Polymers
Properties and Applications

2

Hans-Henning Kausch

Polymer Fracture

2nd revised and enlarged edition

With 180 Figures

Springer-Verlag
Berlin Heidelberg New York
London Paris Tokyo

Professor Dr. rer. nat.
Hans-Henning Kausch-Blecken von Schmeling

Laboratoire de Polymères, Département des Matériaux,
École Polytechnique Fédérale de Lausanne, Suisse

This volume continues the series *Chemie, Physik und Technologie der Kunststoffe in Einzeldarstellungen,* which is now entitled *Polymers / Properties and Applications.*

ISBN 3-540-13250-3 2. Aufl. Springer-Verlag Berlin Heidelberg New York
ISBN 0-387-13250-3 2nd ed. Springer-Verlag New York Heidelberg Berlin

ISBN 3-540-08786-9 1. Aufl. Springer-Verlag Berlin Heidelberg New York
ISBN 0-387-08786-9 1st ed. Springer-Verlag New York Heidelberg Berlin

Library of Congress Cataloging in Publication Data:

Kausch, H. H.: Polymer fracture. (Polymers, properties and applications; 2) Includes bibliographies and index. 1. Polymers and polymerization – Fracture. I. Title: II. Series.
TA 455.P 58 K 38 1987 620.1'920426 84-26876
ISBN 0-387-13250-3 (U.S.)

Typesetting: Schwetzinger Verlagsdruckerei GmbH, 6830 Schwetzingen. Printing: Mercedes-Druck, Berlin. Bookbinding: Helm, Berlin
2154/3020-543210

Preface to the Second Edition

The first edition of this book had been written with the special aim to provide the necessary information for an understanding of the deformation and scission of chain molecules and its role in polymer fracture. In this field there had been an intense activity in the sixties and early seventies. The new results from spectroscopical (ESR, IR) and fracture mechanics methods reported in the first edition had complemented in a very successful way the conventional interpretations of fracture behavior. The extremely friendly reception of this book by the polymer community has shown that the subject was timely chosen and that the treatment had satisfied a need.

In view of the importance of a molecular interpretation of fracture phenomena and of the continued demand for this book which still is the only one of its kind, a second edition has become necessary.

The aims of the second edition will be similar to those of the first: it will be attempted to reference and evaluate completely the literature on stress-induced chain scission, now up to 1985/86. References on other subjects such as morphology, viscoelasticity, plastic deformation and fracture mechanics, where the treatment was never meant to be exhaustive, have remained selective, but they have been updated.

Some sections where major progress has been achieved during 1978 to 1985 will be revised (such as those on impact fracture, on the failure of pipe materials and on crazing) or newly written (that on crack healing). Smaller additions to the text will be introduced by the phrase "In recent years", which refers to the time interval from 1978 to 1986 and permits to identify information not available at the time of the first publication of this book.

As far as possible, the author has tried to consider the many constructive comments which he had received from his colleagues in the past few years. Unfortunately, some of these suggestions would have gone much beyond the intentions of author and publisher to bring out the principal mechanisms and the role of molecular chains in polymer fracture. A certain abstration therefore, has become absolutely necessary. Thus a detailed treatment of the anelastic and plastic deformation of polymers and of the influence of processing parameters would have meant to at least double the volume of this book. Nevertheless, in view of the undeniable importance of these parameters an increased attention has been given to them this time.

Undoubtedly the manuscript still contains points which should have been treated more extensively; the author apologizes in advance for any inconsistencies and omissions. He hopes again that the special viewpoint of this book will be useful in the study of the different aspects of fracture of polymers, from rapid crack propagation to long time loading and fatigue, from crazing to rupture of highly oriented fibres.

Lausanne, July 1986 H. H. Kausch

Preface to the First Edition

This book on "Polymer Fracture" might as well have been called "Kinetic Theory of Polymer Fracture". The term "kinetic theory", however, needs some definition or, at least, some explanation. A kinetic theory deals with and particularly considers the effect of the existence and discrete size, of the motion and of the physical properties of molecules on the macroscopic behavior of an ensemble, gaseous or other. A kinetic theory *of strength* does have to consider additional aspects such as elastic and anelastic deformations, chemical and physical reactions, and the sequence and distribution of different disintegration steps.

In the last fifteen years considerable progress has been made in the latter domains. The deformation and rupture of molecular chains, crystals, and morphological structures have been intensively investigated. The understanding of the effect of those processes on the strength of polymeric materials has especially been furthered by the development and application of *spectroscopical methods* (ESR, IR) and of the *tools of fracture mechanics*. It is the aim of this book to relate the conventional and successful statistical, parametrical, and continuum mechanical treatment of fracture phenomena to new results on the behavior of highly stressed molecular chains.

The ultimate deformation of chains, the kinetics of the mechanical formation of free radicals and of their reactions, the initiation and propagation of crazes and cracks, and the interpretation of fracture toughness and energy release rates in terms of molecular parameters will be discussed. Although the prime interest had been given to these subjects it was not possible to cover the literature completely. The author apologizes in advance for all intentional and unintentional omissions that will have occurred. In any event this book draws on the existing literature on morphology, viscoelasticity, deformation and fracture of high polymers. It is hoped that it forms a useful supplement of such monographs in view of a molecular interpretation of polymer fracture.

Lausanne, September 1978 H. H. Kausch

Acknowledgements

The author gratefully acknowledges the great help he has received in many different forms from his friends and colleagues in the preparation of this revised version. Thus he is deeply indebted to the late Prof. P. J. Flory (Stanford) for his comments on the first two chapters and the long discussions on the notion and importance of "entanglements". Valuable suggestions came from:

J. Bauwens (Bruxelles),
W. Döll (Freiburg),
C. G'Sell (Nancy),
N. Heymans (Bruxelles),
A. Keller (Bristol),
T. Kinloch (London),
E. J. Kramer (Ithaca),
V. A. Marichin (Leningrad),
G. Michler (Halle),
L. P. Mjasnikova (Leningrad),
L. Monnerie (Paris),
R. Porter (Amherst),
N. Rapoport (Moscow),
W. Retting (Ludwigshafen),
J. A. Sauer (Piscataway),
G. Strobl (Freiburg),
F. Szöcs (Bratislava),
M. Tirrell (Minneapolis),
I. M. Ward (Leeds),
J. Wendorff (Darmstadt),
J. G. Williams (London),
H. Wilski (Wiesbaden),
R. Wool (Urabna),
and G. Zhaikov (Moscow),

who have been kind enough to read larger sections or whole chapters of this book.

These colleagues and Drs. R. Casper (Leverkusen), E. Gaube (Frankfurt-Hoechst), and F. Ramsteiner (Ludwigshafen) as well as the Tate Gallery, London, have kindly provided quite a number of the newly incorporated figures. Some of the figures are reproduced from the litterature by permission of the publishing houses

Butterworth & Co.,
Chapman and Hall, Ltd.,
Hütig & Wepf Verlag,
Society of Plastics Engineers,
Dr. Dietrich Steinkopf Verlag,
J. Wiley and Sons, Inc.,

permission which is also gratefully acknowledged.

Last but certainly not least, the author would like to thank his collaborators, Ph. Béguelin, Privat-Docent Dr. M. Dettenmaier, Drs. T. Q. Nguyen, A. C. Roulin-Moloney and B. Stalder for their constructive and valuable assistance, Mme A. Bolanz for her patience and long-standing experience in splendidly organizing a text full of repetitive and intercalating corrections, and the publisher for an excellent and speedy production of the manuscript.

Table of Contents

Chapter 1

Deformation and Fracture of High Polymers, Definition and Scope of Treatment

The importance of a thorough understanding of the deformation behavior and the strength of polymeric engineering materials need not be emphasized. It is obvious to anyone who wants to use polymers as load bearing, weather-resisting, or deformable components or who wants to grind or degrade them. "Strength" and "fracture" of a sample are the positive and negative aspects of one and the same phenomenon, namely that of stress-biased material disintegration. The final step of such disintegration manifests itself as macroscopic failure of the component under use, be it a water pipe, a glass fiber reinforced oil tank, or a plastic grocery bag. The preceding intermediate steps – nonlinear deformation, environmental attack, and crack initiation and growth – are often less obvious, although they cause and/or constitute the damage developed within a loaded sample.

Efforts of fracture research are directed, therefore, towards three goals:

1. To understand the cause and development of a failure and to assess the importance of any one process contributing to a failure or delaying it

2. To see whether, how, and with what effect the most damaging process can be controlled or even eliminated, and how the strength-determining material properties can be improved

3. With the first two goals only partly attainable, one would like to have knowledge of the endurance limit of a part in terms of stress or strain level, environment, number of loading cycles, or of loading time.

The first goal is equivalent to the ability to predict the material response under all kinds of foreseeable conditions of mechanical, thermal, or environmental attack, to characterize the constantly changing state of the material, and to know at any moment which conditions of attack are critical for the state the material just attained. To achieve this goal in all generality is hardly possible, in view of the complexity involved in characterizing polymeric structures, let alone their deformation. A conclusive discussion of those two subjects and of their interrelation with and dependency on other aspects of polymer science could perhaps be achieved in the framework of a polymer science encyclopedia [1, 2] and partially in specialized textbooks on the viscoelastic [3, 4], mechanical [5], and physical [6–8] behavior of polymers. Although the first goal cannot be achieved in all generality, the literature on fracture (cf. textbooks [9–12] and, recently, [46–48]) deals extensively and successfully with interpretations of the fracture event which focus on just those processes which account for the most obvious changes of sample morphology during fracture development.

The previous statements on the goals of fracture research are well illustrated by the fracture patterns of rigid polyvinylchloride (PVC) shown in Figures 1.1–1.3. The samples, water pipes internally pressurized, fractured in a brittle manner when the circumferential stress level was high, in a partially ductile manner and only after prolonged times at medium stresses, and through thermal crack growth (creep crazing) after very long times at low stress levels. The three processes leading to pipe failure in these three cases are respectively the rapid extension of a flaw, the flow of matter, and the thermally activated growth of a flaw. In all three cases the volume element within which the fracture is initiated is finite, consequently nonhomogeneous deformation must have occurred locally. Later Chapters deal with the nature and interpretation of such heterogeneous deformation of a supposedly homogeneous material.

Fig. 1.1. Brittle fracture of an internally pressurized water pipe (polyvinylchloride, PVC), taken from Ref. [13]

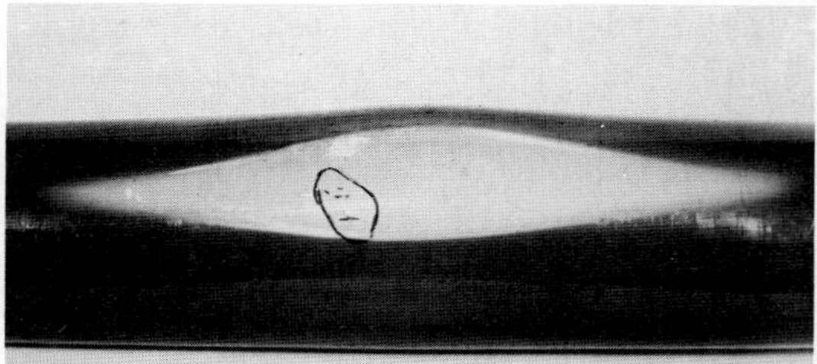

Fig. 1.2. Ballooning, i.e. ductile failure of a PVC pipe at medium stress levels (rupture as a result of the reduction of wall thickness at a defect at the indicated position), taken from Ref. [13]

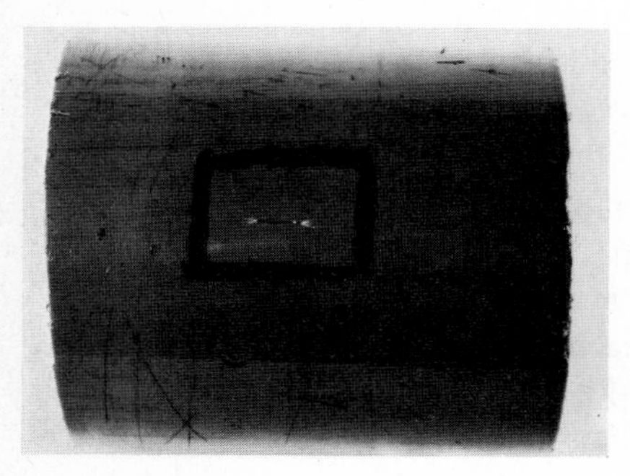

Fig. 1.3. Long-time failure of a PVC pipe through development of a creep craze (from [13])

One also understands the importance of the first goal with respect to achieving the second. In the described example of pressurized PVC pipes it was known that in the medium time range the performance of a pipe was limited by its creep behavior. One would wish, therefore, to decrease the creep rate by increasing the rigidity of the material; in doing so, however, one would facilitate crack propagation and, possibly, craze initiation and thus reduce the permissible stress levels at short and very long times of application.

The third goal, the knowledge of the endurance limit of a part in terms of stress or strain level, is generally directly attainable within a certain experimental region. Figure 1.4 summarizes, for instance, the test results on PVC pipes at temperatures between 20 and 60 °C, circumferential stresses between 10 and 50 MN/m^2 and a range of times to fracture of 0.3 to 10^4 h. Such representations, which in nongraphi-

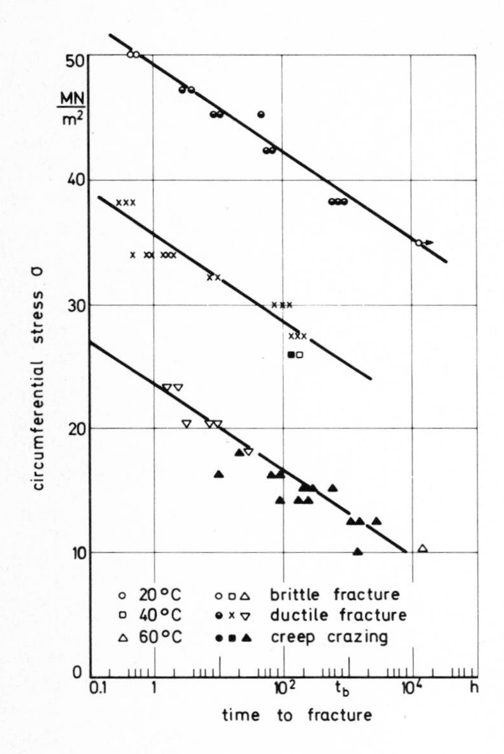

Fig. 1.4. Times-to-failure of internally pressurized PVC water pipes at different stresses and temperatures (after [13])

cal form can be extended to interrelations between more than three parameters, form an important basis of our understanding of material behavior within the multi-dimensional space covered by the parameters investigated. Material selection and optimization of its properties through parameter variation become possible *to the extent the mathematical framework of parameter-property coefficients holds.* With regard to fracture this means that one would like to define a fracture criterion in terms of the relevant structural and environmental parameters.

A good example for the occurrence of unpredicted events is offered by the behavior of polyethylene (PE) pipes – and of other semicrystalline materials – under constant circumferential stress; they showed a rapid drop of sustained stresses after long loading times (Fig. 1.5). With PE – as with PVC – an initial time range is observed where the times to failure are only a weak function of stress. Depending on the temperature, the failure mode is either brittle (Fig. 1.1) or ductile (Fig. 1.2 and 1.6). Both materials are also comparable insofar as the thermally activated growth of a creep crack (Fig. 1.3, 1.7, and 1.8) can be held responsible for the fail-

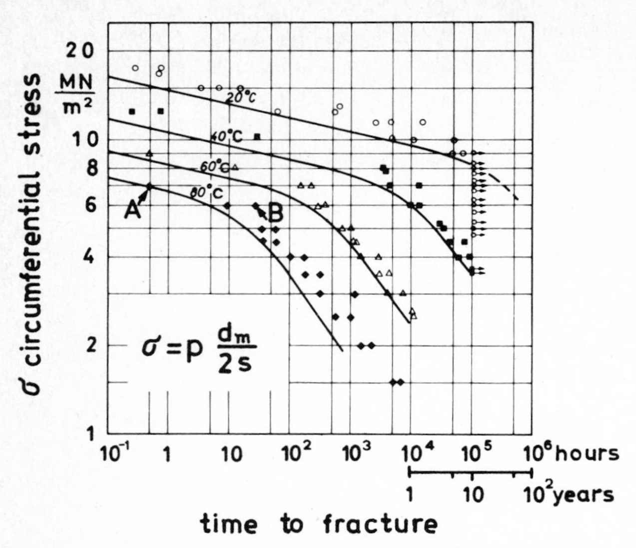

Fig. 1.5. Times-to-failure of high-density polyethylene (HDPE) water pipes under internal pressure p at different stresses and temperatures. d_m: average diameter, s: wall thickness; A: ductile failure, B: creep crazing (specimens are shown in Figs. 1.6–1.8; after [14])

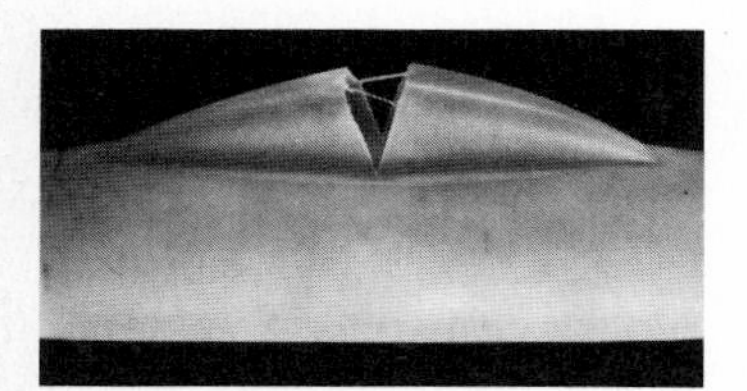

Fig. 1.6. Ductile failure of a HDPE water pipe under the conditions shown as point A in Fig. 1.5: $\sigma_v = 7$ MN/m^2, $T = 80$ °C (after [14])

Fig. 1.7. Surface of a creep craze formed in HDPE under conditions shown as point B in Fig. 1.5: $\sigma_v = 6$ MN/m^2, $T = 80$ °C (after [14])

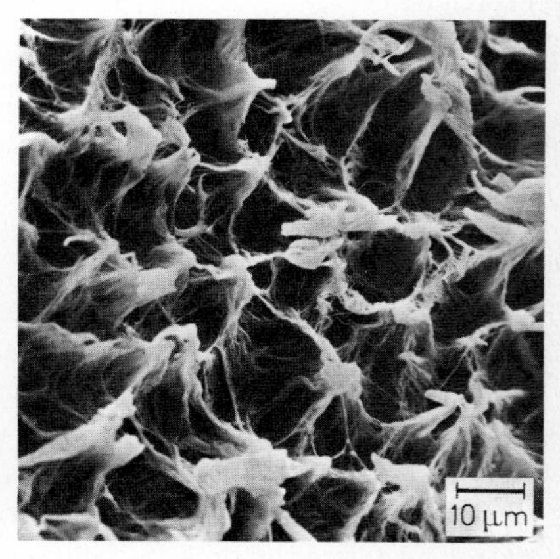

Fig. 1.8. Detail of the fracture surface close to the upper center of the mirror zone, the point of craze initiation (after [14])

ure of a pipe after long times of service. Both materials are different insofar, as one finds in PE – but practically not in PVC – that the kinetics of creep crack development differs markedly from that of ductile failure (Fig. 1.5). This finding only underlines that it is necessary to deal directly with the physical nature of material defect development in order to predict reliably material behavior – especially for new applications – and/or improve properties through additional components or modified production.

In this monograph we wish to discuss the physical nature of defect development, and we will focus on linear thermoplastic and elastomeric polymers (Table 1.1.). We recognize that these materials cover a wide range of properties, although they are composed of molecules which exhibit a number of similarities: they are predominantly linear, flexible, highly anisotropic (nonextended) chains having molecular weights between 20000 and more than 1000000. Figure 1.9 gives a represen-

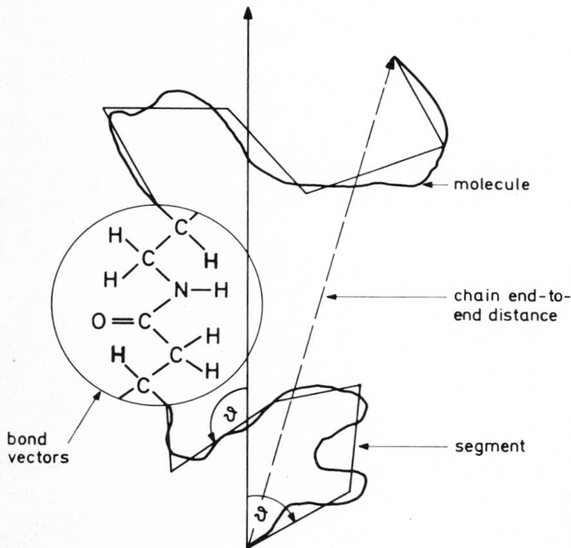

Fig. 1.9. Representation of linear flexible chain molecule (polyamide 6), (after [21])

tation of a (polyamide 6) chain showing the nonextended *conformation*, an arbitrary division into *segments*, and – in the insert – the primary *bonds* forming the *backbone chain*. The relative sizes of the atoms and the proportion of volume occupied by them within the chain is illustrated by the Stuart model of a polyamide segment (Fig. 1.10). In actual size this segment has an extended length of 1.97 nm. If such a segment could be stressed in the chain axis direction, the bending and stretching of primary bonds would result in a chain stiffness of about 200 GN/m^2 [15], whereas the intermolecular attraction of segments through the much weaker van der Waal's forces entails a stiffness of "only" 3 to 8 GN/m^2 in a direction perpendicular to the chain axis. *The fact that polymer molecules are long, anisotropic, and flexible chains accounts for the characteristic properties of polymeric solids,*

Table 1.1. Mechanical Properties of Chain Molecules and Polymeric Solids*)

Polymer	Chain molecule				Oriented Fiber [17]		Isotropic Solid		
	Abbrev. (ISO/R 1043–1969)	Monomer molecular weight M_0 (g/mol)	Monomer length L (10^{-10}m) (from unit cell)	Young's modulus E_c (MN/m^2) (Refs. [15, 16])	Young's modulus E (MN/m^2)	Strength σ_b (MN/m^2)	Young's modulus E (MN/m^2)	Strength σ_b (MN/m^2)	Critical stress σ_c (MN/m^2) (Ref. [18])
Semicrystalline Polymers									
1. Polyethylene	PE	28		235,000					
high density			2.534		8000–50.000	480	1000	27	160
low density			2.54		[19]		200	13	
2. Polypropylene	PP	42	6.5/3	41,000	6000–9000		1250–2400	33	98
3. Polytetrafluorethylene	PTFE	100	2.60	153,000			400–980	23	117
4. Poly(oxymethylene)	POM	30	1.92	53,000		1200–2550	3500	68	216
5. Poly(ethylene oxide)	(PEO)	44	19.3/9	9800			530		
6. Poly(propylene oxide)		58	7.15/2				133		
7. Polyethylene terephthalate	PETP	192	10.77	74,000–137,000	20,000–25,000	800–1400	3000	53	155
8. Polyamide									
polycaprolactam	PA 6	226	17.2	24,500	5000–8500	390–870	1250	61	
polyhexamethylene adipamide	PA 66	226	17.3			420–930	1500–4100	80	179
aromatic	(Kevlar 49)	119	6.6	180,000	144,000	3000			
Amorphous Polymers									
9. Polystyrene	PS	104	2.21	9000–11,800			3250	50	
10. Polyvinyl chloride	PVC	62.5	2.55	(160,000–230,000)			2500–3000	55	142
11. Polyvinyl alcohol	PVAL	44	2.52	250,000	35,000–50,000	1200–2800		28	

Table 1.1. (continued)

Polymer	Chain molecule				Oriented Fiber [17]		Isotropic Solid		
	Abbrev. (ISO/R 1043–1969)	Monomer molecular weight M_0 (g/mol)	Monomer length $L(10^{-10}m)$ (from unit cell	Young's modulus $E_c(MN/m^2)$ (Refs. [15, 16])	Young's modulus E (MN/m^2)	Strength σ_b (MN/m^2)	Young's modulus E (MN/m^2)	Strength σ_b (MN/m^2)	Critical stress σ_c (MN/m^2) (Ref. [18])
Amorphous Polymers									
12. Polyvinyl acetate	PVAC	86					1300–2800	29–49	
13. Polymethyl methacrylate	PMMA	100	2.11				3000–4000	70	68
14. Polycarbonate (bisphenol A)	PC	254	10.75				2350	53	145
(Cross-linked Rubbers, for comparison)									
Natural rubber	NR	136	8.1				1–2	17–25	
Butyl rubber	IIR	56	4.7				0.7–1.5	18–21	
Polyurethane	PUR						0.50	20–30	
Silicon rubber	SIR						0.15	2–8	

*) Reference values of technical polymers at room temperature (critical stresses refer to temperature of ductile-brittle transition); data taken from D. W. van Krevelen [20] and Polymer Handbook [20a] unless otherwise indicated.

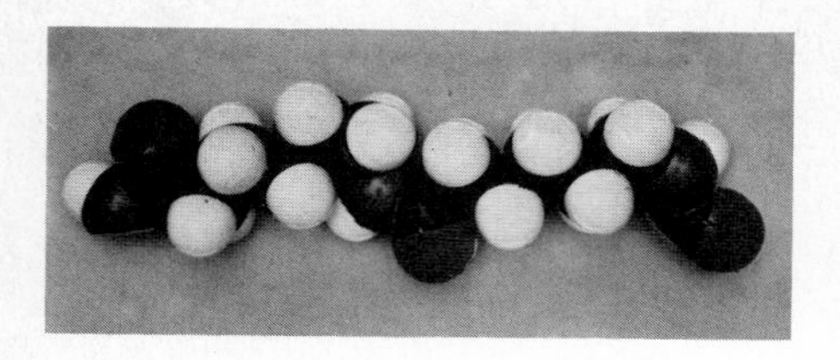

Fig. 1.10. Arbitrary segment of polyamide 6 chain (extended length 1.97 nm, cross section 0.176 nm^2)

namely the anisotropy of macroscopic properties, the microscopic heterogeneity and the nonlinearity and strong time-dependency of polymer response to mechanical excitation.

All solid plastics, from rubbery networks to rigid glasslike bodies, are aggregates of such molecular chains, which cohere through physical (temporary) or chemical (permanent) "cross-links". The wide variety of the macroscopic mechanical properties of these materials is caused by differences in aggregation (crystallinity or superstructure) and in molecular parameters, as for instance by differences in chain structure, intermolecular segmental attraction, segment mobility, molecular weight (distribution), or cross-link density. Unfortunately these parameters are not fixed values attached to a molecular species or a polymeric body, but depend on polymer processing and on environmental conditions. In fact by simply changing the time or temperature scale of the mechanical excitation of one and the same polymer a wide range of responses will be obtained: from that of a brittle glass to that of a highly viscous liquid. This general behavior is depicted in a time-temperature diagram, given by Retting [22] for hard PVC (Fig. 1.11). Within the relatively small temperature interval from 0 to 100 °C (rapid loading) or –100 to +60 °C (creep) the indicated changes from glass to rubber, from – macroscopically – brittle to highly duc-

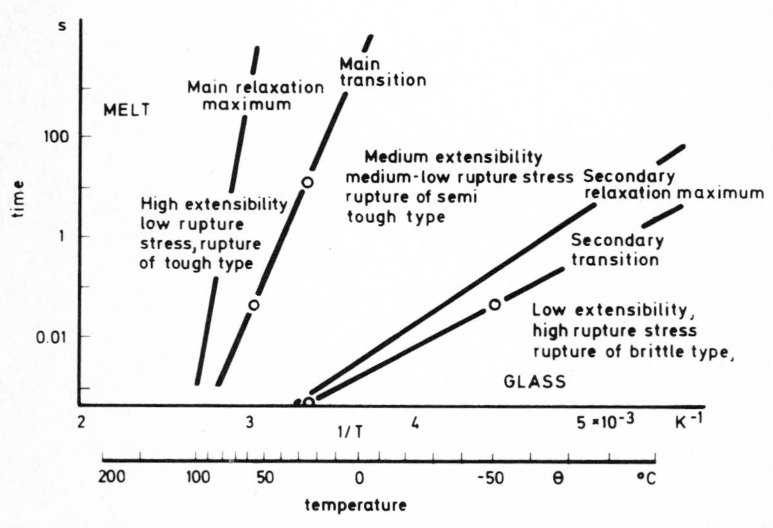

Fig. 1.11. Time-temperature diagram of the mechanical properties of PVC (after Retting [22])

tile deformation and fracture are observed. The changes in the type of response of a given aggregate of chains clearly indicate that different deformation processes must be dominant in different regions of the time-temperature regime.

In discussing the physical nature of defect development it is necessary to characterize the molecular and microscopic structure and to specify which form of local deformation eventually leads to a disintegration of the polymeric solid or to a localized heterogeneous deformation and to crack initiation. In previously proposed kinetic theories quite different processes were held responsible for the development of macroscopic failure. *Viscous flow* (within cotton fibers) was suggested by Busse and his colleagues [23] to determine the fatigue life of these fibers. The *"slippage" of secondary bonds* and the net decrease of the number of bonds was considered by Tobolsky and Eyring [24] for the two cases where repair of bonds under stress is possible or impossible. A model involving two independent networks was discussed by Stuart and Anderson [25] who argued that only *fracture* within the first network with unsymmetrical potential barriers between broken and unbroken states contributes to failure of glasses. A theory of *correlated Brownian motion* of groups of atoms (*Vielfachplatzwechsel*) proposed by Holzmüller [26] is based on Boltzmann statistics of phonon energy distribution where the phonons are the independent statistical units. *Local rupture at chain ends* had been suggested by Spurlin [27] to explain the dependence of the ease of failure on the number of ends of cellulose molecules. The *loss of entanglements* or the partial *disentanglement* of a network may also be cited here since it constitutes a weakening mechanism that influences deformation and failure [3, 6].

In 1945 Flory [28] stated that the tensile strength of a (cross-linked) rubber appears to be a linear function of the percentage of the chains permanently oriented by stretching. This implies that macroscopic failure is caused by breakage of chains at/or between cross-links. This idea that the *breaking of primary chemical bonds* plays a decisive role in fracture was later further developed by Zhurkov [29] and Bueche [30]. Bueche, and Halpin [32, 33] and Smith [34], had been concerned with the strength of rubbery polymers, and found that the time-temperature dependence of the ultimate properties could be explained by the deformation behavior and the *finite extensibility of molecular strands.* Prevorsek and Lyons [35] emphasized that the random thermal motion of chain segments leads to *nucleation and growth of* voids without breakage of any load-carrying bonds. The lifetime of a sample then is determined by the time it takes one flaw to grow to a critical size. Niklas and Kausch [13] discussed the effect of the *dissociation of dipole associates* on the strength of PVC.

To the molecular processes specified above one has to add the unspecified ones: internal destruction, damage, or crack formation probability. In analogy to the molecular description of deformation used by Blasenbrey and Pechhold [38] all of these molecular processes can be referred to the four physical rearrangements that can occur between neighboring chain segments with parallel chain axes: change of conformation (segment rotation, *gauche-trans* transformation), cavitation, slip, and chain scission. In Figure 1.12 these rearrangements of segment pairs are shown. Of these chain scission and – to some extent – cavitation and slip are potentially detrimental to the load-bearing capacity of a polymer network. Conformational changes on the other hand appear to be "conservative" processes, which by them-

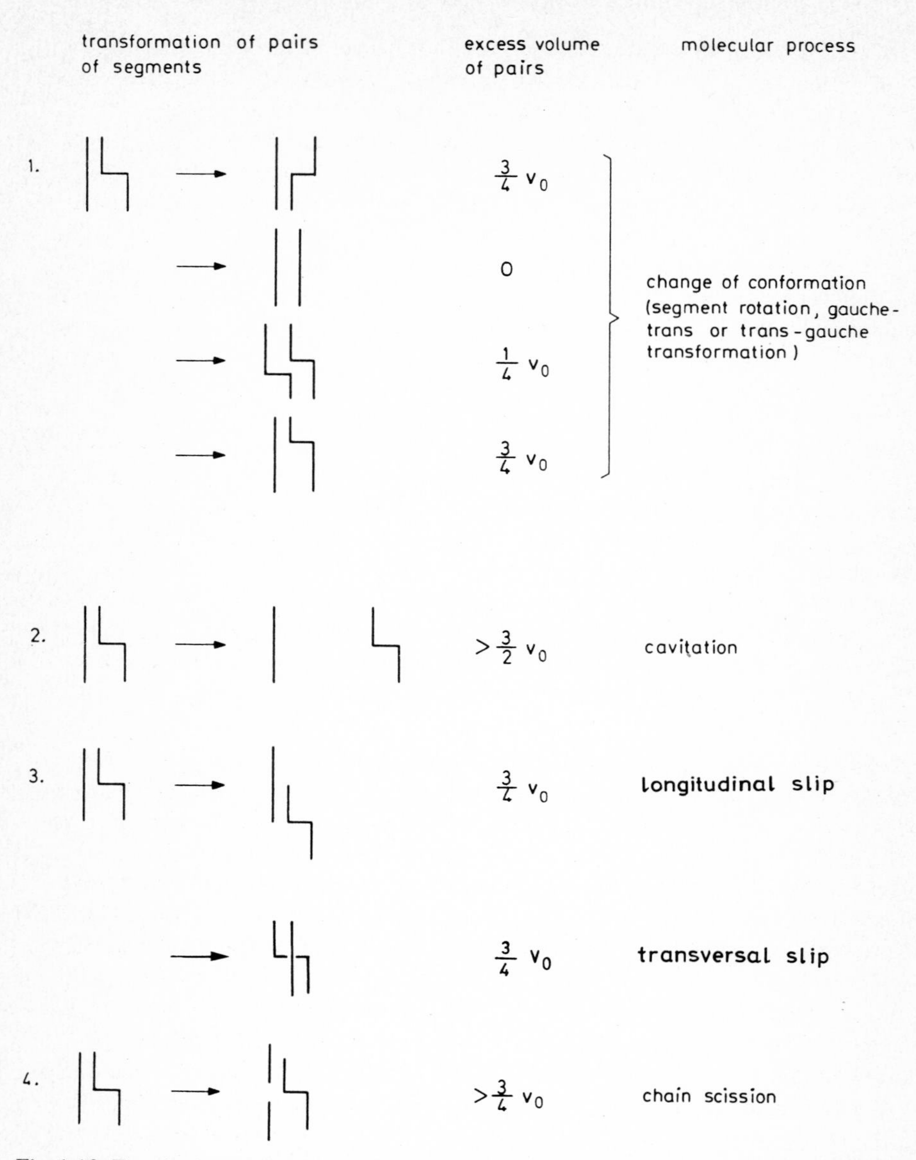

Fig. 1.12. Four basic transformations of pairs of linear chain segments. Excess volume of pairs refers to PE with v_0 being the volume of a CH_2-group [38]. In this scheme segments with parallel axes have been shown. Evidently similar transformations are also possible if the chain axes are not parallel to each other, and/or if the angle of inclination changes during a transformation

selves will modify and delay but never generate any progress in the development of a fracture process.

Here it seems to be necessary to say a word about the meaning of the terms fracture, rupture, and failure as they will be used within this text. Andrews [9] says, "Fracture is the creation of new surfaces within a body . . . The term rupture may be regarded as truly synonymous but not the term failure since the latter indicates only that some change has taken place in the body (generally an engineering component of some kind) which renders it unsuitable for its intended use. Fracture is only one of many possible modes of failure". We will supplement this definition by saying that as a result of the creation of new surfaces either holes have to open up or formerly connected parts of the fractured specimen have to become detached from each other.

In mathematical terms we may define a fracture process as one by which the sum of the degrees of connectedness of the parts of a sample or a structural element is increased by at least one. This definition includes the piercing of films and tubes, but excludes minute damages as caused by crazing or chain scission.

The term fracture process is meant to describe the whole development of a fracture from its initiation through crack growth and propagation to completion of the fracture. In contrast to thermal or environmental degradation and failure the fracture process is understood as stress-biased material disintegration. It carries on if and only if a driving force exists, which has literally to be a force. Forces, however, will give rise to deformations. This means that fracture initiation will always be preceeded by specimen deformation.

From Figure 1.12 it could be concluded that, in the absence of flow, chain scission and chain strength determine the mechanical properties of polymers. In fact, there are a number of observations which at first glance support this assumption:

- sample strength is a function of molecular weight; with increasing number n of chain atoms, a saturated linear hydrocarbon will turn from gas to liquid ($n \geqslant 5$) and to solid ($n \geqslant 18$); the solid will attain technically useful strength (10 MN/m^2) if n is around 2000; with increasing molecular weight and chain orientation that strength may still be increased 50-fold (see Figs. 1.13 to 1.15 and Refs. [19, 27, 28, 36, 37, 41, 42, 49]),
- the mechanical loading and rupture of virtually all natural and synthetic polymers leads to the scission of molecular chains and the formation of free radicals [39];
- the strength and the hardness of solids increase with volume concentration of primary bonds [40];
- in a number of uniaxial tensile fracture experiments the energy of activation of sample fracture coincided with the activation energy for main chain bond scission [39].

The observation that sample strength increases with molecular weight and chain orientation is illustrated by Figures 1.13 and 1.14. In low molecular weight material chain slip occurs readily, so that the sample strength depends solely on the strength of *intermolecular* attraction. Noticeable macroscopic strength is only obtained once the molecular weight is sufficient to permit physical cross-links through entanglements or chain folds between several chains [36, 37]. In the case of PMMA (Fig. 1.13) a gradual increase in flexural strength is observed in the molecular weight range of 30000 to 150000.

The increase in strength is much more rapid for the semi-crystalline PA6 (Fig. 1.14). In the range of molecular weights between 1.5 and 3×10^4, fiber strength increases with molecular weight, whereas at still higher molecular weights the effect of the increasing number of defects introduced in processing such a fiber, tends to outweigh

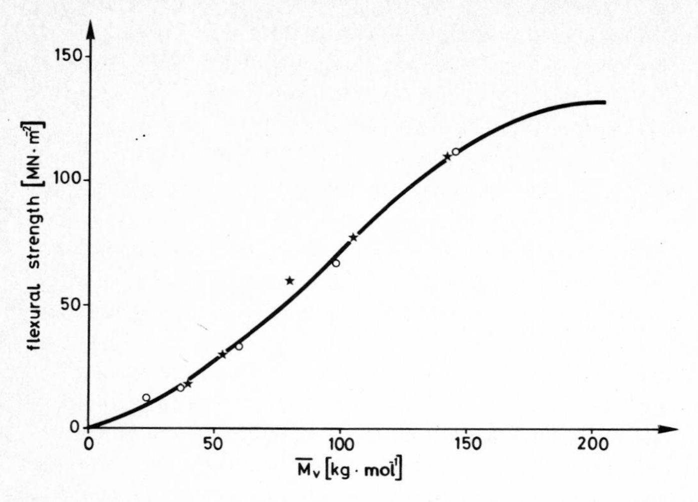

Fig. 1.13. Dependence of flexural strength of two commercially available grades of PMMA on molecular weight (after [37])

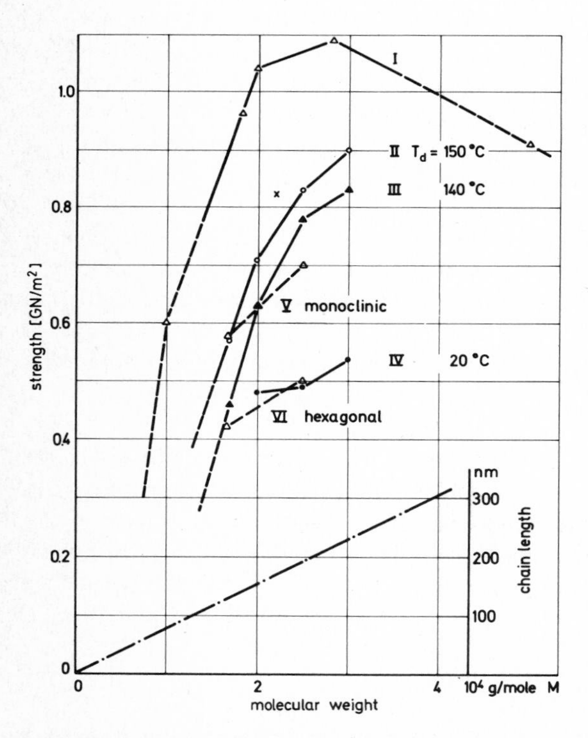

Fig. 1.14. Effect of molecular weight and temperature T_d of the fiber orientation process (drawing) on the tensile strength of polyamide 6. I: low-denier filaments. T_d = 170 °C [41], II–IV: effect of drawing temperature. T_d = 150 °C, 140 °C, 20 °C [42], V, VI: effect of crystal structure [43]

the effect of the decreasing number of chain ends (Fig. 1.14, Curve I). It is also indicated in Figure 1.14 (Curves I–IV) that the realization of high strength through a high degree of orientation (e.g., obtained at higher drawing temperatures T_d) requires high molecular weights [41, 42]. The effects of crystal structure and molecular weight on macroscopic tensile strength are shown by Curves V and VI [43]. This simple diagram already shows that multiple interrelations between strength and molecular, morphological and processing parameters exist.

Also the data of Table 1 do not support a simple relation between the strength values of individual chains and of the solids formed thereof. Young's moduli of bulk specimens are one to two orders of magnitude smaller than the corresponding chain moduli, bulk strengths at room temperature are two to three orders of magnitude smaller than chain strengths. The critical stress, which is the fracture stress at that temperature where fracture changes from ductile to brittle, i.e., at a temperature well below the glass transition temperature, is also one to two orders of magnitude smaller than the corresponding chain strength. The tensile strengths observed at different morphological levels are indicated in Figure 1.15.

In the cited monographs [2–10] very little has been said about the details of how molecular destruction processes effect ultimate properties. In the years 1965 to 1975, however, considerable progress has been made in precisely this area, in the investigation of the deformation, strength, and scission of chain molecules. Part of the progress achieved by Russian scientists working in this area has been covered by the book of Regel, Slutsker, and Tomashevskii [11] which is available in Russian; meanwhile an

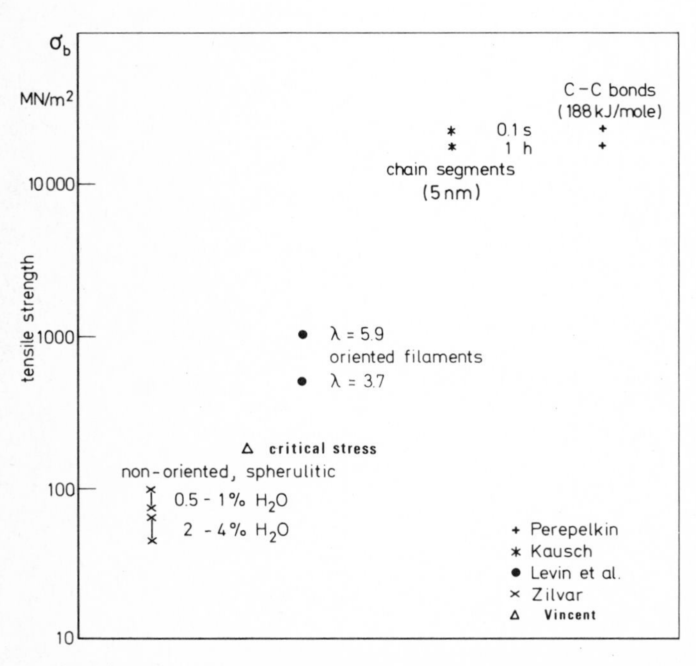

Fig. 1.15. Tensile strengths of polyamide 6 at different levels of morphologic organization (using data from Refs. [17, 18, 21, 44, 45])

English translation of a newer Russian text on this subject has appeared, the book by Kuksenko and Tamush [46].

With these words of introduction we are in a position better to define the scope of this book. The central theme is the behavior of chain molecules under conditions of extreme mechanical excitation and the likely role of chain scission in the initiation and development of macroscopic fracture. We will study the structure and deformation of solid polymers (Chapter 2) in an attempt to define the conditions under which elastic straining of chains occurs. We will then discuss existing non-morphologic fracture theories (Chapter 3) in view of our central theme. The molecular description of fracture begins with a brief account of the strength of primary bonds (Chapter 4) and of the thermomechanical excitation and scission of chain segments (Chapter 5). The study of free radicals formed as a consequence of stress-induced chain scission (Chapters 6–8) constitutes the main part of this volume. It is the only part of this book where a comprehensive coverage of the existing literature has been attempted. In Chapter 9 the role of molecular chains in heterogeneous fracture will be analyzed. For this purpose a brief introduction is given into the mechanics of fracture (stress analysis of cracks in elastic and elastic-plastic materials, slow crack propagation). The fracture mechanical functions critical energy release rate G_{Ic} and fracture toughness K_{Ic} are interpreted from a noncontinuum viewpoint. The same approach is applied to the phenomenon of crazing. A discussion of molecular and morphological aspects of crack propagation and of possible failure criteria concludes that section. A final chapter is devoted to the new area of crack healing.

As mentioned in the introduction, a kinetic theory of fracture deals with and particularly considers the existence and discrete size, the motion, and the response of molecules or their parts in a stressed sample. It thus requires a molecular description of polymer structure and deformation. Such a description is available in the cited books [1–12], [20] or papers [15, 38]. In Chapter 2 only a very brief summary is offered which is not meant to be self-supporting or complete.

References for Chapter 1

1. H. A. Stuart (ed.): Physik der Hochpolymeren. Berlin – Heidelberg: Springer-Verlag 1952
2. H. F. Mark, N. G. Gaylord, N. M. Bikales (eds.): Encyclopedia of Polymer Science and Technology. New York: Interscience 1964, sec. ed. in preparation
3. J. D. Ferry: Viscoelastic Properties of Polymers (2nd ed.). New York – London: Wiley 1970; (3rd ed. 1980)
4. N. G. McCrum, B. E. Read, G. Williams: Anelastic and Dielectric effects in Polymeric Solids. London – New York: Wiley 1967
5. I. M. Ward: Mechanical Properties of Solid Polymers. London: Wiley-Interscience 1971, Sec. Ed. 1983
6. A. D. Jenkins (ed.): Polymer Science, Amsterdam – London: North Holland 1972
7. R. N. Haward (ed.): The Physics of Glassy Polymers. London: Applied Science Publishers 1973
8. J. A. Manson, L. H. Sperling: Polymer Blends and Composites, New York – London: Plenum 1976
9. E. H. Andrews: Fracture in Polymers. Edinburgh London: American Elsevier 1968
10. G. M. Bartenev, Yu. S. Zuyev: Strength and Failure of Viscoelastic Materials. Oxford – London: Pergamon 1968

11. V. R. Regel, A. I. Slutsker, E. E. Tomashevskii: Kinetic Nature of the Strength of Solids (in Russian). Moscow: Nauka 1974
12. R. Hertzberg: Deformation and Fracture Mechanics of Engineering Materials. New York: Wiley 1976, Sec. Ed. 1983
13. H. Niklas, H. H. Kausch: Kunststoffe *53,* 839 and 886 (1963)
14. E. Gaube, H. H. Kausch: Kunststoffe *63,* 391–397 (1973)
15. J. A. Sauer, A. E. Woodward, In: Polymer Thermal Analysis. Vol. II. P. E. Slade, L. T. Jenkins (eds.). New York: Marcel Dekker 1970. pp. 107–224.
16. I. Sakurada, T. Ito, K. Nakamae: J. Polym. Sci. *15,* 75 (1966)
17. K. E. Perepelkin: Angew. Makrom. Chemie *22,* 181–204 (1972)
18. P. I. Vincent: Polymer *13,* 558 (1972)
19. N. I. Capiati, R. S. Porter: J. Polym. Sci., Polym. Phys. Ed. *13,* 1177–1186 (1975)
20. D. W. van Krevelen: Properties of Polymers, Correlations with Chemical Structures, Amsterdam – London: Elsevier 1972
20a. J. Brandrup, E. H. Immergut: Polymer Handbook (2nd ed.). New York: Wiley 1975
21. H. H. Kausch: J. Polymer Sci. *C32,* 1–44 (1971)
22. W. Retting: Kolloid-Z *213,* 69 (1966)
23. W. F. Busse, E. T. Lessing, D. L. Loughborough, L. Larrick: J. Appl. Phys. *13,* 715 (1942)
24. A. Tobolsky, H. Eyring: J. Chem. Phys. *11,* 125 (1943)
25. O. L. Anderson, D. A. Stuart: Ind. Eng. Chem. *46,* 154 (1954)
26. W. Holzmüller: Z. Phys. Chemie (Leipzig) *202,* 440 (1954) and *203,* 163 (1954), Kolloid-Z. *203,* 7 (1965) and *205,* 24 (1965)
27. H. M. Spurlin: Cellulose and Cellulose Derivatives. Emil Ott (ed.). New York: Interscience Publishers 1943, pg. 9 and 935–936
28. P. J. Flory: Am. Chem. Soc. *67,* 2048 (1945)
29. S. N. Zhurkov, B. N. Narzulaev: J. Techn. Phys. (USSR) *23,* 1677 (1953)
S. N. Zhurkov: Z. Phys. Chemie (Leipzig) *213,* 183 (1960)
30. F. Bueche: J. Appl. Phys. *26,* 1133 (1955) and *28,* 784 (1957)
31. F. Bueche: J. C. Halpin: J. Appl. Phys. *35,* 36 (1964)
32. J. C. Halpin: Rubber Chem. Technol. *38,* 1007 (1965)
33. J. C. Halpin, H. W. Polley: J. Composite Materials *1,* 64 (1967)
34. T. L. Smith: J. Polym. Sci. *32,* 99 (1958); J. Polym. Sci. A, *1,* 3597 (1963)
35. D. C. Prevorsek, W. J. C. Withwell: Textiles Res. J. *33,* 963 (1963)
D. C. Prevorsek, M. L. Brooks: J. Appl. Polym. Sci. *11,* 925 (1967)
36. A. N. Gent and A. G. Thomas: J. Polym. Sci., Part A-2 *10,* 571 (1972)
37. D. J. T. Hill and J. H. O'Donnell: J. Chem. Ed. *58,* 174 (1981)
38. W. Pechhold, S. Blasenbrey: Kolloid-Z. Z. Polymere *241,* 955 (1970)
39. S. N. Zhurkov: Int. J. of Fracture Mechanics *1,* 311–323 (1965)
40. L. Holliday, W. A. Holmes-Walker: J. Appl. Polym. Sci. *16,* 139–155 (1972)
41. D. Prevorsek: J. Polym. Sci. *C32,* 343–375 (1971)
42. V. M. Poludennaja, N. S. Volkova, D. N. Archangelskij, A. G. Zigockij, A. A. Konkin: Chim. Volokna *5,* 6–7 (1969)
43. H. Kraessig: Textilveredelung *4,* 26–37 (1969)
44. B. Ya. Levin, A. V. Savitskii, A. Ya. Savostin, E. Ye. Tomashevskii: Polymer Science USSR *13,* 1061–1068 (1971).
45. V. Zilvar, I. Boukal, J. Hell: Trans. J. Plastics Inst. *35,* 403–408 (1967)
46. V. S. Kuksenko and V. P. Tamuzs: Fracture micromechanics of polymer materials. The Hague: Martinus Nijhoff Publishers 1981
47. A. J. Kinloch and R. J. Young: Fracture Behaviour of Polymers. Barking: Applied Science Publishers 1983
48. J. G. Williams: Fracture Mechanics of Polymers. Chichester: Ellis Horwood Publishers, New York: John Wiley & Sons 1984
49. R. W. Nunes, J. R. Martin, J. F. Johnson: Polym. Eng. Sci. *22,* 205 (1982)

Chapter 2

Structure and Deformation

I. Elements of the Superstructure of Solid Polymers

A. Amorphous Regions

The basic structural elements of high polymer solids are the chain molecules. The large variety of their chemical structures and their flexibility permit widely different modes of organization and mechanical interaction, thus giving rise to a variety of deformation mechanisms and fracture patterns. At this point the characteristic elements of structure and superstructure of amorphous and semicrystalline polymers will be introduced. The interrelations between chain parameters (structure and regularity), crystal or superstructural parameters (degree of crystallinity, lattice structure, nucleation, growth kinetics, and defects of crystals), and external parameters are extensively discussed in the literature and are not the principal object of this book [see Ref. 1–3, 107, for an introduction and for further references].

If no regular organization within a solid aggregate of chains becomes apparent one speaks of an *amorphous state.* In the last two decades the interpretation of chain conformation in the amorphous state of solid high polymers has been subject to lively discussions and careful studies [4–12, 100–104, 108–125, 212]. Until about 1960 the general view prevailed that the chain molecules in isotropic, noncrystalline polymers

*) In the annex (Table A. 1) a list of standardized abbreviations of those polymer names is given which are used in this monograph. The abbreviations are in accordance with the international standards proposed by ASTM, DIN, and ISO.

as in many rubbers, glassy polymers such as PS*), PVC, PMMA, PC, or quenched "semicrystalline" polymers such as PCTFE, PTFE, PETP have a random conformation which is described by the random coil or spaghetti model . In this model it is assumed that the presence of long chain molecules does significantly affect the *close range order* observed in low molecular weight amorphous substances and that strong intermolecular orientation correlations do not appear. In the seventies, the concept that intermolecular correlation does occur in high molecular weight glassy polymers had found some support which was thought to derive from a comparison of segment volume and amorphous density, electron microscopic observation of structural elements, calorimetric investigations, from a discussion of crystallisation kinetics, and a study of network orientation. Thus, Hosemann, Pechhold, and Yeh [9–12] propose that amorphous states and defect-rich, originally ordered states are commensurable. They thus retain elements of anisotropic close range order. Their amorphous state is not principally but only gradually different from a well ordered state. Figure 2.1 shows model representations of the various interpretations of the nature of the amorphous state.

On the other hand Flory [7, 8, 213] summarized his extended observations on polymers in solution, rubbery networks, and amorphous one- and two-phase materials in favor of the random coil structure of chains in the bulk. Also Kirste, Schelten, Fischer, Be-

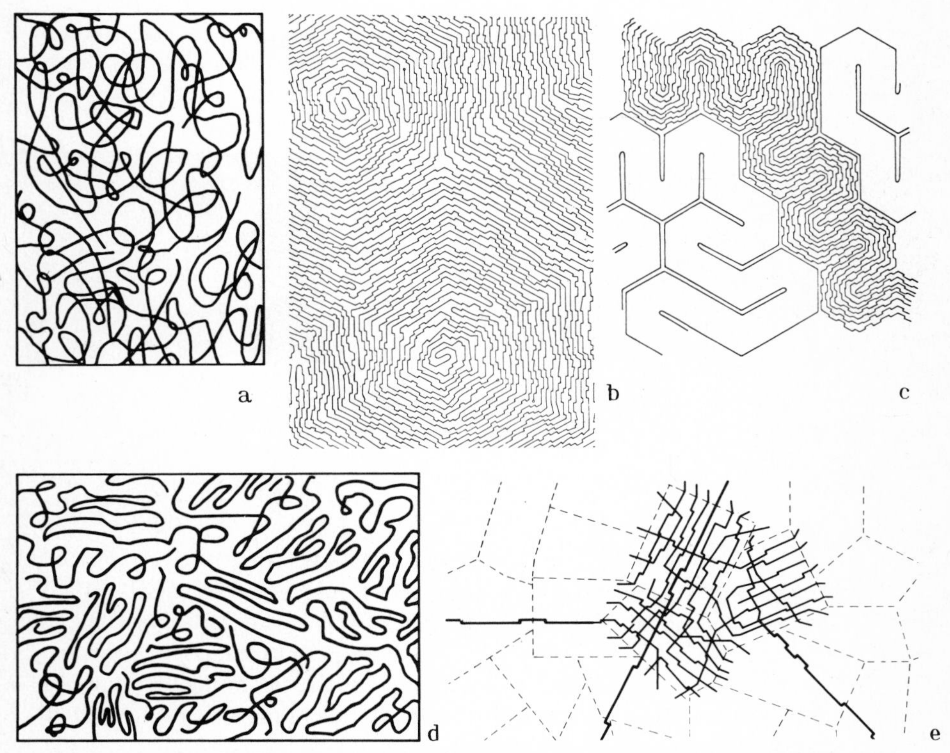

Fig. 2.1 a–e. Model representations of the amorphous state. **(a)** interpenetrating random coils [4, 7], **(b)** and **(c)** honeycomb and meander model [10, 11], **(d)** folded chain fringed micellar grains [12], **(e)** fringed micellar domain structure [13]

Fig. 2.2a. "In that moment", painting by Bernard Cohen, created in 1965 and discovered by the author of this book in the Tate Gallery, London. Whether coincidental or not, the painting gives an artistic representation of the "structural" features of the amorphous state as discussed in the early sixties: randomly oriented chain segments showing a certain close range order and the presence of some chain folds

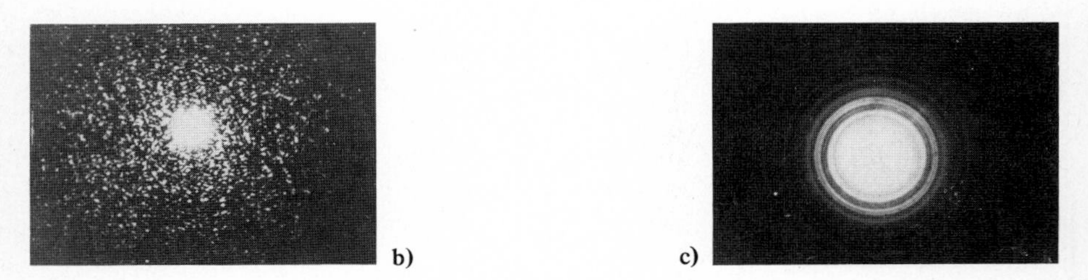

Fig. 2.2b u. c. The randomness of the segment orientation in Cohen's painting had kindly been asserted by Prof. G. Bodor in Budapest. He obtained the laser light scattering pattern **(b)** and, by rotation of that pattern about its center point, Figure **(c)**. The latter reveals the random (planar) orientation of the segment axes but a non-random distribution of their lateral distances, i.e. a close range order

noit and their associates [4–6, 115–116, 118–120] studied the structure of amorphous polymers using scattering techniques such as neutron-, X-ray and light-scattering. The results they obtained do not indicate the existence of strong intermolecular orientation correlations in conventional polymer melts and glasses. For example, the neutron scattering of mixtures of protonated and deuterated molecules in the low- and high-angle region is in excellent agreement with the scattering of unperturbed chains calculated on the basis of the rotational isomeric state theory by Yoon and Flory [117–121]. Thus, it will be considered in this monograph that the *random coil model* is correct and that Figure 2.1a gives the most appropriate representation of the *amor-*

phous state. Nevertheless, some *weak* intermolecular orientation correlations may exist as shown for paraffin melts by various authors [113–115] employing depolarized lightscattering and several other techniques. Occasionally, the effect of annealing on the mechanical properties of glassy polymers has been rationalized in terms of increasing anisotropic close range order. However, depolarized light-scattering experiments performed by Dettenmaier and Kausch [125] have demonstrated that appreciable intermolecular orientation correlations do not exist in PC nor are created by annealing samples below Tg.

In the following discussion ordered structural elements are to be characterized. In doing so one should keep in mind that for the interpretation of ordered structures in real polymer solids one finds a situation somewhat inverse to the interpretation of the amorphous state, i.e. one should keep in mind that in any real polymer well ordered regions are of limited size and finite degree of perfection.

B. Crystallites

From the beginning of polymer science it had been known that native as well as synthetic high polymers crystallize [14a]. Analysis of x-ray diffraction patterns has revealed the lattice structure and unit cell dimensions of high polymer crystallites. Up to 1957 it was believed that the crystallites are of the fringe micelle type. A typical micelle was supposed to bundle some tens of hundreds of different molecules which – after leaving the micelle and passing through amorphous regions – would at random join other micelles. In 1957 Fischer [15], Keller [16], and Till [17] independently from each other discovered and proposed that high polymers form single crystal lamellae within which chain folding occurs. Figure 2.3 a shows an electron micrograph of a stack of PE single crystals [18] grown from dilute solution and Figure 2.3 b the arrangement of the chain molecules in such lamellar crystals. Here the PE chain is folded – with adjacent reentry of the chain after each fold – within the (110) plane of the ortho-rhombic PE crystal. The unit cell dimensions have been determined [19] to be

a = 0.74 nm
b = 0.493 nm
c = 0.353 nm (chain axis direction).

The largest of the lozenge-shaped single crystals shown has an edge length of 4700 nm and a thickness of about 10 to 15 nm (originally the crystal seems to have been a hollow pyramid which became flattened during preparation). Lamellar crystals – including extended chain crystals – of various degrees of perfection and size are the product of crystallization from solution or from the melt under appropriate conditions. The extreme mechanical anisotropy of the chain molecules is reflected by that of their crystallites. For example the elastic response of PE single crystals in the three axis directions differs by two orders of magnitude. From the shifts in equatorial reflections of the X-ray patterns Sakurada et al. [20] have determined the elastic moduli to be:

$E_a = 3.1\ GN/m^2$ $\qquad E_b = 3.8\ GN/m^2$ $\qquad E_c = 240\ GN/m^2$.

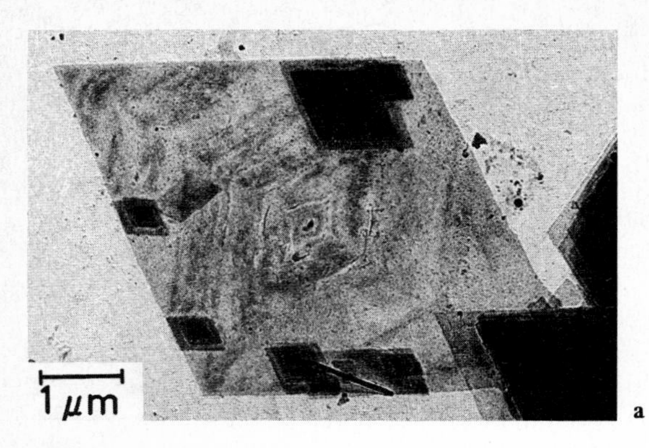

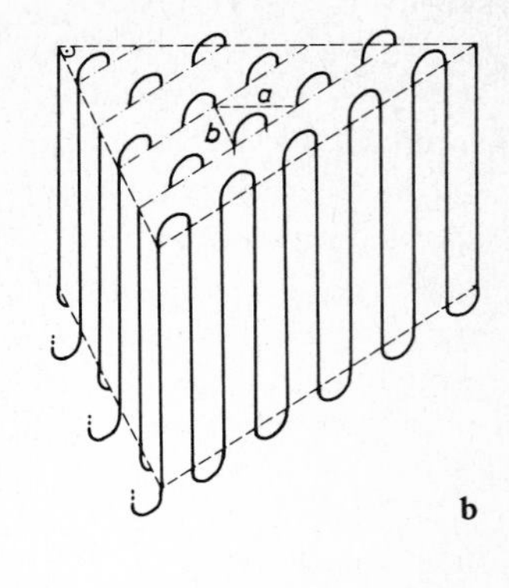

Fig. 2.3 a u. b. Polyethylene single crystal. **a** microphoto of a crystal grown from dilute solution (Courtesy E. W. Fischer, Mainz), **b** model of chain folding

It should be noted that these values may depend on the experimental method used for their determination [56]. Further information on crystal structure, crystallization kinetics and melting is found in the general [2, 3, 107] and specific references [56, 57, 126–131] of this Chapter.

C. Superstructure

At this point the superstructure or morphology of unfilled solid polymers is reviewed with a desire to identify and define elements of their heterogeneity. Heterogeneities can evidently arise from phase separation in copolymers or blends. Structural inhomogeneities can also be due to particularities of the polymerization and/or processing technique as is the case with globular structures and variations in cross-link density. However, the most important features of polymer superstructure certainly arise from the tendancy of many linear macromolecules to crystallize and to form polycrystalline aggregates.

1. Spherulitic Structure

Spherulites are the most common type of organized superstructure found in semicrystalline polymers. In a simplified manner a spherulite can be viewed as an aggregate of twisted crystal lamellae (Fig. 2.4). The radially symmetric growth of such lamellae from several adjacent nuclei leads to the typical spherulitic structure shown in Figure 2.5. Spherulites are the largest and dominant feature of melt crystallized specimens. Their size has a strong influence on sample strength, coarsely spherulitic material being rather fragile [129–130]. This influence and the means to control spherulite size through nucleation agents will be discussed in Section 9 III.

In the seventies it was discussed [91, 92, 100] whether and to what extent the lamellar crystals of a solid semicrystalline polymer show the same regularly folded structure as solution grown single crystals. Schelten et al. [91] concluded that the results of the small-angle neutron scattering of deuterated PE molecules (PED) in a

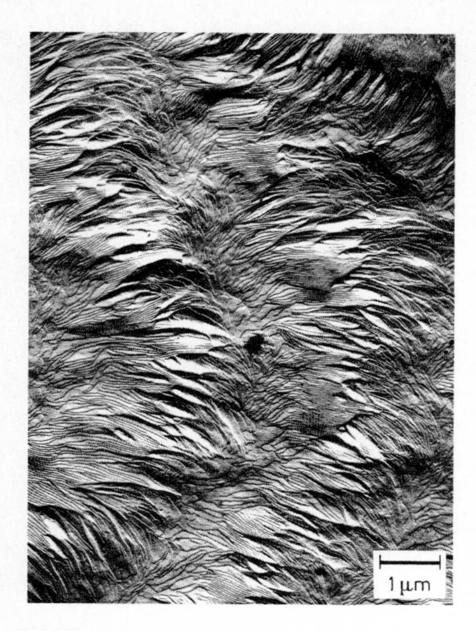

Fig. 2.4. Detail of spherulite in polyethylene showing stacks of crystal lamellac (Courtesy E. W. Fischer, Mainz)

Fig. 2.5. Spherulitic structure of meltcrystallized polyoxymethylene

protonated PE host (PEH) are incompatible with a regularly folded morphology. Yoon and Flory [92] supported Schelten's conclusions. They compared the intensity distributions calculated for several morphological models with the experimental scattering data [91]. In their model they recognized the parallel arrangement of stems within the crystalline layers, the existence of an *interfacial region* which contains the folded sections interconnecting two stems of the same crystalline layer, and an amorphous region. The best fit with the experimental data could be obtained by Yoon and Flory [92] for a model where 70% of the stems forming a given crystalline layer were connected with each other by *irregular, non-adjacent re-entry* of the chains from the interfacial layer (switchboard model). The remaining portion (30%) of stems was connected to the amorphous region and, subsequently, to the adjacent crystalline layer. In a recent series of papers the same authors [214, 215] have developed a lattice theory to describe in detail the transition from perfect crystalline order to an isotropic amorphous state. Assuming infinite molecular weight (absence of chain ends) the authors find that of all the chains emanating from a crystal 70% undergo reversal without proceeding beyond the third lattice layer (which corresponds to an interphase thickness of 1–1.2 nm). The fraction of chains that become engaged in adjacent folds should be 0.2 or less for most semicrystalline polymers. For PP however, the probability for adjacent reentry seems to be close to zero [213]. After extended analysis of the available data, Fischer et al. [100, 126, 127] stated that the radii of gyration (of quenched PE, of PEO crystallized by slow cooling, and of isothermally crystallized isotactic PP) were only slightly affected by crystallisation or annealing even though drastic changes of the long spacing (12–25 nm) had occurred. On this basis Fischer [126–127] advanced a "solidification-model of crystallization" which excludes large scale diffusional motion of chain segments during crystallization.

A different situation may arise if the crystallization growth rate becomes comparable or lower than the rate of conformational rearrangement in the melt. In fact, Guenet and Picot [128] studied the chain conformation in isotactic polystyrene and observed large crystallized sequences in samples of relatively low molecular weight and at high crystallization temperatures.

The boundaries between spherulites resemble grain boundaries; these grain boundary zones are enriched in low molecular weight material, impurities, chain ends, and defects. The deformation and strength of such a "composite" structure naturally depend on the compliances of all of its components. In such a composite large compliances (small constants of elasticity) have to be assigned to the cohesion across grain boundaries and across lamellar fold surfaces. The cohesion between chains within a crystal lamella is much stronger than the intercrystalline interaction; this renders a certain stability to the lamellar elements in sample deformation. The deformation behavior of such a non-oriented semicrystalline polymer will, therefore, depend much more on the nature of the secondary forces between the structural elements than on the length or strength of the chain molecules.

2. Oriented Lamellar Structures

Oriented lamellar structures are obtained through cold or hot drawing of semicrystalline polymers and by spinning from solutions or gels. These structures comprise those found in synthetic fibers known and used for a long time [20, 35, 132–135], the shish-kebab or row-nucleated structures [2, 3, 36, 136–139] and the springy or hard-elastic polymers [46–51, 105].

The morphology of fibers is extensively discussed in the literature (see for instance 141). The essential feature of many fibers is their *microfibrillar, sandwichlike structure.* The role of this microfibrillar structure in fiber deformation and fracture will be discussed in more detail in Section B and in Chapters 7 and 8. In Figure 2.6 models of

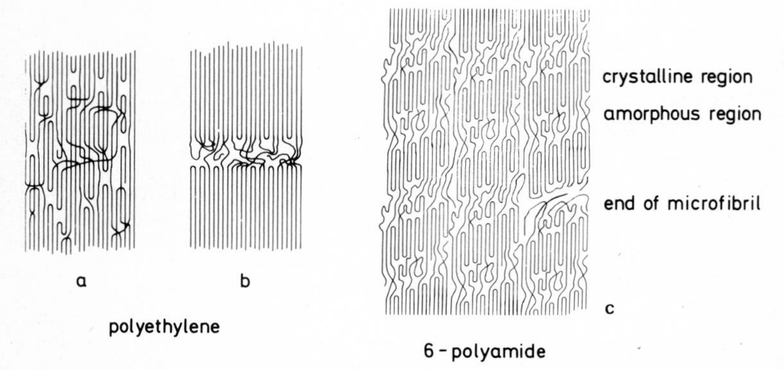

Fig. 2.6 a–c. Models of the structure of drawn semicrystalline polymers: **a** polyethylene as drawn [131]; **b** annealed [131]; **c** 6-polyamide as drawn

such microfibrils are shown. The very high degree of chain orientation in the crystalline *but also* in the amorphous regions of the microfibrils should be noted. Comprehensive reviews on these microfibers within fibers, especially of their dimensions as a function of fiber formation technique, have been given by Tucker and George [35–36]. According to them the transverse dimensions of these microfibers are of the order of 0.8 to 20 nm in various celluloses, and between 7 and 50 nm in most thermoplastic fibers; microfiber lengths are given as 18 to 1000 nm [35].

For a study of the effect of chain length and strength obviously structures which contain a high degree of oriented chains are more promising than others. Among the most widely investigated ones are the semicrystalline fibers. Their deformation and fracture behavior will be discussed in connection with the electron spin resonance investigations of chain scission especially observed in these fibers.

Fibrous crystals or shish-kebabs have first been observed in the late fifties, subsequently they have been studied intensively, especially by Keller [136], Lindenmeyer [137] and Pennings [138–139]. They are readily grown in supercooled, stirred or agitated solutions and are characterized by strings of platelets which are nucleated by and growing on the fibrous backbones previously formed within the solution (Fig. 2.7). In 1976 Zwijnenburg and Pennings [37] report on a technique to obtain PE macrocrystals of virtually infinite length. There, a seed crystal and – subsequently – the growing crystalline PE fiber is pulled under steady-state conditions from a flowing solution. The diameter of the obtained fiber, an agglomeration of shish-kebab fibrils, varied between 10 and 40 μm [37]. These fibers may already be considered as belonging to the *ultra-high modulus* polymers, where an extremely high degree of orientation, stiffness and strength has been obtained [143–154].

In the seventies it had been discovered that by controlled drawing and annealing treatment of a number of fiber forming thermoplastics a highly crystalline morphology with "springy" or hard elastic response could be obtained [46–51, 105, 140]. Springy behavior, the characteristics of which will be discussed in the next section, has been found so far in PP, POM, PE, PMP, PES, and even in PA 66 [47, 105]. The morphology is proposed to consist of stacks of flat, highly regular, c-axis oriented (folded) chain lamellae with the lamellae interconnected by amorphous chains [46–51, 105, 140].

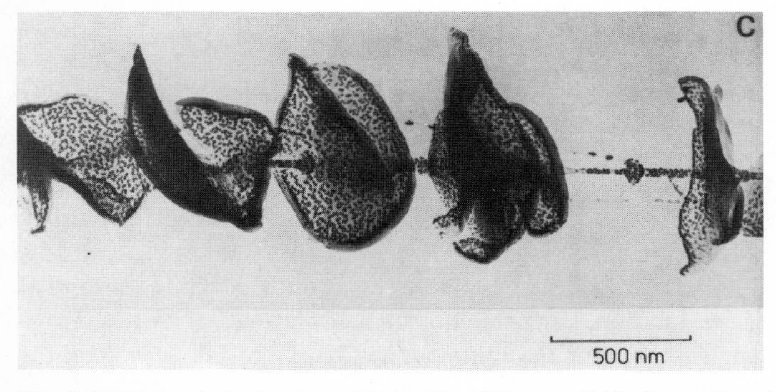

Fig. 2.7. Shish-kebab structure obtained by Hill et al. [142] from ultra high molecular weight polyethylene after growing in stirred xylene solution at 106 °C and storage at 96 °C for 66 hours; the average distance between overgrowths in this sample is 250 nm (Courtesy Prof. A. Keller)

3. Ultra-high Modulus Polymers

During the past decade several techniques to obtain ultra-high modulus fibers have been developed [143–144]. Porter et al. [40, 41, 43, 44], Takayanagi et al. [38, 145], and Capaccio and Ward [143, 146, 153] employed the solid state extrusion of thermoplastic polymers involving very high pressure (up to 5 kbar = 0.5 GN/m^2), temperatures between 30 and about 250 °C, and an extensional flow field. In the case of PE such a treatment led to a new high-pressure, high-temperature phase of hexagonal symmetry at a high level of chain extension. In analogy to the extended-chain crystals observed by e.g. Anderson [45] in the fracture surfaces of low-molecular weight PE the term *extended-chain* crystals is being used for pressure crystallized PE also. Weeks and Porter found that highly oriented strands of such material (M_w = 58000) have an exceptional stiffness of 70 GN/m^2 at room temperature which is comparable to that of E-glass [40]. In addition a good tensile strength of 500 MN/m^2 is reported [41]. Lupton and Regester [39] worked with ultrahigh-molecular-weight (UHMW) HDPE with M_w between 2 and $3 \cdot 10^6$ g/mol and obtained a stiffness of 2 to 3 GN/m^2 and a strength of 33–39 MN/m^2, but an excellent tensile-impact resistance of 300 to 600 kJ/m^2. Capaccio and Ward [93–98, 146, 153] prepared ultra oriented polymers (PE, PP, POM) by means of cold drawing and hydrostatic extrusion. A fibrillar structure seemed to be retained in those samples. With increasing draw ratio the morphology would be characterized primarily by a continuum of oriented material in which a statistical distribution of chain ends would constitute the only discontinuities. Thus an increasingly smaller fraction of the material would show the conventional morphology comprising crystals and oriented amorphous chains, including tie molecules, in series.

For linear polyethylene Capaccio [146] finds that the fiber elastic modulus (between 5 and 60 GPa) is a unique function of draw ratio (λ = 1.0 to 40). He proposes a network superstructure in which entanglements and crystalline regions provide the junction points. Arridge and Barham propose for cold drawn PE in the module range of 4 to 90 GPa a fiber composite model [147]. Ultrasonic measurements of the elastic constants [148] and X-ray studies of the annealing behavior of ultra-oriented HDPE [149] permit to characterize the physical network structure even further.

The pressure crystallization of polymide 6, the morphology of the ensuing material and its fracture properties have been extensively investigated by Gogolewski et al. [150] and are discussed in Section 9 III A.

A recent review by Ohta [144] on the processing of ultra-high tenacity fibers treats six different flexible polymers with strength values ranging from 1 GPa (PP at λ = 28) to 1.7 GPa (PA 6, draw ratio not indicated), 3.2 GPa (Kevlar®, commercial fiber), and 4.3 GPa (PE at λ = 35).

A drastically enhanced effective drawability of high molecular weight polyethylene can be obtained by spinning or casting from *semidilute solutions* [151, 152, 154]. Smith, Lemstra and Booij have argued that a relatively low entanglement density is present within a semidilute solution. Quenching of such a solution transforms it into a *gel* and fixes the entanglement network. Spinning (at 120 °C) of such a gel permits to attain draw ratios up to 72.

The structure, deformation, and fracture of high strength fibers from rigid rod molecules (such as Kevlar®) will be discussed in Chapter 8.

4. Globular Aggregates

Some submicroscopic structural features have been observed in amorphous and semi-crystalline polymers, which are due to particularities of the polymerization technique. As an example some fibrous or globular structures are shown (Fig. 2.8) which had been observed by Wristers [21] after the initial phase of polymerization of a poly-olefine.

These structures are obtained after polymerization in gaseous or liquid environment and at low or high efficiency of various Ti-, V-, Cr-, or Al-based catalysts. In Figure 2.8 electron micrographs of such nascent polymer structures are reproduced [21]. Polypropylene at low catalyst efficiency is formed in globules of 0.5 μm diameter (a), at high catalyst efficiency fibers several μm long are observed (b); the fiber diameter is closely correlated with the lateral face dimension of the primary catalyst crystal and varies between 0.37 and 2 μm for $TiCl_3$-crystal breadths of between 5 and 50 nm. Polyethylene samples prepared with $TiCl_4$-$Al(i\text{-}Bu)_3$ or other colloidal catalysts show less regular surface structures (c).

In the course of the bulk polymerization of lactams it was observed [22] that the resultant polymers precipitated in the form of spherulites. Other examples are the fibrous, globular, or star-shaped aggregates existing in PVC [23–26] which have sizes

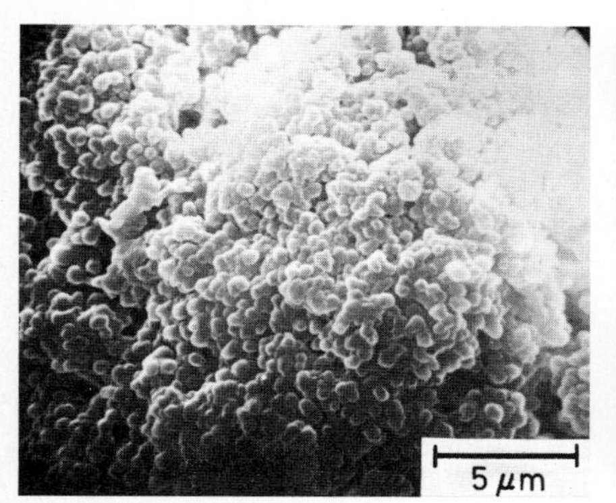

Fig. 2.8a. Nascent polymer morphology: polypropylene prepared at low catalyst efficiency

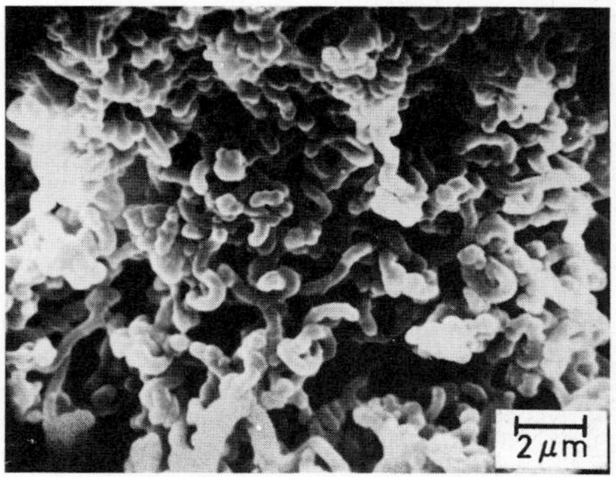

Fig. 2.8b. Polypropylene prepared at high catalyst efficiency

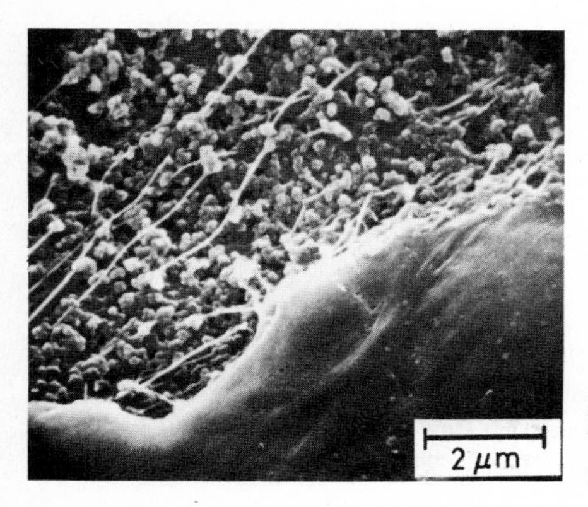

Fig. 2.8c. Polypropylene prepared with $TiCl_4$-$Al(i\text{-}Bu)_3$. Figs. a–c were taken from Ref. [21]

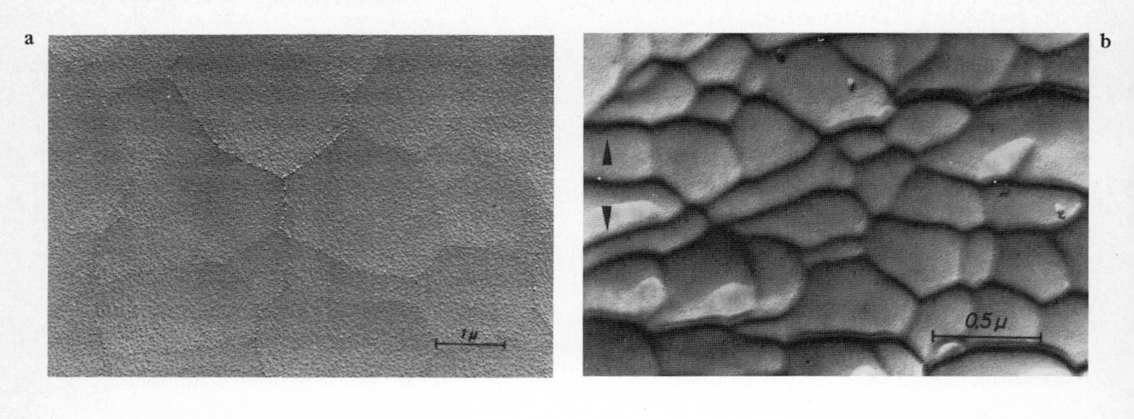

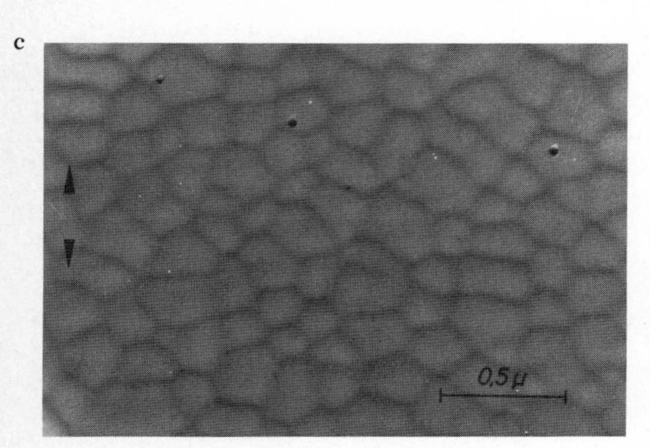

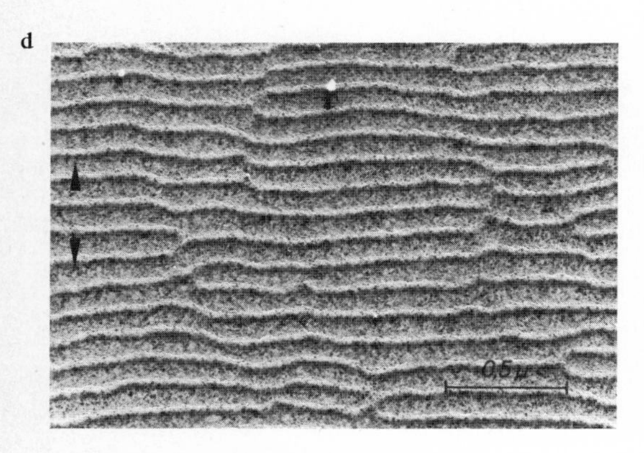

Fig. 2.9 a–d. Ion-etching patterns obtained by Grosskurth [158] on surfaces of oriented polystyrene: **a** no orientation; **b** drawn 25%; **c** drawn 50%; **d** drawn 500%. (Transmission electron micrograph of surface replica, courtesy Prof. K. P. Grosskurth)

of between 10 and about 500 nm. In good solvents the largest aggregates disintegrate into single molecules but only at temperatures higher than 200 °C [25].

Whereas the above globular superstructures can fairly unambigously be identified and related to the polymerization technique, there are others which can only be seen by specific methods and where an explanation is not obvious. From the evaluation of SAXS curves Reichenbach et al. [155–156] derive a superstructure in PVC which changes with the degree of polymerization. They infer a tetrahedral arrangement of 1.2 nm small PVC particles at 1% conversion which turns into the hexagonal arrangement of 7.8 nm globules at a conversion of 8.9%. The latter should form the globules of 0.1 to 1 μm generally visible by electron microscopy [24, 155].

Grosskurth [157–159], Kämpf and Orth [160], and Schwarz [161] have made the interesting observation that ion etching of oriented PS, SAN, PMMA, PC, PETP, and PP led to a globular surface structure which was unambigously related to the degree and the direction of orientation (Fig. 2.9).

The question arose whether in this case ion-etching had revealed a preexisting morphology or whether it had created a surface structure. If the globular structures existed in the solid polymer it should be possible to trace these structures by other, very sensitive means (light scattering for instance). Pursuing this interesting problem further Grosskurth et al. [162] made the following observations:

- a polystyrene sample which in light scattering had proven to be "absolutely clean", i.e. free from density fluctuations other than thermal fluctuations [125], did not show any systematic globular structures after prolonged oxygen etching (only in some surface areas traces were observed which might well have been caused by the polishing procedure);
- in isotropic PMMA which was free from molecular orientation and internal tensions a very weak, grain boundary-like network with grain sizes of between 300 and 500 nm was obtained [159];
- in non-oriented PC a certain globular structure was observed on ionetched surfaces [162]; the characteristic distances between boundaries (150 to 300 nm) were somewhat larger than the correlation lengths inferred from light scattering (90 to 140 nm of unetched bulk samples [125, 181]); in oriented PC globular and row structures appeared;
- globulit structures have also been seen by transmission electron microscopy in highly oriented OsO_4-stained ultramicrotome PS slices *not* subjected to ion etching [159];

Recently, Neppert et al. [163] studied the sputtering mechanisms in the ion etching of isotropic and oriented semicrystalline and amorphous polymers (PS, PC, PE; PMMA); they also observed highly excavated as well as weakly modulated sample surfaces; they related these structures to the fact that target sputtering yields depend on the angle of incidence of the ions (and thus on local surface curvature), on chain orientation, on the gradient of the chemical potential μ and on secondary fluxes of ejected atoms. They state that chains at the surface cut during ion etching would relax and diffuse. The effect of this directed surface diffusion is to roughen the surface into rows which are subsequently modulated by the incoming ions [163].

According to the latter hypothesis, the strong dependence of chain relaxation and diffusion on the structure of the network and on its frozen-in anisotropy would explain the excellent correlation between the degree of sample orientation and the characteristic surface features. This correlation has been verified by Grosskurth [157–159] for a large variety of molecular and processing parameters of PS and PMMA.

5. Block Copolymers and Blends

A much more distinct structure is found in the large group of multi-phase polymeric systems formed by the block copolymers, graft copolymers, random copolymers, and poly-blends. This group of polymers has received increasing attention (e.g. [27–29, 164–167]) because of their improved, controllable, "tailor-made" properties which have made them excellent engineering materials. Their superstructure is determined by the relative weight, sequential order, and limited miscibility or even incompatibility

- of the covalently connected molecular segments of hard (glassy, thermoplastic) and soft (elastomeric, rubbery) chains as within the (two- or three-) block copolymers and random copolymers
- of the (two) phases grafted onto each other within graft-copolymers or
- of the molecules mixted together in poly-blends.

The variety of ensuing multi-phase structures cannot be discussed here in any detail; it ranges from the extremely ordered hexagonal patterns of spherical domains of one phase embedded within the continuous phase (mostly of the dominant component) to the partial phase separation of solution-cast poly-blends occurring only after heat treatment [27–29]. In Figure 2.10 an electron micrograph of the morphology of a butadience styrene diblock copolymer is reproduced [31]. At 17% by weight the styrene forms the discrete phase of spherical or cylindrical aggregates in a continuous butadiene phase. Owing to the homogeneity of block lengths the aggregates are arranged in highly regular lattices. The sizes of the aggregates correspond very well to the volume occupied by the styrene blocks. The spherical or cylindrical aggregates must be larger than one diameter of the coil formed by the PS segment and must be smaller than two such diameters to permit phase separation while maintaining material continuity [28–30].

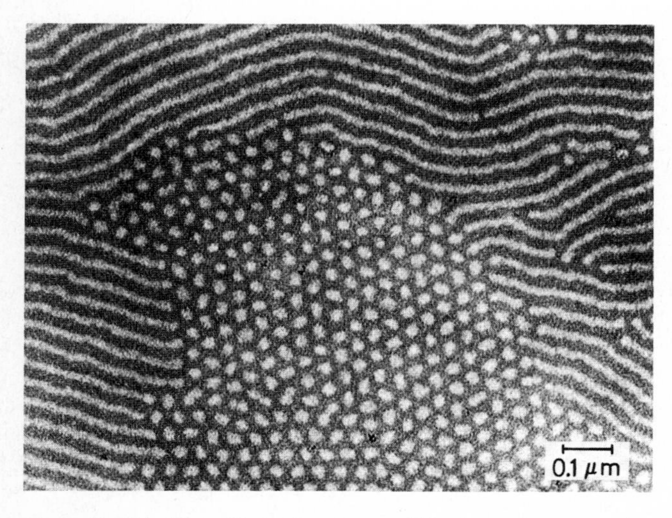

Fig. 2.10. Thin film of butadien-styrene diblock copolymer cast from petrol ether solution and annealed 1 h at 100 °C. At 17% by weight the styrene aggregates in the form of spheres and of cylinders (Courtesy G. Kämpf, Uerdingen)

In view of their interesting properties and their large technical importance block copolymers and blends have been extensively discussed in the literature. At this point be cited the "classical" books by Bucknall [164], Manson and Sperling [28], and Paul and Newman [165]. Aspects of the thermodynamics and miscibility were treated by Lipatov [168] and Prud'homme [169], advances in the theory of polymeric alloys were reviewed by Noolandi [170].

A rather intimate combination of two polymers in network form are the *interpenetrating polymer networks (IPN).* They are defined as a material containing two polymers, each in network form, one of which is generally synthesized in the presence of the other. The IPN's can be distinguished from simple polymer blends, blocks or grafts in that they swell, but do not dissolve in solvents and that creep and flow are suppressed [171–173]. A monograph on their structure, behavior and application has been publishers recently by L. H. Sperling [171].

With respect to another interesting area, the structures of *liquid crystallike polymers,* the reader is referred to the following pertinent references [174].

D. Characterization

The experimental methods available to determine or characterize the structure of polymer chains or of their aggregates are discussed in the general references of Chapter 1, by Hoffmann et al. [175] and Kämpf [176], and in [177]. In addition specific information on X-ray diffraction [178], neutron scattering [179, 182], electron- and light-scattering [4, 52, 53, 180–182], optical and electron microscopy [3, 14b], thermal [3, 54] and viscoelastic behavior [14c, 55–57], and nuclear magnetic resonance (NMR) technique [180, 182] can be obtained from the references indicated. In his book [183] Koenig describes how these methods can be employed to analyse the chemical microstructure of polymer chains (composition, stereo regularity, crosslinks, branches). Recent progress and new methods to characterize the solid state are discussed in two volumes of the Advances in Polymer Science, edited by Kausch and Zachmann [184]: Synchrotron radiation, solid state ^{13}C-NMR (including the magic angle spinning technique), deuteron-NMR, quantitative electron microscopy, fluorescence spectroscopy.

Electron spin resonance (ESR) and infrared absorption (IR) techniques will be introduced in Chapters 6 and 8 respectively. Except for the microscopic methods all of the mentioned techniques provide an average information, an information to which all structural elements within the active volume of an investigated sample contribute according to their individual response, concentration, and position. Any macroscopic information which is not solely taken as such but is used to characterize an inhomogeneous solid generally involves elemental response functions and orientation distributions which are not completely known. For this reason a certain ambiguity is introduced into the deconvolution of such experimental data. This leaves some room for individual interpretations as to the concentration and nature of structural elements. This ambiguity has been inherent in the discussion of the existence or non-existence of close range order in amorphous polymers and of the nature and distributions of defects within crystallites. Of even more concern for the structural interpretation of mechanical properties is the fact that the nature of the connecting members between structural elements is much less well known than the nature of the latter. This is true on all levels: for the arrangement and interaction of non-ordered chain segments, for the fold surfaces of lamellar crystals and the connections between them, for the amorphous regions between crystallites in fibers or unoriented polycrystalline samples and for the boundaries of microfibrils or spherulites. This will be further illustrated in the next section.

II. Deformation

A. Phenomenology

The deformation of the complex structures described in the last section involves a number of different phenomena ranging from a linear elastic to a highly viscous response.

In the following three sections will be outlined the characteristic deformation behavior of the various structures treated above. The molecular mechanisms and the principal laws of deformation will be indicated. They will be more extensively treated in Chapters 8 to 10. For any general information concerning the deformation of single crystals [3, 57, 141, 185], amorphous [1, 188], semicrystalline [188], or oriented polymers [132, 133], ultra-drawn fibers [143], elastomers [186], and toughened plastics [28, 164–166, 187] or interpenetrating networks [171] the reader is also referred to the indicated literature.

1. Thermoplastic Polymers

The characteristic quasi-static uniaxial-deformation behavior of different *thermoplastic polymers* in the non-rubbery temperature region is shown in Figure 2.11. The stress-strain curve a) is that of a brittle polymer (polystyrene at room temperature) characterized by its very limited extensibility and the steep and monotone increase of stress with strain. Curve b) represents a hard elastic polymer, lamellar polypropylene film [58], which combines strong elastic behavior with good extensibility at a high stress level and almost complete recovery (within a period of days). Curve c) is that of a ductile polymer. The initial monotone increase of the engineering stress generally is less steep than that of a brittle polymer, i.e. the secant modulus is smaller. The engineering stress σ reaches a maximum at the yield point (Streckgrenze) which marks the beginning of the so-called cold drawing, a notable reduction in cross-section of the strained specimen. If the stress-strain curve extends considerably beyond the yield point one speaks of a tough polymer; the shape of the curve depends on the strain hardening behavior of the polymer and on the tendency to neck formation. Curve d) is characteristic for soft polymers, e.g. for thermoplastics at a temperature close to their glass transition temperature or plasticised polymers, which do not yet show rubber elastic behavior.

As indicated earlier the different stress-strain curves are not characteristic for particular, chemically defined species of polymers but for the physical state of a polymeric solid. If the environmental parameters are chosen accordingly transitions from one type of behavior (e.g. brittle, curve a) to another (e.g. ductile, curve c) will be observed. These phenomenological aspects of polymer deformation are discussed in detail in [14], [52–53], [55–57], and in the general references of Chapter 1. A decrease of rate of strain or an increase of temperature generally tend to increase the ductility and to shift the type of response from that of curve a) towards that of curves c) and d). At small strains (between zero and less than one per cent) the uniaxial stress σ and the strain ϵ are essentially linearly related (Hooke's law):

$$\sigma = E \cdot \epsilon \tag{2.1}$$

Even at these small deformations the apparent Young's modulus E is a function of the rate of strain. This shows that E is not solely determined by the energy elastic deformation of bond angles, bond lengths, and intermolecular distances but also involves *time-dependent* displacements of atoms and small atom-groups. In the following region (1 to about 5% of strain) stress and strain are no longer proportional to each other, structural and conformational rearrangements occur that are mechanically although not thermodynamically reversible, one speaks of *anelastic* (viscoelastic in the narrow sense) or *paraelastic* behavior. Beyond the yield point large scale reorientation of chains and lamellar crystals begins, a process generally termed plastic deformation. A *true plastic deformation* can be understood as the transformation from one equilibrium state to another equilibrium state without the creation of frozen-in tensions. The latter point is particularly important in view of the fact

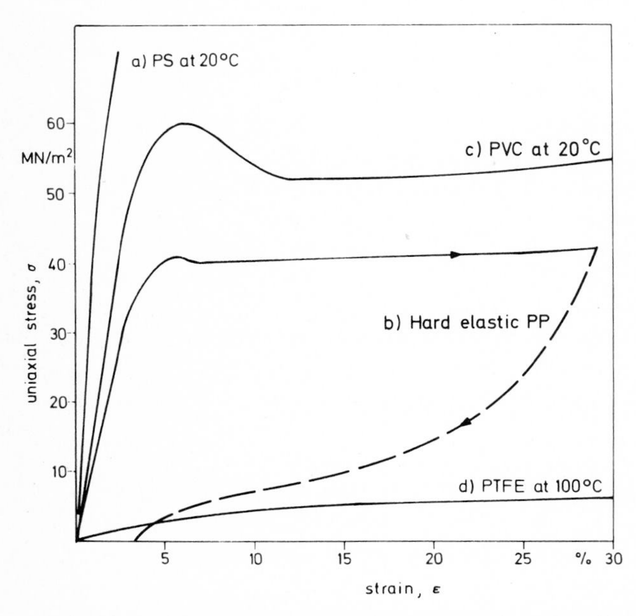

Fig. 2.11 a–d. Characteristic uniaxial deformation behavior of polymers. **a.** brittle (PS at room-temperature), **b.** hard elastic (lamellar PP film), **c.** ductile (PVC at room-temperature), **d.** soft (PTFE at 100 °C)

that the average post-yield deformation is brought about largely by mechanically reversible anelastic conformational changes of the molecules and not by flow of molecules past each other. Unless a state of equilibrium had been established through proper annealing treatment highly drawn samples may retract considerably. Within the context of this book emphasis should be placed, however, not so much on the processes leading to or accompanying molecular reorientation – which is basically a reinforcing effect – but on the damaging processes of chain scission, void formation, and flow. The latter processes gradually occur in the strain region from just before the yield point up to final fracture. Among the damaging processes one has also to include the phenomenon of normal stress yielding or crazing of glassy polymers which will be discussed in Chapter 9.

In *oriented fibers* generally a slightly non-linear anelastic deformation up to brittle fracture is observed, shish-kebabs seem to deform by transformation of lamellar into fibrillar material [217]. *Ultra-high modulus filaments* (of PE) show a strongly strain-rate dependent yield point. At strain rates of $\dot{\epsilon} > 3 \cdot 10^{-3}\ s^{-1}$, however, the PE filaments break in a brittle manner at stresses of up to 1.2 GPa [190]. This subject, the fracture of oriented or ultra-drawn fibers will be resumed in Chapter 8.

2. Elastomers

At the time of this writing it is exactly 50 years since Kuhn [191] presented his famous concept on the conformation of chain-like molecules. The representation of a chain by statistical links Kuhn had applied for the first time, has established the theoretical basis in such different fields as rubber elasticity, flow- and stress-induced optical anisotropy or, as already discussed in Section A, of the structure of amorphous polymers. Kuhn [191, 192], Guth and Mark [193], and Meyer and Ferri [194] also realized that the elastic response of an ideally flexible chain depends on the change of entropy accompanying its elongation and not on the change of internal energy.

Application of the Gauss approximation of chain-end distances to a rubbery network (which is then called a "Gaussian Network") yields the following relation between uniaxial stress σ, extension ratio λ and shear modulus G:

$$\sigma = G(\lambda - 1/\lambda^2). \qquad (2.2)$$

The shear modulus G is proportional to the absolute temperature T, Boltzmann constant k, and the number N of subchains between cross-links. For polymer chains of number average molecular weight M which are subsequently cross-linked, one obtains:

$$G = NkT = \rho RT(1 - 2M_e/M)/M_e \qquad (2.3)$$

where M_e is the number-average molecular weight of the subchains and ρ is the polymer density.

The stress-strain curve as given by Eq. (2.2) is plotted in Figure 2.12 using a value of G of 0.4 MPa; it is compared with the experimental data of Treloar [59]. The Gaussian approximation, which breaks down at higher strains, has been refined in

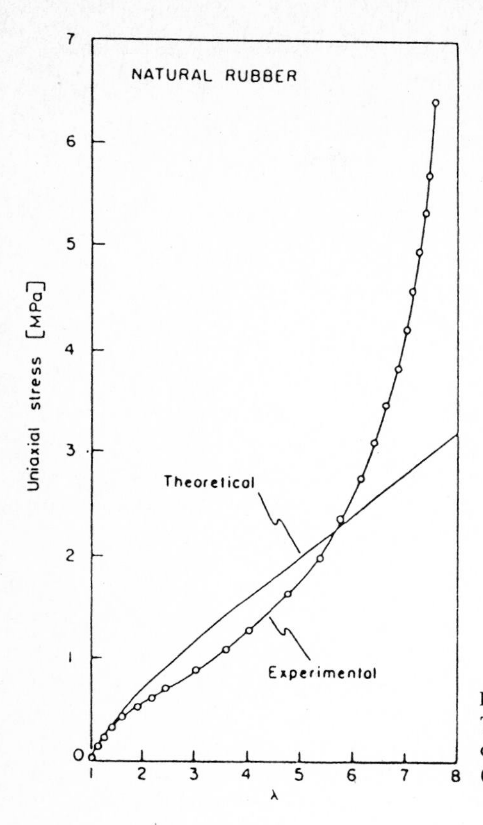

Fig. 2.12. Stress-strain curve of natural rubber. The theoretical curve is calculated on the basis of Eq. 2.2 using a shear modulus G of 0.4 MPa (after Treloar, 59)

several ways to account for this discrepancy [see for instance the general references 59, 186, 195]. In Chapter 5 it will be necessary to analyze in particular the elastic behavior of flexible chains *at very high elongations* where one has to expect notable energy-elastic contributions to the axial chain stresses (and thus to stress-induced chain scission).

3. Phenomenological Theory of Viscoelasticity

A purely elastic deformation is mechanically completely reversible and does not involve chain scission or creep. A *real* rubber or thermoplastic polymer, however, like any viscoelastic solid, shows on a molecular scale energy *and* entropy elastic deformation accompanied by viscous flow. Thus it shows stress relaxation at constant strain, creep at constant load, and energy dissipation in dynamic excitation. Representation of the macroscopic mechanical response of viscoelastic solids – even in a strain region where no large-scale reorientation occurs – requires, therefore, the use of damped elastic elements consisting of springs (modulus G) and rate sensitive damping elements (dashpots characterized by a viscosity η). The simplest elements are the Maxwell element with spring (G) and dashpot (η) in series and the Voigt (Kelvin) element with spring G and dashpot η in parallel. The Maxwell element has a relaxation time $\tau = \eta/G$, the Voigt-Kelvin element has the same relaxation time which here more precisely is called retardation time. The phenomenological theory of viscoelasticity [55] describes the mechanical response of a body in terms of a distribution of basic viscoelastic elements primarily characterized by their relaxation times τ_i. If the spectra of the molecular relaxation times, H (lnτ), are known, then the viscoelastic modulus functions are, in principle, derivable therefrom [14c, 14d, 55]. The time-dependent stress relaxation modulus in shear G(t) becomes in terms of a continuous spectrum of relaxation times:

$$G(t) = G_e + \int_{-\infty}^{+\infty} H \exp(-t/\tau) \, d \ln\tau \quad (2.4)$$

where G_e is the modulus at infinite times which – in the case of a solid – is supposedly larger than zero. The dynamic modulus functions are conveniently treated as complex quantities consisting of a (recoverable) storage and an (irrecoverable) loss component; the ratio of loss (e.g. G'') and storage component (G') is called the loss tangent or loss factor, i.e. $\tan \delta = G''/G'$. The phenomenological theory of viscoelasticity is based on a linear system of differential equations, i.e. on the linear additivity of the effects of mechanical history. It should be emphasized that the aim of that theory is not so much to derive the form of the relaxation time spectra from detailed structural considerations but rather to utilize the response functions obtained in one experiment, e.g. creep, to predict the response of the same material under different excitations, e.g. dynamic loading [14, 55]. The general application of this approach is restricted to the linear range of response, i.e. to strains smaller than 0.5% [196–200]. Beyond that limit strain-softening occurs. Thus, Koppelmann et al. [198] and Rahaman et al. [199], observed a small decrease of G', an increase in G'' and in the activation volume for creep for PMMA, PEMA, PnBM, and PA 66. In fact, Wendorff [201] and Jansson [202] associate with the deviation from a linear behavior the appearance of the first *defects*. In a formal manner the theory of viscoelasticity has been extended into the non-linear range [14d], it has also been adapted to account for limited changes in the structure and/or of the orientation of the stressed system [14e]. In the context of this book use will be made of the established results of the phenomenological theory as well as of the "molecular theory" of viscoelasticity [55–57], particularly of the molecular friction coefficient introduced in it.

B. Molecular Description

A kinetic theory of fracture intends to interrelate the motion and response of molecules to the ultimate properties of a stressed sample. A kinetic theory, therefore, entails a molecular description of the deformation of the microscopically heterogeneous and anisotropic aggregates of chains to such an extent that critical deformation processes can be identified. The macroscopic deformation of any aggregate potentially involves the deformation, displacement, and/or reorientation of so different substructural elements as bond vectors, chain segments or crystal lamellae. The molecular origin of observed deformation mechanisms have been elucidated by various spectroscopic methods – including mechanical relaxation spectroscopy – and by the previously listed tools which characterize the morphology. They are supplemented by dynamic scattering and diffraction techniques [52, 53].

The following molecular description of the uniaxial straining of unoriented semicrystalline polyethylene characterizes the ductile deformation of fiber-forming spherulitic thermoplastics and may serve to illustrate the large variety of deformation mechanisms involved. At small strains of less than one per cent one observes the anisotropic elastic response of the (orthorhombic) PE crystallites [57] and of the amorphous material [53]. At the same time those anelastic deformations of CH_2-groups and chain segments occur that characterize the low temperature β-, γ-, and δ-relaxation mechanisms [10, 56]. At larger strains (1 to 5 per cent) additional chain segments change their relative position and conformational changes are initiated (bond rotation). A detailed study on the behavior of the chains within amorphous regions had been carried out by Petraccone et al. [53]. In crystalline regions subjected to strains of this order, dislocations and dislocation networks are generated (in lamellar crystals observed by means of moiré patterns). Depending on environmental parameters and the type of the dislocations their movement leads to crystal plastic deformation through twinning or slip or to a phase transformation from the orthorhombic to a monoclinic unit cell. An extensive review on the deformation behavior of polymer *single crystals* has been given by Sauer et al. [57] and in the book of Wunderlich [3]; a detailed account of the contribution of the various structural elements and defects to the deformation of semicrystalline polymers is found in the literature in a large number of papers of which only a few have been cited here [47–62]. Although the above mentioned effects lead to a non-linear stress-strain response, the initially present substructure is still maintained. Such a deformation is called non-disruptive.

Still larger stresses cause a destruction of the substructure, involving reorientation of chain segments and of lamellar crystals (crystal rotation, chain tilt and slip), opening-up of voids, and the first breakages of chains take place. These processes account for the ductile deformation. At this stage the mechanical energy input is largely dissipated as heat as shown in later Chapters. Since the deformation proceeds at an almost constant level of engineering stress it has generally been termed plastic deformation – although on a molecular basis it does not correspond to the plastic deformation of metals. The initial post-yield deformation of PE is predominantly not brought about by mechanically irreversible flow. In a strain region of up to about 50 per cent the lamellar crystals slip and/or orient their a-axes perpendicular to the draw direction. The c-axes (chain axes) assume a preferred angle of 35 to 40° with respect to the draw direction. The latter orientation, however, is reversible and changes into a random distribution if external stresses are removed [61]. In a strain region of between 100 and 400 per cent an increasing c-axis orientation in the draw direction is observed. In this strain region the polymer undergoes the important transition from a spherulitic to a fibrillar structure. According to Peterlin [62] microfibrils with lateral dimensions of 20 to 40 nm are formed which contain nearly unmodified blocks of folded chains broken away from crystal lamellae but still interconnected by unfolded sections of tie molecules. A model of the fibrillar structure of a highly drawn fiber is shown in Fig. 2.6. The originally random distribution of chain segments and crystal blocks has become highly oriented. The load carrying capability of such a structure is – as will be studied in considerable detail – many times larger than that of the unoriented, spherulitic sample.

The principle mechanism permitting the large deformation of the springy or hard-elastic semi-crystalline polymers is the "adhesive fracture" between contacting lamellar crystals and their bending. The retractive forces are provided by stretched interconnecting fibrils and, as pointed out by Göritz [50, 140] and Wool [203], by the adhesive free energy.

A specific mechanism during large deformation of elastomers, to which attention must be drawn at this point is the strain-incluced crystallization [59, 186, 204]. The basic deformation behavior of elastomeric networks will be discussed in relation with that of a single chain. Excellent general introductions into rubber elasticity are given by Treloar [59] and Eirich [186].

Most of the above results have been obtained by analysis of X-ray diffraction patterns. In recent years considerable additional information on the behavior of individual chain segments or whole chains has become available through neutron scattering studies [179, 182], Fourier transform infrared (FTIR) spectroscopy [180, 206] and fluorescence spectroscopy [180, 207]. Thus, it results from recent SANS studies that on the scale of the radius of gyration the molecular coils in an isotropic amorphous polymer (PMMA) deform affinely if the polymer is stretched below or above T_g to between 60 and 170% [205]. Detailed information on a segmental level, e.g. on rotational transitions between isomers *during* the plastic deformation has become available through the FTIR technique [206]; its results will be discussed and used in Chapters 7 and 8.

The described molecular deformation processes affect the macroscopic fracture behavior of a given high polymer. The relative magnitudes of the different deformation modes completely determine:
- the level of stress and strain to be attained in nondisruptive deformation
- the volume density of elastically stored energy
- the rates of energy dissipation and local heating
- chain elongation, network reorientation and possible strain hardening
- chain scission, void formation, and other forms of structural weakening.

III. Model Representation of Deformation

For the interpretation of the complex mechanical response of a highly anisotropic polymeric network it is indispensable to have a simplifying model representation of the arrangement and interaction of the structural elements and of their deformation. Such model representations will be useful in the further investigations which necessarily focus on some aspects while averaging or neglecting a large number of others. In this section one would like to characterize the assumptions made on the geometry and mode of interaction of structural elements in basic theories of the deformation of a composite body. These theories had been developed to account for the behavior at small strains. They can be extended into theories of strength *if and only if* failure criteria are introduced which become effective within the range of validity of the particular deformation theory.

There is a large body of literature on the purely *elastic* interaction of non-prismatic structural elements within a deformed "composite" solid (e.g. 63–70); this literature also forms part of the theoretical foundation of the behavior of *viscoelastic* composite solids. While not generally discussing these foundations three groups of model descriptions of multi-phase materials will be singled out:

- those of Voigt [63], Reuss [64], or Takayanagi [71] which assume a homogeneous distribution of strain *or* stress,
- the calculations of Kerner [65], Hashin [66], van der Poel [67], or Uemura et al. [72] which minimize the elastic energy of a system containing assumedly spherical inclusions, and
- the empirically modified models of Eilers [68], Guth [69], Mooney [70], and Nielsen [73] which are based on the original calculations of Einstein [74], where the disturbances of an elastic field in the vicinity of – and caused by – completely rigid, spherical inclusions is considered.

It is to be investigated which *chain* properties are effectively recognized by these groups of model descriptions of polymer deformation. One knows that the semicrystalline samples discussed previously are polycrystalline solids containing dispersed amorphous regions with frequently ill-defined boundaries and equally ill-defined interaction between amorphous and crystalline regions. In a direct approach Takayanagi [63] represented semicrystalline polymers by "diagrams" (Fig. 2.13) involving straight boundaries and conditions of plane stress or strain. He obtained a meaningful interpretation of the relative contributions of crystalline and amorphous phases to the complex moduli of these materials. The Takayanagi approach can well accomodate information on the deformation behavior of individual crystalline [57] or amorphous regions [53] that becomes available from other sources. In the analysis of e.g. the response of a fiber to dynamic loading through a Takayanagi diagram the nature of chain segment orientation is considered in detail but likely distributions of chain length and axial chain stresses do not enter the model.

The two-shell model of Kerner [65] conforms to the conditions of the second group of models. The dilatation of a spherical inclusion surrounded by a homogeneous medium is derived subject to the condition that displacements and tractions at the surface of the inclusion are continuous. The homogeneous medium is supposed to have the elastic properties of the composite as a whole. The model interrelates

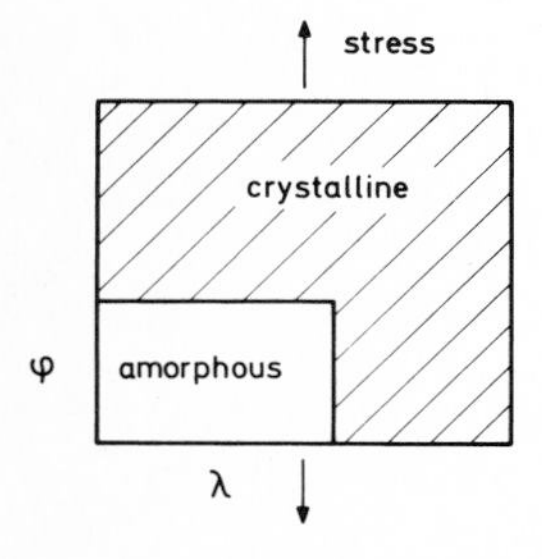

Fig. 2.13. Takayanagi diagram of semicrystalline polymer; content of amorphous material corresponds to $\lambda \cdot \varphi$

shear (G_i) and compressive (K_i) moduli (or Poisson's ratios ν_i) of an arbitrary number of isotropic elements with the macroscopic moduli G_c and K_c.

In the case of isotropic inclusions of an elastic shear modulus G_f comparable to that of the matrix, G_m, Eq. (2.5) describes the variation of the complex shear modulus, G_c^*, of the composite as a function of the volume content V_f of the discrete phase:

$$\frac{G_c^*}{G_m^*} = \frac{(1 - V_f)\, G_m^* + (\alpha + V_f)\, G_f^*}{(1 + \alpha V_f)\, G_m^* + \alpha (1 - V_f)\, G_f^*} \tag{2.5}$$

The quantity α is derived from the Poisson's ratio ν_m as:

$$\alpha = \frac{(8 - 10\, \nu_m)}{(7 - 5\, \nu_m)} \tag{2.6}$$

As will be noted no molecular anisotropies and no effects due to size and size distribution of the particles of the discrete phase are recognized. Through $E^* = 2(1 + \nu)G^*$ Eq. (2.5) can be used to predict also the complex tensile modulus. A good example for the applicability of Eq. (2.5) is furnished by the experimental data obtained by Dickie et al. [75]. For the dynamic tensile modulus of a physical mixture (polymer blend) of 75% by weight polymethylmethacrylate (PMMA, continuous phase) and 25% butylacrylate (PBA, discrete phase) within experimental error correspondence of calculated and measured data was obtained (Fig. 2.14, solid curves). A copolymer of equal volumetric composition, where 25% by volume of the elastomeric butylacrylate had been grafted onto PMMA, yielded the experimental data points shown. In the region between the glass transition temperatures of the PBA (−20 °C) and of the PMMA (at these conditions 160 °C) a difference between the response of the mixture and of the graft copolymer can be observed. This difference can be related to the existence of the primary bonds which connect matrix and inclusion and which lead to frozen-in stresses.

A three-shell model of a two-phase solid was proposed by van der Poel ([67], Fig. 2.15). An interior sphere representing a hard component (G_f, K_f, V_f) is surrounded by a soft shell (G_a, K_a, V_a), and by a continuum, for which the elastic moduli G and K are being calculated in terms of the mentioned elastic constants and of the volume concentration (V_f) of the hard component. The condition of

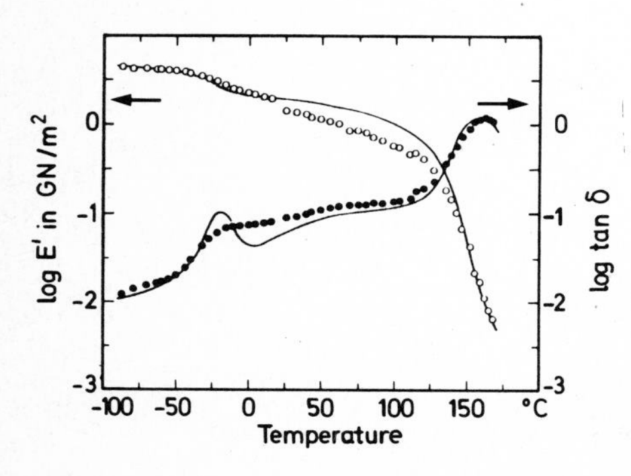

Fig. 2.14. Complex dynamic tensile modulus of blends (solid line) and graft copolymers (circles) respectively of polymethyl-methacrylate (75% by weight) and polybutylacrylate (25% by weight) at 110 Hz (after [75, 76])

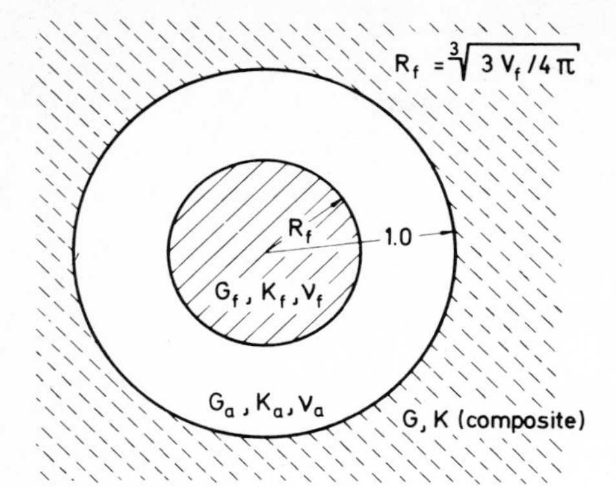

Fig. 2.15. Arrangement of phases in the three-shell model of Van der Poel [67]; bulk and shear moduli (K, G) and Poisson's ratios ν of filler (index f) and matrix (index a) and volume concentration V_f enter calculated composite moduli K and G (after [76])

mechanical equilibrium under shear and hydrostatic pressure leads to an explicit expression for K and to a complicated determinant for G which has to be solved numerically and for each individual system. The van der Poel approach had yielded the best results in describing the elasticity of elastomers highly filled (up to 50 vol.%) with mineral fillers (see e.g. [77]). Again no orientation distributions or physical anisotropies of the constituent particles enter the model.

As mentioned frequently the mechanical and optical response of molecules – and of their crystallites – is highly anisotropic. Depending on the property under consideration the carriers of the molecular anisotropy are the bond vectors (infrared dichroism), chain segments (optical and mechanical anisotropy), or the end-to-end vectors of chains (rubber elastic properties). For the representation of the ensuing macroscopic anisotropies one has to recognize, therefore, the molecular anisotropy and the orientation distribution of the anisotropic molecular units (Fig. 1.9.). Since these are essentially one-dimensional elements their distribution and orientation behavior can be treated as that of rods; such a model had been used successfully to explain the optical anisotropy [78], and the anisotropies of thermal conductivity [79], thermal expansion or linear compressibility [80], and Young's modulus [59, 65, 80, 81] of rubbers and oriented thermoplastics. The rod models focus on the large anisotropies which exhibit properties such as the polarizability or the retractive elastic forces; only two components of such a property, namely that in chain axis direction and that perpendicular to it enter the model as independent parameters.

For example the macroscopic elastic response of an aggregate of randomly, partly, or well oriented chains can be represented by that of a corresponding distribution of one-dimensional finite elements (Fig. 2.16) if the following conditions are fulfilled:

- external forces P cause a homogeneous local strain which determines the axial stress ψ (ϑ, φ, t) of a chain segment
- forces between chains are not transferred by shear or bending moments
- the state of stress at any arbitrary point A is given as the integral over all the traction forces exerted by the chain segments intersecting the surface of a volume element enclosing the point A [82].

The components of stress of such a homogeneously strained, partly oriented aggregate of elastic elements with axial Modulus E_k, length L, volume concentration

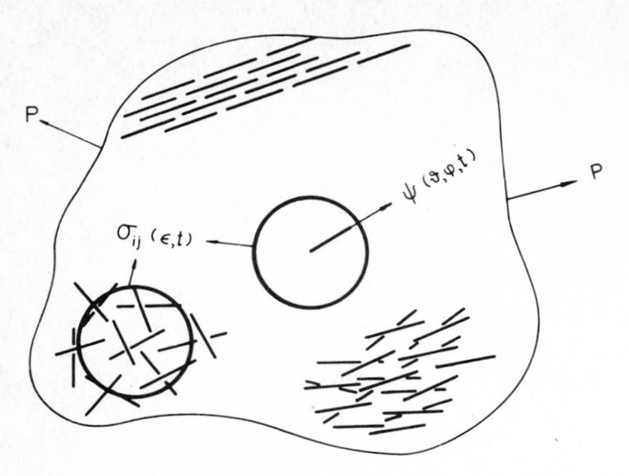

Fig. 2.16. Application of the condition of homogeneous strain ϵ (P, t) to a partially oriented system of finite, one-dimensional elements. The stress tensor σ_{ij} is obtained by space averaging over the traction vectors Ψ (ϑ, φ, t)

λ, and orientation distribution ρ was derived by Hsiao [82] as a special case of the Voigt-model employing a homogeneous distribution of strain:

$$\sigma_{ij} = \int E_k L^2 \lambda \rho \epsilon_{mn} s_m s_n s_i s_j d\omega. \tag{2.7}$$

Here the summation convention of repeated indices is used. The s_i are the components of the unit vector in the principal axis directions and ω the solid angle. Into this model only one molecular elastic component enters: the axial chain modulus. Any interactions by shear or in a direction perpendicular to the chain axes cannot be accounted for. It should only be applied, therefore, if these interactions can be meaningfully neglected.

A much more general representation of either different or differently oriented molecular domains can be achieved, of course, through three-dimensional elements. In the case of transverse symmetry the molecular elements must be characterized by 5 elastic constants (compliances), the orientation of one or two axes, and the condition of stress and strain at the boundary of an element. Voigt [63] based his calculations on the assumption that there be no discontinous change of strain at any boundary; Reuss [64] made the assumption of homogeneous stress. Following either Voigt or Reuss the space averaging over the elastic constants c_{ijmn} or elastic compliances s_{ijmn} of the molecular domains leads to the upper or lower bound respectively of the macroscopic moduli [83]. For an affinely deforming aggregate of such elements Ward [84] and later Kausch [85] have calculated the macroscopic elastic moduli as a function of domain orientation. The calculated curves for the change of the elastic moduli with draw ratio are particularly characterized by the rate of change and by the extent of change ultimately attained. If only reorientation of otherwise unchanged molecular domains occurs in drawing, the properties of a "completely" oriented sample must correspond to the properties of the molecular domains. In Figure 2.17 the calculated changes of Young's modulus in the direction of draw as a function of draw ratio and domain anisotropy are represented in

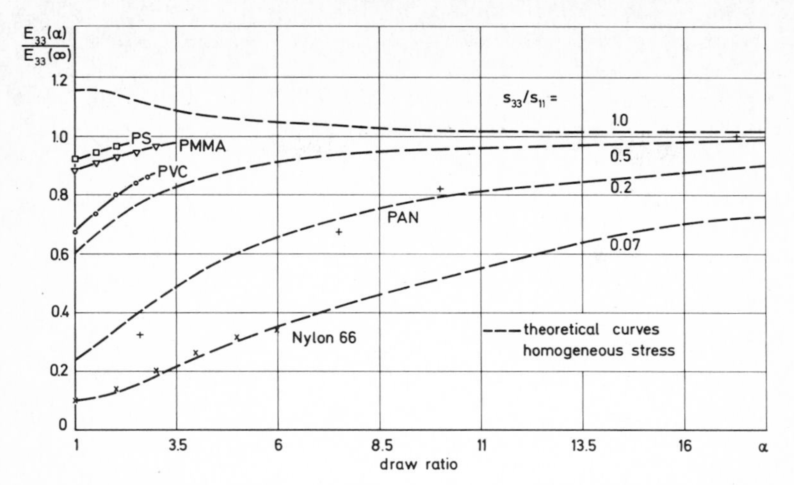

Fig. 2.17. Change of Young's modulus $E_{33} = \sigma_{33}/\epsilon_{33}$ as a function of draw ratio α and of the mechanical anisotropy $s_{33}/s_{11} = s_{3333}/s_{1111}$ of the molecular domains (after [13], [85])

comparison to experimental data [13, 85]. The findings of Ward and Kausch may be summarized by saying:

- the orientation behavior of amorphous thermoplastics (PS, PMMA, PVC) could be represented by the affine deformation of domains (cf. also [205])
- the observed data for PS, PMMA, and PVC were in agreement with the assumption of homogeneous stress at the boundaries of the orienting domains, the data for PC were better in agreement with a condition of homogeneous strain
- the orienting molecular domains have comparatively small elastic anisotropies (s_{3333}/s_{1111}-values ranging from 0.5 to 1), the axial compliances being more than an order of magnitude larger than those of the chains; this indicates that the domains are not equivalent to chains, micro-crystallites, or fringe micelles; if such domains do have a physical significance in amorphous polymers then they must be considered to be sections or aggregates of chain segments with a stiffness about twice as large as that of the sample as a whole [13]
- the concept of *affinely* orienting lamellae in semicrystalline polymers is only a crude approximation [13, 87–89, 106].

The mathematical representation of the elastic behavior of oriented heterogeneous solids can be somewhat improved through a more appropriate choice of the boundary conditions such as proposed by Hashin and Shtrikman [66] and Sternstein and Lederle [86]. In the case of lamellar polymers the formalisms developed for reinforced materials are quite useful [87, 88, 147]. An extensive review on the experimental characterization of the anisotropic and non-linear viscoelastic behavior of solid polymers and of their model interpretation had been given by Hadley and Ward [89]. In order to describe polymer microstructure and deformation models will generally be used which are based on the rotational isometric state theory (see IA and C and [117–121, 214–216]). Some other concepts such as the theory of paracrystals [9, 90] and the meander model [10, 11] have also been proposed. They will briefly be presented here although it seems that in the context of this book they do not offer additional insight.

The theory of paracrystals recognized that chain molecules form three-dimensional lattices and establish a certain long range order between members of the same lattice while postulating at the same time that in a real solid the lattice will be distorted. The degree of lattice distortion is measured by the variation in length that the three distance vectors $\mathbf{a}_i$ between corresponding lattice points encounter if moved in the three lattice directions. If the dimensionless relative mean fluctuations g_{ik} of the distance vectors $\mathbf{a}_i$ are all zero the structure is crystalline and if all $g_{ik} > 0.1$ it is amorphous. The g_{ik} provide a quantitative measure of the colloidal structure of microheterogeneous solids. If for instance g_{13} and g_{23} are large compared with the other g_{ik} we deal with the nematic state (segments parallel but at random distances), if g_{31} and g_{32} are large compared with the other g_{ik} we have a smectic state where the segments arrange in layers [9]. The relative paracrystalline distance fluctuation was shown to be inversely related to the maximum number of network planes within one microdomain [9]. The g_{ik} are obtained from line profile measurements in small angle X-ray scattering (SAXS).

Hosemann's schematic representation of a two-dimensional paracrystalline lattice is given in Figure 2.18. If we interpret a superstructure (e.g. the fibrillar structure shown in Fig. 2.6) in terms of paracrystals rather than crystals we arrive at the very same size and orientation distribution of the scattering elements – be they crystallites or paracrystals. The differences in interpretation become apparent, however, if intercrystalline rearrangements in annealing, during deformation, or after irradiation are discussed [9].

The general concept that there are only gradual differences between the more or less ordered regions of polymers is also employed by Pechhold and his co-workers in their quantitative microstructural theory (*Kinkenmodell*) of deformation [10, 208–211]. Pechhold recognizes that chain molecules exist in energetically different states of rotational isomers between which – cooperative and mostly rapid – transitions can be accomplished. He bases his quantitative thermodynamical calculations on kink isomers, which show only a small departure from the extended planar chain and are most likely to form bundles ([11], Fig. 2.1 c). The neighbouring chains within such bundle form various pairs of segments (cf. Fig. 1.12) which have different volumes, axial lengths and interaction energies. The concentration of kinks determines the axial extension of the bundle. In the partition function of a bundle of chains all these quantities enter; the partition function relates, therefore, the kink concentration with bundle geometry and internal and free energy of the isomers. Consequently the elastic compliances of a transversely isotropic bundle of chains could be determined as a function of kink concentration, kink block size, width of kink steps, and temperature [10, 11, 208–211]. The anelastic and plastic deformation may be interpreted in terms of successive dislocation motion and slip of bundle surface areas. The kink model offers in principle a *molecular* description of polymer structure. Again there are only gradual differences between ordered and non-ordered regions. Pechhold estimates that an apparently perfect PE-crystallite may contain up to 4 kinks per 1000 CH_2-groups whereas in a melt-like structure this concentration is about 200 in 1000. Although this concentration is so large as to annihilate all close and long range order some logical principles should govern the space filling through chain molecules. Pechhold generated appropriate patterns, the honeycomb- and the meander model

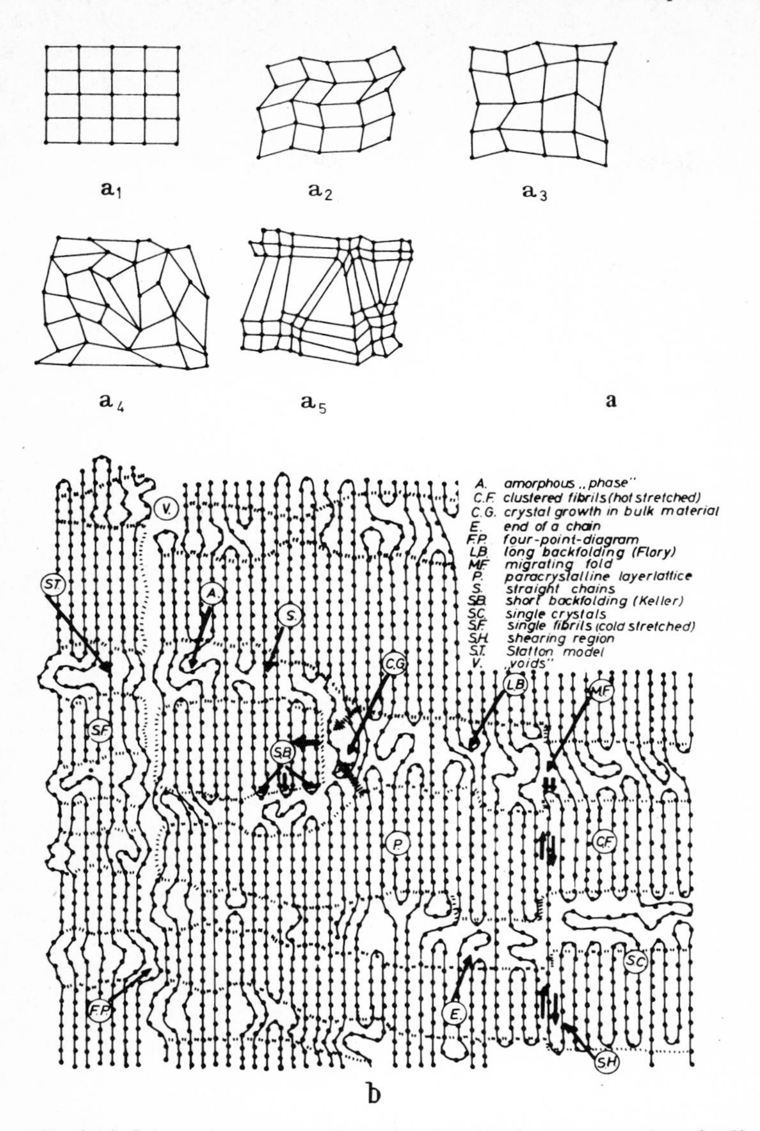

Fig. 2.18. Schematic paracrystalline (**a**) and molecular representation of different two-dimensional lattices and superstructures (after [9]). a_1 = crystal, a_2 = ideal paracrystal, a_3 = real paracrystal, a_4 = amorphous state, a_5 = micro-paracrystallites (function of "micellar cross-links"), **b** molecular model of linear polyethylene

(Fig. 2.1 c) and, more recently, the *superfolded meander topology* (Fig. 2 19). Taking as a basis the bundle concept, he proposes that chain segments gain orientational entropy by sharply bending or by *superfolding.* The most symmetric and simplest topology in which a melt bundle can tightly superfold, in order to fill densely 3-dimensional coarse grains (for the bulk phase) or plane grains (for the thin surface films), is an arrangement of *meander cubes* [208–211] which are shown in Figure 2.19. In

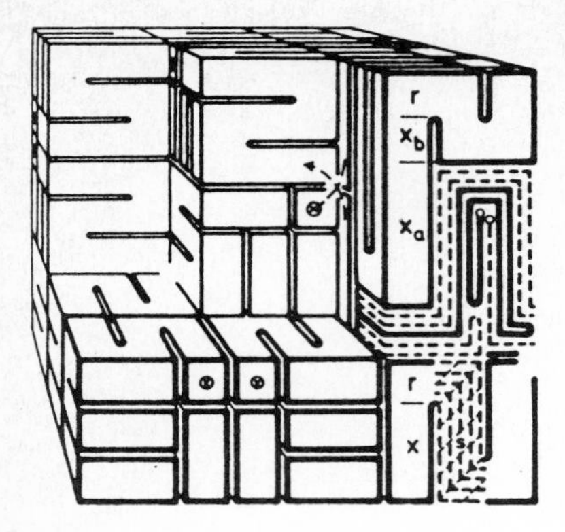

Fig. 2.19. Superfold meander topology according to Pechhold [211] showing tightly folded bundles of diameter r, consisting of chains with segments of length s and distance d; a minimum of free energy is expected if the length x of the superfolds is approximately 2 r (from ref. 211)

this model *cube rotation* and *shear deformation of molecular* layers are forming the elementary viscoelastic deformation mechanisms [111, 208–211]. Although meanwhile the neutron scattering experiments [100–104] strongly disprove the existence of a distinct meander arrangement of chains, Pechhold's considerations have greatly fertilized the efforts to study the structure of amorphous regions.

The close geometrical description of the arrangement of chain molecules within bundles offered by the kink-model have been used by Kröner and Anthony [14g, 99] to develop a quantitative non-linear deformation theory based on the structure defect "disclination". The meander model, then, is reduced to a particular arrangement of disclinations.

With regard to the elucidation of phenomena occurring during crack propagation in polymers it is generally not necessary to consider intracrystalline processes in detail. If, however, the intracrystalline segments of tie molecules are concerned then one has to be aware of the fact that the intermolecular interaction determines what forces can be transmitted onto these chain segments and whether or not these forces are large enough to cause chain scission.

Thus far model representations of polymer deformation have been discussed. Each of those could be converted into a model representing the fracture behavior if it were possible to formulate an adequate fracture criterion within the range of validity of these models. Having dealt with deformation the fracture criteria to be formulated would have to involve finite extensibility, critical load, or limited volume concentration of stored or dissipated energy. Dealing with fracture one will find that strain, stress, and energy are not sufficient as variables and that one will have to add at least two new dimensions: time and structural discontinuity. This will be explained in the following Chapter.

References for Chapter 2

1. R. N. Haward ed.: The Physics of Glassy Polymers, London: Applied Science Publishers, 1973.
2. A. Keller: Polymer crystals, Rep. Progr. Phys. *31*, 623–704 (1968).

3. B. Wunderlich: Macromolecular Physics: Crystal Structure, Morphology, Defects (Vol. 1), Crystal Nucleation, Growth, Annealing (Vol. 2), New York and London: Academic Press 1973 Crystal Melting (Vol. 3) 1980.
4. E. W. Fischer: Structure of amorphous organic polymers in bulk, Proc. Conf. Non-Crystalline Solids, Clausthal-Zellerfeld, Sept. 1976.
5. R. G. Kirste, W. A. Kruse, J. Schelten: Makromol. Chem. *162,* 299 (1972).
6. J. P. Cotton, D. Decker, H. Benoit, B. Farnoux, J. Higgins, G. Jannink, R. Ober, C. Picot, J. des Cloizeaux: Macromolecules *7,* 863 (1974).
7. P. J. Flory: Statistical Mechanics of Chain Molecules, New York: Wiley 1969.
8. P. J. Flory: Spatial configuration of macromolecular chains, The 1974 Nobel Lecture, Brit. Polymer J. *8,* 1–10 (1976).
9. R. Hosemann: Makromol. Chem., Suppl. *1,* 559–577 (1975). R. Hosemann: Ber. Bunsenges. phys. Chem. *74,* 755–767 (1970).
10. W. Pechhold: Kolloid-Z. Z. Polymere *228,* 1 (1968).
11. W. Pechhold: J. Polymer Sci. *C32,* 123–148 (1971).
12. G. S. Y. Yeh: Polymer Prepr. *14/2,* 718 (1973).
13. H. H. Kausch: Kolloid-Z. Z. Polymere *234,* 1148–1149 (1969), *237,* 251–266 (1970) and J. Polymer Sci. *C32,* 1–44 (1971).
14. a) H. F. Mark: p. xiii. b) H. Gleiter, R. Hornbogen, J. Petermann: p. 149. c) J. D. Ferry: p. 27. d) R. S. Rivlin: p. 71. e) R. F. Landel, R. F. Fedors: p. 131. f) W. Pechhold: p. 301. g) K. H. Anthony, E. Kröner: p. 429. Deformation and Fracture of Polymers: H. H. Kausch, J. A. Hassel, R. I. Jaffee, eds., New York/London: Plenum Press 1973.
15. E. W. Fischer: Z. Naturforsch. *12a,* 753 (1957).
16. A. Keller: Phil. Mag. *2,* 1171 (1957).
17. P. H. Till: J. Polymer Sci. *24,* 301 (1957).
18. Courtesy of E. W. Fischer: Mainz 1974.
19. C. W. Bunn: Trans. Faraday Soc. *35,* 482 (1939).
20. I. Sakurada, T. Ito, K. Nakamae: J. Polym. Sci. *C15,* 75 (1966).
21. J. Wristers: J. Polymer Sci., Polym. Physics Ed. *11,* 1601–1617 (1973).
22. T. Komoto, M. Iguchi, H. Kanetsuna, T. Kawai: Makromol. Chem. *135,* 145 (1975).
23. K. C. Tsou, H. P. Geil: Int. J. Polym. Mater. *1,* 223 (1972).
24. T. Hattori, K. Tanaka, M. Matsuo: Polym. Eng. Sci. *12,* 199 (1972).
25. A. H. Abdel-Alim, A. E. Hamielec: J. Appl. Polym. Sci. *17,* 3033–3047 (1973).
26. A. H. Abdel-Alim: J. Appl. Polym. Sci. *19,* 2179–2185 (1975).
27. N. A. J. Platzer: Copolymers, Polyblends, and Composites, Adv. Chem. *142* (1975).
28. J. A. Manson, L. H. Sperling: Polymer Blends and Composites, New York and London: Plenum Press 1976.
29. Mehrphasensysteme, Spring Meeting of the German Chemical and Physical Societies, Bad Nauheim, March 29 – April 2, 1976, papers published in Angew. Makromol. Chem. *58/59* and *60/61* (1977).
30. G. Kämpf, M. Hoffmann, H. Krömer: Ber. Bunsen-Ges. *74,* 851 (1970).
31. Courtesy to G. Kämpf: Uerdingen, 1976.
32. A. Keller: J. Polymer Sci. *15,* 32–49 (1955).
33. A. J. Pennings, A. M. Kiel: Kolloid-Z. Z. Polymere *205,* 160–162 (1965).
34. T. Kawai, T. Matsumoto, M. Kato, H. Maeda: Kolloid-Z. Z. Polymere *222,* 1–10 (1968).
35. P. Tucker, W. George: Polym. Eng. Sci. *12,* 364–377 (1972).
36. P. Tucker, W. George: Text. Res. J. *44,* 56–70 (1974).
37. A. Zwijnenburg, A. J. Pennings: Kolloid-Z. Z. Polymere *254,* 868–881 (1976).
38. M. Takayanagi, In: Deformation and Fracture of High Polymers, H. H. Kausch, J. A. Hassel, R. I. Jaffee, eds., New York: Plenum Press 1973, p. 353–376.
39. J. M. Lupton, J. W. Regester: J. Appl. Polymer Sci. *18,* 2407–2425 (1974).
40. N. E. Weeks, R. S. Porter: J. Polymer Sci., Polymer Physics Ed. *12,* 635–643 (1974).
41. N. J. Capiati, R. S. Porter: ibid. *13,* 1177–1186 (1975).
42. R. B. Morris, D. C. Bassett: ibid. *13,* 1501–1509 (1975).
43. N. E. Weeks, R. S. Porter: ibid. *13,* 2031–2048 (1975).
44. N. E. Weeks, R. S. Porter: ibid. *13,* 2049–2065 (1975).

45. F. R. Anderson: J. Polymer Sci. *C3,* 123 (1963).
46. E. S. Clark, C. A. Garber: Int. J. Polym. Mater. *1,* 31 (1971).
47. W. O. Statton: Report UTEC MSE 73–100, University of Utah, Salt Lake City, 1973.
48. H. D. Noether, W. Whitney: Kolloid-Z. Z. Polymere *251,* 991 (1973).
49. I. K. Park, H. D. Noether: Colloid and Polymer Sci. *253,* 824–839 (1975).
50. D. Göritz, F. H. Müller: Colloid and Polymer Sci. *253,* 844–851 (1975).
51. B. Cayrol, J. Petermann: J. Polymer Sci., Polymer Phys. Ed. *12,* 2169–2172 (1974).
52. R. S. Stein: J. Polymer Sci. *C32,* 45–68 (1971).
53. V. Petraccone, I. C. Sanchez, R. S. Stein: J. Polymer Sci., Polymer Phys. Ed. *13,* 1991–2029 (1975).
54. B. Wunderlich, H. Baur: Adv. Polymer Sci. *7,* 151–368 (1970).
55. J. D. Ferry: Viscoelastic Properties of Polymers, New York: Wiley 3rd ed. 1980.
56. J. A. Sauer, A. E. Woodward: Polymer Thermal Analysis, II. P. E. Slade, Jr., L. T. Jenkins, eds., New York: Dekker, Inc. 1970, p. 107.
57. J. A. Sauer, G. C. Richardson, D. R. Morrow: J. Macromol. Sci. – Revs. Macromol. Chem. *C9 (2),* 149–267 (1973).
58. H. D. Noether: personal communication.
59. L. R. G. Treloar: The Physics of Rubber Elasticity, London: Oxford Univ. Press 1958.
60. M. Maeda, S. Hibi, F. Itoh, S. Nomura, T. Kawaguchi, H. Kawai: J. Polymer Sci., A-2, *8,* 1303–1322 (1970).
S. Nomura, A. Asanuma, S. Suehiro, H. Kawai: ibid. *9,* 1991–2007 (1971).
Rheo-Optical Studies of High Polymers, Onogi Laboratory, Kyoto University, Kyoto, Japan, 1971.
61. N. Kasai, M. Kakudo: J. Polymer Sci. A-2, 1955 (1964).
62. A. Peterlin: Adv. Polymer Science and Engineering, K. D. Pae, D. R. Morrow, and Yo Chen, eds., New York: Plenum Press 1972, p. 1.
63. W. Voigt: Lehrbuch der Kristallphysik, B. G. Teubner Leipzig (1910).
64. A. Reuss: Z. angew. Math. Mech. *9,* 49 (1929).
65. E. H. Kerner: Proc. Phys. Soc. B *69,* 802 (1956), Proc. Phys. Soc. B *69,* 808 (1956).
66. Z. Hashin: J. Mech. Phys. Solids *10,* 335 (1962). Z. Hashin, S. Shtrikman: J. Mech. Phys. Solids *10,* 343 (1962).
67. C. van der Poel: Rheol. Acta *1,* 198 (1958).
68. H. Eilers: Kolloid-Z. *97,* 313 (1941).
69. E. Guth: J. Appl. Phys. *16,* 20 (1954). E. Guth, O. Gold: Phys. Rev. *53,* 322 (1938).
70. M. Mooney: J. Colloid Sci. *6,* 162–170 (1951).
71. M. Takayanagi, K. Imada, T. Kajiyama: J. Polymer Sci., C*15,* 263–281 (1966).
72. S. Uemura, M. Takayanagi: J. Appl. Polymer Sci. *10,* 113 (1966).
73. L. E. Nielsen: Mechanical Properties of Polymers and Composites, New York: Marcel Dekker, 1974, Vol. 2, p. 387 ff.
74. A. Einstein: Annalen Phys. *19,* 289 (1906), Annalen Phys. *34,* 591 (1911).
75. R. A. Dickie, M. Cheung: J. Appl. Polymer Sci. *17,* 79–94 (1973).
76. H. H. Kausch: Angew. Makromol. Chem. *60/61,* 139 (1977).
77. F. R. Schwarzl, H. W. Bree, C. J. Nederveen: Proc. IV. Int. Congr. Rheology (Providence 1963). E. H. Lee, ed., New York: Interscience, 1965, Vol. 3, p. 241–263.
78. W. Kuhn: Kolloid-Z. *87,* 3 (1939). W. Kuhn, F. Grün: Kolloid-Z. *1,* 248 (1942).
79. K. Eiermann: Kunststoffe *51,* 512 (1961). K. Eiermann, K. H. Hellwege: J. Polymer Sci. *1,* 99 (1962).
80. J. Hennig: Kunststoffe *57,* 385 (1967). J. Hennig: Kolloid-Z. Z. Polymere *202,* 127 (1965).
81. W. Hellmuth, H. G. Kilian, F. H. Müller: Kolloid-Z. Z. Polymere *218,* 10 (1967). F. H. Müller: J. Polymer Sci. C *20,* 61 (1967).
82. C. C. Hsiao: J. Polymer Sci. *44,* 71 (1960).
83. J. Bishop, R. Hill: Phil. Mag. *42,* 414 and 1298 (1951).
84. I. M. Ward: Proc. Phys. Soc. *80,* 1176 (1962).
85. H. H. Kausch: J. Appl. Phys. *38,* 4213 (1967).
86. S. S. Sternstein, G. M. Lederle, In: Polymer Networks: Structure and Mechanical Properties, Newman and Chompff, eds., New York: Plenum Press, 1971.

87. J. C. Halpin, J. C. Kardos: J. Appl. Phys. *43,* 2235 (1972).
88. E. H. Andrews: Pure Appl. Chem. *39,* 179–194 (1974).
89. D. W. Hadley, I. M. Ward: Rep. Progr. Physics, *38,* 1143–1215 (1975).
90. R. Bonart, R. Hosemann: Kolloid-Z. Z. Polymere *186,* 16 (1962).
91. J. Schelten, D. G. H. Ballard, G. D. Wignall, G. W. Longman, W. Schmatz: Polymer *17,* 751 (1976).
92. D. Y. Yoon, P. J. Flory: Polymer *18,* 509–513 (1976).
93. G. Capaccio, I. M. Ward: Polymer *16,* 239 (1975).
94. G. Capaccio, T. A. Crompton, I. M. Ward: J. Polymer Sci. A*214,* 1641 (1976).
95. G. Capaccio, T. J. Chapman, I. M. Ward: Polymer *16,* 469 (1975).
96. G. Capaccio, T. A. Crompton, I. M. Ward: Polymer *17,* 645 (1976).
97. I. M. Ward: Seminar on Ultra-High Polymers, St. Margherita Ligure, May 1977.
98. G. Capaccio: Seminar on Ultra-High Polymers, St. Margherita Ligure, May 1977.
99. K. Anthony: Habilitationsschrift, Universität Stuttgart, November 1974.
100. E. W. Fischer: Europhysics Conf. Abstracts 2 E, 71–79 (1977).
101. J. Schelten, G. D. Wignall, D. G. H. Ballard, G. W. Longmann: Polymer *18*/11, 1111 (1977).
102. H. Benoit, D. Decker, R. Duplessix, C. Picot, P. Rempp, J. P. Cotton, B. Farnoux, G. Jannink, R. Ober: J. Polym. Sci., Polym. Phys. Ed. *14,* 2119–2128 (1976).
103. C. Picot, R. Duplessix, D. Decker, H. Benoit, F. Boue, J. P. Cotton, M. Daoud, B. Farnoux, G. Jannink, M. Nierlich, A. J. deVries, P. Pincus: Macromolecules *10,* 436–442 (1977).
104. Europhysics Conf. Abstracts *2F* (1978).
105. S. L. Cannon, G. B. McKenna, W. O. Statton: J. Polymer Sci., Macromol. Rev. *11,* 209–275 (1976).
106. H. G. Kilian, M. Pietralla (Polymer *19,* 664–672, 1978) derive from the anisotropy A of the thermal diffusivity $\alpha = \lambda/c_p\rho$ of oriented polyethylenes that the intrinsic anisotropy A_i of the orienting, partly crystalline lamellar clusters increases with the degree of crystallinity ($A_i = 7$ to 26, linear extrapolation yields $A_i = 2$ for the fully amorphous and $A_i = 50$ for a completely crystalline cluster); the average degree of orientation of the lamellae, $\langle\cos^2\vartheta\rangle$, as determined from thermal measurements agrees very well with X-ray data; the observed increase of orientation with draw ratio is more rapid than it would be in affine deformation.
107. L. Mandelkern: An Introduction to Macromolecules, 2nd Ed., New York–Berlin–Heidelberg: Springer 1983.
108. see J. Macromol. Sci. *B12,* (1976).
109. see Faraday Disc. Chem. Soc. *68,* (1979).
110. J. H. Wendorff: Polymer *23,* 543 (1982).
111. W. R. Pechhold, T. Gross, H. P. Grossmann: Coll. Polymer Sci. *260,* 378 (1982).
112. I. G. Voigt-Martin, J. H. Wendorff: Encyclopedia of Polymer Science and Technology, 2nd ed. New York: Wiley-Interscience 1985.
113. G. D. Patterson, P. J. Flory: J. Chem. Soc. Faraday Trans. II *68,* 1098 (1972).
114. P. Bothorel, G. Fourche: J. Chem. Soc. Faraday Trans. II *69,* 441 (1973).
115. E. W. Fischer, G. R. Strobl, M. Dettenmaier, M. Stamm, N. Steidle: Faraday Disc. Chem. Soc. *68,* 26 (1979).
116. R. G. Kirste, W. A. Kruse, K. Ibel: Polymer *16,* 120 (1975).
117. D. Y. Yoon, P. J. Flory: Macromolecules *9,* 299 (1976).
118. M. Dettenmaier: J. Chem. Phys. *68,* 2319 (1978).
119. E. W. Fischer, M. Dettenmaier: J. Noncryst. Solids *31,* 11 (1978).
120. W. Gawrisch, M. G. Brereton, E. W. Fischer: Polymer Bulletin *4,* 687 (1981).
121. D. Y. Yoon, P. J. Flory: Polymer Bulletin *4,* 693 (1981).
122. R. Lovell, A. H. Windle: Polymer *22,* 175 (1981).
123. H. R. Schubach, E. Nagy, B. Heise: Coll. Polymer Sci. *259,* 53 (1981).
124. M. Meyer, J. van der Sande, D. R. Uhlmann: J. Polym. Sci., Polym. Phys. Ed., *16,* 2005 (1978).
125. M. Dettenmaier, H. H. Kausch: Coll. Polymer Sci. *259,* 209 (1981).
126. M. Stamm, E. W. Fischer, M. Dettenmaier: Faraday Disc. Chem. Soc. *68,* 263 (1979).
127. M. Dettenmaier, E. W. Fischer, M. Stamm: Coll. Polym. Sci. *258,* 343 (1980).

129. E. H. Andrews: Fracture in Polymers, Edingburgh, London: Oliver & Boyd, 1968.
130. G. Menges, D. Kirch, J. Nordmeier, E. Winkel, J. Wortberg: Kunststoffe *73,* 258 (1983) Polymere Werkstoffe H. Batzer, Ed. Bd. 3, Stuttgart–New York: Georg Thieme, 1984.
131. E. W. Fischer, H. Goddar, G. F. Schmidt: Makromol. Chem. *119,* 170 (1968).
132. I. M. Ward Ed.: Structure and Properties of Oriented Polymers, London: Applied Science Publishers, 1975.
133. I. M. Ward Ed.: Developments in Oriented Polymers-1, London–New Jersey: Applied Science Publishers, 1982.
134. V. A. Marichin: Acta Polymerica *30*/8, 507 (1979).
135. J. L. White: Pure and Appl. Chem. *55*/5, 765 (1983).
136. A. Keller: Kolloid.-Z. *165,* 15 (1959).
137. P. H. Lindenmeyer: SPE Transactions *4,* 1 and 157 (1964).
138. A. J. Pennings: Proc. Int. Conf. on Crystal, Boston: Oxford: Pergamon, 1966, p. 389.
139. A. J. Pennings: J. Polymer Sci. C *16,* 1799 (1967).
140. D. Göritz: Habilitationsschrift, Universität Ulm, 1978.
141. J. W. S. Hearle: Polymers and their Properties, Vol. 1 Fundamentals of Structure and Mechanics, Chichester: Ellis Horwood Publishers, 1982.
142. M. J. Hill, P. J. Barham, A. Keller: Coll. & Polym. Sci. *258,* 1023 (1980).
143. A. Ciferri, I. M. Ward, Eds.: Ultra-High Modulus Polymers, London: Applied Science Publishers, 1979.
144. T. Ohta: Polymer Eng. Sci. *23*/13, 697 (1983).
145. K. Imada, M. Takaynagi: Intern. J. Polymeric Mater. *2,* 89 (1973). S. Maruyama, K. Imada, M. Takayanagi: ibid 125.
146. G. Capaccio: Coll. & Polymer Sci. *259,* 23 (1981). G. Capaccio, I. M. Ward, ibid *260,* 46 (1982).
147. R. G. C. Arridge, P. J. Barham: Polymer *19,* 654 (1978).
148. J. G. Rider, K. M. Watkinson: ibid 645.
149. A. Tsuruta, T. Kanamoto, K. Tanaka: Polymer Eng. Sci. *23*/9, 521 (1983).
150. S. Gogolewski: Polymer *18,* 63 (1977). S. Gogolewski, A. J. Pennings: ibid, 647.
151. P. Smith, P. J. Lemstra, H. C. Booij: J. Polymer Sci., Polym. Phys. Ed. *19,* 877 (1981). P. J. Lemstra, J. P. L. Pijpers: Europhysics Conf. Abstracts 6 G, 1982, 88.
152. P. J. Barham: Polymer *23,* 1112 (1982).
153. G. Capaccio: Pure & Appl. Chem. *55*/5, 869 (1983).
154. A. J. Pennings, J. Smook, J. de Boer, S. Gogolewski, P. F. van Hutten: ibid, 777.
155. H. Behrens, G. Griebel, L. Meinel, H. Reichenbach, G. Schulze, W. Schenk, K. Walter: Plaste und Kautschuk *22*/5, 414 (1975).
156. H. Reichenbach, H. Behrens, G. Schulze: Faserforsch. & Textiltechn. *29*/2, 86 (1978).
157. K. P. Grosskurth: Gummi, Asbest, Kunstst. *25,* 1159 (1972).
158. K. P. Grosskurth: Coll. & Polym. Sci. *255,* 120 (1977).
159. K. P. Grosskurth: Progress in Coll. & Polym. Sci. *66,* 281 (1979); Coll. & Polym. Sci. *259,* 163 (1981) and Schadensanalyse an Kunststoff-Formteilen, Düsseldorf: VDI-Verlag, 1981, 79 and 95.
160. G. Kämpf, H. Orth: J. Macromol. Sci. – Phys., B *11*/2, 151 (1975). G. Kämpf: Progr. Coll. & Polym. Sci. *57,* 249 (1975).
161. G. Schwarz: Coll. & Polym. Sci. *258,* 807 (1980).
162. M. Dettenmaier, K. P. Grosskurth, H. H. Kausch: Lausanne and Braunschweig, unpublished results, 1982.
163. B. Neppert, B. Heise, H.-G. Kilian: Coll. & Polym. Sci. *261,* 577 (1983).
164. C. B. Bucknall: Toughened Plastics, London: Applied Science Publishers, 1977.
165. D. R. Paul, S. Newman Eds.: Polymer Blends, Vol. 1 and 2, New York–San Francisco–London: Academic Press, 1978.
166. E. Martuscelli, R. Palumbo, M. Kryszewski: Polymer Blends – Processing, Morphology, and Properties, New York: Plenum Publishing Corp., 1980.
167. Polymer Alloys: Structure and Properties, 16th Europhysics Conference on Macromolecular Physics, Europhys. Conf. Abstr. *8c* (1984).

168. Yu. S. Lipatov: Polymer Sci. USSR *20,* 1 (1978).
169. R. E. Prud'homme: Polymer Eng. & Sci. *22*/2, 90 (1982).
170. J. Noolandi: Polymer Eng. & Sci. *24*/2, 70 (1984).
171. L. H. Sperling: Interpenetrating Polymer Networks and Related Materials, New York–London: Plenum Press, 1981.
172. G. Meyer: 13e Colloque National "Systèmes polymères tridimensionnels", Strasbourg 1983, 61.
173. H. L. Frisch: Europhys. Conf. Abstr. *8c,* 55 (1984);
H. X. Xiao, K. C. Frisch, H. L. Frisch: J. Polymer Sci., Polymer Chem. Ed. *22,* 1035 (1984).
174. A. S. Nowick, G. G. Libowitz, eds.: Polymer Liquid Crystals, New York: Academic Press, 1982.
P. J. Flory et al.: Liquid Crystal Polymers I–III, Advances in Polymer Science *59–61,* Berlin–Heidelberg–New York: Springer, 1984.
175. M. Hoffmann, H. Krömer, R. Kuhn: Polymeranalytik I und II, Stuttgart: Georg Thieme 1977.
176. G. Kämpf, Charakterisierung von Kunststoffen mit physikalischen Methoden, München–Wien: Hanser, 1982.
177. R. A. Fava, ed.: Methods of Experimental Physics: Polymers, Vol. 16, A–C, New York–London: Academic Press, 1980.
178. A. Guinier: Théorie et Technique de la Radiocristallographie, Paris: Dunod, 1964;
O. Glatter, O. Kratky Eds.: Small Angle X-Ray Scattering, New York: Academic Press, 1982.
179. L. H. Sperling: Polymer Eng. Sci. *24,* 1 (1984).
180. W. Klöpffer: Introduction to Polymer Spectroscopy, Berlin–Heidelberg–New York–Tokyo: Springer, 1984.
181. M. Dettenmaier: Progr. Colloid & Polymer Sci. *66,* 169 (1979).
182. J. V. Dawkins Ed.: Developments in Polymer Characterisation-2, London: Applied Science Publishers, 1980.
183. J. L. Koenig: Chemical Microstructure of Polymer Chains, New York–Chichester–Brisbane–Toronto: John Wiley & Sons, 1980.
184. H. H. Kausch, H. G. Zachmann, eds.: Polymers in the Solid State, Advances in Polymer Science *66/67* (1984).
185. D. H. Reneker, J. Mazur: Polymer *24,* 1387 (1983);
ibid *23,* 401 (1982).
186. F. R. Eirich Ed.: Science and Technology of Rubber, New York–San Francisco–London: Academic Press, 1978.
187. A. S. Argon, R. E. Cohen, O. S. Gebizlioglu, C. E. Schwier: Advances in Polymer Science *52/53,* H. H. Kausch ed., Berlin–Heidelberg–New York: Springer, 1983, 275.
188. I. M. Ward: Mechanical Properties of Solid Polymers, London: Wiley-Interscience 1971, 2nd ed. 1983.
189. J. C. Seferis: J. Macromol. Sci. – Phys., B *13*/3, 357 (1977).
190. D. L. M. Cansfield, I. M. Ward, D. W. Woods, A. Buckley, J. M. Pierce, J. L. Wesley: Polymer Communic. *24,* 130 (1983).
191. W. Kuhn: Kolloid-Z. *68,* 2 (1934).
192. ibid *76,* 258 (1936).
193. E. Guth, H. Mark: Monatsh. d. Chem. *65,* 93 (1934).
194. K. H. Meyer, C. Ferri: Helv. Chim. Acta *18,* 570 (1935).
195. M. Shen, W. F. Hall, R. E. DeWames: Rubb. Chem. and Techn. *45*/3, 638 (1972).
196. I. V. Yannas, A. C. Lunn: J. Macromol. Sci. Phys. *B4,* 603, 620 (1970);
I. V. Yannas, N. H. Sung, A. C. Lunn: ibid *B5,* 487 (1971);
I. V. Yannas: ibid *B6,* 91 (1972).
197. I. V. Yannas, R. R. Luise: J. Macromol. Sci.-Phys. *B21*(3), 443 (1982).
198. J. Koppelmann, R. Hirnböck, H. Leder, F. Royer: Colloid & Polymer Sci. *258,* 9 (1980).
199. M. N. Rahaman, J. Scanlan: Polymer *22,* 673 (1981).

200. R. H. Boyd, M. E. Robertsson, J. F. Jansson: J. Polymer Sci., Polym. Phys. Ed. *20,* 73 (1982); M. E. Robertsson, PhD thesis, The Royal Inst. of Technology, Stockholm 1984.
201. J. H. Wendorff: Progr. Coll. & Polym. Sci. *66,* 135 (1979).
202. J. F. Jansson: personal communication, to be published.
203. R. P. Wool: J. Polymer Sci., Polym. Phys. Ed. *14,* 603 (1976).
204. L. Jarecki, A. Ziabicki: Polymer *20,* 411 (1979).
205. M. Dettenmaier, H. H. Kausch, T. Q. Nguyen, J. S. Higgins, A. Maconnachie: Meeting of the American Physical Society, Detroit, March, 26–30, 1984, Bulletin Americ. Phys. Soc. *29,* 532 (1984), Macromolecules *19,* 773 (1986).
206. H. W. Siesler, K. Holland-Moritz: Infrared and Raman Spectroscopy of Polymers, New York–Basel: Marcel Dekker Inc., 1980.
207. J. P. Jarry, L. Monnerie: J. Polym. Sci. – Polym. Phys. Ed. *16,* 443 (1978).
208. W. R. Pechhold, H. P. Grossmann: Roy. Soc. Chem., Faraday Discussions *68* (1979).
209. W. R. Pechhold: Coll. & Polym. Sci. *258,* 269 (1980);
W. R. Pechhold, H. P. Grossmann, W. v. Soden: ibid 248.
210. W. R. Pechhold, B. Stoll: Polymer Bulletin *7,* 413 (1982).
W. R. Pechhold: ibid 615.
211. W. R. Pechhold: Makromol. Chem. Suppl. *6,* 163 (1984).
212. G. R. Mitchell and A. H. Windle (Polymer *25,* 906 (1984)) propose an interesting model of the structure of amorphous PS in which the phenyl groups segregate on a molecular scale to form stacks.
213. P. J. Flory: Seminar at the Polymer Laboratory, Swiss Federal Institute of Technology, Lausanne, August 31, 1984.
214. P. J. Flory, D. Y. Yoon, K. A. Dill: Macromol. *17,* 862 (1984).
215. D. Y. Yoon, P. J. Flory: ibid., 868.
216. D. N. Theodorov and U. W. Suter: Macromolecules *18*, 1467 (1985) and ibid. *19*, 139 (1986) have developed a method for the atomistic modeling of the mechanical properties of (amorphous) glassy polymers. Predicted elastic constants (for atactic polypropylene) are within 15% of the experimental values without the use of adjustable parameters.
217. P. F. van Hutten, C. E. Koning, A. J. Pennings: Coll. & Polym. Sci. *262*, 521 (1984).

Chapter 3

Statistical, Continuum Mechanical, and Rate Process Theories of Fracture

I. Introduction

Test specimens loaded under laboratory conditions, but also regular engineering components fracturing in service, provide many different data which lend themselves to evaluation of the fracture process. These data are, for instance, the time to onset and completion of fracture, details of the fracture pattern (ductile or brittle fracture, appearance of the breaking specimen and of the fracture surface), crack dynamics, and change in physical or chemical properties. Naturally the most straightforward evaluation of a test or a set of data is the direct correlation of the property of interest (e.g., time under stress) to the environmental parameter(s) of interest (e.g., stress and temperature). In Figure 1.4 a set of data with just these variables has been plotted (PVC pipes under internal pressure). If one uses such a plot a number of questions arise:

- what is the statistical significance of an individual data point
- what are the failure conditions and how do they depend on external parameters and material properties
- what is the likely cause of failure
- what conclusions can be drawn with respect to extrapolating the set of curves into hitherto non-accessible regions of time, pressure, or temperature?

The first of these points will be commented upon subsequently in connection with an interpretation of the statistical variability of fracture events. To answer this and the second question the highly developed tools of statistical analysis (e.g., 1–2) are available; their application forms a backbone of industrial quality control and material design. Understandably, the complex input (mechanical, thermal, and environmental attack) acting upon a complex system (e. g., a highly structured polymer) has a complex, non-deterministic output. The determination and evaluation of only a small number of different data will necessarily reveal only a partial aspect of the fracture process. For this reason the questions as to cause and kinetics of fracture development can rarely be answered unambiguously. *Any mathematical interrelations established between variables* (e.g., time and stress) *are valid for extrapolation only if the basis does not change.*

In this chapter an overview of conceptually different fracture theories is presented which have in common that they do not make explicite reference to the characteristic properties of the molecular chains, their configurational and supermolecular order and their thermal and mechanical interaction. This will be seen to apply to the classical failure criteria and general continuum mechanical models. Rate process fracture theories take into consideration the viscoelastic behavior of polymeric materials but do not derive their fracture criteria from detailed morphological analysis. These basic theories are invaluable, however, to elucidate statistical, non-morphological, or continuum mechanical aspects of the fracture process.

II. Statistical Aspects

Experimentally determined quantities (observables) such as strength, time to fracture, or concentration of free radicals show a wide scatter of values; they are stochastic variables. As an extreme example of a stochastic variable in Figure 3.1 the service lives t of 500 HDPE-tubes under identical test conditions are plotted in the form of a histogram [3]. The distribution shown can be represented by a logarithmic normal distribution (Fig. 3.2) with a mean of lg t/h equal to 2.3937 and a variance of $s = 0.3043$. The expectation value of the time to fracture of a specimen subject to the test condition employed is the time which corresponds to the mean value of the (logarithmic) distribution, here 247.6 h. The actually determined values of t evidently scatter widely around this expectation value. Nevertheless even such a distribution can be characterized by testing just a few randomly selected specimens. For a normally distributed observable any three random values cover on the average a range of 1.69 s – which contains in 86% of all cases the mean value of the (infinite) distribution also [3]. This extreme example has been chosen to emphasize the meaning of stochastic variables with which one is dealing subsequently. It was also intended to show the significance of small samples – containing 3 to 10 members – which are generally used in practice to establish correlations between stochastic variables and environmental or material parameters. At this point we are not going to talk further about the statistical reduction of such a set of data. Here we are interested in the information which may be drawn from such a statistical analysis.

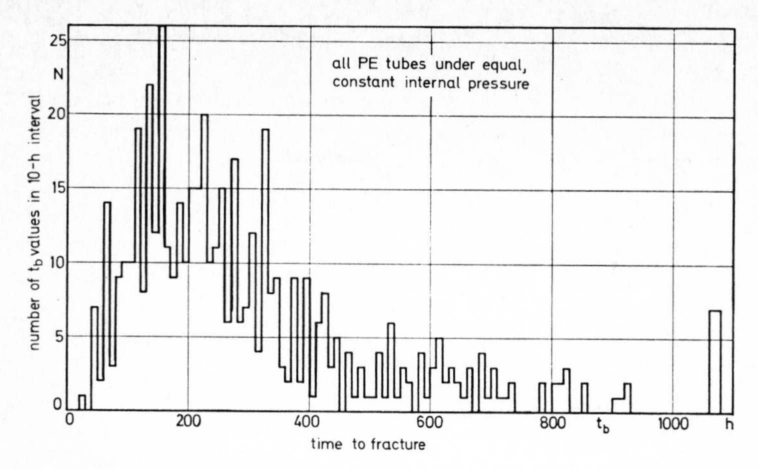

Fig. 3.1. Scatter of times to fracture of 500 PE tubes under identical loading conditions. 80 °C, $\sigma_V = 4$ MN/m^2 (after [3], from [78]).

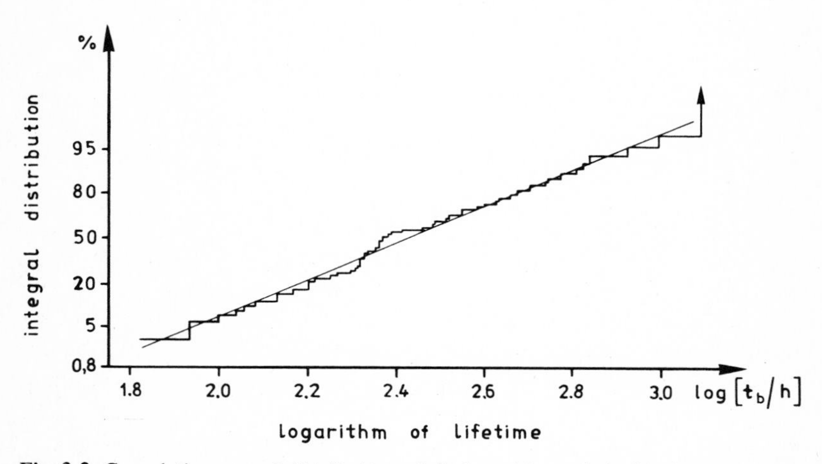

Fig. 3.2. Cumulative normal distribution of the logarithms of the lifetimes plotted in Fig. 3.1.

From a statistical standpoint, the variability of a material property, e.g., fracture strength, is explained by referring to one or more of the following three arguments:

1. The beginning of fracture is a statistical event, the occurrence of which is described by probability laws.

2. Apparently identical specimens of a set are inherently different, e.g., they contain a large number of flaws of different size or severeness and the most severe of these flaws determines the fracture strength.

3. The load to which a sample is subjected until it fractures activates a large number of molecular processes. The uncertainty of the cumulative event, e.g., mac-

roscopic failure, then results from the uncertainty of the molceular events and the mode of correlation of these events.

The first argument (fracture is a statistical event) would lead to the following general equation for the rate of change of the number N of members belonging to one ensemble:

$$dN/dt = -N K \tag{3.1}$$

where K may be a function of N, stress, stress history, temperature, or other environmental parameters. For a number N of statistically equal, independent and separately loaded units the rate function K in Eq. (3.1) is independent of N, but generally a function of time. Equation (3.1) is readily solved with respect to the number N(t) of surviving units:

$$N(t) = N_0 \exp\left[-\int_0^t K(\tau)\, d\tau\right] \tag{3.2}$$

where N_0 is their initial number. The fraction of units broken after some time t_b is obviously $1 - N(t_b)/N_0$ and the cumulative distribution function of lifetimes, $Q(t_b)$, therefore is:

$$Q(t_b) = 1 - \exp\left[-\int_0^{t_b} K(\tau)\, d\tau\right]. \tag{3.3}$$

The average lifetime, $< t_b >$, is solely determined by K and becomes for constant K:

$$< t_b > = \frac{1}{N_0} \int_0^\infty t N_0 K \exp(-Kt)\, dt = K^{-1}. \tag{3.4}$$

It may be noted that in this case the life expectancy of a surviving unit is at any time independent of N and t. This is equivalent to the assumption that the residual strength is equal to the initial strength, i.e., that *no aging or deterioration has taken place* in the stressed sample. The rate K of failure occurrence can be determined experimentally from the scatter of t_b [4]. Using this assumption Kawabata and Blatz [5] developed a simple stochastic theory of creep failure. They argued that the first moments of t_b for an ensemble of ruptured specimens should reveal whether or not K is independent of time. If K depends only on initial load and is independent of N the equation

$$K = (m!/< t_b^m >)^{1/m} \tag{3.5}$$

should hold for any m, that is for any order of moments of t_b. An investigation of the variability of fracture times in the special case of a rubber vulcanizate gave excellent agreement with the assumption of a time-independent K as used in Eq. (3.4).

The same conclusion was drawn by Narisawa et al. [82] from the exponential decrease of N(t) with t of uniaxially stressed PA 6 monofilaments. The general independence of K from time seems to be doubtful, however, especially if the samples are exposed to accelerating environments [87].

In other experiments Kawabata et al. [6] investigated the failure statistics of carbon-filled styrene butadien rubber (SBR). They arrived at the conclusion that either the coefficient relating stress and rate of failure is increasing with time or that several local fracture events must be independently initiated before the rubber specimen breaks. The best fit of their theory with experiments was obtained for a critical number of 3 to 4 microscopic fracture events as nucleus for unstable crack formation. For the asymmetric distribution of times-to-fracture shown in Figure 3.2 the relation Eq. (3.5) is also not valid for higher orders of m (m equal to or larger than two). This means either that the probability density K for pipe failure is smaller for specimens having a longer service life or that K is a function of loading time. In the first case it must be assumed that from the beginning the specimens had not been statistically identical, in the second that they underwent structural changes affecting K. The first assumption seems to be valid for pipe failure through creep crack development. The investigations of Stockmayer [83] recently have shown that nature and location of material inhomogeneities largely determine the creep lives t_b of LDPE pipes. Morphological parameters thus account for more than half of $<t_b^2>$ [84]. The conclusion that structural differences exist and affect K must be drawn for other materials as well which exhibit a log-normal distribution of t_b values [3, 83–85]. These aspects will have to be discussed in later sections in terms of detailed structural models of the failing specimens.

Without reference to any such models and based purely on statistical considerations, Coleman and Marquardt developed their important theory of *breaking kinetics of fibers* (reviewed in ref. 7). They especially investigated the distribution of lifetimes of fibers under constant or periodic load and the effects of fiber length, rate of loading, and bundle size on the strength of filaments or bundles thereof (Figs. 3.3 and 3.4). Two statistical effects should be noted: the lower strength of a bundle as compared with that of a monofilament (due to the accelerating increase in failure probability K after rupture of the first filament within a bundle) and the increase of strength with loading rate here derived from the reduced time each filament spends at each load level. The authors [8] determined the average tensile strengths of 15-denier 66 nylon monofils and that of an infinite bundle; the strength values are plotted as a function of loading rate in Figure 3.3.

In recent years, the evaluation of the statistical aspects of fracture has been greatly furthered by Phoenix [88–91]. He extends the *breaking kinetics of bundles* to random variation in fiber strength and imperfect fiber migration, obtains the asymptotic distribution for the time to failure of large bundles, and applies his model to the breakdown of *fibrous materials* (twisted yarns and cables, unidirectional composites). In the latter approach he partitioned the fibrous composite into a series of m short sections called "bundles" each containing n fiber elements of length L_t, where L_t was to correspond to the *effective stress transfer length* (see Chapter 5 for definition). Thus, the composite is modelled as a *weakest-link arrangement* of its m independent bundles; it has a total length mL and a volume mnL. Phoenix assumed that single fiber lifetimes could be described by a *Weibull* distribution with shape parameter β. As an example, it may be cited that the lifetime distribution for a Kevlar 49/epoxy vessel is quite well represented by this approach.

The above considerations on the *influence of specimen geometry* and *fluctuation of material properties* play an important role in the transformation of laboratory results into real-life expectation values. The statistical implications of such transformations and the models used are discussed by Juillard et Mozzo [92].

The large scatter of strength values and the dependency on sample geometry has also been explained by introducing the *concept of flaws of different degrees of severity*. This means substituting one statistical variable (e.g., underlying strength) for another one (e.g., macroscopic strength or time to failure). Even though such a substitution leaves the nature of the microscopic defects out of consideration, it still allows one to obtain, from statistical analysis, some information on the size, number, location, and severity of these defects (or flaws). The term flaw needs some deliberation. First, a flaw can be considered as an *ellipsoidally shaped void* which may act as a stress concentrator and as the eventual cause of instability and failure. Secondly, it should also be understood as a *weak region* containing molecular irregularities. When being stressed the weak region may fail, leading to a void, a craze, and/or a propagating crack and macroscopic failure. Following Epstein's statistical theory of extreme

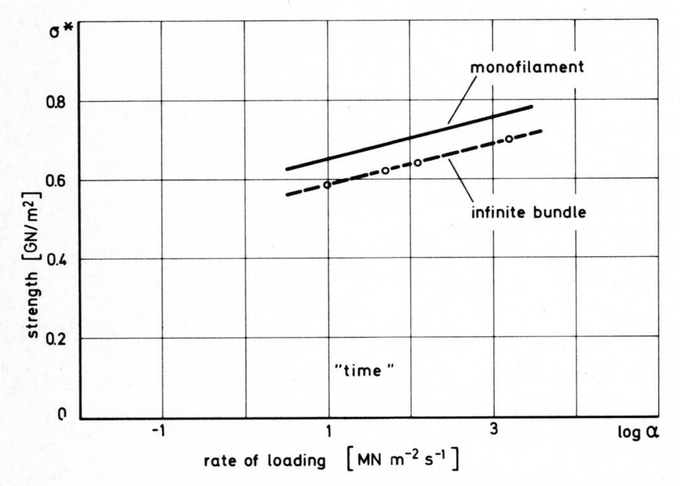

Fig. 3.3. Strength of 66 polyamide monofilaments and bundles of filaments as a function of loading rate (after Coleman et al., [8]).

values [9] the strength of a material volume element depends on the "severeness" of the most critical flaw being present within that volume element. The macroscopic strength distribution g(L, x) of fibers of different length L but uniform cross section can then be expressed as

$$g(L, x) = Ln_0 f(x) \left[1 - \int_{-\infty}^{x} f(y)\,dy\right]^{Ln_0 - 1} \tag{3.6}$$

where n_0 is the number of flaws per unit length of fiber and f(x) the underlying strength distribution. For f(x) being a Gaussian distribution with mean μ and variance v, the most probable strength, σ^*, depends on Ln_0 as shown by the solid line (theory) in Figure 3.4. Also plotted are the experimental data of Levin and Savi-

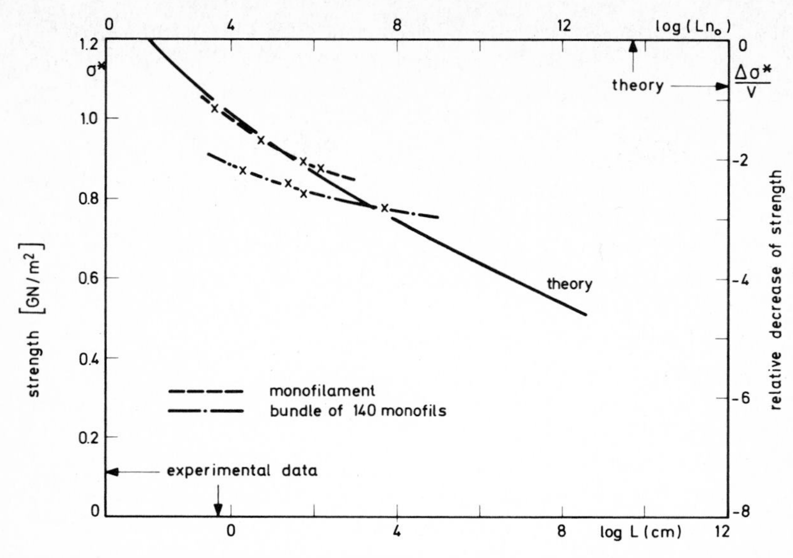

Fig. 3.4. Effect of specimen length L on 6-polyamide fiber strength (after [11]); theoretical curve from Eq. 3.6, experimental data from [10].

ckij [10]. Even though it appears as if the experimental data would be fitted by the theoretical curve in the range of $10^3 < Ln_0 < 10^6$, some caution is required. If indeed the inherent distribution of flaws is a Gaussian distribution, then the decrease of $\sigma^*(Ln_0)$ with L and the variance of $\sigma(Ln_0)$ are both related to v and $n(= Ln_0)$ in a known manner and subject to proof. A proof is also possible for other underlying distributions. For the data of Levin and Savickij [10] neither a Gaussian distribution of x, nor a Laplacian, nor a Weibull distribution describes $\sigma^*(Ln_0)$ and $v(Ln_0)$ simultaneously [11]. As a result of this analysis one may conclude that either the underlying distribution f(x) is truncated at the left-hand side (small values of x), or the variance of σ is much larger than to be expected from the number of flaws. The truncation at small strength values could be caused by a multi-step crack initiation process which would require a certain time. The effect of sample volume on strength was also not in agreement with extreme value statistics for viscose filaments, carbon fibers, and cotton yarn [12]. Fedors and Landel [13] obtained a doubly exponential distribution of strength values for eight SPR gum vulcanizates. Such a distribution derives, for example, from an underlying exponential distribution of flaw cross sections [14]. Here the volume effect was not studied.

When dealing with underlying distributions it should be recognized that different regions (core, surface) of a (seemingly) homogeneous specimen may have different underlying distributions. Evidence of this behavior was for instance found in polyamide fibers [15] and LDPE pipes [84]. Depending on processing conditions volume-related and surface-related breakages with quite different stress-lifetime characteristics could be distinguished. At this point the many attempts to determine underlying distributions are not discussed any further but reference is made to the publications on this subject [11–17]. If any conclusion may be attempted from statistical investigations of uniaxial failure of thermoplastics it is that the underlying distribution, f(x), of natural flaws could be a Laplacian distribution possibly skewed at the left-hand side (small values of x). Moreover, it appears unlikely that a deterministic view may be adopted associating a particular flaw with a particular strength (or time to failure) but rather necessary to allow for the uncertainty of multistep flaw growth.

The third statistical argument (*fracture as the result of a large number of molecular processes*) is very old. Its application to polymers definitely was stimulated from many sides including fracture and fatigue of metals [4] or glasses [17] and the thermodynamics of reactions [19]. The interpretation of fracture as a sequence of individual steps will be mainly used in the following. It necessitates a consideration of the correlation of such steps and of the final failure criterion.

The discussion of the three different statistical arguments can be summarized by saying that in the first case (fracture as a statistical event) a property (probability of fracture) is attached to the material body as a whole. In the second case one flaw, i.e., one micro-heterogeneity (out of many), within the body is considered to dominate failure; in the third case many events interact, influence, and determine each other. *The very same lines of thought used here to explain the variability of strength,for example, are found in the continuum mechanics, fracture mechanics, and molecular approach to theories of strength, respectively.*

III. Continuum and Fracture Mechanics Approach

A. Classical Failure Criteria

In classical continuum mechanics homogeneous isotropic materials are taken into consideration. Failure criteria are established on the basis that there is indeed one strategic material property such as uniaxial tensile strength, shear strength, elastic or total extensibility, or energy storage capability that determines the failure of the stressed sample. If, in establishing such criteria, all environmental parameters (e.g., temperature T, strain rate, $\dot{\epsilon}$, or surrounding medium) are taken to be constant, then failure has to be expected whenever the components of multiaxial stress – usually the three principal stresses σ_1, σ_2, and σ_3 are considered – combine in such a way that the strategic quantity reaches a critical value C. For different T or $\dot{\epsilon}$ C may assume different values. The condition $f(\sigma_1, \sigma_2, \sigma_3) = C(T, \dot{\epsilon})$ represents a two-dimensional "failure surface" in three-dimensional stress space. The stable states of stress form a continuous body bounded by the failure surface against the unstable stress points.

It has already been indicated that failure in polymeric materials may occur in a number of phenomenologically different ways, e.g., as brittle fracture through propagating cracks, as ductile fracture through plastic deformation following shear yielding, or as quasi-brittle fracture after normal-stress yielding (crazing). It had correctly been anticipated that these different phenomena respond differently to the magnitude and the state of stress. This means that for different fracture phenomena separate failure surfaces exist which may overlap and penetrate each other. These facts are extensively investigated and discussed in the monographs by Ward [20] and Bardenheier [93] and in the general literature (e.g., 21–34).

The *classical criteria* were conceived to describe the failure of isotropic, elastic-plastic solids. The physical content of the six principal criteria is in brief the following:

1. The *maximum principle stress theory* (Rankine's theory) states that the largest principle stress component, σ_3, in the material determines failure regardless of the value of normal or shearing stresses. The stability criterion is formulated as

$$\sigma_3 < \sigma^*. \tag{3.7}$$

In this case σ^* is considered a basic material property which can be determined from, for example, a tension test. In stress space the surface of failure according to the maximum stress theory is a cube. A modified form of this theory allows for a larger critical value σ^* if one of the stresses is compressive. Then the surface of failure is a cube again, but with the center displaced from origin.

2. The *maximum elastic strain theory* (St. Venant's theory) states that inception of failure is due if the largest local strain, ϵ_3, within the material exceeds somewhere a critical value ϵ^*. The failure criterion, therefore, is derived as

$$\epsilon_3 = \frac{1}{E}\,[\sigma_3 - \nu(\sigma_1 + \sigma_2)] < \epsilon^* \tag{3.8}$$

where E is the isotropic Young's modulus and ν Poisson's ratio. The term in square brackets thus has to be smaller than an equivalent stress, σ^*.

3. The *theory of total elastic energy of deformation* was first proposed by Beltrami. Since under high hydrostatic pressure large amounts of elastic energy may be stored without causing either fracture or permanent deformation, the elastic energy as such seems to have no significance as a limiting condition.

4. *Theory of constant elastic strain energy of distortion* (Huber, v. Mises, Hencky). The consideration of the poor significance of the elastically stored energy in hydrostatic compression or tension led to the idea of subtracting the "hydrostatic" part of the energy from the total amount of elastically stored energy. Thus, it is assumed that only the energy of distortion, W, determines the criticality of a state of stress. For small deformations one obtains:

$$\frac{6E}{1+\nu}\,W = [(\sigma_1 - \sigma_2)^2 + (\sigma_2 - \sigma_3)^2 + (\sigma_3 - \sigma_1)^2] < 2(\sigma^*)^2. \tag{3.9}$$

5. If a *limiting octahedral shearing stress,* τ^*, is postulated as a failure criterion the same mathematical expression as in Eq. (3.9) is obtained with

$$9(\tau^*)^2 = 2(\sigma^*)^2. \tag{3.10}$$

In this case the limiting condition of small deformations does not pertain.

6. The *Coulomb yield criterion* states that the critical stress, τ^*, for shear deformation to occur in any plane increases linearly with the pressure, σ_n, applied normally to that plane:

$$\sigma_3 - \sigma_1 < \tau^* = \tau_0 + \mu\sigma_n. \tag{3.11}$$

Here the material constant τ_0 is given by the cohesion of the material and μ is a "coefficient of friction". If friction is neglected the Tresca yield criterion is obtained.

The above failure (or stability) criteria do not explicitly contain time as a variable. One may apply them to rate sensitive materials, however, if one recognizes that σ^* and τ^* will depend on stress history. The classical approach of Eyring and other rate theories will be discussed in Section IV of this chapter. The limits of the applicability of the classical criteria and of their extension to anisotropic materials are analyzed by Ward [20] and Bardenheier [93]. Numerous papers (e.g., 21–28) and review articles [29–34] have appeared which concern the geometry of failure surfaces of polymeric materials. Multiaxial failure experiments on elastomers which seem to corroborate the St. Venant criterion have been carried out [21] by Gent and Lindley on endbonded penny-shaped samples of a natural rubber gum vulcanizate, by Ko on polyurethan rubber, by Oberth and Bruenner on castable elastomers containing solid inclusions, and by Lim [22] on tubular samples of polyurethane and of a butadiene/acrylic acid copolymer. Figure 3.5 depicts Lim's data and a failure envelope for $\nu = 0.37$ resulting from the St. Venant criterion.

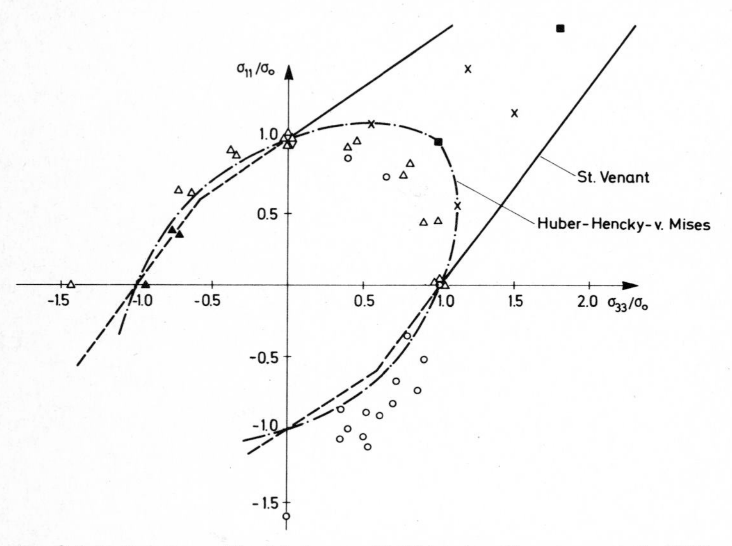

Fig. 3.5. Failure in multiaxial stress, ○ PMMA tubes (Broutman et al., [23]), △ 6 PA tubes, ▲ buckling (Ely, [24]), x PUR tubes (Lim, [22]), ■ SBR membranes (Dickie et al., [25]) – – – maximum strain failure criterion, – · – · – octahedral shear stress failure criterion.

Of the other theories, the Huber-v. Mises-Hencky criterion and the Coulomb yield criterion have received the most experimental support. The failure surface in the stress space described by Eq. (3.9) is a cylinder with its axis along the space diagonal. Traces of this cylinder with the (σ_1, σ_2)-plane are ellipses symmetric with respect to the point of origin. One such ellipse is shown in Figure 3.5. The failure surface corresponding to the Coulomb criterion is a cone. Its traces with the (σ_1, σ_2)-plane are also ellipses but displaced with regard to the origin. Elliptic traces seem to form the best available failure criterion in the biaxial yielding experiments with tubular specimens of PMMA [23, 33], PVC [26, 27], PC [27] and of PS, PE, PP, PA, ABS, and CAB [33]. In the case of multiaxial fracture without yielding (Trennbruch) elliptic and parabolic traces are obtained which pertain to a cone or a paraboloide in stress space [30, 33]. In any event *there is no strict distinction between the form of failure surfaces describing failure through flow and the form of the surfaces referring to failure through material separation (brittle fracture).* This indicates that in these isotropic polymers the *initiation* of failure was strongly influenced by shear stresses irrespective of the *ultimate* material behavior.

The careful and comprehensive analysis meanwhile carried out by Bardenheier [93] in terms of various stress potentials for brittle and ductile fracture of more than a dozen polymers has essentially confirmed the above results.

For obvious reasons there is a strong interest in the deformation and fracture behavior of plastics under large hydrostatic pressures [31–32]. One should expect – and one observes – that the rigidity of a polymer increases with pressure. Sauer et al. [32] report that a pressure of 3.5 kbar raises the initial Young's moduli of amorphous thermoplastics (PC, PI, PSU, PVC, CA) by a factor of 1.2 to 1.9, that of crystalline polymers by 1.4 (POM) to 7.5 (PUR). Despite the increased rigidity, ductile fracture occurs. The effects are not yet understood in all generality. Following the two major review articles on this subject by Radcliffe [31] and Sauer and Pae [32] the Coulomb criterion corresponds best to most pressure-yield experiments. In the case of PTFE and PC with their nonlinear increase of yield strength with hydrostatic pressure ($\sigma_n = p$) even the consideration of a second-order term of p appeared to be necessary [32]. Whereas the application of pressure restricts the cold drawing of many ductile polymers, it enables yielding of normally brittle polymers, e.g., of PS at room temperature at a pressure of 2 to 3 kbar, and of PSU and polyimides at 3 to 7 kbar [32].

In this chapter no molecular interpretations of fracture phenomena are given. The above discussion has already shown, however, that a consideration of the three-dimensional state of stress is not sufficient to elucidate the possible role of chain scission or disentanglement in failure of polymers. This is particularly true if the phenomenon of crazing ("normal stress yielding") is included. Before investigating the molecular aspects, however, in the later chapters, the presentation of general, non-molecular fracture theories should be continued.

B. Fracture Mechanics

Real solids are inhomogeneous. Even if they are one-phase materials they will contain defects, voids, inclusions, flaws, cracks, and/or other inhomogeneities which are able to distort an otherwise homogeneous stress field. A continuum mechanical approach that analyzes these (singular) stress-strain fields in the vicinity of flaws or cracks and derives its criteria of crack stability from an energy balance consideration is the fracture mechanics approach. Griffith [35] was the first to find wide acclaim for his considerations of the energetic changes occurring as a consequence of the widening of a crack (of length $2a$). He equated the resulting decrease dU of elastic energy stored within a stressed specimen with the surface energy γ_cdA necessary to enlarge the crack by an infinitesimal area dA:

$$-\frac{dU}{da} = \gamma_c \frac{dA}{da}. \tag{3.12}$$

For a plate containing an elliptical hole and stressed at right angles to the major axis a, he obtained the stability criterion

$$\sigma^* = (2\gamma_c E/\pi a)^{1/2}. \tag{3.13}$$

The size of the largest material defect together with the material constants modulus of elasticity (E) and surface work parameter (γ_c), thus, supposedly determine the strength of the material, i.e., that macroscopic value σ^* of uniaxial stress at which irreversible crack propagation begins. Equation (3.13) gives a mathematical form to the earlier used concept that flaw size (or severity) determines the inherent strength of a sample. It also explains that the actually observed macroscopic strength is much smaller than the theoretical strength of flawless specimens.

The application of fracture mechanics to viscoelastic media is caught between the limitations that are brough about by the deviation from infinitesimal strain condition, by molecular anisotropy, local concentration of strains, and the time dependency of stress and strain functions. It has been notably successful in investigations of crack propagation. The analytical extension of Griffith's work to linearly viscoelastic materials has been reviewed by Williams [36], Knauss [37], and Nikitin [94]. In Chapter 9 a more detailed account on the application of fracture mechanics to crack propagation is given. There the morphological aspects and the effect of time-dependent plastic deformation, craze initiation and growth, and chain scission on the energy of cohesive fracture of high polymers will be considered.

At this point attention will be drawn to an application of the Griffith approach to the effect of matrix orientation on strength. Such an interpretation has been forwarded by Mikitishin et al. [79] and Sternstein et al. [80]. These authors suggest that during the orientation process a rotation, elongation, and widening of possible defects takes place. The change of shape and orientation changes the effect of these flaws as stress concentrators. They become less dangerous in a direction parallel to the draw direction (z) and more critical in the perpendicular direction (x). If only the increase in length of (elliptical) voids which are oriented parallel to the draw direction is considered, the lateral strength σ_x is derived from Eq. (3.13) as:

$$\sigma_x = [2\,\gamma_c\,E/(1+\lambda)\pi a]^{1/2} \tag{3.14}$$

where λ is the draw ratio. Sternstein et al. [80] have made comprehensive calculations in two dimensions of the flaw-rotation and widening effect on axial and lateral strength for a statistical distribution of flaw orientations and various flaw geometries. They compare their predictions of the lateral strength of polystyrene with the experimental data of Retting [81]. For moderate molecular weights and draw ratios up to 1.5 they find a correspondence to within 20%. For higher molecular weights and larger draw ratios the discrepancies are above 70%.

If a crack travels at the interface of an *adhesive joint* one observes an adhesive debonding rather than cohesive failure. Except for minor differences in boundary condition the analgy to the fracture mechanical treatment of cohesive failure is complete [38–40] and the normal debonding stress σ_a^* is obtained as:

$$\sigma_a^* = (2\gamma_a\,E/\pi a)^{1/2}. \tag{3.13a}$$

The interpretation of γ_a and E is rather involved. These quantities not only depend on the static surface free energies and elastic moduli of the interacting (polymeric) materials but also on the work of deformation which the adhesive layer undergoes in debonding and on the surface activity of a possible third phase having access to the debonded area [40].

C. Continuous Viscoelastic Models

Under the heading of this section belongs the very fruitful model of a common "failure envelope" proposed by Smith [41]. It is based on a consideration of the viscoelastic behavior of continuous high polymer bodies, i.e., on the concept that a reduction according to the time-temperature superposition principle of the environmental parameters stress, strain-rate, and temperature should lead to corresponding molecular states. If a fracture criterion refers indeed to unique limits of molecular load-carrying capability, for example, then the plotting of reduced stress at break versus strain at break for different experimental conditions should lead to one master curve, the *failure envelope* (Fig. 3.6). For a large number of natural and synthetic rubbers and vulcanizates under similar modes of mechanical excitation, this concept was found to hold. It thus generally allows the extrapolation of the ultimate behavior into unknown regimes of time, strain-rate, or temperature. An excellent review on

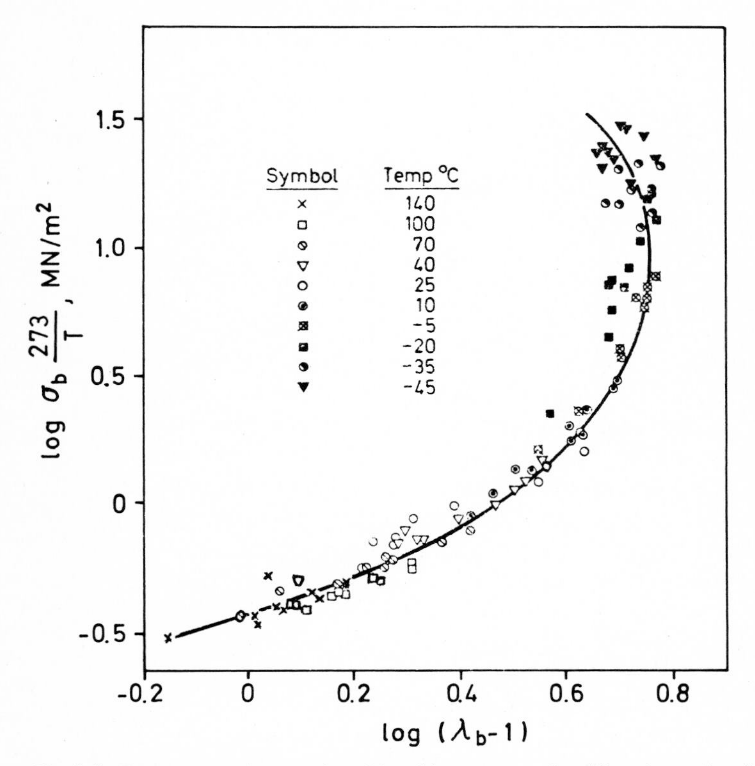

Fig. 3.6. Fracture envelope: reduced breaking stress vs. breaking elongation for an unfilled butyl rubber vulcanizate (after Smith, [41]).

the uniaxial rupture of elastomers has been given by Fedors [111]. This model, however, does not allow prediction of the shape of the fracture envelope or the conversion of data obtained under different modes of excitation (uniaxial, equal biaxial, or other multiaxial loading).

An argument to resolve the discrepancy between the failure envelopes obtained for different modes of straining is indicated by the work of Blatz, Sharda, and Tschoegl [42]. These authors have proposed a generalized strain energy function as constitutive equation of multiaxial deformation. They incorporated more of the nonlinear behavior in the constitutive relation between the strain energy density and the strain. They were then able to describe simultaneously by four material constants the stress-strain curves of natural rubber and of styrene-butadiene rubber in simple tension, simple compression or equibiaxial tension, pure shear, and simple shear.

The continuous viscoelastic models discussed in this section make use of the theory of rubber elasticity and of the time-temperature superposition principle. They implicitly very well recognize the molecular origin of the viscoelastic behavior of a material – but explicitly they do not refer to noncontinuous quantities such as the discrete size, structure, and arrangement of molecules, the anisotropy of molecular properties, or the distribution of molecular stresses or stored strain energy. If individual molecular events or discontinuities of stress or strain are either not discernible or not of particular importance, a representation of a solid as a continuum is quite adequate.

IV. Rate Process Theories of Fracture

A. Overview

As opposed to the continuum mechanical theories, "molecular" rate process theories of fracture recognize the presence of discrete particles or elements forming the material body. Rate process theories intend to relate in a straightforward manner breakage, displacement, and reformation of these elements to deformation, defect development, and fracture of the structured material.

All theories to be described subsequently have in common the assumption that macroscopic failure is a rate process, that the basic fracture events are controlled by thermally activated breakages of secondary and/or primary bonds, and that the accumulation of these events leads to crack formation and/or to breakdown of the loaded sample. The basic events referred to in this Section are defined in a general manner and not experimentally related to particular morphological changes. The "morphological" theories based on the physical evidence of molecular damage obtained from spectroscopy and X-ray scattering methods will be discussed in Chapters 7 and 8. A third approach considers the role of cracks and craze initiation and is treated in Chapter 9. In the following some non-morphological fracture theories will be presented which concern respectively the calculation of the time to failure as a function of flow or damage accumulation (Sections B–D), the determination of sample strength (E–F) and time to failure (G–J) as a function of molecular weight or concentration of primary bonds.

One of the early applications of the theory given by Glasstone, Laidler, and Eyring [43] was the prediction of the flow and diffusion behavior of liquids. From the original concept of flow, the thermally activated jump of molecules across an energy barrier, various fracture theories of solids have emerged. Thus Tobolsky and Eyring [44] considered the general decrease of *secondary* bonds. A more general approach, the *rule of cumulative damage* [95–98], does not explicitely specify the nature of the damage incured during loading but tries to account for the influence of load history on sample strength. The importance of *primary bonds* for static strength had been deduced very early from the dependences of sample strength on molecular weight [99–103] and on volume concentration of primary bonds [70, 71]. Later Zhurkov [45–47] and Bueche [48–50] took up this idea and developed their kinetic theories based on the *breakage of primary bonds.* The Russian school has established and reviewed a considerable amount of experimental data on the effect of time and temperature on the fracture behavior of polymers of different composition and structure and under various environments [51, 53, 104].

Following the earlier works of Eyring, Zhurkov, and Bueche, these and other authors have continued to develop different aspects of the kinetic theory of fracture. Particularly in the USSR fracture data have been interpreted in terms of the failure of regular, nonmorphological model lattices (e.g., 54–58). Gubanov et al. [54], Bartenev [55], and Perepelkin [56] focussed on the potential energy of interaction between adjacent members of a polymer chain, Mikitishin et al. [57] on the defects introduced into the lattice by chain ends, and Dobrodumov [58] on the increase of the load and – consequently – of the rate of scission of bonds adjacent to a broken bond.

Hsiao and Kausch [59–61] studied the effect of orientation-dependent local strain on the total rate of chain scission. Holzmüller [62], Bartenev et al. [63], and Salganik [64] analyzed the amount of thermal energy and the direction of relative motion imparted by statistically fluctuating phonons to adjacent segments. Different statistical aspects of the accumulation of molecular defects were treated by Orlov et al. [65], Goikhman [66], and Gotlib [67] who simulated the isolated production, growth, interaction, and coalescence of defects. An energy probability theory was forwarded by Valanis [68] who combined the role of strain energy density, the stochastic nature of fracture, and the fracture hypothesis of Zhurkov.

Whereas all of the above theories [54–68] refer in one way or another to the thermally activated breakage of one kind of molecular bonds as primary fracture event, the Bueche-Halpin theory for the tensile strength of gum elastomers [69] employs the idea that the viscoelastic straining of rubber filaments together with the degree of their ultimate elongation determine the kinetics of crack propagation in an elastomeric solid.

In the following the characteristic aspects of these generalized, nonmorphological rate process fracture theories are to be discussed, i.e. the nature of the basic fracture events, their distribution in space and time, the law of the accumulation of these events, and resulting fracture criteria.

B. Eyring's Theory of Flow

Flow of a liquid, melt, or solid is the result of a stress-biased thermodynamically irreversible displacement of molecules past each other. A molecule in thermal equilibrium with its surroundings is in thermal motion, which in the case of a liquid or solid is predominantly a vibration around a temporary equilibrium position. The amplitude of vibration is constantly changing. Eyring [43] postulated that a displacement (or jump) of a molecule from an initial to a neighboring equilibrium position may occur if the thermal energy of the molecule is sufficient to reach the "activated" stage, i.e., the top of the energy barrier separating the initial and the final equilibrium positions. The rate of decay of activated states towards the final position was obtained by him as

$$k_0 = \kappa \frac{kT}{h} \exp(-u_0/kT) \exp \Delta s/k. \tag{3.15}$$

Here κ is the transmission constant which indicates how many of the activated complexes actually disintegrate, k is Boltzmann's, h Planck's, and R the gas constant, T is absolute temperature, u_0 the height of the potential barrier, and Δs the difference in entropy between ground state and activated state (the quantities u and s in small letters refer to one particle, the capital quantities to a mole of them). In the absence of external forces initial and final equilibrium states are assumed to have the same potential energy. Then the rates of flow of particles across the dividing potential barrier are the same in the forward and backward directions. Under the influence of external forces, however, local stress fields are generated. A particle thermally activated to move in the direction of, say, stress Ψ by a distance $\lambda/2$ then gains an amount of energy w:

$$w = \Psi \lambda \, q/2 \tag{3.16}$$

where q is the average cross section occupied by the particle normal to the direction of motion. A particle moving in the $-\Psi$ direction will loose the same amount of energy. The first particle, therefore, needs less thermal energy to reach the activated state than the second particle. Consequently the rates of flow of particles across an energy barrier in opposite directions are biased by the local stress:

$$k_1 = k_0 \exp(w/kT) \tag{3.17}$$

$$k_2 = k_0 \exp(-w/kT). \tag{3.18}$$

The net flow rate, therefore, is:

$$K = k_1 - k_2 = 2\, k_0 \sinh(w/kT). \tag{3.19}$$

It should be noted very carefully that w is equal to the product of local stress times an activation volume; this is the volume drawn up by particle cross section and step width in the process of activation. Activation volume and activation energy necessarily must refer to one and the same activation process.

C. Tobolsky-Eyring

The concept of slipping of secondary bonds was applied by Tobolsky and Eyring in 1943 to the breaking of polymeric threads under – uniaxial – load [44]. The rate of decrease of the number N of such bonds per unit area under constant uniaxial stress Ψ_0 is obtained from Eqs. (3.1) and (3.19):

$$-\frac{1}{N}\frac{dN}{dt} = 2\,k_0 \sinh(\Psi_0\lambda/2\,NkT). \tag{3.20}$$

For large values of stress the flow of bonds takes place almost exclusively as breakage and not as reformation. In that case one may use Eq. (3.17) instead of (3.19). If one substitutes there $\Psi_0\lambda/2$ NkT by y one obtains a solution in terms of the exponential integral $-\mathrm{Ei}(-y)$:

$$-\mathrm{Ei}(-y) = \int_{y_0}^{\infty} dy'\exp(-y')/y' = k_0 t_b. \tag{3.21}$$

The lower limit of integration is $y_0 = \Psi_0\lambda/2\,N_0kT = w/kT$. Equation (3.21) offers an implicit relation between the expectation value of the time to fracture of the thread, t_b, the rate of flow of unstressed secondary bonds, k_0, and the stress Ψ_0 acting on the N_0 initial bonds. With $w \gg kT$ the exponential integral may be approximated by:

$$-\mathrm{Ei}(-y) = [(\exp - y)/y](1 - 1/y). \tag{3.22}$$

Within the range of validity of this approximation the logarithm of lifetime, $\log t_b$, is almost linearly related to w, i.e., to the applied stress. Exactly this behavior is exhibited by a large number of stressed metals, ceramics, or polymers. Together with Eyring's original interpretation this has laid the foundations of the kinetic theory of fracture, to which – as indicated above – subsequently a large number of researchers have contributed.

D. Rule of Cumulative Damage

The above concept of Tobolsky and Eyring can be considered as a specific form of the more general rule of cumulative damage. This rule stipulates that fracture of a specimen occurs if the time- and stress-dependent "damage" D(t) has accumulated to unity:

$$D(t,\sigma) = \int_0^t dt'/t_b(\sigma[t']). \tag{3.23}$$

This relation was originally formulated by Bailey [59] in order to compare the fracture behavior of plate glas subjected to different load histories (constant stress, constant rate of stressing, load cycles of different amplitude). An exactly analogous concept has been applied by Miner [96] to cyclic fatigue. He used the ratio of the number of cycles $n(\sigma)$ spent at stress amplitude σ to the endurance limit $N_b(\sigma)$ at that stress as integration variable in Eq. (3.23) (*Miner's rule*).

It should be mentioned, however, that according to this hypothesis a certain load element $\sigma(t)\ \Delta t$ creates the same damage whether it occurs in the beginning or at the end of the specimen life. Thus, no stress concentration through crack growth or defect interaction is foreseen in this model. This, however, is contrary to experience. The apparent confirmations of the linear rule of cumulative defects (LRCD) concern practically always increasing stress histories (in fatigue or in tension, see [95–98]. Such a behavior is readily explained by the fact that "the damage" developes mainly in "the last moment". To give a quantitative example, McKenna and Penn [98] have measured the time to fracture, t_b, of PMMA under constant and under increasing load σ. For constant load they obtained:

$$t_b = t_0 \exp -\gamma\sigma/RT \tag{3.24}$$

with $t_0 = 7.78 \cdot 10^9$ s

and $\gamma/RT = 0.25\ \mathrm{MPa}^{-1}$ at 297 K.

The rate of damage formation, $\dot{D}(t)$, is then defined as:

$$\dot{D}(t)dt = dt/t_b = t_0^{-1} \exp \gamma\sigma/RT. \tag{3.25}$$

According to the LRCD a specimen breaks once

$$\int_0^t \dot{D}dt' = \frac{t}{t_b} = 1, \tag{3.26}$$

thus, evidently at $t = t_b$.

In a tensile test at constant load rate one has:

$$\dot{D}(t) = t_0^{-1} \exp [\dot{\sigma}(t)\gamma t/RT] \tag{3.27}$$

and:

$$\int_0^t \dot{D}(t)dt = (t_0\gamma\dot{\sigma}/RT)^{-1}[\exp (\gamma\dot{\sigma}t/RT) - 1] \tag{3.28}$$

which gives:

$$t_b = [\ln (t_0\gamma\dot{\sigma}/RT + 1)]/(\gamma\dot{\sigma}/RT). \tag{3.29}$$

Figure 3.7 shows the damage accumulation. It is readily visible that in McKenna's experiments more than 95% of the time under stress elapses before one half of the damage is created.

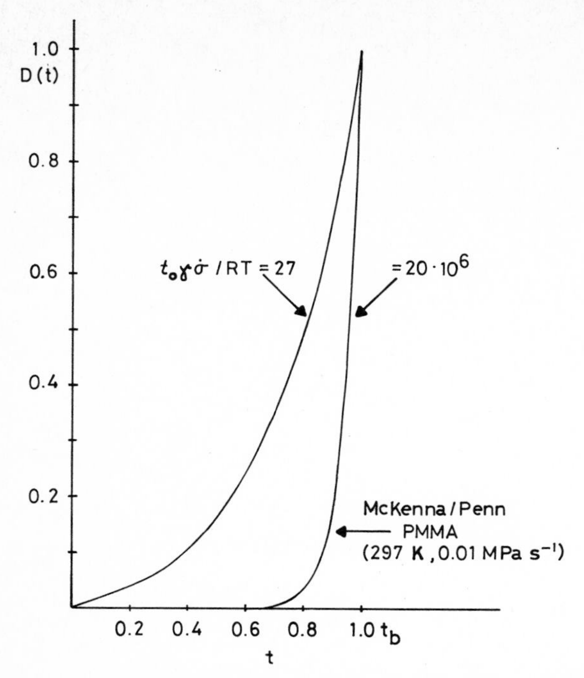

Fig. 3.7. The damage accumulation function (Eq. 3.28) according to the hypothesis of Miner/Bailey for a tensile test ($\dot{\sigma}$ = constant = 10^{-2} MPa s^{-1}, t_0 and γ/RT as indicated)

For decreasing loads the LRCD underestimates the service life and leads to *over-design.* For statistically varying loads notable derivations are also observed [97, 98]. The same is true for the smaller exponential arguments which are often obtained (see Chapter 8).

E. Early Theories of the Molecular Weight Dependence of Strength

The fact that the static strength of polymer samples depends on their molecular weight had been noticed rather early ([99–102], a comprehensive review of the literature up to 1946 has been given by Haward [103]). Working with polyvinylchloride-acetate copolymer fractions Douglas and Stoops [99] and with cullulose acetate fractions Sookne and Harris [100] found that sample strength σ_b increased with number average molecular weight M_n according to

$$\sigma_b = A - B\,M_n^{-1} \tag{3.30}$$

but independent of M_w/M_n. Spurlin [101] concluded from this independence that the chain ends may be the weak spots of the structure giving rise to slip of adjacent molecules.

Referring to the work of Bueche [105] some authors [106, 107] recently proposed an apparently morphological entanglement network model (Fig. 3.8). In their calculations it was assumed that the ideal strength of an isotropic entanglement network is given by the force nf_b it takes to break those (n) chain segments which intersect a unit area oriented perpendicularly to the stress axis:

$$\sigma_b = nf_b = \left(\frac{\rho N_A}{3M_e}\right)^{2/3} f_b \left(1 - \frac{2M_e}{M}\right). \tag{3.31}$$

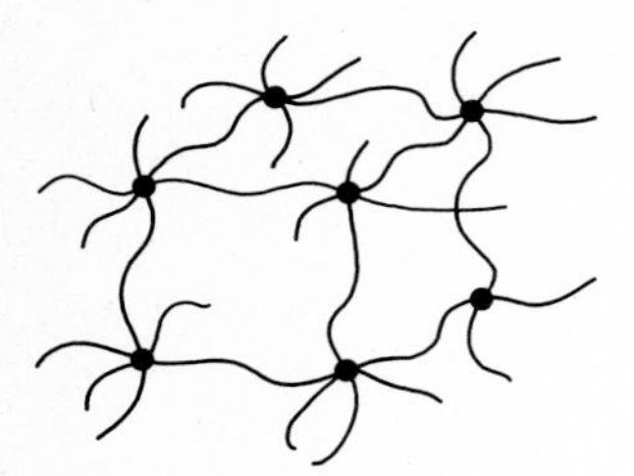

Fig. 3.8. Bueche's model of the strength of an ideal network of crosslinked chains

Whereas n and f_b have been quite correctly estimated, it has completely been neglected that the chain segments in an isotropic polymer matrix have a statistically coiled conformation. Thus, before being loaded by a force of the order of f_b a chain segment of molecular weight M_e would have to be stretched by the natural draw ratio. This point had already been underlined by Flory in 1945 [102] and by Taylor and Darin [108] who indicated that only the (elastomer) chains oriented at less than a very small angle to the direction of stress *at the instant of breaking* can be highly elastically stressed. This is also true for thermoplastic polymers. This means that the physical network of entanglements must be heavily deformed before forces of the order of f_b can be supported. The energy of plastic deformation dissipated in this process should (at least implicitely) enter Eq. (3.31). In order to predict reliably the ideal brittle strength of a polymer, it is therefore necessary *to know* the critical conditions for the onset of large deformations, *to identify* their limits (for instance in terms of the natural draw ratio λ_{nat}), and *to ascertain* the role of the entanglement network. This will be attempted in the final chapters.

F. Effect of Volume Concentration of Primary Bonds

At this point one would like to check further arguments that the existence (and volume concentration) of the strong primary bonds determines the rigidity and strength of (polymeric) solids. Holliday et al. [70] have looked under this aspect on the degree of connectivity CN (average number of main network bonds per chain atom, e.g., 2 in hydrocarbon chains, 3 in graphite, and 4 in diamond) and on the relative volume density of main bonds NCR (which here is density times number of network atoms in the repeat unit divided by molecular weight of a repeat unit). The term NCR varies between less than 0.01 for bulky polymeric chains and 0.58 for diamond. The bulk moduli, Young's moduli, hardness values, and cubical coefficients of thermal expansion of the more than 50 organic and inorganic polymers investigated at room temperature exhibit a strong correlation with NCR. In Figure 3.9 the best fit curves drawn by Holliday for Young's modulus and Brinell hardness values BHN are reproduced (the actual NCR values – except for those of the graphitelike structures – fall into a narrow band along the curves having a width of ± 0.05). The nearly identical shapes of the upper portions of both curves suggest that hardness (compressive yield strength) and elastic modulus depend in the same way on the concentration of primary bonds. In fact for the points on the curve with NCR > 0.08 (mostly organic thermosetting resins or inorganic polymers) the ratio of BNH value and E has a constant value of 0.0625.

For values of NCR < 0.08 (thermoplastic materials) the correlation between mechanical properties and bond density is much less pronounced although the steep slopes of the shown curves tend to make the scatter of the actual data points less apparent. The main reason for the lack of good correlation certainly is the fact that *in high polymer solids the stiffness and strength can be utilized only to that extent permitted by intermolecular attraction and segment orientation* (cf. Table 1.1 for the differences in elastic moduli of chains, highly oriented fibers, and isotropic solids; in section E below and in Chapter 5 the role of orientation and intermolecular attraction on chain strain is discussed in detail).

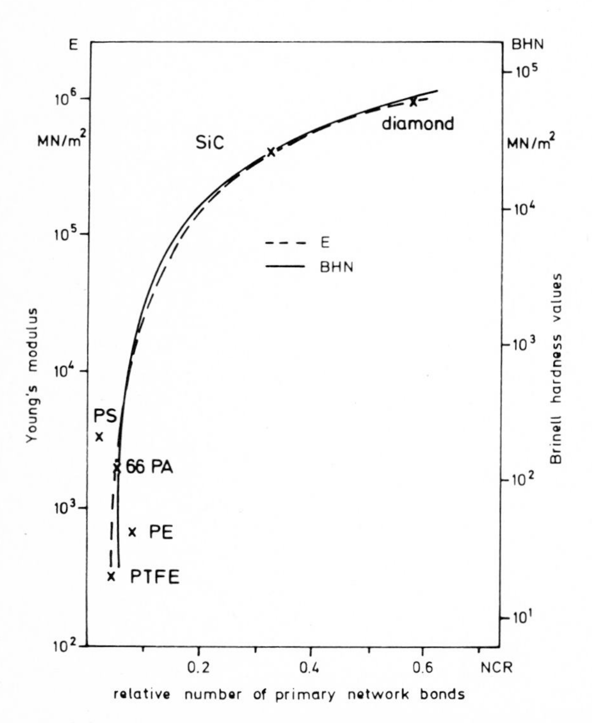

Fig. 3.9. Young's modulus E and Brinell hardness values BHN of various solids vs. volume concentration of primary bonds (after Holliday [70])

In order to investigate the effect of the density of main chain bonds on the mechanical properties it is necessary to select experimental conditions such that comparable mechanical excitation of said bonds does indeed take place. According to Vincent [71] commensurable reference temperatures to measure a critical tensile strength are the brittle point temperatures, where brittle failure changes to a ductile failure. As indicated in Figure 3.10 he has related the number of backbone bonds per unit area (i.e., sample density times length of repeat unit over weight of repeat unit) to the critical tensile strength of (isotropic) polymers. A linear relationship exists between critical sample strength and bond packing density which does not seem to depend on the chemical nature of the bond. From the slope of the straight line the critical strength (in MN/m^2) of the isotropic polymers is obtained as 36.8 times the number of backbone bonds per nm^2. In highly oriented textile fibers this figure is much larger and amounts to between 100 and 250; in aromatic polyamide fibers (Kevlar 49) it reaches about 650. This should be compared with the theoretical value for fully oriented chains of infinite length which is about 3000, since the axial load to break a molecular chain of any one of the species listed in Figure 3.9 is of the order of 3 nN.

In other words, in isotropic thermoplastics at the point of brittle failure only a minute fraction – less than 1% – of all main chain bonds is fully strained. Under these conditions, onset of unstable crack propagation is determined by the magnitude of intermolecular attraction. Reinterpreting Vincent's data one must say that it is not the number of backbone bonds per unit area and their accumulated strength

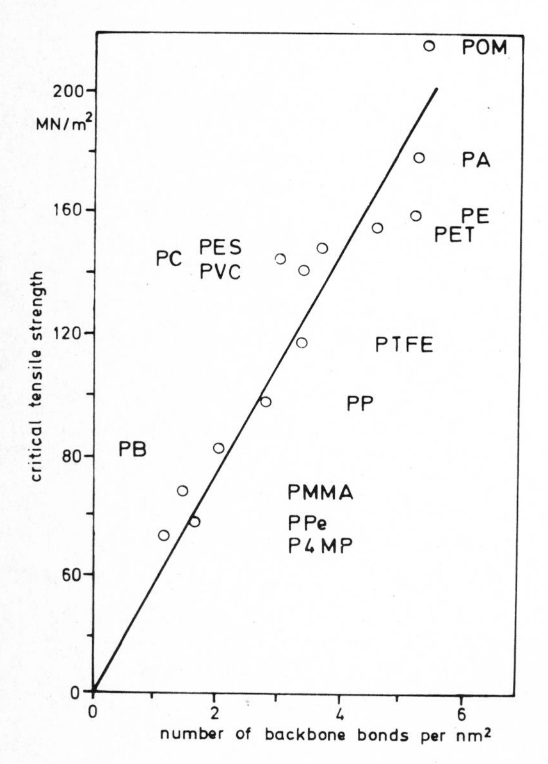

Fig. 3.10. Critical tensile strength (strength at ductile-brittle transition) vs. chain cross-section (after Vincent [71])

which determine macroscopic strength but the level of intermolecular forces; the latter benefit from a closer packing (see also [112, 113] and Chapter 10).

G. Zhurkov, Bueche

Independently from each other Zhurkov et al. [45–47] in the USSR and Bueche [48–50] in the USA reactivated the idea that in fracture of polymers – and also of metals for that matter – the breakage of primary (chemical) bonds plays an important role. They found that the time to fracture under uniaxial stress σ_0 of Zn, Al, and for example, PMMA and PS below their glass transition temperatures could be expressed by an exponential relationship involving three kinetic parameters:

$$t_b = t_0 \exp (U_0 - \gamma\sigma_0)/RT. \tag{3.32}$$

The three parameters were – based on the available data – interpreted as energy for activation of bond scission (U_0), as the inverse of molecular oscillation frequency (t_0), and as a structure sensitive parameter γ. In Figure 3.11 experimental data and the theoretical curves according to Eq. (3.32) are reproduced.

Equation (3.32) can be considered as a general expression of the kinetic nature of material disintegration. A closer discussion of the meaning of the molecular oscillation frequency, of the structure-sensitive parameter γ, and also of the activation

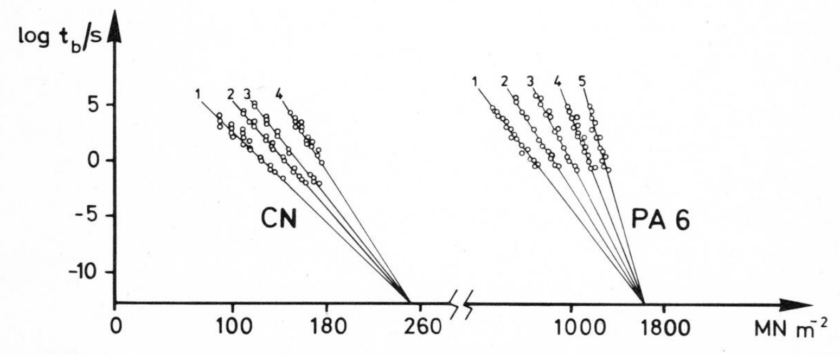

Fig. 3.11. Times to fracture under constant uniaxial load: cellulose nitrate (CN) at T = 70 (1), 30 (2), –10 (3), and –50 °C (4); polyamide 6 (PA 6) at T = 80 (1), 30 (2), 18 (3), –60 (4), and –110 °C (5) (after Savitskii et al. [86].

energy term is necessary in view of the molecular structure and of the sequence of molecular processes [87, 109, 110]. It should already be indicated, however, that the energy term U_0 is not a priori identical with the thermal dissociation energy [87, 109].

In the following section it will be discussed in more detail to what extent the orientation distribution of chain segments, the ensuing distribution of axial chain stresses, and the gradual accumulation of defects through successive breakage of chains can be held responsible for a discrepancy between actual and theoretical strength.

H. Hsiao-Kausch

In the fracture theory of Hsiao-Kausch [60, 61] the state of orientation of a polymeric solid is explicitly recognized. The theory combines the kinetic concept of Zhurkov or Bueche and the theory of deformation of anisotropic solids developed by Hsiao [59] for a model made up of rodlike elastic elements. The fracture theory is based on the simplifying assumption that the mechanical properties of the anisotropic solid are predominantly determined by the state of orientation and the properties of the (one-dimensional) elastic elements. The intermolecular interaction is neglected. In Chapter 2 the mathematical model (Fig. 2.12) has been described and the constitutive stress-strain relations (Eq. (2.5)) have been derived. The kinetic aspects are introduced into this model through the assumption that the rodlike elastic elements can undergo breakage, that the probability of failure of an element is determined by the axial stress acting upon that element, and that an immediate redistribution of the load carried up to then by the breaking element occurs. Again the model is characterized by the fact that it *distinguishes between the macroscopic state of stress (σ) and the microscopic (axial) stress Ψ of an element*. The axial stress of a one-dimensional elastic element deformed in a homogeneous strain field ϵ_{mn} is determined by its axial elongation:

$$\Psi = E\, \epsilon_{mn} s_m s_n . \tag{3.33}$$

The summation convention for repeated indices is used. E is the modulus of elasticity of the elastic elements, and s_m, s_n are components of the unit vector in the direction of orientation identified by ϑ and ϕ ($s_1 = \sin\vartheta \cos\phi$, $s_2 = \sin\vartheta \sin\phi$, $s_3 = \cos\vartheta$). For uniaxial stress, $\sigma_{33} = \sigma_0$, and transverse symmetry about the 33-axis Eq. (3.33) becomes:

$$\Psi(\vartheta, t) = E\,(\cos^2\vartheta - \nu \sin^2\vartheta)\,\epsilon_{33}(t) \tag{3.34}$$

where ν is Poisson ratio $\epsilon_{11}/\epsilon_{33}$.

In a partially or nonoriented system, the microscopic axial stresses are larger than σ_0 by a factor of up to E/E_λ where E_λ is the longitudinal modulus of elasticity of the subvolume containing the elastic element. If all subvolumes of a sample behave identically, then E_λ is the longitudinal modulus of the (partially) oriented sample. Naturally the largest axial stresses are experienced by those elements oriented in the direction of uniaxial stress. In a completely oriented system E and E_λ are equal by definition, and macroscopic and microscopic stress (in the direction of orientation) are equal also.

As stated above it is then first assumed in accordance with Zhurkov's kinetic considerations that elements break and secondly that the orientation-dependent rate of breakage of elements is given by Eq. (3.17) as

$$-\frac{d\,f(\vartheta, t)}{f\,d\,t} = k(\vartheta) = k_0 \exp(\beta\Psi/RT) \tag{3.35}$$

where β is the activation volume for breakage of the elastic elements in question and $f = N(\vartheta, t)/N(\vartheta, 0)$ the fraction of unbroken elements.

Thirdly it is assumed that after breakage of an element mechanical equilibrium is established within the affected subvolume and that strain ϵ_{33} increases. This means that the elastic elements of all orientations take over an appropriate share of the load carried previously by the broken element to balance the load acting on the boundaries of the subvolume. It has been shown in Chapter 2 that the stress tensor acting on a small volume element of the elastic network can be calculated once the orientation distribution function, ρ, is known

$$\sigma_{ij} = \int \rho(\vartheta)\, f(\vartheta, t)\, s_i\, s_j\, \Psi(\vartheta, t)\, d\Omega \tag{3.36}$$

where $d\Omega$ is the infinitesimal solid angle.

In uniaxial stressing, mechanical equilibrium within the elastic network is established if

$$\sigma_{11} = \sigma_{22} = 0 = \int_0^{\pi/2} E\rho(\vartheta)\, f(\vartheta, t)\, [\epsilon_{11} \sin^2 \vartheta + \epsilon_{33} \cos^2 \vartheta] \sin^3 \vartheta\, d\vartheta \tag{3.37}$$

and

$$\sigma_{33} = 2\pi \int_0^{\pi/2} E\rho(\vartheta)\, f(\vartheta, t)\, [\epsilon_{11} \sin^2 \vartheta + \epsilon_{33} \cos^2 \vartheta] \cos^2 \vartheta \sin \vartheta\, d\vartheta. \tag{3.38}$$

The system of Eqs. (3.35), (3.37), and (3.38) has been evaluated for a completely oriented and an unoriented network under constant uniaxial stress σ_0. The case of a completely oriented network is that treated by Tobolsky and Eyring. All elements are thought to be subjected to the same stress Ψ which increases inversely proportional to the decreasing number of unbroken elements. Breakdown of the subvolume occurs with breakage of the last element within the subvolume, i.e., after f has reached zero. The time to breakdown of the subvolume, t_b, is taken from Eqs. (3.20), (3.21), and (3.35) as

$$k_o t_o = - \mathrm{Ei}\,(-\beta\Psi_o / RT) \tag{3.39}$$

where Ψ_0 is the initial stress carried by elements of orientation $\vartheta = 0$.

In the case of random network orientation the system of Eqs. (3.35), (3.37), and (3.38) cannot be solved analytically. An approximation method has been proposed, however [60]. Without repeating details of the method it may be said that the changes of f for a time interval Δt as calculated from Eq. (3.35) are being fed back into Eqs. (3.37) and (3.38) giving the corresponding changes of $\epsilon_{11}(t)$ and $\epsilon_{33}(t)$, which in turn determine the changes of $\Psi(t)$. The iterative integration of Eqs. (3.35), (3.37), and (3.38) can be carried out numerically.

Two results of these calculations seem to be of interest with respect to the effect of network orientation on defect accumulation and strength: that the range of orientation angles within which elements break preferentially is narrow and that the increase in strength resulting from improved uniaxial orientation is limited. The first effect is illustrated in Figure 3.12 for a randomly oriented network where the initial distribution $\rho = 1/2\,\pi$, of elements is independent of ϑ. Elements oriented in the direction of uniaxial stress ($\cos \vartheta = 1$) experience the largest initial stress (Ψ_0), and they break preferentially. The angular distribution of f ($\cos \vartheta$, TIME) is

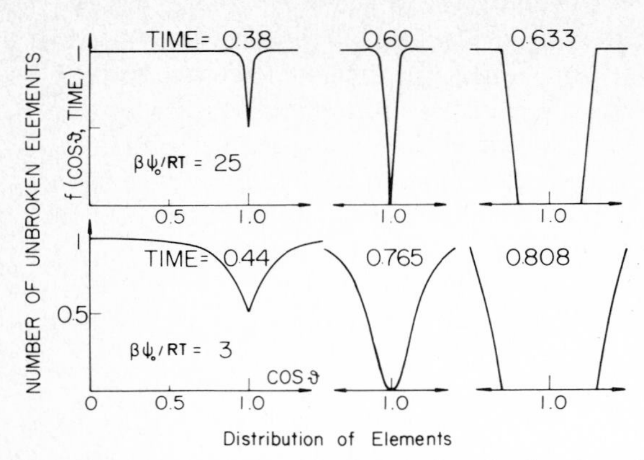

Fig. 3.12. Decay within a population of randomly oriented elements; at high stresses ($\beta\Psi_0 > 10$ RT) essentially those elements oriented in the direction of uniaxial stress break (upper graph). TIME = linear time scale according to Eq. (3.40) (Kausch, Hsiao [60]).

shown for three different stages of defect development, namely after f (1, TIME) has dropped to 0.5 and zero, respectively, and for the point of impending network failure – where the numerical calculations had been terminated because breakage of the remaining elements occurred at too high a rate. For ease of comparison the time scale is given in terms of TIME where

$$\text{TIME} = t/t_z = k_0 t \exp(\beta\Psi_0/RT) \tag{3.40}$$

and t_z the time to failure according to Zhurkov's Eq. (3.32). It should be noted that more than one-half of the loading period has to pass before one-half of the most heavily stressed elements is broken. More than 90% of the loading period has to elapse before all elements within a small angular section are broken. At this second stage the majority of all elements is still unbroken. The step from the second to the third stage of defect accumulation is rather short (< 6% of loading time) and from there to network failure practically negligible. It should also be noted that TIME (failure) of a randomly oriented network composed of successively breaking interacting elements is only somewhat smaller than the time given by the empirical Eq. (3.32) where TIME (failure) for a stress Ψ_0 is unity; on the other hand it is larger than $RT/\beta\Psi$, the value predicted by the exponential integral (Eq. (3.22)) for a completely oriented network loaded to such an extent that again the elements experience the same initial stress Ψ_0.

The numerical calculations according to the above model resulted in an estimate of the time to failure of arbitrarily oriented networks under constant load:

$$k_0 t_b = -[\text{Ei}(-\beta\Psi_0/RT)]C \tag{3.41}$$

where C is a slowly varying correction term the limits of which are expressed by:

$$1 \leq C \leq 1 + 0.63\,\beta\Psi_0/RT. \tag{3.42}$$

The lower limit refers to completely oriented, the upper to randomly oriented networks. With C being determined it has become possible to compare Eqs. (3.32) and (3.41) and to express the structure sensitive kinetic parameter γ in terms of the element properties. In the range of validity of Eq. (3.22) one obtains for a fully oriented network:

$$\gamma = \beta \frac{\Psi_0}{\sigma_0} \left(1 - \frac{RT}{\beta\Psi_0} \ln \frac{RT}{\beta\Psi_0}\right) \qquad (3.43)$$

and for a randomly oriented one

$$\gamma = \beta \frac{\Psi_0}{\sigma_0} \left(1 - \frac{RT}{\beta\Psi_0} \ln 0.63\right) . \qquad (3.44)$$

The term in parentheses is an orientation-dependent and minor kinetic contribution derived from the assumption of successive failure of many elements (in the technically relevant range of $\beta\Psi_0$ values from 40 to 200 kJ/mole the term in parentheses assumes values between 1.17 and 1.0). If the nature of the breaking elements is not changed during sample treatment or fracture development, then β can be considered to be constant. The major effect on γ is derived from the local stress concentration Ψ_0/σ_0 which here is equal to the stiffness ratio E/E_λ. This theory essentially predicts, therefore, that the increase in strength is equal to the increase in stiffness; it is based on the presumption that the elements break indeed at a critical local strain (kinetic version of St. Venant's criterion of maximum strain). A different interpretation of the strength of oriented samples which is based on the rotation of flaws has been presented in an earlier section of this chapter.

In Figure 3.13 the experimental data obtained by Regel and Leksovskii [75] for the lifetimes of partially oriented PAN fibers are compared with theoretical curves derived from Eq. (3.41). It should be emphasized that the gain in strength

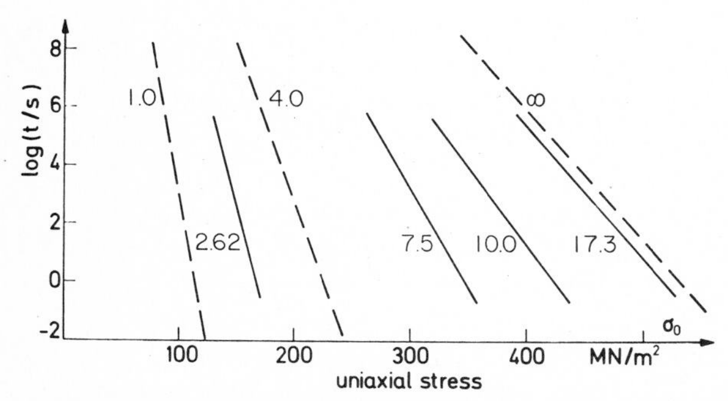

Fig. 3.13. Time to fracture of differently oriented PAN fibers. —— experimental curves (Regel et al. [75]), ---- theoretical curves (Eq. 3.32), parameter: draw-ratio λ.

through improved orientation of the PAN fibers (or of their model representation) amounts to a factor of $\Psi_0/\sigma_0 = 5$. Similar values of between 2 and 5 for the increase in strength with orientation have been reported for PE, PP, PS, PVC, PMMA, or PA [51, 54]. It should be pointed out that the rather limited increase in stiffness in these experiments indicates that the orienting "elements" are not simply highly extended chain segments but larger molecular domains of limited anisotropy. This does not preclude the possibility that breakage of an element is essentially identical to breakage of the most highly stressed chain molecule(s). This will be the case if the breaking chain(s) carry practically the total load of an element or initiate its breakage.

A comparison of different reaction rate models was made by Henderson et al. [76] who found that the bond rupture theories (Tobolsky-Eyring, Zhurkov-Bueche) gave better agreement with fracture data on filled polybutadiène and filled and unfilled PVC than the bond slippage hypothesis. Primary uncertainties exist, however, with respect to a repair reaction and to the temperature dependence of β and γ. The question of the true meaning of the activation volumes β and γ, of the stress ratio Ψ_0/σ_0, and of the role of chain stretching and chain scission in macroscopic fracture will be resumed in Chapters 7 to 9.

J. Gotlib, Dobrodumov et al.

In none of the previously treated molecular fracture theories (B – H) allowance has been made for possible stress concentration in the neighborhood of a breaking element. The first ruptures in a large ensemble of initially equally stressed molecules certainly will occur at random within the stressed volume. The breakage and retraction of a particular element must lead, however, preferentially to an increase in axial stresses of those elements to which it is coupled (through secondary forces). The probability density for rupture of those elements will, therefore, be slightly higher than that for others. But the higher probability density applies only to the small number of elements in the vicinity of the existing fracture points. So the (integrated) probability for rupture of one of these elements is initially smaller than the probability for rupture of one of the large number of chains hitherto not affected by a ruptured element. The rupture events, therefore, will continue for some time to occur at random [67]. With their growing number, chances are increasing that rupture events occur in an immediate vicinity of each other thereby forming the nucleus of a crack and enhancing the fracture probability of neighboring elements.

Whenever the random accumulation of defects in one site has grown to such a size that subsequent fractures will preferentially occur at that site, one has reached the end of the phase of fracture initiation and the beginning of (thermal) crack growth.

In Figure 3.14 a computer simulation by Dobrodumov et al. of the fracture development within a regular model lattice is shown [58]. The authors have taken the ratio of the elastic constants in and perpendicular to the direction of orientation as 10:1 and an element fracture probability as given by Eq. (3.32). The distribution of local stresses was determined by a numerical iteration procedure. It is readily observed that for small loads ($\beta\Psi_0/RT = 5$) a large number of isolated defects appear before (from the accidental accumulation of a few defect points) a "crack" starts growing. At higher loads ($\beta\Psi_0/RT = 20$) a smaller number of defect points initiate

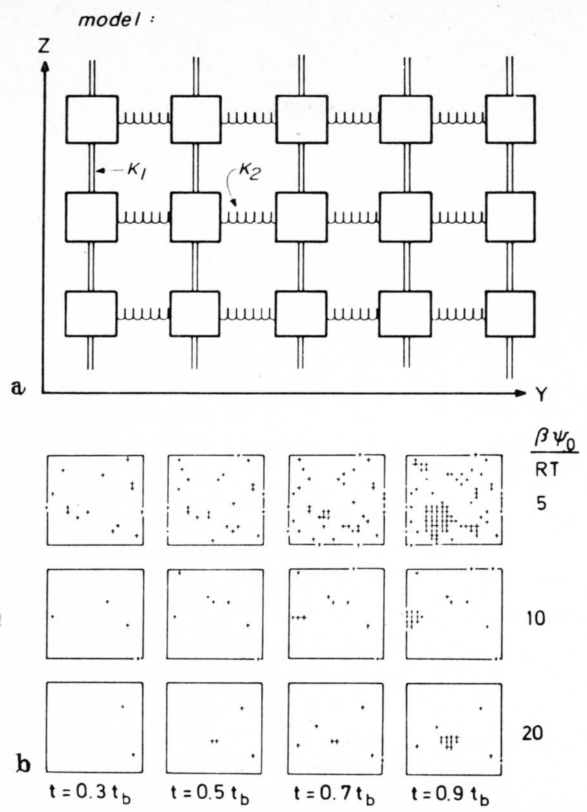

Fig. 3.14 a and b. Simulation of brittle fracture of a three-dimensional network model (after [58]). Upper part: model representation, elastic constants $K_1:K_2 = 10$. Bottom part: successive fracture events in x-y plane as a function of t/τ (τ = time to complete failure); note the increase in cooperativeness with $\beta\Psi_0/RT$ similar to that in Figure 3.12.

a crack and very little damage is done to the remainder of the plane during fracture initiation. These results are qualitatively similar to those of Hsiao and Kausch shown in Figure 3.12 From these studies the authors [57, 67] conclude that in polymers only weak stress concentration on a molecular scale can exist (axial stresses adjacent to a site comprising j defects should be larger by a factor of $\sqrt[8]{j}$ than the stress further away). It should be noted that this model – as the previous one – strictly operates with elastic elements and with their rupture as the only way to relieve local stress. The application of these models, therefore, seems to be restricted to the brittle fracture of organic glasses or to other systems not involving large anelastic or plastic deformation in fracture initiation.

K. Bueche-Halpin

The limited viscoelastic extensibility of rubber strands determines according to the theory of Bueche and Halpin [69] the fracture of elastomers. The authors assume that a large number of filaments at the tip of a growing crack have to be strained successively to their critical elongation λ_c. The sample will fracture at a gross strain λ_b after q filaments have been broken in a time $t_b = qt'$. The values λ_b and λ_c are connected through the creep function of the material and the stress concentration factor. The theory allows one to calculate the elongation at break, λ_b, from the knowledge of the creep function. It does not take into consideration any change of stress concentration with growing crack length or the decrease of t', the time necessary to fracture one filament, with continuing creep of the sample. It is assumed that all filaments will have to be stretched from practically zero elongation to λ_c. That will, in the first place, affect any numerical values of q which may be calculated from fitting experimental failure envelopes. A group of q filaments subject to the statistical condition that fracture of one filament may start once the fracture of the preceding one is completed has an average lifetime, $\langle t_b \rangle$, of qt' and a Poisson distribution of t_b:

$$p(t_b)\,dt_b = \frac{dt_b}{t'(q-1)!}\left(\frac{t_b}{t'}\right)^{q-1} \exp(-t_b/t'). \tag{3.45}$$

The same concept of filament deformation was independently employed by Bartenev et al. [77] to the opening and growth of crazes (silver cracks) in PMMA.

The molecular theories treated in this chapter have considered the thermally activated breakage of elements or filaments as a source of crack initiation and macroscopic failure. It is the purpose of the following chapters to investigate the mechanical strength of primary bonds and of molecular chains and to study the occurrence of chain breakages. With that information available it will be possible to resume the discussion on the nature of the "elements" used in molecular theories of fracture.

References for Chapter 3

1. R. M. Bethea, B. S. Duran, T. L. Boullion: Statistical Methods for Engineers and Scientists, New York: Marcel Dekker 1975.
2. A. Hald: Statistical Theory with Engineering Applications, New York: Wiley 1952.
3. H. H. Kausch: Materialprüf. *6,* 246 (1964).
4. T. Yokobori: Kolloid-Z. *166,* 20 (1959).
5. S. Kawabata, P. J. Blatz: Rubber Chem. Technol. *39,* 923 (1966).
6. S. Kawabata, S. Tatsuta, H. Kawai: Proc. Fifth Internat. Congr. Rheology, Kyoto, Japan *3,* 111 (1968).
7. B. D. Coleman: J. Appl. Phy. *29,* 968 (1958).
8. B. D. Coleman, D. W. Marquardt: J. Appl. Phys. *28,* 1065 (1957).
9. B. Epstein: J. Appl. Phys. *19,* 140 (1948).
10. B. Ya. Levin, A. V. Savickij: Chim. Volokna *5,* 48 (1968).
11. H. H. Kausch: Proc. Internat. Conf. Mech. Behavior of Materials, Kyoto, Japan, Aug. 1971, Vol. III, pg. 518.

12. L. J. Knox, J. C. Whitwell: Text. Res. J. *41,* 510 (1971).
13. R. F. Fedors, R. F. Landel: Trans. Soc. Rheol. *9,* 195 (1965).
14. S. Kase: J. Polym. Sci., *11,* 424 (1953).
15. S. Kaufmann: Faserforsch. u. Textiltechn. *22,* 406 (1971) and ibid. 463 (1971).
16. K. E. Hansen, W. C. Forsman: J. Polym. Sci., A-2, *7,* 1863 (1969).
17. D. A. Stuart, O. L. Anderson: J. Am. Ceramic Soc. *36,* 416 (1953).
18. H. H. Kausch: J. Polym. Sci., C. (Polymer Symposia) *32,* 1 (1971).
19. A. Tobolsky, H. Eyring: J. Chem. Phys. *11,* 125 (1943).
20. I. M. Ward: Mechanical Properties of Solid Polymers, London: Wiley-Interscience 1971, p. 280, 2nd Ed. 1983.
21. A. N. Gent, P. B. Lindley: Proc. Roy. Soc. (London) *A249,* 195 (1968).
W. K. Ko, Ph. D. Dissertation, California Institute of Technology, 1963.
A. E. Oberth, R. S. Bruenner: Trans. Soc. Rheology *9,* 165 (1965).
22. C. K. Lim, Ph. D. Thesis: The Pennsylvania State University, University Park, PA, 1967.
23. L. J. Broutman, S. M. Kirshnakumar, P. D. Mallick: J. Appl. Polym. Sci. *14,* 1477 (1970).
24. R. E. Ely: Polymer Engng. and Sci. *7,* 40 (1967).
25. R. A. Dickie, T. L. Smith: J. Polym. Sci. A-2, *7,* 687 (1969).
26. J. C. Bauwens: J. Polymer Sci. A-2 *8,* 893–901 (1970).
27. R. Raghava, M. Caddell, G. S. Y. Yeh: J. Materials Sci. *8,* 225–232 (1973).
28. R. G. Dong: Trans. Soc. Rheol. *18,* 45 (1974).
29. N. W. Tschoegl: J. Polym. Sci. *C 32,* 239–267 (1971).
30. T. L. Smith: ibid., 269–282.
31. S. V. Radcliffe: Deformation and Fracture of High Polymers, H. H. Kausch, J. A. Hassell, R. I. Jaffee, eds. New York: Plenum Press 1973, p. 191–208.
32. J. A. Sauer, K. D. Pae: Colloid a. Polymer Sci. *252,* 680–695 (1974).
33. W. Schneider, R. Bardenheier: Z. Werkstofftechnik *6,* 269–280 and 339–348 (1975).
34. W. Retting: Angew. Makromol. Chem. *58/59,* 133 (1977).
35. A. A. Griffith: Phil. Trans. Roy. Soc. *221,* 163 (1921).
36. M. L. Williams: Fracture of Solids, D. C. Drucker and J. J. Gilman, eds. New York: Interscience, 1963, p. 157.
37. W. G. Knauss: Appl. Mechanics Rev., January 1973, p. 1.
38. M. L. Williams: J. Adhesion *4,* 307 (1972).
39. G. P. Anderson, K. L. DeVries, M. L. Williams: Int. J. Fracture *9,* 421 (1973).
40. D. H. Kaelble: J. Appl. Polymer Sci. *18,* 1869–1889 (1974).
41. T. L. Smith: Rheology, Vol. 5, F. R. Erich, ed., New York: Academic Press 1969, p. 127.
T. L. Smith, IV, V, and VI Microsymposia on Macromolecules, Main Lectures, London: Butterworths, 1970, p. 235–253.
42. P. J. Blatz, S. C. Sharda, N. W. Tschoegl: Trans. Soc. Rheol. *18,* 145 (1974).
43. S. Glasstone, K. J. Laidler, H. Eyring: The Theory of Rate Processes, New York: McGraw Hill 1941.
44. A. Tobolsky, H. Eyring: J. Chem. Phys. *11,* 125 (1943).
45. S. N. Zhurkov, B. N. Narzullayev: Zhur. Techn. Phys. *23,* 1677 (1953).
46. S. N. Zhurkov, E. Ye. Tomashevskii: Zhur. Techn. Phys. *25,* 66 (1955).
47. S. N. Zhurkov, T. P. Sanfirova: Dokl. Akad. Nauk SSSR *101,* 237 (1955).
48. F. Bueche: J. Appl. Phys. *26,* 1133 (191955).
49. F. Bueche: J. Appl. Phys. *28,* 784 (1957).
50. F. Bueche: J. Appl. Phys. *29,* 1231 (1958).
51. G. M. Bartenev, Yu. S. Zuyev: Strength and Failure of Viscoelastic Materials, Oxford: Pergamon Press 1968, p. 143ff.
52. E. H. Andrews: Fracture in Polymers, Edinburgh: Oliver and Boyd 1968.
53. V. R. Regel, A. I. Slutsker, E. E. Tomashevskii: Kinetic Nature of Durability of Solids (in Russian), Isdat. Nauka, Moscow 1974.
54. A. I. Gubanov, A. D. Chevychelov: Fiz. Tverd. Tela *4,* 928–933 (1962), Sov. Phys.-Solid State *4,* 681–684 (1962).

55. G. M. Bartenev: Mekh. Polim.(USSR) *6,* 458–464 (1970), Polym. Mech. (USA) *6,* 393–398 (1970).
56. K. E. Perepelkin: Fiz. Khim. Mekh. Materialov *6,* 78–80 (1970).
57. S. I. Mikitishin, Yu. N. Khonitskii, A. N. Tynnyi: Fiz. Khim. Mekh. Materialov *5,* 69–74 (1969).
58. A. V. Dobrodumov, A. M. El'yashevich: Fiz. Tverd, Tela *15,* 1891–1893 (1973). Sov. Phys. – Solid State, *15,* 1259–1260 (1973).
59. C. C. Hsiao: J. Appl. Phys. *30,* 1492 (1959).
60. H. H. Kausch, C. C. Hsiao: J. Appl. Phys. *39,* 4915–4919 (1968).
61. H. H. Kausch: Kolloid-Z. Z. Polymere *236,* 48–58 (1970).
62. W. Holzmüller: Kolloid-Z. Z. Polymere *203,* 7–19 (1965).
63. G. M. Bartenev, I. V. Razumovskaya: Fiz. Khim. Mekh. Materialov *5,* 60–68 (1969).
64. R. L. Salganik: Int. J. Fracture Mech. *6,* 1–5 (1970).
65. A. N. Orlov, V. A. Petrov, V. I. Vladimirov: phys. stat. sol. (b) *47,* 293–303 (1971).
66. B. D. Goikhman: Strength Mater. (USA) *4,* 918–922 (1972).
67. Yu. Ya. Gotlib, A. V. Dobrodumov, A. M. El'yashevich, Yu. E. Svetlov: Fiz. Tverd. Tela *15,* 801 (1973), Sov. Phys. – Solid State, *15,* 555–559 (1973).
68. K. C. Valanis: Report G 378 – ChME – 001 (1974), University of Iowa, Iowa City, USA.
69. F. Bueche, J. C. Halpin: J. Appl. Phys. *35,* 36–41 (1964)
70. L. Holliday, W. A. Holmes-Walker, J. Appl. Polym. Sci. *16,* 139–155 (1972).
71. P. I. Vincent: Polymer *13,* 557–560 (1972).
72. D. C. Prevorsek: J. Polym. Sci. *C32,* 343–375 (1971).
73. V. M. Poludennaja, N. S. Volkova, D. N. Archangel'skij, A. G. Zigockij, A. A. Konkin: Chim. Volokna *5,* 6–7 (1969).
74. H. Krässig: Textilveredelung *4,* 26–37 (1969).
75. V. R. Regel, A. M. Leksovsky: Int. J. Fracture Mechanics *3,* 99 (1967).
76. C. B. Henderson, P. H. Graham, C. N. Robinson: Int. J. Fracture Mechanics *6,* 33–40 (1970).
77. G. M. Bartenev, I. V. Razumovskaya: Fiz. Tverd. Tela *6,* 657–661 (1964), Sov. Phys. – Solid State *6,* 513–516 (1964).
78. H. H. Kausch: J. Polym. sci. *C32,* 1–44 (1971).
79. S. I. Mikitishin, A. N. Tynnyi: Soviet materials sci. *4*/3, 187–188 (1968).
80. S. S. Sternstein, J. Rosenthal: Toughness & Brittleness of Plastics, Adv. Chemistry Series, A.C.S., Washington, D. C. (1974).
81. W. Retting: Orientation in Polymers Programme, Polystyrene, Rep. Results of BASF, 2nd part, March 1972, Macromol. Division, I.U.P.A.C.
82. I. Narisawa, T. Kondo: J. Polymer Sci. – Polymer Phys. Ed. – *11,* 1235–1245 (1973).
83. P. Stockmayer: Stuttgarter Kunststoff-Kolloquium 1977, Kunststoffe *67,* 470 (1977).
84. H. H. Kausch: Materialprüfung. *20,* 22 (1978).
85. J. W. S. Hearle: J. Textile Inst. *68*/4, 155–157 (1977).
86. A. V. Savitskii, V. A. Mal'chevski, T. P. Sanfirova, L. P. Zoshin: Polymer Sci. USSR *16,* 2470–2477 (1974), Vysokomol. Soyed. *A 16,* 2130–2135 (1974).
87. R. Cook: to be published in Polymer.
88. S. L. Phoenix: Adv. Appl. Prob. *11,* 153 (1979).
89. S. L. Phoenix: Text. Res. J. *49*/7, 407 (1979).
90. S. L. Phoenix: Proc. 9th U.S. National Congress of Applied Mechanics, New York: The American Society of Mechanical Engineers, 1982, 219.
91. S. L. Phoenix, R. L. Smith: Int. J. Solids Struct. *19*/6, 479 (1983).
92. M. Juillard, G. Mozzo: Kunststoffe *69*/6, 334 (1979).
93. R. Bardenheier: Mechanisches Versagen von Kunststoffen – Anstrengungsbewertung mehraxialer Spannungszustände, Kunststoffe-Fortschrittsberichte Bd. 8, München–Wien: Carl Hanser Verlag, 1982; an abridged version of this text is found in: Kunststoffe *72*/11, 729 (1982).
94. L. V. Nikitin: Int. J. Fract. *24,* 149 (1984).
95. J. Bailey: Glass Industry *20,* 95 (1939).

96. M. A. Miner: J. Appl. Mech. *12*, A–159 (1945).
97. D. C. Prevorsek, G. E. R. Lamb, M. L. Brooks: Polym. Eng. Sci. *10*, 1 (1967).
98. G. B. McKenna, R. W. Penn: Polymer *21*, 213 (1980).
99. S. D. Douglas, W. N. Stoops: Industr. Eng. Chem. *28*, 1152 (1936).
100. A. M. Sookne, M. Harris: Industr. Eng. Chem. *37*, 478 (1945).
101. H. M. Spurlin: Cellulose and Cellulose Derivatives, E. Ott, ed., New York: Interscience Publishers Inc., 1943, 930.
102. P. J. Flory: J. Am. Chem. Soc. *67*, 2048 (1945).
103. R. N. Haward: The Strength of Plastics and Glass, London: Cleaver-Hume Press, 1949, Ch. II.
104. V. S. Kuksenko, V. P. Tamuzs: Fracture micromechanics of polymer materials, The Hague–Boston–London: Martinus Nijhoff Publishers, 1981.
105. F. Bueche: Rubber Chem. Techn. *32*, 1269 (1982).
106. B. H. Bersted: J. Appl. Polymer Sci. *24*, 37 (1979).
107. D. T. Turner: Polymer *23*, 626 (1982).
108. G. R. Taylor, S. R. Darin: J. Polymer Sci. *17*, 511 (1955).
109. E. Sacher: J. Macromol. Sci.-B *15*/1, 171 (1978).
110. H. A. Papazian: J. Appl. Polymer. Sci. *28*, 2623 (1983).
111. R. F. Fedors: The Stereo Rubbers, W. M. Saltman, ed., John Wiley & Sons, Inc., New York 1977, 679.
112. S. M. Aharoni: Macromolecules *18*, 2624 (1985) has compiled data from 72 different polymers concluding that tensile failure is brittle if the chain volume between entanglements is larger than the activation volume for stress-induced glass transition changes.
113. T.-B. He: J. Appl. Polymer Sci. *30*, 4319 (1985) also relates chain cross-sections to Tg.

Chapter 4

Strength of Primary Bonds

I. Covalent Bonds

In the last chapter, in the section on molecular theories of fracture, almost throughout an Arrhenius equation has been used to describe the activation of element breakage. The energy of activation, U_0, frequently turned out to be equal to (or was assumed to be equal to) the dissociation energy of the weakest main chain bond. Before further analyzing the kinetics of element – and possibly chain – breakage a definition of the *mechanical strength of a bond* and of a chain have to be given. In order to do this basic results of quantum chemistry [1, 2] are recalled in this chapter which concern the "strengths" of intramolecular bonds and factors influencing the binding potential such as electronic excitation or ionization.

A. Atomic Orbitals

Covalent chemical bonds between atoms of the same or a different species rely on the interaction of the outermost – or valence – electrons. Even though one speaks of electrons one should rather think of electron clouds, i.e. of electronic density distributions. The radial and angular distribution of the electron density is described by "one electron" wave functions Ψ – also called atomic orbitals – which are derived as a solution of the quantum mechanical Schrödinger equation:

$$\left(-\frac{h^2}{8\pi^2 m_r}\nabla^2 + V\right)\Psi = E\Psi. \tag{4.1}$$

Here h is Planck's constant, m_r the reduced mass of an electron with charge e, ∇ the Nabla operator, and E the energy eigenvalue of an electron moving within a potential V. The term in brackets constitutes the Hamiltonian operator **H**.

The Schrödinger equation has a solution only for certain discrete energy values E. Different values of E mean different forms of the electronic density distributions which are characterized by the principal quantum number n, the angular quantum number ℓ and the magnetic quantum number m.

For the simplest atomic system, the hydrogen atom, which consists of one proton and one electron the potential V is $-e^2/r$. In this case energy levels are obtained, which are degenerate, i.e. the energy levels for different l and m coincide. The corresponding wave functions, however, do depend on the three quantum numbers n, l and m. It should be noted that the radial distribution, R(r), and the angular distribution $\Theta(\theta, \phi)$ may be separated:

$$\Psi_{n\ell m} = \Theta_\ell(\theta, \phi)\, R_{n\ell}(r). \tag{4.2}$$

The quantum numbers are subject to the condition that

$$\ell = 0, 1, 2 \ldots n-1$$
$$m = 0, \pm 1, \pm 2 \ldots \pm \ell.$$

It is customary to denote the atomic orbitals by the spectroscopical terms s, p, d . . . if ℓ is 0, 1, 2 . . . respectively.

The first few wave functions for a hydrogen-like atom with a charge of +Ze of the nucleus then have the following form:

$$1s \quad R_{10}(r) = (Z/a_0)^{3/2}\, 2 \exp(-Zr/a_0) \tag{4.3}$$
$$2s \quad R_{20}(r) = (Z/a_0)^{3/2}\, (2 - Zr/a_0)\, (\exp - Zr/2a_0)\, (1/2\sqrt{2}) \tag{4.4}$$
$$2p \quad R_{21}(r) = (Z/a_0)^{3/2}\, (Zr/2a_0\sqrt{6}) \exp(-Zr/2a_0) \tag{4.5}$$

and

$$s \quad \Theta_{00}\ (\theta, \phi) = 1/2\sqrt{\pi} \tag{4.6}$$
$$p_x \quad \Theta_{1x}\ (\theta, \phi) = (\sqrt{3}/2\sqrt{\pi}) \cos\phi \sin\theta \tag{4.7a}$$
$$p_y \quad \Theta_{1y}\ (\theta, \phi) = (\sqrt{3}/2\sqrt{\pi}) \sin\phi \sin\theta \tag{4.7b}$$
$$p_z \quad \Theta_{1z}\ (\theta, \phi) = (\sqrt{3}/2\sqrt{\pi}) \cos\theta. \tag{4.7c}$$

Here a_0 is the radius of the first Bohr orbit in hydrogen ($0.529 \cdot 10^{-10}$ m). According to the Pauli exclusion principle every atomic orbital may be occupied by one or two electrons which in the latter case have to have their spins aligned antiparallel to each other (paired electron spins).

For many-electron atoms an exact analytical solution of the Schrödinger equation (Eq. 4.1) is not possible. Nevertheless atomic orbitals may be obtained by an iterative procedure, which starts by placing the appropriate number of electrons on hydrogen-like one-electron orbitals. With the potential obtained from such a charge distribution a better approximation of the first orbital can be calculated which in turn is considered in the recalculation of the second orbital and so on until no further corrections become necessary. Again these orbitals are denoted by 1s, 2s, 2p, 3d if the quantum numbers n and ℓ are 1 and 0, 2 and 0, 2 and 1, 3 and 2 and so on.

In a many-electron atom – as for instance Li – one finds a particular electron in any of the accessible orbitals (1s, 2s) and in any of the two possible states of spin ($m_s = +1/2$ or $-1/2$). Since electrons are indistinguishable the energy of the atom must not be affected by interchanging the coordinates of any pair of electrons. The wave function belonging to one eigenvalue E will have to encompass, therefore, all terms resulting from an interchange of electron numbering.

In the case of Li the ground state has the wave function:

$$\Psi = \begin{vmatrix} 1s\,\alpha\,(1) & 1s\beta\,(1) & 2s\,\alpha\,(1) \\ .. & .. & .. \\ . \ . & . \ . & 2s\,\alpha\,(3) \end{vmatrix} \quad \text{(Slater's determinant)} \tag{4.8a}$$

or

$$\Psi = 1s\,\alpha(1)\,1s\beta\,(2)\,2s\,\alpha\,(3) - 1s\,\alpha\,(2)\,1s\beta\,(1)\,2s\,\alpha\,(3) + \ldots \tag{4.8b}$$

Each of the six terms of the fully written determinant (4.8b) contains an orbital and a spin function for each of the three electrons. The spin function $\alpha(1)$ may denote that the electron 1 has a spin quantum number $m_s = +1/2, \beta(1)$ corresponds to $m = -1/2$.

B. Hybridization

It can easily be shown that an atom with one electron in each of the 2s-, $2p_x$-, $2p_y$-, and $2p_z$-orbitals has a total charge density distribution which is completely spherical, i.e. independent of Θ and ϕ. The s- and p-atomic orbitals are not the only functions that lead to a spherical distribution of the total charge. A large number of suitable linear combinations of the s- and p-orbitals do have the same property. These linear combinations are called hybrid orbitals or hybrids. Depending on the number and kind of atomic orbitals used in hybridization the hybrids are denoted as sp, sp^2, sp^3, dsp^3 etc. The four normalized and orthogonal sp^3-hybrids (of a C-atom) have the form:

$$\Psi_1 = \frac{1}{\sqrt{4}}(2s + 2p_x + 2p_y + 2p_z) \tag{4.9a}$$

$$\Psi_2 = \frac{1}{\sqrt{4}}(2s + 2p_x - 2p_y - 2p_z) \tag{4.9b}$$

$$\Psi_3 = \frac{1}{\sqrt{4}}(2s - 2p_x - 2p_y + 2p_z) \tag{4.9c}$$

$$\Psi_4 = \frac{1}{\sqrt{4}}(2s - 2p_x + 2p_y - 2p_z) \ . \qquad (4.9d)$$

The individual hybrid is, of course, non-spherical, i.e. the electron density distribution has a characteristic shape. In this particular case the directions of maximum electron density point into the corners of a regular tetrahedron.

The electronic energy of a C-atom with electrons in hybrid orbitals is higher than that of the ground state. In carbon the ground state will be formed by two electrons in each of the lowest orbitals: $1s^2\ 2s^2\ 2p^2$. Hybridization involves the change from pyramidal orbitals (p^3) to tetragonal ones (sp^3); it requires, that three p-orbitals be occupied. The first step is, therefore, that one electron be promoted from 2s to 2p which necessitates an energy of promotion. The effect of promotion and hybridization is, however, that four rather than two electrons are available for bonding. The gain in binding energy of the two additional C–H bonds (90 to 100 kcal/mol per C–H bond) more than balances the energy expended for promotion and hybridization.

C. Molecular Orbitals

Within the limited scope of this book a mere terminology of molecular orbital theory can be supplied – intended only for the discussion of bond energies. Rigorous quantum mechanical calculations for any molecule being more complex than the hydrogen molecule are not feasible. Therefore, approximate methods for calculation of the orbitals of the bonding electrons have been developed [1, 2]. The two most important ones are the *molecular orbital* (MO) method relating to the work of Hund, Mulliken, and Hückel [1a] and the *bond orbital* method, the Heitler-London-Slater-Pauling (HSLP) method [1b]. In the MO approach electrons are placed in *one-electron*-wave functions which belong to the molecule as a whole with each nucleus in its proper configuration. In its crudest form these wave functions are obtained as a *linear combination of the atomic orbitals* (LCAO) of all atoms forming the molecule [2a].

In bond orbital theory the wave functions of the molecule are derived from the wave functions associated with the different bonds of a molecule, i.e. from the bond orbitals. In the LCBO model the molecular orbitals are a *linear combination of bond orbitals* which in turn are a combination of the atomic orbitals or hybrids forming the bond in question [2b]. In the HLSP-method the *many-electron* wave functions are a sum of product functions which contain a Heitler and London type of factor (space function and spin function) for each bond of the molecule.

Bond orbital and molecular orbital approach lead to the common conclusion that bonding will occur only at directions where the atomic orbitals have large values and only if the atoms are close enough to each other so that the atomic orbitals overlap, i.e. if the "electron clouds" partly penetrate each other.

A covalent bond is associated with an electron pair and hence subject to the quantum mechanical phenomenon of electron exchange: in a molecule A–B the electron from A may also be found near the nucleus of atom B and vice versa. It is this additional exchange energy term which significantly deepens the interatomic potential and leads to a distinct potential minimum – prerequisite for a stable binding. Thus a bond can be formed if a bonding molecular orbital exists and if it is occupied by an electron or an electron pair. In the ground energy state of a molecule paired electrons occupy the molecular orbitals with the lowest energy.

D. Multiple Bonds

So far multiple bonds between atoms of the same or a different species such as carbon, nitrogen, and oxygen have not been explicitly discussed). At this point attention should be drawn to the fact that the two (or three) molecular orbitals involved in forming a double (or triple) bond are usually not of the same kind but of different character. In ethylene ($H_2C{=}CH_2$) all six atoms lie in one plane. Therefore, one may take three sp^2-hybrid orbitals of each C-atom to form the four C–H bonds and the one axially symmetric C–C bond, the σ bond.

The two remaining p_z orbitals are combined into the so called (molecular) π bond orbital. Since the p_z orbitals are directed perpendicular to the π bond which they are forming their overlap is smaller than that of orbitals forming a σ bond. Also the regions of maximum overlap are not in the xy plane between the C-atoms but rather above and below this plane. Because of this orientation of the p_z orbitals the interaction between conjugated π bonds is so strong that in fact it leads to a delocalization of the π electrons. In view of the geometry of the π bond it is easily realized that rotation of a group around multiple bonds is forbidden [1c].

II. Bond Energies

A. Electronic Energy and Heat of Formation

The term bond energy requires a definition. The potential energy-scheme of an arbitrary bond A–B in a polyatomic molecule given in Figure 4.1 serves to illustrate this point. As pointed out previously one has to take into account that in polyelectronic atoms the valency state of the atom may lie above the respective ground state. If this is the case no binding can occur between two atoms in their ground states; if they encounter each other their potential energy will rise. At some interatomic distance the potential energy of the system will approach the energy of the atoms in their valency states (Fig. 4.1, dashed line) and a transition to the binding state can occur. The intrinsic binding energy (wahre Bindungsenergie) E is, therefore, the difference between the energy of the molecular ground state and that of the valency states at infinite interatomic distance. The energy of dissociation D is smaller than E, namely by the zero point energy (Nullpunktsenergie) $h\nu/2$ of vibration and by the sum P of energies of promotion, hybridization, and polar and steric rearrangements necessary to attain the valency state. The difference between zero point energy and the maximum of the potential energy curve is the activation

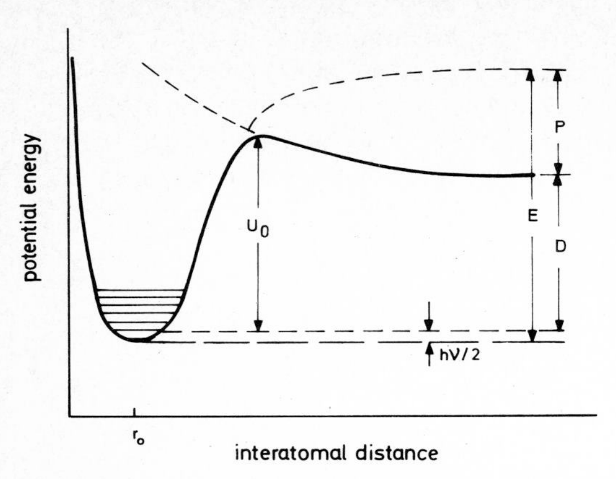

Fig. 4.1. Scheme of the potential energy of an arbitrary bond in a polyatomic molecule. D energy of dissociation, E intrinsic binding energy, P energies of promotion, hybridization and of polar and steric rearrangements, U_0 activation energy of bond dissociation, r_0 equilibrium bond length, $h\nu/2$ zero point energy of vibration.

energy U_0 of bond dissociation. In diatomic molecules U_0 and D generally have the same value and represent the bond energy.

In molecules with three or more atoms the assignment of a specific increment of energy to a particular bond is complicated by the presence of the above mentioned energy term P for promotion and electron reorganization, by the occurrence of delocalized electrons, and by the fact that the dissociation energies of successively broken, initially identical bonds are different. The latter statement may be illustrated by the disintegration of CH_4 using the energy values given by Stuart [5]. It must be assumed that the four hydrogen atoms in CH_4 are equivalent, all C–H bonds are formed by hydrogen 1 s orbitals and carbon 2 sp^3 orbitals. Abstraction of the first hydrogen atom requires an energy of 425.8 kJ/mol (101.7 kcal/mol) and changes the methane atom into the methyl radical CH_3. From symmetry considerations confirmed by ESR investigations it has been concluded [4] that the methyl radical is planar, with the unpaired electron undoubtedly in a 2 p orbital of the C-atom. The bond character of the C–H bonds remaining in what now is the methyl radical has changed, therefore, from the tetrahedral configuration of sp^3 hybrid orbitals into the planar one of sp^2 hybrids. The dissociation energy of one of these C–H bonds is 477.3 kJ/mol (114 kcal/mol). The third hydrogen has a dissociation energy of 418.7 (100 kcal/mol), the fourth of 341.2 kJ/mol (81.5 kcal/mol). It should be mentioned at this point that the literature data on the energies of the different C–H bonds vary slightly (cf. for instance [15–17]), they do agree, however, on the total energy gained in forming CH_4 from its constituent atoms, i.e. 1663 kJ/mol (397.2 kcal/mol), the *average* C–H bond energy is, therefore, 415.7 kJ/mol (99.3 kcal/mol) [5]. It is this average from the energy of formation of the molecule which is sometimes simply termed bond energy (mittlere Bindungsenergie).

The bond energies of saturated and unsaturated homologues of *hydrocarbons* may be treated in some detail since they will give us some understanding of the correlation of bond energies in *polyolefins.* In saturated hydrocarbons with localized electrons it is reasonable to equate the energy of a bond with the energy of the electrons forming that particular bond. Theoretically the electronic energy of a molecule may be calculated using any of the mentioned approximations, for instance the LCBO method [2b, 6]. In that case the *molecular* orbitals for the 2N σ-electrons of a hydrocarbon with N bonds may be written as a linear combination of N bond orbitals ϕ_i:

$$\Psi_j = \sum_{i=1}^{N} c_{ij} \phi_i . \qquad (4.10)$$

One assumes that the wave functions Ψ_j are a solution of the equation

$$\mathbf{H} \Psi_j = E \Psi_j \qquad (4.11)$$

or

$$(\mathbf{H} - E) \Psi_j = 0. \qquad (4.12)$$

If one defines the notations

$$H_{ij} = \int \phi_i \mathbf{H} \phi_j \, dv \qquad \text{(resonance integral)} \qquad (4.13a)$$
$$S_{ij} = \int \phi_i \phi_j \, dv \qquad \text{(overlap integral)}, \qquad (4.13b)$$

introduces Eq. (4.10) into Eq. (4.12), and multiplies that equation from the left hand side by ϕ_j one obtains

$$\sum_{i=1}^{N} c_{ji} (H_{ji} - ES_{ji}) = 0 \quad j = 1, 2, \ldots N. \qquad (4.14)$$

The system (4.14) of equations has a nontrivial solution only if the determinant of the quantities in brackets is zero. Since in principle the coefficients H_{ij} and S_{ij} are derivable from the electron wave functions the *secular determinant* of the system (4.14) may be solved to yield the energy eigenvalues E_1 to E_N. The total electronic energy E of the molecule then is

$$E = \sum_{j=1}^{N} 2 E_j . \qquad (4.15)$$

Following the treatment of Brown [6] it is assumed that the *Coulomb integral*, H_{mm}, of an orbital ϕ_m depends only on the type of bond in which it resides, thus all C–H bond orbitals will be assigned the same value α of the Coulomb integral. The term is one half of the binding energy of an isolated C–H bond. The resonance integral, H_{mn}, will be assumed to be zero unless m and n denote adjacent bonds. In that case H_{mn} still depends on the type of neighboring bonds. In Table 4.1 the assignments of Brown for the bond interaction parameters in saturated hydrocarbons are listed.

Table 4.1. Bond Interaction Parameters

Coulomb parameter H_{mm}		Resonance parameter H_{mn}		Overlap integrals S_{mn}	
Bonds	Parameter	Bonds	Parameter	Bonds	Parameter
C–H	α	C–H, C–H	β	C–H, C–H	S
C–C	$\alpha + h\gamma$	C–H, C–C	$\Theta\beta$	C–H, C–C	ΘS
		C–C, C–C	$\eta\beta$	C–C, C–C	ηS

Table 4.2. Electronic Energies of Paraffins

Paraffin	Electronic Energy E	E_{theor} ($S_\gamma = 6.5$) kJ/mol	$E_{exp} + W$ kJ/mol
Methane	$8\alpha - 24S\gamma$	−1765	−1665
Ethane	$14\alpha + 2h\gamma - (24 + 24\theta^2)S\gamma$	−2923	−2828
Propane	$20\alpha + 4h\gamma - (28 + 40\theta^2 + 4\eta^2)S\gamma$	−4088	−4001
Butane	$26\alpha + 6h\gamma - (32 + 56\theta^2 + 8\eta^2)S\gamma$	−5253	−5175
iso Butane	$26\alpha + 6h\gamma - (36 + 48\theta^2 + 12\eta^2)S\gamma$	−5260	−5182
Pentane	$32\alpha + 8h\gamma - (36 + 72\theta^2 + 12\eta^2)S\gamma$	−6418	−6350
iso Pentane	$32\alpha + 8h\gamma - (40 + 64\theta^2 + 16\eta^2)S\gamma$	−6425	−6358
neo Pentane	$32\alpha + 8h\gamma - (48 + 48\theta^2 + 24\eta^2)S\gamma$	−6438	−6370
Hexane	$38\alpha + 10h\gamma - (40 + 88\theta^2 + 16\eta^2)S\gamma$	−7583	−7525
2-Methylpentane	$38\alpha + 10h\gamma - (44 + 80\theta^2 + 20\eta^2)S\gamma$	−7590	−7532
3-Methylpentane	$38\alpha + 10h\gamma - (44 + 80\theta^2 + 20\eta^2)S\gamma$	−7590	−7529
2,3-Dimethylbutane	$38\alpha + 10h\gamma - (48 + 72\theta^2 + 24\eta^2)S\gamma$	−7597	−7535
2,2-Dimethylbutane	$38\alpha + 10h\gamma - (52 + 64\theta^2 + 28\theta^2)S\gamma$	−7603	−7543

In Table 4.2 and 4.3 the electronic energies of a few hydrocarbons and of their radicals are given in terms of Brown's parameters. It will be noted that to a zeroth approximation (S = 0) the total electronic energy E of a molecule or a radical is given by the sum of the energies of the corresponding number of isolated C–C and C–H-bonds. In the first approximation also the energetic interactions of neighboring bonds are considered (terms to the first power of S as listed in Tables 4.2 and 4.3) and higher powers of S are neglected. Such terms would be necessary, however, to account for the effect of third neighbors within the molecule. That effect is clearly revealed by analysis of thermochemical data [7].

If it were possible to determine the numerical values of the bond interaction parameters by simply using molecular geometry and atomic constants one would have a direct and independent way to calculate bond energies. So far, however, it has proven very cumbersome to derive the explicit values of the bond energy parameters α, h, γ, Θ^2, η^2, and S directly and sufficiently correctly from molecular wave functions. Usually thermochemical data are employed to obtain the electronic energy of a molecule. For example the electronic energy E of a hydrocarbon can be

Table 4.3. Energy* of Hydrocarbon Radicals R

Radical R	Electronic Energy	$E(R-H)-E(\dot{R})$	D(R–H)* kJ/mol	ΔH_f(298 K)* kJ/mol
Methyl $\dot{C}H_3$	$6\alpha - 12S\gamma$	$F_1 + 3F_2$	436	134
Ethyl $H_3C-\dot{C}H_2$	$12\alpha + 2h\gamma-(16 + 20\theta^2)S\gamma$	$F_1 + 2F_2$	414	110
n-Propyl $H_3C-CH_2-\dot{C}H_2$	$18\alpha + 4h\gamma-(20 + 36\theta^2 + 4\eta^2)S\gamma$	$F_1 + 2F_2$	411	86
iso-Propyl $H_3C-\dot{C}H-CH_3$	$18\alpha + 4h\gamma-(24 + 32\theta^2 + 4\eta^2)S\gamma$	$F_1 + F_2$	394	72
n-Butyl $H_3C-(CH_2)_2-\dot{C}H_2$	$24a + 6h\gamma-(24 + 52\theta^2 + 8\eta^2)S\gamma$	$F_1 + 2F_2$	411	67
iso-Butyl $H_3C-CH(CH_3)-\dot{C}H_2$	$24\alpha + 6h\gamma-(28 + 44\theta^2 + 12\eta^2)S\gamma$	$F_1 + 2F_2$	411	57
sec.-Butyl $H_3C-CH_2-\dot{C}H-CH_3$	$24\alpha + 6h\gamma-(28 + 48\theta^2 + 8\eta^2)S\gamma$	$F_1 + F_2$	394	50
tert.-Butyl $(CH_3)_3\dot{C}$	$24\alpha + 6h\gamma-(36 + 36\theta^2 + 12\eta^2)S\gamma$	F_1	381	28

* Energy values from Refs. [8a] and [9].

calculated from the enthalpies H_a of atomization of hydrogen, L of sublimation of graphite, and ΔH_f of formation of the hydrocarbon from H_2 and C_n:

$$E(C_nH_{2n+2}) = -n\,L(C) - (n+1)H_a + \Delta H_f - W. \tag{4.16}$$

The term W accounts for the fact that the minimum of a binding potential is by the zero point vibrational energy (1/2 $h\nu$) lower than the corresponding dissociation energy (Fig. 4.1). Since the relevant frequencies are of the order of $2.5 \cdot 10^{13}\ s^{-1}$ (C–C skeleton) and $9 \cdot 10^{13}\ s^{-1}$ (C–H) the zero point vibrational energies range between 5 kJ/mol and 18 kJ/mol.

A quantitative analysis of the quantum mechanical calculations through thermochemical data will be attempted, although considerable uncertainties exist. These are caused by the limitations of the used first order approximation, the lack of consideration of steric interaction of atoms, and of the difficulties to determine W. The analysis is firstly based on the fact that the differences between the electronic energies of the listed hydrocarbons and their radicals can be expressed by only four quantities F_i:

$$F_1 = -2\alpha + 12\,\Theta^2\,S\gamma \tag{4.17}$$
$$F_2 = 4(1-\Theta^2)\,S\gamma \tag{4.18}$$
$$F_3 = -4(1 - 2\,\Theta^2 + \eta^2)\,S\gamma \tag{4.19}$$
$$F_4 = -2\alpha - 2\,h\gamma + 24\,\eta^2\,S\gamma. \tag{4.20}$$

Brown [6] determined F_2 to be 15.5 kJ/mol (from the radical dissociation energies listed in Table 4.3) and F_3 to be –6.7 kJ/mol (from the energy differences found for 5 pairs of isomers listed in Table 4.3).

Since these difference values are not so much affected by the systematic uncertainties mentioned above they will be adopted here along with the value of 300 kJ/mol for F_4; F_4 is the electronic energy of a C–C bond interacting solely with C–C bonds as in $(CH_3)_3C–C(CH_3)_3$. Further the standard values (at 298.15 K and 1 bar) L = 717.2 kJ/mol and H_a = 436.2 kJ/mol [7] and a $\Delta H_f\,(CH_4)$ = = –75 kJ/mol have been used [6]. The zero point vibrational energy of the 9 vibrational modes of the –C–H and H–C–H bonds of methane is estimated at 100 kJ/mol. The total electronic energy of methane then is derived from Eq. (4.16) as $E\,(CH_4)$ = –1765 kJ/mol. From Eqs. (4.17) and (4.18) it follows that $6.2 < S\gamma < 8.4$ kJ/mol. Employing $E\,(CH_4)$, F_2, F_3, and F_4 as given above and a value of

$S\gamma$ = 6.5 kJ/mol one obtains

α = –200 kJ/mol
$h\gamma$ = 55 kJ/mol
Θ^2 = 0.4 kJ/mol
η^2 = 0.06 kJ/mol .

On the basis of these values the electronic energies of the hydrocarbons have been determined and listed as E_{theor} in Table 4.2. For comparison the thermochemically determined enthalpies of formation are referenced as E_{exp} + W. The latter values are consistently larger than E_{theor} although the difference decreases rather than to increase with the number of vibrational modes available to a molecule.

Table 4.4. Dissociation Energies of Primary Bonds in Organic Molecules and Polymer Chains*

	$-CH_3$	$-CH_2-CH_3$	$-CH(CH_3)_2$	$-C(CH_3)_3$	$-CH=CH_2$	$-C\equiv CH$	$-O-CH_3$	$-C(=O)CH_3$	$-O-C(=O)CH_3$	$-C(=O)H$	$-CH_2-CN$	C_6H_5-	$-C(=O)-C_6H_5$	$-CH_2-C_6H_5$	$-CH(CH_3)-C_6H_5$	$R-CH_2-CH(-)-R'$	
H-	431	406	394	373	436	507	427	360	469	365		427		347		300	
F-	453	444	440	427			245					524		377		448	
Cl-	352	339	339	331	360		205	344				419		285		314	
OH-	381	381	385	381				427		419		469				356	
O=CH-	314	297															
NH_2-	331	327	323	323				411				419					
N≡C-	461											545				451	
$C_6H_5-CH(CH_3)-$	276															355	$-CH_2-R$
$C_6H_5-CH_2-$	302	260	230		293			264		210		323		199		324	$-C(CH_3)_2-R$
$C_6H_5-C(=O)-$							377					364*					$R-CH_2-CH(-)-R'$
C_6H_5-	427	381	348	327	423	499	423			377	355	432	316		335	345	-O-R
$CN-CH_2-$	306													290		311	$-CH_2O-R$
H-C(=O)-	314	297						251		251					336		$-(CH_2)_2O-$
$CH_3-C(=O)-O-$	379*								126							311	-CH(OH)-R
$CH_3-C(=O)-$	344	323						251								328	$-CH(CH_3)-R$
CH_3-O-	335	335	339	327			142									326	-CHCl-R
CH≡C-	419	457	432			461										342	-CHF-R
$CH_2=CH-$	377	377	356	339	423											298	-CH(CN)-R
$(CH_3)_3C-$	335	314	306	293				433							377	301	$-C_6H_4-R$
$(CH_3)_2CH-$	348	327	318													270	$-CH(C_6H_5)-R$
CH_3-CH_2-	356	337					224										$RNH-C_6H_4-NH-$
CH_3-	369										328					254	$-C(CH_3)(C_6H_5)-R$
										272							$RC(=O)-C_6H_4-C(=O)-O-CH_2-$
									285								$-(CH_2)_m-CO-R$
																206	$-CH_2-C_6H_4-R$
							$R-C(=O)-C_6H_4-C(=O)-$	$R-C_6H_4-$	$RNH(CH_2)_nNHCOCH_2-$	$ROCH_2-$	$R-CH_2-CH(-)-R'$	$R-CH_2-CH(CH_3)-$	$R-(CH_2)_4-$	$R-(CH_2)_3-$	$R-(CH_2)_2-$	$R-CH_2-$	

* Upper and left hand side: after van Krevelen [9]
Lower and right hand side: after Stepan'yan et al. [10]

The treatment may be extended to deal with the energy of alkyl radicals if one assumes that the single unpaired electron on the "radicalized" carbon atom occupies of $2p\pi$ molecular orbital and does not change the interaction of other electrons on other orbitals. As mentioned earlier this assumption is in accord with ESR-investigations [4]. In electronic terms the dissociation energy D of a bond $R_1 - R_2$ can then be formulated as

$$D(R_1 - R_2) = E(R_1 - R_2) - E(\dot{R}_1) - E(\dot{R}_2) - 1/2\, h\nu(R_1 - R_2)\,. \tag{4.21}$$

In Table 4.3 the electronic energy of hydrocarbon radicals and R–H dissociation energies are listed together with thermochemically determined values. The difference values of electronic and thermochemical data are in correspondence with each other. The absolute value of $F_1 = D(R\text{–}H) + 1/2\, h\nu(R\text{–}H)$ can be calculated from Eq. (4.17) to be 431 kJ/mol; since D(R–H) for tertiary carbon atoms is 381 kJ/mol a contribution of 50 kJ/mol must be assigned to the zero point vibrations. This contribution is certainly too high by $\simeq$30 kJ/mol. The discrepancies between calculated and measured dissociation energies of C–C bonds, however, lie between 0 and 6 kJ/mol for the bonds covered by Tables 4.2 and 4.3. The experimentally determined dissociation energies of small molecular groups are listed in the upper and left hand side of Table 4.4 [after van Krevelen, 9]. For these molecular fragments the activation energy of bond rupture and the dissociation energy are assumed to correspond. In the lower right hand side we have listed the dissociation energies obtained for sp^2 and sp^3 bonds by Stepan'yan et al. [10]. In their semi-empirical approach the authors calculate the bond dissociation energies R_1–R_2 by extrapolation from those of the series of compounds R–X, X representing F, H, Cl, Br, and I. Additional data on the energy of chemical bonds especially, concerning halogen- and nitrogen-containing groups are given in [16]. Judging the available experimental and theoretical determinations of bond energies of chain molecules it must be stated that still considerable differences with respect to the depth of the binding potentials for main chain bonds exist.

Before leaving the subject of bond energies one should briefly discuss the meaning of the term bond strength. In quantum chemistry the term bond strength, or more explicitly the bonding strength of atomic orbitals, refers to the excentricity of a (hybrid or nonhybrid atomic) orbital and is measured by the maximum angular intensity of the wave function as compared with the spherically symmetrical s-orbital [1]. So the *strength* of an sp^3 hybrid is $\Psi_1(\Theta = 54°\,44'; \phi = 45°)/\left(\frac{1}{4\pi}\right)^{1/2} = 2$. Other hybrids of interest have the following bond strengths: the (pyramidal) p^3 hybrid 1.732, the (plane triangular) sp^2 hybrid 1.991, and the (linear) sp hybrid 1.932.

Throughout our further discussions, however, the term bond strength will have a more technical meaning, i.e. that of the force required to break the bond (cf. Chapter 5).

B. Binding Energy of Excited States and Radicalized Chains

In as much as the covalent binding force between any two atoms rests on the electron density distribution between (and around) these atoms any change of the electron distribution necessarily affects the bond strength. Since the states of different electronic energy of a molecule are states of different electron distribution they are also states of different interatomal forces.

This is also true, of course, for molecular ions, i.e. for molecules where one or more electrons have been removed completely. As already seen for methane the character of a bond and its energy are also affected through the radicalization of the atom group to which it belongs.

1. Electronic Excitation

The electronic excitation of a polymeric network chain can be caused by electromagnetic (light, UV, γ) or particle irradiation. The transfer of energy from the impinging particle or quantum of radiation to the electron requires that sufficient energy is available to reach an excited state and that an interaction mechanism exists. In the case of visible light a photon of, say, a wavelength of 330 nm has enough energy to break a C–C bond. However, the photon will not be absorbed by alkanes and there are no electronic states of that or a smaller energy which could be excited. To be effective a photon has to be absorbed and to interact with a binding electron. This interaction occurs either directly or indirectly through the mechanisms of energy transfer by exciton diffusion, single step transfer or absorption of fluorescence light emitted by the same or another (impurity) molecule [11]. The nature and sequence of these important processes which determine the photochemical stability – or instability – of polymers will not be treated here in detail. One is interested, however, to know the levels of energy where the excitation of electrons – or the ionization of molecules – begins and to know the changes of binding energy caused by excitation – or ionization – respectively.

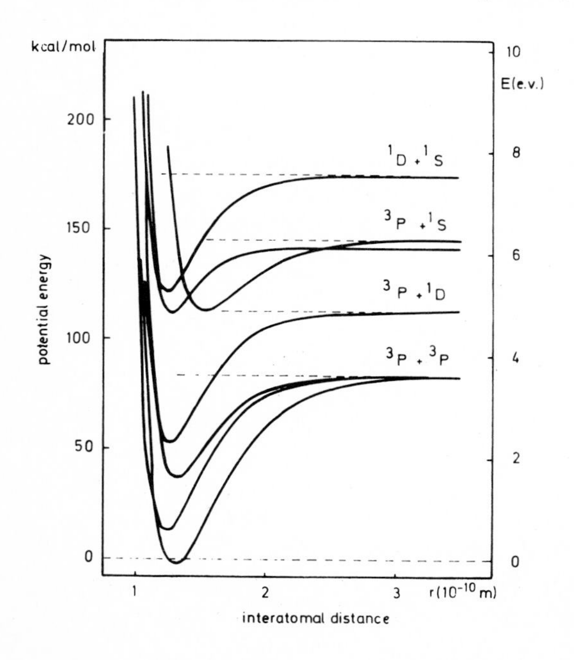

Fig. 4.2. Interatomic energy of excited states of a C_2 molecule [12].

The potential curves of different excited states of the C_2-molecule illustrate the effect of electronic excitation on bond energies (Fig. 4.2). We would like to point out that the *dissociation energies* of the different molecular states shown range from 350 to 119 kJ/mol (84 to 28 kcal/mol). These dissociation energies are generally smaller than the energy differences caused by the electronic excitation. The excited states of the (many-electron) C-atoms are denoted in the usual manner as S, P, D. Since two excited atoms may combine in a number of ways there are more than one molecular state corresponding to any two atomic states.

Transitions from the ground state of a molecule to an excited state usually occur from a bonding orbital (σ or π electron) or a non-binding orbital (n electron) to an antibonding σ^* or π^* orbital. In Table 4.5 are listed the electronic transitions for some common high polymers as given by [13]. If an electronic excitation of a network chain occurs, it leads to decreased bond energies. This will reduce the mechanical stability of a loaded polymeric network and thus contribute to chain scission, fracture initiation or facilitated crack propagation. It is readily seen from Table 4.5,

Table 4.5. Electronic Transitions for some Polymers[a]

Polymer	λ_{max}, nm	$h\nu_{max}$ kJ/mol	kcal/mol	Transition
Polyethylene	<150	>802	>191.5	$\sigma \rightarrow \sigma^*$
Polybutadiene	180	670	160	$\pi \rightarrow \pi^*$
Polystyrene	187–260	461–645	110–154	b
Poly(methyl methacrylate)	214	561	134	$n \rightarrow \pi^*$
Poly(ethylene terephthalate)	290	415	99	$n \rightarrow \pi^*$
Poly(ethylene terephthalate)	240	502	120	b

a As solid films (after Brenner and Kapfer, [12]).
b π-electron system involved.

however, that the energy values of the lowest lying excited states are larger than that of any C—C bond. One does not expect, therefore, that the recombination of any chain radicals will lead to energy quanta which are large enough to cause electronic excitation of the reacting radicals or of neighboring molecules. (This is different, though, if additional reactions with oxygen are involved; this aspect, the *chemiluminescence*, will be treated in Chapter 8).

2. Ionization

The complete removal of an electron from a molecule, i.e. the ionization, requires more energy than electron excitation and has an even stronger effect on the dissociation energy. In Table 4.6 the dissociation and ionization potentials of the first few hydrocarbons have been collected from various sources. It is assumed that ionization is first accomplished through removal of the least bound electron from the highest occupied orbital and also that this orbital embraces the whole molecule.

Table 4.6. Dissociation and Ionization Energies of Hydrocarbons

Molecule (R_1-R_2)	$D(R_1-R_2)$[a] kJ/mol	$I(R_1-R_2)$[b] eV	kJ/mol	Radical Ion (R_2^+)	$I(R_2)$[c] eV	kJ/mol	$I(R_1-R_2) - I(R_2)$ eV	kJ/mol	$D(R_1-R_2^+)$ kJ/mol
Hydrogen H_2	436	15.43	1490	H^+	13.6	1314	1.83	176	260
Ethylene C_2H_6	352	11.76	1134	CH_3^+	9.82	950	1.94	184	167
	414	11.76	1134	$C_2H_5^+$	$\leqslant$ 8.4	$\leqslant$ 812	$\geqslant$3.36	326	88
Propylene C_3H_8	347	11.21	1084	CH_3^+	9.82	950	1.39	134	213
	350	11.21	1084	$C_2H_5^+$	$\leqslant$ 8.4	$\leqslant$ 812	$\geqslant$2.81	$\geqslant$272	$\leqslant$ 75
	406	11.21	1084	$nC_3H_7^+$	< 8.1	< 783	>3.11	>301	<109
	393	11.21	1084	$isoC_3H_7^+$	< 7.5	< 724	>3.71	>360	< 33
n Butane C_4H_{10}	347	10.8	1042	CH_3^+	9.82	950	0.98	92	255
	347	10.8	1042	$C_2H_5^+$	$\leqslant$ 8.4	$\leqslant$ 812	$\geqslant$2.4	$\geqslant$230	$\leqslant$117
	347	10.8	1042	$C_3H_7^+$	< 8.1	< 783	>2.7	>260	< 88

[a] From Table 3.4.
[b] From Landolt Börnstein I/3, 6. Auflage, Springer Verlag, Berlin 1951, p. 363.
[c] From F. A. Elder, C. Giese, B. Steiner, M. Inghram, J. Chem. Phys. *36*, 3292 (1962).

The dissociation energies $D(R_1 - R_2^+)$ in the last column of Table 4.6 can be calculated from the ionization energies according to the relation

$$D(R_1 - R_2^+) = D(R_1 - R_2) - I(R_1 - R_2) + I(R_2). \tag{4.22}$$

The values derived in that (indirect) way are given in parenthesis because of the ambiguities that are still involved in obtaining the ionization potentials from electron impact or photo ionization efficiency measurements. For the molecules listed here it is clearly apparent that the dissociation energies of the positive molecular ions are much smaller than those of the non-ionized molecule. (It should be noted that the dissociation energies of excited states are not necessarily *always smaller* than those of the corresponding ground states. But in the organic compounds here of interest dissociation energies are reduced by electron removal.)

It has been emphasized here that the ionization should lead to a *positive* molecular or radical ion because the electronic energy of negative ions may be quite different. These differences are not expected and not explained in molecular orbital theory. If no electron repulsion term is explicitely introduced then the same amount of energy will be predicted for the removal of an electron and for placing an electron onto such an orbital. These differences can be quite considerable. For H_2 the state $H^+ + H^-$ lies above the state H + H by an amount of 12.8 eV (1230 kJ/mol). The energy difference between $R_1^+ + R_2^-$ and $R_1 + R_2$ is given by $I(R_1) - E'(R_2)$ where $E'(R_2)$ is the *electron affinity* of R_2 (i.e. the energy which is released when the atom or radical R_2 combines with an electron to form a negative ion). As is illustrated by Table 4.7 the electron affinities of saturated hydrocarbons are of the order of 1 eV (96 kJ/mol) or less, the differences between I(R) and E(R) are of the order of 7 eV (672 kJ/mol) or more. The dissociation of a hydrocarbon molecule $R_1 - R_2$ into the two radical ions R_1^- and R_2^+ generally requires an amount of 672 kJ/mol or more *in addition* to the energy required for homolytic scission into the radicals R_1 and R_2. Homolytic chain scission, therefore, is a much more probable mode of molecular disintegration (of saturated hydrocarbons) than ionic dissociation.

3. Radicalization

The removal of a hydrogen atom from a hydrocarbon leads to the formation of a radical. Its electron distribution differs from the original one near the radical

Table 4.7. Electron Affinity and Ionization Potential

Radical R	Electron Affinity* E'(R) [eV]	Ionization potential I(R) [eV]	I(R) – E(R) eV	kJ/mol
H	0.80	13.60	12.8	1230
CH_3	1.08	9.82	8.74	840
C_2H_5	0.89	8.4	7.5	720
n-C_3H_7	0.69	8.1	7.4	710

* Values of E(R) taken from Handbook of Chemistry and Physics, 53 rd Edition, CRC Press, Cleveland 1972/73, p E 55.

C atom. In saturated hydrocarbons the 4 sp^3 hybrid orbitals change to 3 sp^2 hybrids. The radical free electron is considered to be located in a completely separated π eigenfunction which has no interaction with the σ-system. The strength of a bond connecting a C atom with the radicalized C atom, $\dot{C}$, is now mainly determined by the interaction of an sp^2 with an sp^3 hybrid. The energy of a bond formed by an overlap of an sp^3 hybrid and an sp^2 hybrid is not smaller than that formed by two sp^3 hybrids. This is concluded from the dissociation energies of the C–C bonds in RCH_2-CH_2R (337 kJ/mol) and $RCH_2-C_6H_5$ (381 kJ/mol). The assumption that replacing sp^3 orbitals by sp^2 orbitals in bonds will not reduce the bond energy per se is further supported by the observation that the bond energy of the $C_{(sp^2)}-H$ bond in $H_2C{=}CH_2$ is the same as the value of the $C_{(sp^3)}-H$ bond in H_3C-CH_3 (436 kJ/mol). The increased energy of dissociation of the $C_{(sp^2)}-H$ bond in the methyl radical (477 kJ/mol) leads to a similar conclusion.

The C–C bonds in α position to the carbon carrying a free electron are, therefore, slightly stronger than the same bonds in the original molecule. The C–C bonds in β position, however, are adversely affected by the described electronic reorganization. In thermal or mechanical degradation they are found to break preferentially. Reactions involving the scission of a radicalized carbon backbone chain lead to the formation of a double bond:

$$R_1-CHX\underset{\beta}{-}CH_2\underset{\alpha}{-}\dot{C}X-CH_2-R_2 \rightarrow R_1-\dot{C}HX + CH_2{=}CX-CH_2-R_2.$$

The considerable amount of energy gained in transforming a $-CH_2-\dot{C}X-$ bond into a double bond (250–280 kJ/mol) should decrease the activation energy of such

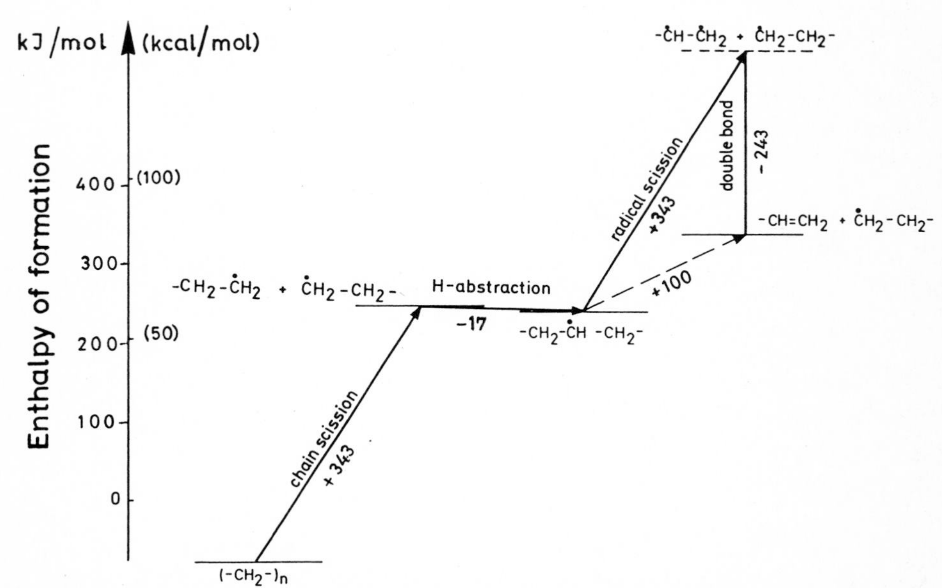

Fig. 4.3. Scheme of enthalpy values observed in the scission of a radicalized PE chain (it should be noted that the rather low enthalpy of 100 kJ mol^{-1} for the step $-CH_2\dot{C}H_2-CH_2- \rightarrow -CH{=}CH_2 + \dot{C}H_2-CH_2-$ is only observed in thermal equilibrium; activation energies in stress induced chain scission of radicalized chains may be closer to 343 kJ mol^{-1})

a reaction correspondingly. Thus one obtains for the reaction n-$C_3H_7 \rightarrow C_2H_4 + CH_3$ an activation energy of 106.8 kJ/mol and for the reaction n-$C_4H_9 \rightarrow C_2H_5 + C_2H_4$ an activation energy of 95 kJ/mol [18]. The enthalpy values of the different states observed in the scission of radicalized PE chains are indicated in Figure 4.3. Kagiya [14] has derived a number of activation energies, D, mostly for vinyl polymers. He uses the assumption that D is equal to the heat of polymerization, Q, and the activation energy E_p of the propagation reaction:

$$D = Q + E_p. \tag{4.23}$$

His values are reported in Table 4.8. Again it must be mentioned that a stress-induced scission reaction may require an energy of activation which is considerably larger than the listed values of D.

Table 4.8. Activation Energy of Main Chain Scission of Radicalized Chains*

Polymer	Activation Energy kJ/mol	Polymer	Activation Energy kJ/mol
PE	118	PVF	116
PP	102	PVDF	131
PIB	76	PTFE	147
PS	85	PVAC	118
Pα-MS	95	PMA	98
PVC	87	PMMA	85
		PA 6	117**

* Ref. 14.
** T. Kagiya, personal communication.

The only experimental indication of the weakening of radicalized chains so far know to the author of this book are the studies of Tomaskevskii et al. [19]. They report that the tensile strength (at 77 K) of PE-, PP- and PA6-fibers decreases with increasing number of (x-ray produced) free radicals (of a concentration of up to 10^{20} cm^{-3}). For these materials weakening coefficients of 1.5, 4 and 5.10^{-20} MPa cm^3 per radical have been determined. As the authors indicate the latter two coefficients may possibly be too high because of unaccounted radiation damage. This means that the normally observed radical concentrations of up to 10^{18} spins/cm^3 reduce by their pure presence the strength of a PA6-fiber by up to 0.05 MPa – a rather insignificant amount (see also 8IC2).

III. Form of Binding Potential

In Figure 4.1 it was *qualitatively* indicated how the potential energy of two bonded atoms changes with their interatomal distance r. Knowledge of the *exact* form of the interatomal potential could – in principle – be obtained from a calculation of the total electronic energy E of a molecule as a function of r. For polyatomic molecules, however, E(r) cannot be calculated by any of the mentioned approximation methods with reasonable accuracy. So the potential functions are described by *empirical,* phenomenological functions. These functions will, of course, necessarily utilize those molecular constants which are otherwise obtainable, e.g. bond length (r_0), bond stretching force constant, and dissociation energy D. One of these functions is the Morse potential function which has the virtue to lead to a solvable Schrödinger Equation:

$$V = D \exp[-2a(r - r_0)] - 2D \exp[-a(r - r_0)]. \tag{4.24}$$

The Morse constant a can be expressed through force constant and anharmonicity constant x of the oscillator formed by the bond atoms. The Morse function yields the energy eigenvalues of an anharmonic oscillator which are correct within a sizeable range of amplitudes (in the case of H_2 for $0.4 < r/r_0 < 1.6$). For large r, however, V becomes progressively too small [8b].

References for Chapter 4

1. K. S. Pitzer: Quantum Chemistry. New York: Prentice Hall, 1953 a) 153, b) 157, c) 163.
2. R. Daudel, R. Lefebvre, C. Moser: Quantum Chemistry, Methods and Applications. New York: Interscience, 1959 a) 48, b) 91, c) 3–24.
3. A. G. Shenstone: Phys. Rev. *72,* 411 (1947).
4. T. Cole, H. O. Pritchard, N. R. Davidson, H. M. McConnell: Mol. Phys. *1,* 406 (1958).
5. H. A. Stuart, Ed.: Molekülstruktur. Berlin – Heidelberg – New York: Springer 1967, 35.
6. R. D. Brown: J. Chem. Soc. 2615 (1953).
7. R. M. Joshi: J. Macromol. Sci.-Chem. *A4* (8), 1819 (1970).
8. J. G. Calvert, J. N. Pitts, Jr.: Photochemistry. New York: Wiley, Inc. 1967, a) 815ff, b) 136f.
9. D. W. van Krevelen: Properties of Polymers, 2nd Edition. Amsterdam: Elsevier 1976.
10. A. Ye. Stepan'yan, Yu. G. Papulov, Ye, P. Krasnov, G. A. Kurakov: Vysokomol. Soyed. *A14,* 2033–2040 (1972), Polymer Sci. USSR *14,* 2388–2397 (1972).
11. M. Heskins: Encyclopedia of Polymer Science and Technology, Supplement Vol. 1. H. F. Mark, N. G. Gaylord, Eds. New York: Interscience 1969, 763.
12. G. Herzberg: Spectra of Diatomic Molecules, 2nd Edit., New York: D. van Nostrand 1953.
13. W. Brenner, W. Kapfer: Encyclopedia of Polymer Science and Technology, Vol. *11,* H. F. Mark, N. G. Gaylord, Eds. New York: Interscience 1969, 763.
14. V. T. Kagiya, K. Takemoto, M. Hagiwara: IUPAC Macromol. Symposium "Long-Term Properties of Polymers and Polymeric Materials". Stockholm, Sweden, Aug. 30 – Sept. 1, 1976.
15. W. R. Kneen, M. J. W. Rogers, P. Simpson: Chemistry, Facts, Patterns, and Principles, London-Reading MA: Addison-Wesley Publ. Ltd., 1972.
16. J. A. Kerr, A. F. Trotman-Dickenson: Handbook of Chemistry & Physics, 58th ed., Cleveland: CRC-Press, 1977–78, Table F 231.
17. H. H. Sisler, R. D. Dresdner, W. T. Mooney, Jr.: Chemistry, a Systematic Approach, New York–Oxford: Oxford University Press, 1980.
18. J. A. Kerr, J. K. Kochi ed.: Free Radicals, New York: John Wiley & Sons, 1973, pp. 31.

19a. E. E. Tomashevskii: personal communication, Leningrad 1985

19b. B. B. Narzullaev, N. G. Kvachadze, E. E. Tamashevskii, A. I. Slutsker: Fiz. Tverd Tela *23,* 429 (1981) (Sov. Phys, solid State *23,* 242 (1981)).

Chapter 5

Mechanical Excitation and Scission of a Chain

I. Stress-Strain Curve of a Single Chain

A chain molecule as part of a thermoplastic body is in thermal contact with other chains and constantly in thermal motion. The atoms vibrate and take part in the more or less hindered rotations of groups and even of chain segments. With no external forces acting all molecular entities try to approach – and fluctuate around – the most probable conformation attainable to them. The action of external forces causes – or maintains – displacements of the chain from those positions and evokes retractive forces. Let us consider a chain or a bundle of chains in thermal contact with the surrounding and at constant volume. The condition of thermodynamic stability of such a system is that the free energy

$$F = U - TS \tag{5.1}$$

assume a minimum. The differential changes of the free energy of the system are given by the changes of its internal energy U and its entropy S. Under isothermal conditions one has:

$$\mathrm{d}F = \mathrm{d}U - T\,\mathrm{d}S. \tag{5.2}$$

An increase of U can be effected by heat $\mathrm{d}Q$ flowing into the system from the outside or by mechanical work $\mathrm{d}W$ done onto the system. The change of entropy may be split up into a portion $-\mathrm{d_e}S$ which is equal to $\mathrm{d}Q/T$ and into the internal entropy production $\mathrm{d_i}S$ caused by irreversible processes as for instance by stress relaxation or generation of heat through friction:

$$dF = dQ + dW - Td_eS - Td_iS = dW - Td_iS. \tag{5.3}$$

From Eq. (5.3) one derives the force f acting upon a chain with end-to-end distance r as:

$$f = \left(\frac{\partial W}{\partial r}\right)_{V,T} = \left(\frac{\partial F}{\partial r}\right)_{V,T} + T\left(\frac{\partial_i S}{\partial r}\right)_{V,T} \tag{5.4}$$

The uniaxial force on a relaxing chain molecule depends, therefore, on the changes of internal energy and entropy associated with a change of its end-to-end distance and on the entropy production term (which is always positive). These three different force contributions will be examined in the following.

A. Entropy Elastic Deformation

The kinetic theory of rubber elasticity was essentially developed between 1930 and 1943. A detailed description of the development of this theory and its present state is given by e.g. the classical treatises of Flory [1] or Treloar [2]. The necessary conditions for the occurrence of rubberlike elastic behavior are the presence of a long chain molecule possessing internal flexibility (freely rotating links) and the absence of strong secondary bonds acting between a segment and surrounding segments of the same or other chains. The stress-strain relation for a single finite chain then is derived from the distribution of chain conformations. Following the treatment of Treloar it will briefly be recalled which chain properties enter the stress-strain relations and what levels of force are attainded.

First the conformations of certain idealized chains are considered. For the *chain of n equal links of length* ℓ *joined entirely at random* (i.e. neglecting valence angles and hindrances to internal rotation) the probability distribution p(x, y, z) to find a chain end at a point (x, y, z) – if the other end is fixed at the origin – is obtained as a Gaussian distribution:

$$p(x, y, z) = (b^3/\pi^{3/2}) \exp[-b^2(x^2 + y^2 + z^2)]. \tag{5.5}$$

The only parameter, b, is the inverse of the most probable chain end separation and is under the above conditions $\sqrt{3/2n\ell^2}$. The extended length of the chain is, of course, equal to $n\ell$, whereas the average chain end separation (the "root-mean-square length" $\sqrt{\overline{r^2}}$ equals $\ell\sqrt{n}$.

More realistic is the model of the *statistically kinked chain with valence angles and hindrances to internal rotation.* Here the mean square length is derived as:

$$\overline{r^2} = n\ell^2 \frac{1 + \cos\alpha}{1 - \cos\alpha} \cdot \frac{1 + a}{1 - a} \tag{5.6}$$

where α is the semi-angle of the cone of bond rotation and a is the mean value of $\cos\phi$, ϕ being the angle of bond rotation counted with respect to the trans-conformation. The (limit for infinite chain lengths of the) ratio $r^2/n\ell^2$ is called the *char-*

acteristic ratio. If the rotation of a bond is considered to be independent of the position of other bonds and free within its cone of rotation then a is zero.

A given (hydrocarbon) chain can be treated as a statistically kinked chain with valence angles and also as freely jointed chain. The root-mean-square lengths and extended lengths of both models correspond to each other then and only then when a random link of appropriate length has been used. This "equivalent random link" of length ℓ_r is rather short and contains only a few chain atoms. For paraffin-type chains Treloar obtained a length ℓ_r of 0.377 nm, for cis-polyisoprene 0.352 nm. These equivalent random links are about two times smaller than the (shortest) segment lengths necessary to form a fold of 180°.

The entropy S of a single chain is according to Eq. (5.5) and the Boltzmann relation

$$S = k \ln p = C_1 - kb^2r^2 \tag{5.7}$$

where C_1 is a constant. An important aspect of the entropy of a chain is – in the words of Treloar [2a] –: "A chain does not possess an entropy by the mere fact of having a specified conformation, but only on the virtue of the large number of conformations accessible to it under specified conditions of restraint". If these conditions consist only in the location of the end-points of the chain at specified positions, then the appropriate expression for the entropy is Eq. (5.7). In the presence of external hindrances to bond rotation the entropy of a chain is not given by Eq. (5.7) since $p(x, y, z)$ must then be corrected for the number of inaccessible states.

The retractive force f acting between the chain ends – and in the direction of the line joining them – is derived from Eqs. (5.2), (5.4) and (5.7). If one considers a reversible deformation and postulates that all conformational states posses the same internal energy U so that U does not depend on conformation then

$$f = 2\,kT\,b^2\,r. \tag{5.8}$$

The random coil in thermal equilibrium behaves as a spring (of length zero) with a spring constant $2\,kT\,b^2$. The level of force exerted by such a spring with its ends separated by the most probable distance $r = 1/b$ is fairly small. For a hydrocarbon chain of molecular weight 14'000 with an extended length of 125.5 nm and a most probable chain end distance of 7 nm, this force is $17 \cdot 10^{-13}$ N which corresponds to a force per cross-section of the chain, f/q, of 10 MN/m^2 (10^8 dyn/cm^2). As one should expect this "chain stress" is very close to the modulus $G = NkT$ of a rubber network having chains of exactly this length between crosslink points and for which G is 17.5 MN/m^2.

The Gaussian theory considers the number of possible conformations of a chain having a specified end-to-end distance. A more accurate non-Gaussian statistical treatment of the freely jointed chain is based on the distribution of $\sin \Theta_1$, i.e. of the angle between the direction of a random link and of the end-to-end vector. From the probability of finding n_1 links in the range $\Delta\Theta_1$, n_2 in $\Delta\Theta_2$ and so on, the entropy of a single chain is derived [2b] as

$$S = C_2 - kn \left[\frac{r}{n\ell} \beta + \ell n \frac{\beta}{\sinh \beta} \right] \tag{5.9}$$

where C_2 is a constant and β the so-called inverse Langevin function $\mathscr{L}^{-1}$ (r/nℓ). Using (5.2) and (5.4) one obtains

$$f = \left(\frac{kT}{\ell}\right) \mathscr{L}^{-1} (r/n\ell) \tag{5.10}$$

which can be expanded into

$$f = \frac{kT}{\ell} \left\{ 3 \left(\frac{r}{n\ell}\right) + \frac{9}{5} \left(\frac{r}{n\ell}\right)^3 + \frac{297}{175} \left(\frac{r}{n\ell}\right)^5 + \ldots \right\}. \tag{5.11}$$

It is of special interest to note that this stress-strain relation can also be derived from a consideration of the potential energy of the statistical links in a unidirectional force field. James and Guth [3] and Flory [1] have used the fact that an amount of work of $f \cdot \ell (1 - \cos \Theta_i)$ must be performed to rotate a link from a position of $\Theta = 0$ to $\Theta = \Theta_i$; if the orientational energy $-f \ell \cos \Theta_i$ is distributed according to Boltzmann statistics then the mean extension of a link in the direction of force is

$$\bar{z}_i = \ell \mathscr{L}(f\ell/kT) \tag{5.12}$$

and the mean total chain end-to-end distance is

$$r = n\bar{z}_i = n\ell \mathscr{L} (f\ell/kT) \tag{5.13}$$

which by inversion directly yields Eq. (5.10). It should be pointed out that so far this result has been derived without introducing – or permitting – any internal or external hindrances to rotation of a link. The absence of external hindrances also means that in longitudinal extension or contraction of the chain or of parts thereof no frictional forces are encountered. The forces exerted by the chain ends are solely the consequence of the imbalance of exchange of momentum between the chain and its surrounding.

If a hydrocarbon chain behaves as described by Eq. (5.13) and if such a chain were to reach a breaking force (of, say, $17 \cdot 10^{-10}$ N) at room temperature then $f\ell/kT$ must exceed 150. For an argument that large the Langevin function is equal to $1 - kT/f\ell$ and f is:

$$f = nkT/(n\ell - r). \tag{5.14}$$

In the chosen example of a hydrocarbon chain of 125.5 nm extended length, breaking stresses will only be reached if r is larger than 124.7 nm. In other words only 2 of the 333 random links of 0.377 nm length may point in a direction perpendicular to the chain end-to-end vector while all the remaining ones then have to be fully aligned. Even at this extreme elongation the Langevin function gives a good approximation of the stress-strain behavior of the randomly kinked chain. This becomes evident by comparison with the so called exact solution of Treloar [2c] which is purely based on geometrical (and combinatorial) considerations and from which one derives for extreme elongations:

$$f = -kT \, d\ell n \, p/dr = -kT/r + (n-2) \, kT/(n\ell - r). \qquad (5.15)$$

The force-elongation relations (5.8), (5.10), and (5.15) have been derived under the condition that all conformational states have the same internal energy, that all states are accessible, and that transitions between different states are easily accomplished. No chain molecule – and especially not within a solid – does fully comply with the above conditions. For one thing there are the *intramolecular* interactions (between nearest and second nearest atoms of one chain) which cause the potential energy of the chain to be a function of the angle of rotation with distinct relative minima in more than one position. All this is well known and documented (e.g. 1–5) and will not be discussed here. The direct or indirect consequences with respect to the retractive forces between (PE or PA) chain ends are the following:

- the internal energy of (in-plane) trans-conformations is lower than that of (out-of plane) gauche-conformations
- the probability to find a conformation with internal energy ΔU is weighted by the Boltzmann factor exp $(-\Delta U/kT)$
- the average end-to-end distance increases with a according to Eq. (5.6), the length ℓ of the equivalent random link also grows in proportion to $(1+a)/(1-a)$, while the number n of random links in a chain of finite length decreases
- the force between chain ends at their most probable distance ($r = 1/b$) is inversely proportional to r and thus also to the square root of the characteristic ratio, the force is the smaller the larger the characteristic ratio is.

The characteristic ratio of PE at room temperature is – according to Abe et al. [4] – 8.0 as compared to a value of 2 calculated under neglect of hindrances to bond rotation. The force between PE chain ends, therefore, is by a factor of 2 smaller if the effect of the potential energy on chain conformation is taken into consideration. For polyamide chains the detailed calculations of Flory and Williams [5] led to characteristic ratios of 6.08 (PA 6) and 6.10 (PA 66) in excellent agreement with the experimental value of 5.95 for PA 66 found by Saunders [6]. The calculated force between PA chain ends, then, will be smaller by a factor of about 1.7.

A second cause to invalidate Eq. (5.7) are strong *intermolecular* interactions between chain segments which do have the effect that only a restricted number of conformational states are accessible to a chain segment, that transitions between different states may not be independent from each other, and that energy is required to activate these (cooperative) transitions. In recent years considerable progress has been made to apply the methods of statistical thermodynamics also to systems subject to the above restrictions in order to derive the distribution of conformational disorder and to explain e.g. the stress-strain behavior or the melting and crystallisation of chain molecules [4–8, 60–63]. With respect to chain loading one is interested in the entropy changes associated with the elongation of a highly extended chain. In the following ΔS will be calculated for PE and PA chain segments of 3 to 6 nm length.

It is convenient to start with Pechhold's [7] treatment of the polyethylene molecule. He derived the partition function of the rotational isomers of a chain with n independent C–C bonds. The chain is allowed to rotate around its C–C bonds, it will, therefore, encounter three minima of potential energy of which the absolute minimum belongs to the extended trans-conformation (t). Two relative minima at an angle ± 120° out of plane allow two gauche-conformations ($g,\bar{g}$) with a conformational energy of + ΔU. The kink-isomers are the subclass of the rotational isomers where only n/2 of the bonds – and no adjacent ones – may have a g or $\bar{g}$ position. For kink isomers entropy, internal energy, and free energy as a function of the average concentration $\bar{x}$ of gauche-conformations are derived [7] as:

$$\bar{x} = \exp(-\Delta U/kT) / [1 + 2\exp(-\Delta U/kT)] \tag{5.16}$$

$$S = -0.5\, nk\, \ln(1 - 2x) + n\bar{x}\,\Delta U/T \tag{5.17}$$

$$U = n\bar{x}\,\Delta U \tag{5.18}$$

$$F = 0.5\, nkT\, \ln(1 - 2\bar{x}). \tag{5.19}$$

The group of kink isomers still comprises chains with quite different deviations from a planar zig-zag conformation, the simplest of those kinks being ttgt$\bar{g}$ttt (2gl kink), tgttg$\bar{}$t (2g2 kink), tgtgtgt (3g2 kink). Each kink causes a contraction in longitudinal direction which is a multiple Δz of $\sqrt{2/3}\,a$ with a the length of a C–C bond (Pechhold accordingly denotes a kink containing q gauche positions and causing a contraction Δz as a qg Δz-kink). A chain with n C–C bonds and m_i qgi-kinks has a projected length in z-direction of $(n - im_i)\sqrt{2/3}\,a$. Vice versa a projected length corresponding to that of n C–C bonds can be realized as well by an extended chain with n C–C bonds, by a chain with n + 1 bonds containing a 2gl kink, by chains with n + 2 bonds containing either two 2gl kinks or a 2g2 kink or by other appropriate combinations. The statistical weight of each conformation is given by the number of ways *that* conformation can be realized and by the Boltzmann factor of its conformational energy. The by far leading term in the partition function for a given *r* is always that listing the number of possible conformations of chains containing only 2gl kinks. The number of ways of arranging $n_g/2$ pairs of g and $\bar{g}$ bonds on n/2 available positions clearly is

$$Z(n, n_g) = 2^{n_g/2} \prod_{i=1}^{n_g/2} \left(\frac{n}{2} - 2i + 1\right) \exp(-2i\Delta U/kT)\,. \tag{5.20}$$

In Table 5.1 the statistical weights of PE chains with a gauche-concentration of up to 20% are given according to Eq. (5.20). Also listed are the statistical weights of 6 PA chains subject to the condition that the two gauche positions of any one gtg$\bar{}$t kink are confined to the (six) C–C bonds of one monomeric unit and that the carbonamide group adheres to a trans-planar conformation [5]. As potential energy ΔU of a gauche position the "preferred value of 0.5 kcal/mol" [5] has been used.

The change of $\ln Z(n, i)$ with *r* is essentially the same as the change of the complete partition function (which accounts for mgi-kinks of all types). One can cal-

Table 5.1. Statistical Weight of (n, n_g) Conformations
Polyethylene chains, $\Delta U = 0.5$ kcal/mol

n	$n_g = 0$	2	4	6	8	10
38	1	6.80	41.1	217	985	3722
40	1	7.18	46.1	261	1282	5329
42	1	7.55	51.4	311	1642	7444
44	1	7.93	56.9	366	2072	10174
46	1	8.31	62.8	427	2580	13646
48	1	8.69	68.9	495	3177	18000

6 Polyamide chains, ΔU of gauche positions of C–C bonds 0.5 kcal/mol

n	$n_g = 0$	2	4	6	8	10
35	1	5.67	25.7	87.3	198	224
42	1	6.80	38.5	174.6	594	1346
49	1	7.93	53.9	305.6	1385	4710
56	1	9.07	71.9	489.0	2771	12560

culate, therefore, from the data of Table 5.1 the entropy elastic contribution to the retractive force as $-kT\,\Delta\,\ell n\,(Z)/\Delta r$. The energy elastic contribution, f_e, derives here from the change of conformational energy and is $\Delta n_g \Delta U/\Delta r$. For the chosen examples of a PE chain with 40 C–C bonds, a 6 PA chain of 6 monomer units and gauche energies ΔU of 0.5 and 0.75 kcal/mol one obtains the force contributions listed in Table 5.2. For comparison the forces derived for the statistically kinked chain according to the Langevin and the "exact" approximation are also given. It is immediately apparent that in kink isomers the energy-elastic contribution f_e is of the same order of magnitude – and opposite sign – as the entropy-elastic force. The molecular stresses, Ψ, are obtained by dividing the axial forces f through the molecular cross-section q, whereas the elastic moduli are understood as:

$$E = r\,\Delta f/\Delta r q. \tag{5.21}$$

This investigation may be summarized by saying that the retractive forces experienced by highly extended chain segments at room temperature due to conformational changes are small in comparison to the ultimate forces that can be carried by a chain. External hindrances to segment rotation have not been treated here. At present is suffices to state that they will only further diminish the –already – small entropy elastic forces.

B. Energy Elastic Chain Deformation

So far the energy-elastic deformation of the (assumedly rigid) chain skeleton has not been considered. In response to axial forces the chain skeleton will deform,

Table 5.2. Entropy-Elastic Forces and Moduli

Approach		Eq.	L Å	r Å	f_e 10^{-10}N	f 10^{-10}N	ψ MN/m^2	E GN/m^2
Langevin	(ℓ = 0.377 nm)	(5.14)	49	45.2	0	1.43	843	5.5
Exact	(ℓ = 0.377 nm)	(5.15)	49	45.2	0	1.20	712	4.6
PE-Kinkisomer	(ΔU = 0.50)	(5.20)	50.3	49	−0.55	0.10	55	1
PE-Kinkisomer	(ΔU = 0.75)	(5.20)	50.3	49	−0.83	−0.45	−247	(0.9)
PA-Kinkisomer	(ΔU = 0.50)	(5.20)	52.1	50.8	−0.55	0.08	49	1.9
PA-Kinkisomer	(ΔU = 0.75)	(5.20)	52.1	50.8	−0.83	−0.47	−278	(1.6)
PE bond stretching and bending:						23.8	14000	290

Table 5.3. Stiffnesses of Molecular Deformation Modes

Mode of Deformation	Polymer	Calculated Modulus [GN/m^2]	Derived from	Ref.
Hindered rotation (helix)	Polyesters	4–7	X-ray diffraction on various polyesters	12
Hindered rotation of kinked chain	PE	$2.6\ n/n_g$	Eq. 5.21, sinusoidal rotation potential	–
Pure bending of C–C bonds	PE	80	–	12
Pure stretching of C–C bonds	PE	740	–	12
Stretching of zig-zag chain	PE	182–340	Force constants	9
	PE	182, 299	Urey Bradley force field	13
	PE	297	Minimum energy electron wave function	11
	PE	272	Deformation of rigid valence tetrapods	47
	PA 6	180–200	Spacing of (070) planes	46
	PA 6	157–196	Force constants	9
	PA 6	244	Urey Bradley force field	13
	PA 66	181	Urey Bradley force field	13
Crystal lattice deformation	PE	235	Spacing of (002) planes	9
	PA 6 (α form)	20, 24.5	Spacing of (020) planes	46, 9

however, through hindered rotation, bending, and stretching of main chain bonds [12]. In Table 5.3 the stiffnesses characterizing these individual modes of deformation and their combination in a loaded chain are listed. The C–C bond bending and stretching moduli have been calculated from the corresponding force constants. The modulus for the hindered rotation of a kinked chain is derived from the rotational potential of CH_2 groups and the associated axial chain elongation as formulated by Pechhold [7] who used an intramolecular rotational barrier of 2 kcal/mol above gauche energy. The resulting modulus is a function of the angle of rotation and ranges between 4 and 7 GN/m^2 in good agreement with a value of Bowden [10] of 5.7 GN/m^2. In axial stretching the *hindered* rotation of gauche bonds contributes to axial elongation, the chain stiffness thus decreases with the concentration of gauche bonds.

Longitudinal chain moduli, defined as in Eq. (5.21), have been determined theoretically and experimentally in a number of ways:

- through calculation from spectroscopically determined bond bending and stretching constants
- from the Raman shifts of extended chain hydrocarbons
- from x-ray observations of load-induced changes of the unit cell dimensions in highly oriented polymers.

These methods, their results, and the arguments to explain a certain disagreement between the numerical results have been reviewed by e.g. Sauer and Woodward [9]. More recently the direct calculation of the total (electronic) energy (of a PE chain) with changing atomic geometry has been attempted by Boudreaux [11]. In his self-consistent field molecular orbital (SCF – MO) calculations he searched for "that set of numbers which, when used as coefficients of the assigned atomic orbitals, produces an approximation to the wave function which minimizes the total energy". Hahn et al. [14] presently investigate a model based on the elastic interaction of essentially rigid sp^3 valence tetrapods (shell-model). The values of PE and PP chain moduli thus obtained by different researchers are represented in Figure 5.1. For PE some preference has been given in this book to the value of $E_c = 290\ GNm^{-2}$ measured by Strobl and Eckel [58], which has recently been confirmed by Kobayashi et al. [58].

From Table 5.3 it is apparent that the weakest mode of skeleton deformation (4.7 GN/m^2) is that derived from the hindered rotation of (non-extended) bonds whereas a fully extended chain segment has a 50 times larger stiffness. The elastic modulus of a partially extended kinked chain, then, is:

$$1/\overline{E} = \sum_i x_i/E_i \tag{5.22}$$

where the E_i are the stiffnesses of the individual (trans-planar, kinked, or helical) segments of a chain and the x_i are the corresponding (volume) fractions.

One is now in a position to determine the change of the free energy F of a partly extended chain as a function of the end-to-end distance r. As an example the quasi-static deformation of PE segments will be treated within the frame of the bond-bending and stretching model. The minimum free energy of a segment containing n C–C bonds and n_k 2g1 kinks is attained at an end-to-end distance

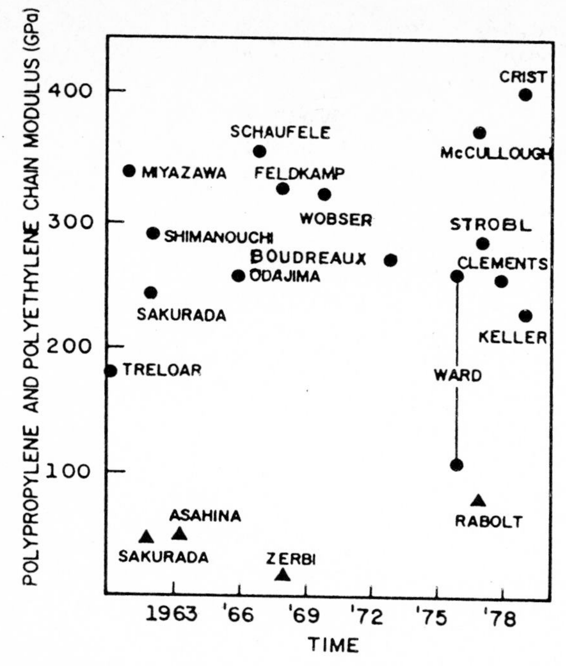

Fig. 5.1. Chain modulus values of polyethylene (•) and polypropylene (▲) (after Wool and Boyd [59]).

$r = (n - n_k)\sqrt{2/3}\,a$. This minimum free energy is equal to $n_k\ \Delta U - RT\ \ell n\ Z$. From the statistical weight of the (n, n_k) conformations of PE segments with $n = 40$ (Table 5.1) the minimum free energy values have been calculated. They are indicated in Figure 5.2 at the appropriate chain-end distances. If the chain ends are axially displaced from these equilibrium positions energy elastic forces arise which have an unsymmetric potential. In *stretching* completely extended sections the trans bonds are deformed in the plane of the zig-zag chain according to the chain modulus E_{str}. The gauche bonds are performing a hindered rotation out of the zig-zag plane (E_{rot}). The stretching modulus $\bar{E}$ of the kinked segment then is derived from Eq. (5.22). The chain is the stiffer the fewer gauche bonds it contains. Employing the rotation potential mentioned above [7] and an extended chain modulus of 200 GN/m^2 the *tensile* sections of the free energy curves have been calculated. The presence of only 5 kinks softens the chain segment quite noticeably ($\bar{E}$ decreases to 9.5 GN/m^2).

The axial *compression* of an extended chain differs in so far as the trans bonds now aquire the freedom to perform a hindered rotation out of the zig-zag plane. This freedom is not exercised in tension because it would be energetically less favorable than bond bending and stretching. The axial compression modulus of a zig-zag chain is essentially determined by the hindered rotation mode of deformation, it corresponds to the extension modules of a highly kinked chain ($n_k = n/4$) and has been taken to be 5 GN/m^2.

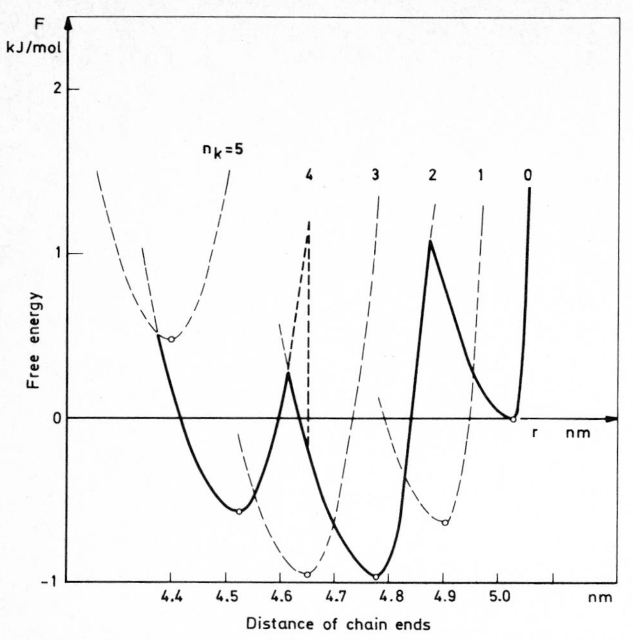

Fig. 5.2. Free energy of uniaxially deformed polyethylene chain segment. Number of C–C-bonds n = 40, number of kinks $n_K = 2\ n_g$, extended length L = 5,03 nm.

For each conformational state an asymmetric parabola is obtained. In quasi-static straining an individual, kinked chain will tend to assume that state of minimum free energy accessible to it. Transitions between different conformational states will occur (solid line in Fig. 5.2). The transition to a more extended conformation will cause an immediate release of axial stress for the chain segment in question – and thus constitute a thermodynamically irreversible process. Macroscopically the step-wise and uncorrelated release of stress in a large number of chains is felt as stress relaxation. The random thermal motion of chain atoms and the existence of external hindrances to bond rotation are the reason that the deformation potentials of individual chain segments deviate from each other and from the solid line – as arbitrarily indicated by the heavy dashed line. The deformation potential shown in Figure 5.2 essentially represents also that of other (planar) zig-zag chains such as PA 6 or 66 which have the same kind and concentration of gauche bonds and about the same length. Recent model calculations are found in refs. [69, 70].

C. Non-Equilibrium Response

The stress-induced conformational transitions of a kinked chain as represented by the solid line in Figure 5.2 are thermodynamically irreversible, internal energy is converted into heat. At this point one is interested in the time-constant of the transition process: if it is small compared to the time within which chain stretching is accomplished then the stress-strain curve will not deviate very much from the solid line, if the time constant is extremely large then transitions may be considered to be impossible, the chains deform energy-elastically. In the intermediate region of time constants, however, the largest stresses of not fully extended chains will depend on the rate at which stress releasing conformational transitions take place. A detailed discussion of this effect requires a consideration of the geometry and interaction of molecular chains, of the principles of irreversible thermodynamics [15], and analysis of the potential of the secondary or van der Waal's bonds between segments [16, 69–71]. It leads to a description of the anelastic *deformation* of polymers which is not intended in this book. Nevertheless one is still interested in some information with respect to the rate of stress relieving transitions between different kink-isomers. Since any transitions leading to the motion of just a single kink generally do not cause any axial chain elongation one has at least to consider the simultaneous activation and annihilation of two kinks. Such process involves the rotation of four gauche bonds and the rotational translation of the segment between the kinks, the energy of activation can be estimated to be 8 kcal/mol for the rotation of gauche bonds [7] and 0.5 to 1.2 kcal per mol of rotating CH_2 groups to overcome the intermolecular barriers [17, 18]. A tgt$\bar{g}$ttg$\bar{}$tgt-kink requires, therefore, an activation energy of 11 to 15.2 kcal/mol to transform into the all-trans conformation. One may assume that in stress-free PE chains such transformations do occur with a rate of at least 1 s^{-1} at a temperature somewhat below the melting point, i.e. at 400 K. One may then calculate the rate of this process at 300 K from Eq. (3.22) as 0.0018 s^{-1}. If the chain is strained the activation energy for segment rotation only decreases and the rate of stress relieving transitions increases [19]. With such a time-scale one may consider stress-induced rotational transitions as being easily accomplished before a chain segment reaches elastic energies sufficient for chain scission at room temperature, i.e. more than 25 kcal/mol (see also [70, 71]).

It remains the entropy production term in Eq. (5.4) to be discussed. Under the conditions of this section – quasi-static axial loading of the ends of a segment and determination of the ensuing retractive forces – this term is essentially zero.

II. Axial Mechanical Excitation of Chains

A. Secondary or van der Waal's Bonds

In Section A the stress-strain behavior of a single chain segment loaded pointwise at its ends has been discussed. In (uncross-linked) thermoplastic polymers, however, large axial forces cannot be applied pointwise to a main chain bond but only be built up gradually along the chain through – the much weaker – intermolecular forces. The forces between molecules are a superposition of the close range (nucleonic) repulsion and the (electronic) van der Waal's attraction, which includes electrostatic forces between ions, dipoles or quadrupoles, induction forces due to polarization of atoms and molecules, and the generally most important quantum-chemical dispersion forces. The van der Waal's attraction is the cause of the solidification and crystallization of high polymers, it is theoretically quite well understood and has been treated in detail by Langbein [16]. With reference to this work and the general references in Chapter 1 it can be stated that the secondary forces are not saturated and are nondirectional, i.e. not confined to precise locations of the neighbor atoms at e.g. tetrahedral bond angles. To the extent that these assumptions are valid the potential of the intermolecular forces acting on a chain or segment can be replaced by the sum of the interaction potentials of all appropriate pairs of atoms. The pair potentials contain an attractive term – which is theoretically derivable and decreases with the 6th power of the interatomic distance [16] and a repulsive term for which only semi-empirical expressions can be given. The total energy of intermolecular interaction, i.e. the cohesive energy of a solid then is to be understood as the sum of the pair interactions. In thermodynamic equilibrium a chain segment attains a position such that the free energy of interaction is a minimum. If a chain segment is displaced with respect to this position the free energy is raised and retractive forces arise. Such axial displacements are realized in quasi-static pulling a chain segment which is part of a crystal lamella against the crystal lattice or of extended chains against a glassy matrix. Dynamic chain stretching occurs in the shear deformation of molecules in solution or the melt. In the following it will be discussed which level of axial chain stress can be attained in these cases, that is in the loading of primary bonds through secondary or van der Waal's forces.

B. Static Displacements of Chains Against Crystal Lattices

The first investigations of the loading of chains displaced with respect to their equilibrium positions within a crystal lattice have been carried out by Chevychelov [20]. He used a continuum approach, i.e. replaced the string of discrete chain atoms by a continuous mass distribution. Each mass point experiences a periodic repulsive crystal potential if displaced from its equilibrium position. Later Kausch and Langbein [21] and Kausch and Becht [22] extended these calculations to treat the static and dynamic interaction of chains of discrete atoms with arbitrary periodic potentials.

The model used in these investigations is represented by Figure 5.3. It consists of a tie molecule (t) which forms part of a crystal lamella (c), leaves it perpendicularly to the crystal fold surface, (I), extends straightly through an amorphous region (a) and enters the adjacent crystallite of which it forms a part. The crystal boundaries

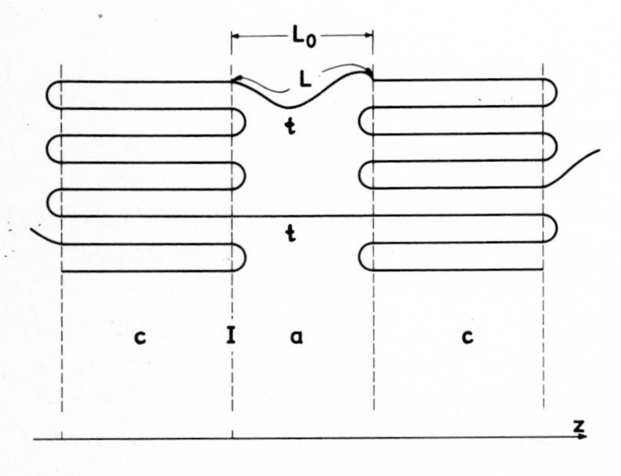

Fig. 5.3. Model of tie molecules (t) connecting folded chain crystal lamellae (c). a = amorphous region, L_0 = length of amorphous region, L = contour length of tie molecule, I = crystal boundary.

are assumed to be ideally sharp. No interaction between the tie chain and the amorphous material outside of the crystallite is taken into account.

A tie molecule thus embedded in two different crystalline layers is streched if the crystal lamellae are separated by macroscopic stresses. Because of the large elastic modulus of the chain molecule very high axial stresses are exerted onto the chain if it experiences the same strain as its amorphous vicinity. This high axial stress tends to pull the tie molecule against the intracrystalline potential and out of the crystal lamella.

In the following calculations the assumedly rigid solid is represented by an interaction potential the depth of which corresponds to two times the average (lateral) cohesive energy U with which the chain segments are bound within lamellar crystals or amorphous regions. Upon axial displacement in z-direction (chain axis) a chain will rotate and translate along a temperature-dependent minimum energy path. For polyethylene (PE) Lindenmeyer [17] has given the maximum potential energy to be overcome by a CH_2 group in twisted translation as ΔU_{tt} = 500 cal/mol at room temperature. The potential energy of a CH_2 group within the amorphous phase can be determined from the difference of the enthalpies of crystalline and amorphous PE at absolute zero temperature, $H_{am,o} - H_{cr,o}$ = 661 cal/mol [23]. It should also be taken into account that the internal energy of a partly oriented PE may be smaller than that of an amorphous one. The energy difference will be comparable to the enthalpy difference $(H_{am} - H_{def})$, which was estimated by Fischer and Hinrichsen [24] to be 140 cal/mol. As function of its z-position within an otherwise undisturbed crystal a CH_2 group would encounter a potential energy as shown in Figure 5.4.

Each point along a tie chain is characterized by its distance z from the crystal boundary, where z refers to the unstressed situation. The displacement of the molecular point z under stress is denoted by u = u (z) and the crystal potential per unit length by V(u). The axial stress, s, acting along the chain decreases with increasing

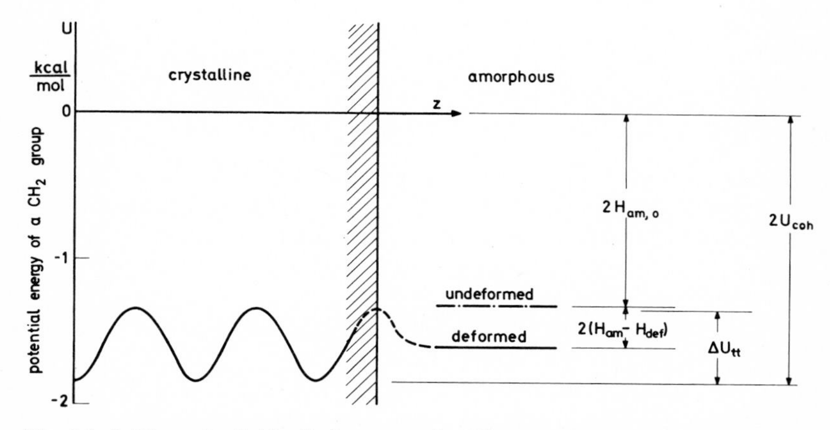

Fig. 5.4. Lattice potential in displacement of a CH_2 group in crystalline and amorphous polyethylene (after [22]). U_{coh} = cohesive energy per CH_2 in PE crystal, ΔU_{tt} = potential maximum in twisted translation, $H_{am,0}$ = enthalpy of PE crystal at 0 K, $H_{am} - H_{def}$ = difference of enthalpies of undeformed and deformed amorphous phase of PE.

$|z|$ due to the forces exerted on the chain through the crystal potential V(u). Equilibrium of forces at each section of the chain is established if

$$\frac{ds}{dz} = \frac{dV(u)}{du} . \quad (5.23)$$

This relation implies no assumptions with respect to the form of V(u) or the response s(u) of the chain.

One does assume, on the other hand, the tie-chain segments within the crystallites to resemble an elastic spring, i.e. one equates the axial stress s to the local chain deformation

$$s = \kappa du/dz \quad (5.24)$$

yielding

$$\kappa \frac{d^2u}{dz^2} = \frac{dV(u)}{du} . \quad (5.25)$$

Here κ is the constant of chain elasticity (not normalized with respect to the chain cross-section).

For a semi-infinite crystal and a sinusoidal potential $V(u) = v(1 - \cos K\, u(z))$ Chevychelov integrated Eq. (5.25) to obtain

$$\frac{1}{2}\left(\frac{du}{dz}\right)^2 = A - \frac{v}{\kappa} \cos K\, u \quad (5.26)$$

and, after having determined $A = v/\kappa$ from $du/dz = 0$ for zero displacement,

$$\frac{du}{dz} = 2\sqrt{v/\kappa}\, \sin K\, u/2 \quad (5.27)$$

and

$$u(z) = \frac{4}{K} \arctan \{C \exp [(K\sqrt{v/\kappa})\, z]\} . \quad (5.28)$$

The integration constant C is tan Ku(0)/4 and K is 2π divided by the intra-chain repeat unit length d. For a polyethylene chain with a spring constant κ of 34 nN Eqs. (5.28) and (5.24) have been evaluated [22]. The crystal potential depth v has been derived as 0.0138 nN from the cohesive energy density δ^2 through

$$v = 2\delta^2\, M/\rho\, d\, N_L \quad (5.29)$$

where N_L is Avogadros number, M the molecular weight of a repeat unit, and ρ the density. In static mechanical equilibrium the stresses and displacements of tie chain atoms within the crystallite are a unique function of the axial tension s_a existing at the crystal boundary (z = 0). In Figure 5.5 tension and displacement of a PE chain within a PE crystallite are shown for maximum axial chain tension s_0 equal

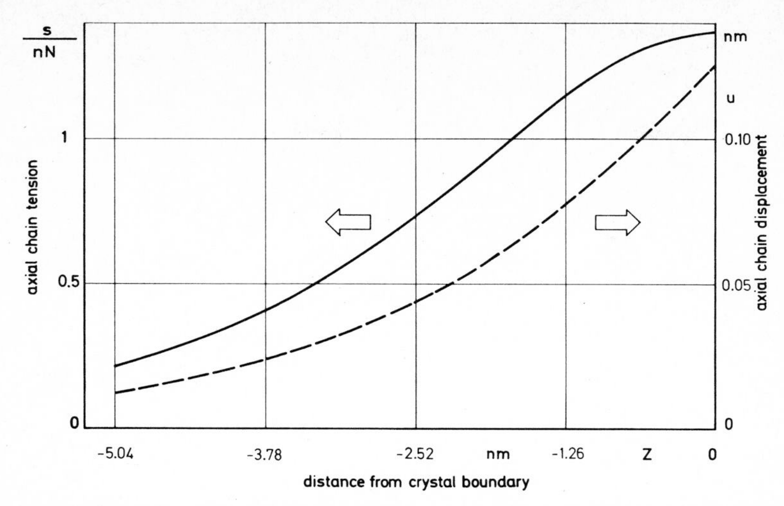

Fig. 5.5. Decrease of tension and displacement of a highly stressed tie chain within a polyethylene folded chain crystal lamella (after [22]).

to 1.372 nN. For any force larger than s_0 no static equilibrium can be established within a defect-free crystallite. It should be noted that this force which pulls out of the crystal a chain of full crystal length is 15.8 times larger than the force κ v d which is necessary to pull out a single monomer unit. If the tension s_0 is divided by the chain cross-section (0.1824 nm^2) one arrives at the maximum axial stress which a perfect PE crystallite is able to exert on a tie molecule: 7.5 GN/m^2. As can be seen from Figure 5.5, the maximum mechanical excitation of a chain penetrates about 5 nm into the crystallite. At a distance of 6 nm from the crystal boundary it is smaller than the average thermal amplitude at room temperature, which is about 0.008 nm.

Kausch and Langbein [21] extended the above calculations to various boundary conditions at the lower stress end, –Z, of the chain. If the chain terminates at –Z or continues into the adjacent amorphous region, it experiences no or only very weak axial forces. This case is covered by putting $A < v/\kappa$. If, on the other hand, a stressed tie-chain folds back at –Z, any axial displacement there implies a deformation of the fold. Such a deformation entails a considerable distortion of the crystal lattice and requires strong axial forces. In this case the integration constant A generally becomes larger than v/κ. After having derived the equations necessary for describing the elastic interaction between a chain and a crystal – represented by one unique potential – for arbitrary boundary conditions the authors [21] took into account also the fact that in a crystal containing hydrogen bonds – like that of 6-polyamide – the intermolecular interaction differs in different crystal regions. It is much stronger at the sites of C=O and N–H chain groups than it is at the sites of CH_2 groups. The problem of calculating the elastic displacement of a continuous chain passing through alternating regions of strong and weak attraction can be solved through alternate application of appropriate boundary conditions. In this case the boundary condition is that chain displacement and chain tension are continuous at all points z_i where the potential changes from weak attraction ($v = v_1$) to strong attraction ($v = v_2$)

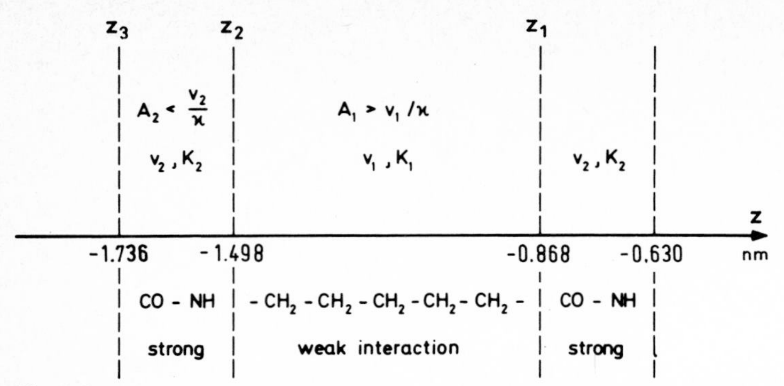

Fig. 5.6. Geometry and notations used in calculating stresses and displacements of 6 polyamide chains.

or vice versa (refer to Fig. 5.6 for notation). Here it was assumed that the difference between the cohesive energies U_{coh} of PA (18.4 kcal/mol) and PE (6.3 kcal/mol) is caused by the presence of hydrogen bonds between the carbonamide groups of neighboring molecules within the grid planes of the (monoclinic) PA crystallites. This excess energy contributes to the intermolecular potential v_2 acting between carbonamide groups. The potential v_2 thus is (12.1 + 0.50) kcal/mol, which is equivalent to 0.369 nN.

Chain tension and displacement of a 6-PA tie molecule within a 6-PA crystal are shown in Figure 5.7. The strong attractive effect of the hydrogen bonds results in a rapid drop of chain tension at the position of the carbonamide groups. The decrease of both tension and displacement is much more rapid than in the case of PE. At a distance of 2.1 nm from the crystal boundary the displacement has already dropped to the average level of thermal vibration at room temperature. With the crystal boundary positioned – as shown in Figure 5.7 – at the end of a $(-CH_2-)_5$ segment the maximum chain tension is s_0 = 3.94 nN. This tension is only 1.7 times the force necessary to remove one carbonamide group from the crystal, i.e. 59% of the maximum chain tension is expended to break the hydrogen bonds of the first CONH group. It should be noted that the value of s_0 depends upon the geometric arrangement of the atoms at the crystal boundary. With a carbonamide group at the crystal boundary the maximum tension s_0 is about 20% smaller than in the case where the crystal boundary runs through the middle of the $(-CH_2-)_5$ segment [21].

If s_0 is divided by the molecular cross section (0.176 nm^2) a stress of 22.4 GN/m^2 is obtained, which may be termed the static strength of a perfect (monoclinic) 6-PA crystal.

The above strength values have been derived for the interaction between a periodic potential and a continuous elastic chain. Kausch and Langbein [21] have shown, however, that there are only insignificant changes if the model using a continuous chain is replaced by the more accurate one using a string of discrete atoms.

In the preceding calculations a lattice potential had been used which did not depend on the applied axial chain stress. Evidently, this condition is not quite correct since each stressed chain will pull on its nearest neighbors which will in turn releave part of the stress by displacement with respect to the second nearest neigh-

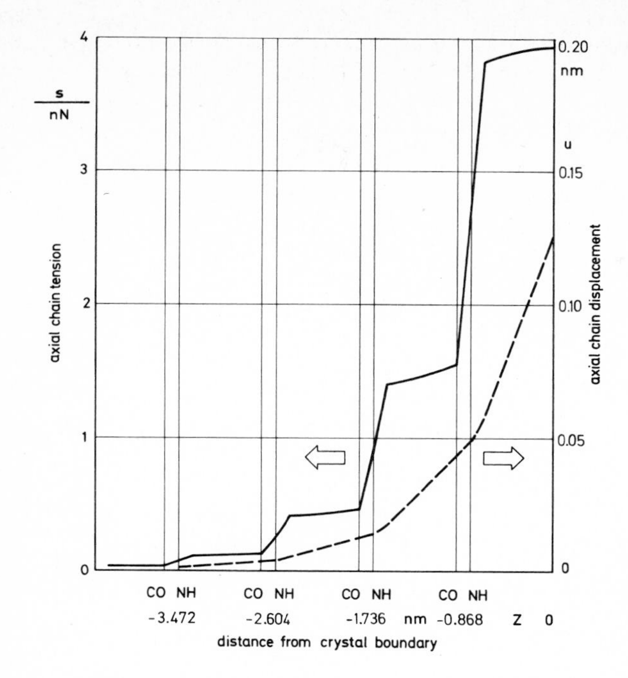

Fig. 5.7. Decrease of tension and displacement of highly stressed PA 6 tie molecules within folded chain crystal lamellae (after [22]).

bors – and so on. Meanwhile, Gibson et al. [48] have analyzed this problem using shear lag theory and finite elements. They find that the factors influencing stress transfer between the ties and the surrounding crystal are the following:

– The ratio of the crystal shear and tensile moduli (G/E)
– The ratio of the tie radius (r_0) to the lamellar crystal thickness (ℓ)
– The proportion of ties spanning the interface (p).

Their calculations for PE crystals show a *lateral* stress-decay half width of the order of the tie radius (~0.2 nm) whereas the axial chain stress decays to one half within about 3 nm. Gibson et al. [48] show that in view of the rather rapid lateral stress decay the Kausch approach (Eq. 5.28), shear lag theory, and finite element analysis of the three-dimensional problem practically coincide. They also note that in view of the chain displacement within the crystal the *effective stiffness* of ties will be only about 0.25–0.7 of the crystal stiffness in most cases.

Crystal defects have been intensively studied for a long time by Reneker et al. He introduced in 1962 the point dislocations created by "double occupancy" of a lattice site [49]. In a recent publication, the structure and energy of other crystallographic defects such as dispirations, disclinations and chain twist are treated [50]. The total excess energy of a dislocation in a PE crystal is given as 27.7 kcal/mol^{-1} of which 14.2 kcal/mol^{-1} are *intra*molecular excess energy.

C. Thermally Activated Displacements of Chains Against Crystal Lattices

As stated repeatedly crystal and chain are in thermal motion; the chain atoms, therefore, vibrate around the new equilibrium positions z + u given e.g. by Eq. (5.28).

The effect of the correlated or uncorrelated motion of the chain atoms on the intermolecular potential energy can be calculated if the dispersion of vibrational amplitudes along the chain is known [22]. A chain subjected to maximum tension s_0 is not resistant to translation. Even the smallest thermal vibration will cause further displacement of the chain and a decrease in axial tension at the boundary. The point of maximum tension will rapidly propagate into the crystallite (Fig. 5.8). The "dis-

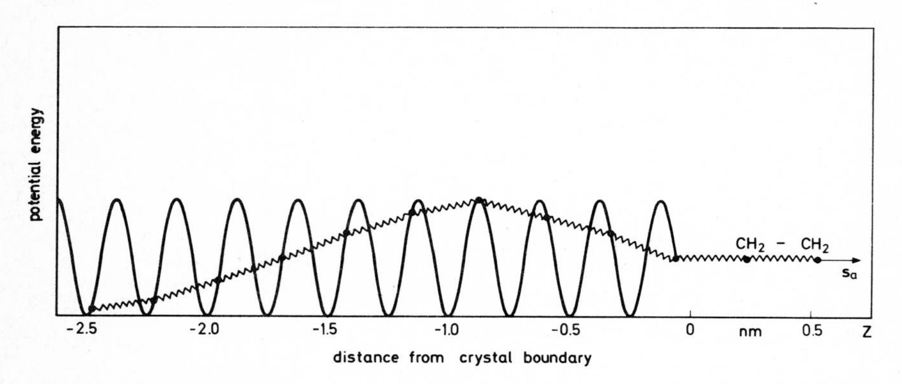

Fig. 5.8. Schematic representation of a stress-induced defect. The point of maximum chain stress travels *into* the crystal as the leading chain group is pulled *out* of the crystal (after [22]).

location" is fully developed once the displacement at the crystal boundary, u_a, has reached the value of the chain repeat unit length d. The energy W_d of the defect introduced into the crystal by displacement of the atoms of one chain is easily calculated as the sum of the potential energies of the chain atoms and of the elastic chain energy. Both are a function of external tension. In Figure 5.9 the dislocation energy has been plotted as a function of displacement u_a and tension s_a at the crystal boundary. The energy of a completed defect (displacement of the chain by d) in PE amounts to 32.85 kcal/mol. This value is in very good agreement with the 27.7 kcal/mol calculated by Reneker et al. [50]. At tensions smaller than s_0 the displacements u_a are smaller than d/2, the defect energy is smaller than $W_d(s_0)$, and the defect has a stable position. At displacements larger than d/2 the system becomes unstable since the chain tension decreases with increasing displacement u_a.

As shown in Figure 5.9, transitions between the stable and the unstable state can be caused by thermal activation. The required activation energy U_d is given by

$$U_d(s_a) = 2\,[W_d(s_0) - W_d(s_a)]. \qquad (5.30)$$

In Table 5.4 the activation energies required for translation of CH_2-CH_2 segments are listed for various displacements u_a and tensions s_a. One observes that at chain tensions which are as little as 4% less than the maximum tension still an activation energy for chain translation of 9.8 kcal/mol is required. One may safely

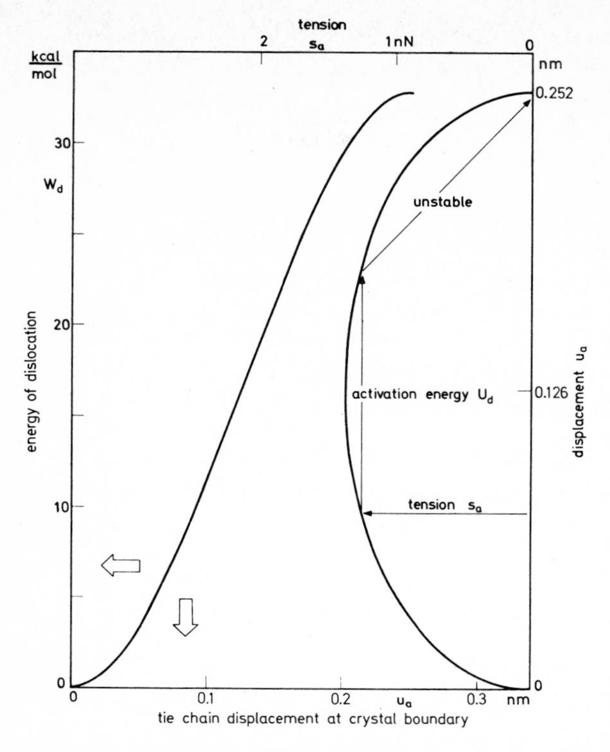

Fig. 5.9. Energy W_d of a chain displacement defect in a polyethylene crystal as function of tie-chain displacement u_a at crystal boundary (after [22]).

Table 5.4. Activation Energies for Chain Translation in PE

u_a Å	s_a 10^{-10} N	U_d 10^{-20} J	U_d kcal/mol
0	0	23.08	33.0
0.104	1.79	22.89	32.7
0.195	3.38	22.38	32.0
0.303	5.18	21.40	30.6
0.412	6.91	20.00	28.6
0.494	8.12	18.69	26.6
0.589	9.41	16.90	24.2
0.699	10.71	14.60	20.8
0.824	11.93	11.68	16.6
0.914	12.63	9.39	13.40
1.009	13.18	6.88	9.82
1.108	13.56	4.19	5.98
1.158	13.67	2.80	4.12
1.209	13.72	1.40	2.06
1.260	13.72	0	0

state, therefore, that even under consideration of longitudinal chain vibrations chain translation will occur at chain tensions close to the specified values.

D. Chain Displacements Against Randomly Distributed Forces

The rupture of chains in amorphous matrices or in solution clearly indicates that large axial forces can also be reached in the absence of periodic potentials. Again forces derive from the displacement of chains or of parts thereof with respect to the surrounding matrix. If one divides a non-extended chain into small sections of identical projected lengths one will obtain sections of quite different conformation. As well the sectional chain moduli E_i as – most probably – the interaction potentials $V_i = -w_i \cdot u_i^2/2$ will vary from section to section. To obtain the axial displacement of an arbitrary sequence of chain sections an interpolation given by Kausch and Langbein [21] may be used who have found that the changes of chain stress σ and displacement u within a heterogeneous matrix are interpolated by an exponential law:

$$u = u_0 \exp \gamma z/L \tag{5.31}$$

$$\text{with } \gamma = \sqrt{n \sum_{i=1}^{n} (w_i L_i/E_i\, q)}. \tag{5.32}$$

The chain stress at Z = 0 is derived from Eq. (5.24) replacing the chain elasticity constant κ by the elastic constant of a series of elastic elements

$$\sigma = u_0\, \gamma/\Sigma(L_i/E_i). \tag{5.33}$$

The above equation shows quantitatively that large chain stresses will only be obtained if the interaction force constants w_i are of about the same order they are in a crystal and if the elastic moduli of the chain sections are large throughout. One "weak" section will greatly enhance average displacement and reduce the stress. It should also be mentioned that it is not the chain length L and the absolute number n of interaction sites which raise the chain stress but the intensity of the interaction forces per unit chain length.

As an example of the static excitation of a chain under these conditions a nearly extended chain of end-to-end distance L embedded in a glassy matrix may be considered. If the matrix (index m) and the chain (no index) were homogeneously strained in chain axis direction they would experience the following stresses:

$$\sigma_m = \epsilon_m E_m \tag{5.34}$$

and

$$\sigma = \epsilon_m L/\Sigma L_i/E_i. \tag{5.35}$$

Due to the different elastic response the free chain would have to act against the lattice potential until chain stress and lattice interaction were in equilibrium at any point of the chain. This interaction can be described by identically the same mathematical formalism developed above. Since the chain ends are assumed to be free they cannot bear any axial stresses. This condition will exactly be met if one superimposes the hypothetical stress σ (Eq. 5.35) by a compressive stress according to Eq. (5.33) where

$$u_0 = \epsilon_m L/\gamma. \tag{5.36}$$

The largest stresses will be found in the center section of the strained chain and here the superposition yields:

$$\sigma(L/2) = \frac{\epsilon_m L}{\Sigma L_i/E_i} [1 - \exp(-\gamma/2)]. \tag{5.37}$$

The distribution of axial stresses along the chain is characterized by the two stress-free chain ends, two terminal sections of length L/γ where the stress increases with a rate determined by γ/L, and a central chain segment subjected to the maximum stress given by Eq. (5.37). The *continuous* straining of a chain segment through interaction with (harmonic) lattice potentials – as required for the terminal sections – is limited to a section of finite length because the largest displacement u may not exceed the range of validity of the intermolecular potential which is of the order of 0.1 nm. For breaking stresses of 20 GN/m^2 to be reached under such a condition the constant γ/L must be larger than 0.7 nm^{-1} *and* the average chain modulus at these loads must be that of a fully extended chain; since the displacements of PE and 6 PA chains within their rigid crystal lamellae led to γ/L-values of 0.4 nm^{-1} and 1.55 nm^{-1} respectively [22] one is forced to the conclusion that the displacements of extended chains within the less tightly packed random aggregates of misaligned chains can only lead to smaller than the above values of γ/L. If chain scission were to occur under the above static conditions it would require a sufficiently stable physical crosslinking of the (then non-extended) terminal chain sections to the matrix through folds, kinks or entanglements.

E. Dynamic loading of a chain

So far the quasi-static interaction between a chain and a surrounding matrix has been investigated. It has been indicated that the level of axial forces to be obtained in that manner is limited by the occurrence of chain slip. In dynamic loading these force levels can be exceeded if frictional forces or forces of inertia become effective.

If the chain displacement does not occur in static equilibrium and not through a single thermally excited step then the translation will not occur reversibly along a path of least energy but will require more energy than in the previous cases. The rate-dependent energy expended per unit distance of forced translation of a chain segment i is equivalent to a frictional shearing force τ_i. The response of chains to shearing forces in solution has extensively been investigated and discussed [25]. A number of different molecular theories of the viscoelastic behavior of polymer chains in solution have been developed. These predict a relation between e.g. molecular weight M (or degree of polymerisation P), solvent viscosity η_s, intrinsic viscosity $[\eta] = \lim (\eta - \eta_s)/c\eta_s$, molecular friction coefficient ζ_0, and mean square end-to-end distance $\overline{r_0^2}$. Following the bead spring model of Rouse [25] one obtains

$$[\eta]\, \eta_s = N_L\, \overline{r_0^2}\, \zeta_0/36\, M_0 . \tag{5.38}$$

The quantity ζ_0 represents the frictional force encountered by each monomeric unit within the chain which is moved with unit velocity through its surrounding. The sectional force f_i, therefore, is:

$$f_i = \zeta_0 \cdot v_i = \zeta_0\, (\dot{u}_{im} - \dot{u}_{ic}). \tag{5.39}$$

In the case of additivity of all sectional forces and in a homogeneous elongation strain field one obtains the following distribution of axial stress ψ within a chain of length L:

$$\psi = E du_c/dz = \frac{\zeta_0\, \dot{\epsilon}\, \rho N_L}{M_0} \left(\frac{L^2}{8} - \frac{z^2}{2} \right) . \tag{5.40}$$

The largest stresses are encountered in the middle (z = 0) and amount to

$$\psi_{max} = \zeta_0\, \dot{\epsilon}\, M^2\, L_{mon}/8\, M_0^2 . \tag{5.41}$$

Equation (5.41) predicts that for a given system (ζ_0, L_{mon}, M_0) axial stresses are proportional to strain rate and square of molecular weight. Chain rupture occurs if ψ_{max} reaches or exceeds the chain strength ψ_b resulting in chain halving.

In a series of intriguing experiments Keller et al. [51–54] have recently studied the conformation of flexible molecular coils in solution subjected to elongational flow. Observing the birefringence of the volume element in the center of a cross-slot device they showed for the first time that as a function of the rate $\dot{\epsilon}$ of elongational strain there is a rapid transition from a coiled conformation to a fully extended one at a critical strain rate $\dot{\epsilon}_c$ (coil-stretch transition). For an atactic polystyrene of $M_w = 4 \cdot 10^6$ in decalin the $\dot{\epsilon}_c$ is of the order of $10^4\ s^{-1}$. For this material the authors verified that the critical strain rate descreased with polymer molecular weight M according to:

$$\dot{\epsilon}_c \sim M^{-1.5} .$$

(The exponent of 1.5 is expected from the theory of Zimm for a non-free draining coil [53]). Lower solvent viscosities would increase the critical strain rate.

If the strain rate of the solution is increased beyond $\dot{\epsilon}_c$ the fully extended chains become *energy*-elastically loaded, and they break at a second critical value, $\dot{\epsilon}_f$, of strain rate (for the above PS at about $1.8 \cdot 10^4\ s^{-1}$). From Eq. 5.41 it can be anticipated that

$$\dot{\epsilon}_f \sim M^{-2}$$

and precisely this result was obtained by Keller et al. [52, 53] for PS in decalin. By molecular weight analysis after accumulation of many chain scission events the authors were also able to state that under these circumstances chain scission occurs in the center: chains are being halved. They estimate [53] that this occurs at axial forces of

$$F_f = 6 \cdot 10^{-9} \text{ N/chain},$$

the strength of a C–C bond in the PS main chain as calculated from a Morse potential function; since the calculations refer to the static strength (at absolute zero) the value of F_f is somewhat higher than the 4 nN given later for dynamic chain scission. Comparing PEO and PS chains of equal length at equivalent viscosities Keller found that $\dot{\epsilon}_f$ (PEO) is 3 times that for PS, the smoother PEO-chain geometry makes the molecule more "slippery" [54] (see also [71] for the dynamics of stretched chains).

In a similar attempt to identify the origin of drag reduction Merrill et al. [55] have studied the flow of dilute solutions through a sudden contraction. They first advanced the interesting hypothesis of intramolecular entanglements in order to explain apparently *random scission* along PS chains [55a]. In a later paper [55b] they confirm, however, *chain halving*; they point out that the width of the lower molecular weight peak in the ensuing distribution will be widened because of the finite probabilities for scission of the non-midpoint bonds. In turbulent flows chains break after they have been extended and aligned with the flow (see also Nguyen and Kausch [72]).

In the above cases of a coil in a strained solution it has to be retained that the coil-stretch transition is relatively sharp, no partly extended chains corresponding to the Langevin region are observed. This is certainly due to the fact that *intra*molecular rotation is achieved in a time which is short compared to $\tau = 1/\dot{\epsilon}_c$. (This does not mean, however, that relaxation from the extended state is equally fast. In fact, Keller and Odell [52] found that chain contraction against the frictional contact of the surrounding liquid is about 50 times slower than chain extension.)

If a solution is circulated repeatedly through the degrading device at a given strain rate chain degradation will stop after all chains longer than the limiting value $M(\infty)$ have been broken:

$$M(\infty) = (8M_0{}^2 \psi_b / \xi_0 \dot{\epsilon} L_{mon})^{1/2}.$$

The Eqs. (5.41), (5.38), and (5.6) together predict that $[\eta]\, \dot{\epsilon}\, M(\infty)$ is constant for a given polymer/solvent system. Kadim [26] obtained for a 0.5 to 1% solution of PIB in decalin $[\eta]\, \tau\, M(\infty) = 1.34$ kJ/mol. He employed shear rates between 25 and 700 s^{-1}. Abdel-Alim et al. [27] report on a relation of $\tau^{0.41}\, M(\infty) = \text{const.}$ for a 2% solution of polyacrylamide in water subjected to shear rates of between 3000 and 100000 s^{-1}.

Within more concentrated solutions the physical-chemical behavior of long chains (size and interpenetration of molecular coils, kinetics of entanglement formation) seems to be much more important than the rheological behavior characterized by ζ_0 and $\dot{\epsilon}$. A striking demonstration of this fact was given by Breitenbach, Rigler and Wolf [28], who prepared solutions of 3.6 to 14.2% by weight of polystyrene in cyclohexane. These systems showed a concentration-dependent phase separation at a temperature T_{tr} between 26.4 and 29.4 °C. Shearing these solutions at a rate of 600 s^{-1} and at temperatures slightly above T_{tr} they observed a dramatic increase in the rate of degradation when approaching T_{tr}. At T_{tr} + 11.6 K no noticeable degradation occurred within 20 h. At T_r + 0.6 K a 13% decrease of the limiting viscosity [η] was measured after one hour already. Within 20 h a reduction of the molecular weight from $\simeq 7 \cdot 10^5$ g/mol to $1.6 \cdot 10^5$ could be achieved.

A related subject of growing interest is the ultrasonic degradation of polymers in solution. In their recent review Basedow and Ebert [29] summarize that ultrasonic chain scission is the result of cavitation, i.e. of the nucleation, growth and collapse of (gas-filled) bubbles. Chain scission occurs in the convergent flow fields in the vicinity of collapsing bubbles and through the ensuing shock waves. The properties of the solvent do not seem to be of particular importance, but the presence of nuclei to initiate cavitation and a minimum intensity of the ultrasonic field (environ 4 W/cm^2) are necessary conditions. Owing to the nature of the degradation process, i.e. to the *shear loading* of chains, the rates of degradation decrease with decreasing molecular weight (degree of polymerization) and become infinitely small at a so-called *limiting molecular weight.* For polystyrene in tetrahydrofuran degraded 88 h at 20 kHz Basedow and Ebert cite one limiting molecular weight of 24000 g/mol and a newer value of 15000 g/mol. In a quite recent publication Sheth et al. [39] report for the same system the appearance of appreciable amounts of polymer material at molecular weight of 1000 g/mol or lower in the gel permeation curve of the degraded sample. They indicate that the limiting degree of polymerization depends on the initial molecular weight distribution and is either much lower than previously reported or perhaps non-existent.

The stresses encountered by a chain displaced with respect to a *solid* matrix can also be described utilizing the concept of the monomeric friction coefficient ζ_0 [25]. The implications of such an assumption are discussed in detail by Ferry [25] who also lists the numerical values of the monomeric friction coefficients for many high polymers. Naturally the coefficients strongly depend on temperature. But even if compared at corresponding temperature, e.g. at the individual glass transition temperatures, the monomeric friction coefficients vary with the physical and chemical chain structure by 10 orders of magnitude. In the upper region one finds the 1740 Ns/m for Polymethylacrylate, the 19.5 Ns/m for Polyvinyl acetate, and the 11.2 Ns/m for Polyvinyl chloride, each at the respective glass transition temperature [25]. This means that a PVC segment pulled at 80 °C through a PVC matrix at a rate of 0.005 nm/s encounters a shearing force of 0.056 nN per monomeric unit. At lower temperatures the molecular friction coefficient increases essentially in proportion to the intensity of the relaxation time spectrum H(τ), the increase being about one to two orders of magnitude with every 10 K temperature difference [25]. If a PVC segment has a length of 2 L = 10 nm and is part of a PVC matrix strained at a rate of 0.001 / s at the glass transition temperature it experiences at the center an axial force of magnitude 1.1 nN. This force is about as large as that statically transferred upon extended PE chains by PE crystallites at room temperature. It may be repeated that the above estimate on the level of axial forces is based on the assumption that in sample straining an extended chain segment is subjected to shearing forces determined by the monomeric friction coefficients ζ_0 and that the segment does not relieve axial stresses through conformational changes. Using the same assumptions one expects forces up to 206 nN for PMMA and up to 0.01 nN for PS each at the respective glass transition temperature and a strain rate of 0.001/s.

This estimate shows that in dynamic straining of glassy polymers very large axial forces can be transferred onto extended chain segments if the straining occurs at a temperature not higher than the glass transition temperature (PMMA) or some 10 K (PVC) to 20 K (PS) lower than that. These forces are sufficient to cause chain breakage.

A loading mechanism not to be investigated in detail in the context of this monograph is the straining of chain molecules through forces of inertia, e.g. through propagating stress waves. Brittle thermoplastic materials (PS, SAN, PMMA) behave "classically" at velocities of uniaxial straining below 3 m/s or strain rates below 50 s^{-1} [30]. In this region increase the strength properties and decrease the elongation with increasing strain rate. At strain rates between 50 and 66 s^{-1} a transition to stress wave-initiated fracture was observed which was accompanied by a tenfold reduction of the apparent load-carrying capability [30]. Skelton et al. [40] studied PA 6, PETP, and aromatic polyamide (Nomex). At ambient temperatures and in the range of loading rates between 0.01 and 140 s^{-1} these fibers also behaved classically. At −67 °C and −196 °C a decrease in strength with loading rate was observed at a loading rate of 30 s^{-1}.

III. Deexcitation of Chains

The mechanisms of deexcitation of stressed chains have been discussed en passant while investigating bond strength and chain loading. These mechanisms are chain slip with respect to a surrounding matrix (enthalpy relaxation), change of chain conformation (entropy relaxation), or chain rupture.

The retraction of a stressed chain through slippage leads to an exponential stress relaxation

$$\psi = \psi_0 \exp(-t/\tau). \tag{5.42}$$

If the slippage is adequately described by the monomeric friction coefficient ζ_0 the relaxation time τ will be given by:

$$\tau = \zeta_0 L_{mon}/E\, q \tag{5.43}$$

where q is the chain cross-section. For a PVC chain at the glass transition temperature one obtains a relaxation time of 0.05 s.

The effect of a change of conformation on the elastic chain energy had been demonstrated in Figure 5.1. The annihilation of 4 gauche conformations within a PE segment of 5 nm length corresponds to an increase in length of the segment of

0.25 nm. This increase of 5% will reduce the axial elastic forces acting on the chain by 0.05 E, i.e. by $\simeq$ 10 GN/m^2. In static straining this will correspond to complete unloading since the largest elastically transmitted forces yielded axial stresses of only up to 7.5 GN/m^2. The stress-biased change of chain conformation will lead, therefore, to a considerable reduction of axial stresses. The rate of conformational changes – even of high T_g polymers and in a stress-free state – is sufficiently large and it will be further increased under stress [19, 31].

Conformational changes under stress have been observed repeatedly (e.g. 41–45). Gafurov and Nowak [41] studied the IR absorption of PE fibers in the range of 1200 to 1400 cm^{-1}. The extinctions of four bands (at 1370, 1350, 1305, and 1270 cm^{-1}) observed there and related to the (wagging) vibrations of methylene groups in gauche conformations decreased with strain. The authors estimated that at 4% sample strain (which corresponded to much larger amorphous strains) the number of gauche isomers in the amorphous regions was decreased by 20%.

Zhurkov et al. [42] derived the degree of chain orientation from the dichroism of "deformed-IR bands" of PETP (cf. Chapter 8 I B and Fig. 8.5); they observed that for highly stressed segments an initial orientation with $\overline{\cos^2\theta} = 0.75$ had turned into practically complete orientation ($\overline{\cos^2\theta} \simeq 1.0$). The problems posed by an application of this method are discussed in detail by Read et al. [43]. Bouriot [44] deduced from IR measurements a reversible cis-trans transformation of the ethylene glycol segment in PETP fibers under the effect of tensile stresses. In polyamide 66 he observed a reversible increase of free (non-associated) NH groups and an increase in the average chain orientation with chain tension.

Conformational changes due to straining of highly uniaxially oriented PETP films have been reported by Sikka et al. [45]. These changes have been inferred from a comparison of the IR spectrum of unloaded and loaded PETP films. Employing a highly sensitive Fourier Transform IR spectrometer Sikka noticed frequency shifts and splittings of the absorption bands in the subtraction spectra (stressed PETP IR spectrum minus unstressed PETP IR spectrum). At a stress of the order of 20% of the breaking stress those IR bands were affected which are assigned to the gauche conformations of the C–O bond next to the C–C bond. These conformations are most likely to exist in the amorphous regions of the highly uniaxially oriented semi-crystalline polymer. In recent years numerous FTIR studies have amply confirmed stress-induced gauche-trans transitions in PETP and PBTP [64–68].

The third mechanism of releasing axial stresses is the breakage of a chain through homolytic bond scission, ionic decomposition or degrading chemical reactions.

It is assumed that chain scission is a thermo-mechanically activated process. A chain segment containing n_c weakest bonds of energy U_0, and carrying at its ends a constant stress ψ_0 will, on the average, break after time τ_c. One refers to ψ_c as the *time-dependent strength* of a chain *segment:*

$$\psi_c\,(U_0, n_c) = (U_0 - RT\,\ln\omega_0 n_c\tau_c)/\beta. \tag{5.44}$$

The quantities ω_0 and β have been defined as bond vibration frequency and activation volume for bond scission respectively [32]. A 5 nm PA 6 segment, for instance, at room temperature, has an instantaneous strength of 20 GN/m^2 (taking U_0 = 188 kJ/mol, $\omega_0 = 10^{12}$ s^{-1}, n_c = 12, τ_c = 4 s, $\beta = 5.53 \cdot 10^{-6}$ m^3/mol). If in straining a segment, a stress of 20 GN/m^2 is maintained for a period of 4 s the segment has a probability of failure of $1 - 1/e = 63\%$. A stress of 20 GN/m^2 is equivalent to a force of ~ 4 nN acting on a chain in the direction of its axis. For PE Boudreaux [11] has calculated breaking forces of 3.3. nN/chain. In view of the different interpretations found in the literature concerning the significance of ω_0, β, and even U_0 the assumptions made in deriving Eq. (5.44) will briefly be discussed.

A chain in thermal contact with its surrounding can be represented by a system of coupled oscillators. The degree of excitation of individual oscillators (modes of vibration) is statistically changing. In the absence of external mechanical forces chain scission at a C–C bond will occur whenever an oscillator representing a C–C stretch vibration becomes excited beyond a critical level, U_0, the "strength" of the C–C bond. In Figure 4.1 the potential energy scheme was given illustrating different states of vibrational energy, the bond strength U_0, and the dissociation energy D. One is interested in the rate of dissociation events of oscillators activated beyond U_0.

To evaluate this problem it will be assumed that Boltzmann statistics can be applied to the system of, say, s oscillators representing the C–C stretch vibrations within a volume element of the polymer. The fraction of energy states in which this system s will have a total energy level $E_r = r\epsilon$ is proportional to the degeneracy g_r of the energy level E_r and the Boltzmann factor $\exp(-E_r/RT)$. (ϵ is the – constant – spacing between oscillator energy levels). The partition function, therefore, will be given by:

$$Q_t = \sum_r g_r \exp(-E_r/RT). \tag{5.45}$$

The number of states in which at least one oscillator has an energy $E_i \geqslant U_0$ is given by Q_a:

$$Q_a = \sum_r g'_r \exp[-(E_r + U_0)/RT]. \tag{5.46}$$

For thermodynamic equilibrium within the volume element comprising s oscillators the fraction K^+ of activated states, i.e. states with $E_i \geqslant U_0$, is given by:

$$K^+ = Q_a/Q_t. \tag{5.47}$$

This can be approximated [33] if $r \gg 1$ by:

$$\frac{Q_a}{Q_t} = \frac{g'_m}{g_m} \exp(-U_0/RT) \tag{5.48}$$

where g_m is the degeneracy of the most probably energy level. Expressing g_m by the entropy S according to:

$$S = R \ln g_m \tag{5.49}$$

one obtains

$$K^+ = \exp(-U_0/RT) \exp \Delta S/R \tag{5.50}$$

which is identical with the relation $\ell n\, K^+ = -\Delta F^+/RT$. The rate of dissociation, k_b, then follows from K^+ [34]:

$$k_b = \kappa \frac{kT}{h} K^+ = \kappa \frac{kT}{h} \exp(-U_0/RT) \exp \Delta S/R \tag{5.51}$$

where κ is the transmission constant which indicates how many of the activated complexes actually disintegrate. By combining all front factors and $\exp \Delta S/R$ one may rewrite this equation as:

$$k_b = \omega_0 \exp(-U_0/RT). \tag{5.52}$$

The quantity ω_0 is proportional to T; this dependency is generally negligible against the T-dependency caused by the exponential term.

The rate k_c for the breakage of one out of n_c equal and independent bonds is n_c times the rate k_b for the breakage of an individual bond. If the bonds are not independent it will mean that more than one oscillator has to acquire enough vibrational energy for dissociation. Depending on the coupling such a *cooperative region* may count some 5 or more oscillators [35]. In this case an experimentally determined term U_0 stands for the total vibrational energy required by all (coupled) atoms which are involved in one chain rupture event. This could mean that the potential barrier between neighboring sites of an uncoupled atom is much less than U_0. This meaning of U_0 has to be kept in mind if the numerical value U_0 is being compared with other activation energies.

Another problem in connection with U_0 arises from the fact that in a kinetic experiment such as the determination of the rate of chain scission one derives the bond strength U_0, that is the maximum of the potential barrier (cf. Fig. 4.1). In calorimetric measurements the reaction enthalpy (dissociation energy D) is determined.

By application of external forces the free energy of the activated state is changed with respect to the ground state. One may say that ΔF^+ has to be replaced by $\Delta F^+ + \Delta F_{ext}$. The term ΔF_{ext} will be some function f of the stress ψ which is acting on the molecular chain along its axis; Equation (5.52) then becomes:

$$k_c = n_c\, \omega_0 \exp - \frac{U_0 + f(\psi)}{RT}. \tag{5.53}$$

The form of f is subject to discussion [35–37]. Linear and quadratic relationships have been proposed. A linear relationship follows if one assumes that the force acting on a particular chain is not changed by the displacement of the chain under action of the force (comparable to a dead weight loading). The correction term $f(\psi)$ will be quadratic in ψ if the stored elastic energy within a small chain section to either side of the point of chain scission is thought to be responsible for the rate of chain fracture. From the Morse potential one obtains a difference of square root expressions which, however, results in an almost linear relationship between force and activation energy in the force region of interest [36]. The effect of a *linear* stress potential on the intramolecular binding energy is shown in Figure 5.10.

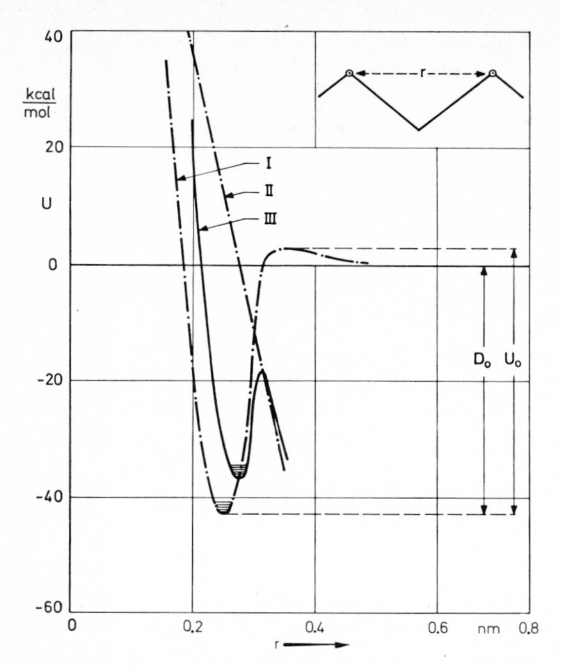

Fig. 5.10. Effect of axial stress on intramolecular binding potential (from [19]). I unstressed chain, II stress potential (20 GN m^{-2}), III resulting intramolecular binding potential of a stressed C–C-bond

Some information concerning the form of $f(\psi)$ and the relation between molecular stress ψ and macroscopic stress σ has been obtained by Zhurkov, Vettegren et al. [37, 32]. These authors have studied the effect of macroscopic stress σ on infrared absorption of oriented polymers. The frequencies of skeleton vibrations of polymer molecules show the effect of external axial forces on molecular binding force-constants. The shift and change of shape of suitable infrared absorption bands should reveal, therefore, the intensity and distribution of actual molecular stresses ψ. Experiments carried out with polypropylene and polyamide 6 have shown that for a stressed sample the maximum of the investigated absorption band decreases slightly and shifts towards lower frequencies [37, 32]. In addition a low frequency tail of the absorption band appears. In agreement with existing structural models the authors interpreted their results in assigning the shifted symmetrical part of the absorption band to the crystalline phase of the sample, the low frequency tail to the amorphous phase. The shift of the symmetrical part of the absorption band, $\Delta \nu_s$, turned out to be linear with macroscopic stress in all performed experiments:

$$\Delta \nu_s = \nu_0 - \nu_\sigma = \alpha \, \sigma. \qquad (5.54)$$

Following a theory of Bernstein [38] the shift of the stretching frequency of a bond can be related to a shift in dissociation energy D of the same bond:

$$\nu_0^2 - \nu_\sigma^2 = c\,(D_0 - D_\sigma). \qquad (5.55)$$

Since $\Delta \nu_s \ll \nu_0$ one may substitute $\nu_0^2 - \nu_\sigma^2$ by $\Delta \nu_s \cdot 2\nu_0$. A consequence of this condition is that also $(\nu_0^2 - \nu_\sigma^2)$ is proportional to σ. The choice of ν or ν^2 in Eq. (5.54) only affects α, not the power of σ. If one further assumes that the activation energy U_0 is changed by a similar amount as D_0 one obtains:

$$U_\sigma = U_0 - \frac{2\nu_0}{c} \alpha \sigma. \qquad (5.56)$$

In order to derive the form of $f(\psi)$ the relation between σ and ψ for oriented crystalline phases has to be established. According to experimental and theoretical evidence it is assumed that a condition of homogeneous stress exists within the crystalline layers of a regular sandwich structure and that ψ is proportional to σ. Equation (5.56) thus yields that the activation energy for scission of an oriented chain molecule is linearly decreased by stresses acting upon this molecule. Equation (5.53) with (5.56) may be written as:

$$k_c = n_c \omega_0 \exp [(-U_0 + \beta \psi)/RT] \qquad (5.57)$$

which is equivalent to Eq. (5.44). It is believed that the rate of fracture of chains *all subjected to the same local stress* ψ can correctly be described by Eq. (5.57).

There are other treatments of coupled anharmonic oscillators [56, 57]. It seems, however, as if the boundary conditions chosen in the first model [56] are incompatible with the situation of overstressed chains. The skeleton atoms within such chains are in rapid vibration and it has to be born in mind that *mechanical stresses* and *thermal motion* contribute to the bond scission energy. For PA6 at room temperature an activation energy of 188 kJ/mol^{-1} is observed to which the term $\beta\psi$ contributes 111 kJ/mol^{-1} (see also Chapter 7). The remaining 77 kJ/mol^{-1} are coming from the fluctuating thermal energy (this is more than 60 times the *average* thermal energy per degree of freedom at that temperature). Just before scission of a PA6 segment at a stress of 20 GN/m^{-2} the segment, which has an axial stiffness of 200 GN^{m-2}, is evidently strained by 10%. An instant later *one bond* is subject to a concentration of fluctuating thermal energy. That bond deforms by notable amounts (in PE for instance by the 35% bond fracture strain indicated by Boudreaux [11]). As discussed in detail by Kausch [22] and Melker [57] the yielding of one bond is accompanied by an in-phase motion of the adjacent bonds in outward direction. Thus, it may be retained that in a stretched PA6 segment in the instant of rupture all main chain bonds except the failing one are equally strained by ~ 10%; *segment* strains of ~ 35% will not be supported.

On the basis of the foregoing considerations the response of two chains of different conformation to axial stresses has been evaluated in Figure 5.11. An extended chain segment ($L = L_0$) stretches elastically up to the time-dependent fracture stress. The calculations have been carried out employing a modulus of elasticity for the polyamide 6 chain of 200 GN/m^2. As indicated in Table 5.3 it is still open to discussion whether this value can be unambiguously accepted. The times to fracture have been calculated from Eq. (5.57) for a segment comprising 12 bonds of low strength (U_0 = 188 kJ/mol). The partially extended chain ($L = 1.1\ L_0$) contains initially four kinks (8 gauche bonds). It responds in region A (Fig. 5.11) mostly by the hindered rotation of its gauche bonds, which, in this region, accounts for 90% of the axial deformation. The resulting combined modulus amounts to 22 GN/m^2. If geometrically possible a transition to the fully extended conformation will occur in region B. Region C represents the energy elastic straining of the then extended chain segment of length 1.1 L_0.

The degradation of chains through ionic decomposition or chemical reactions under stress will be discussed where appropriate in the following chapters.

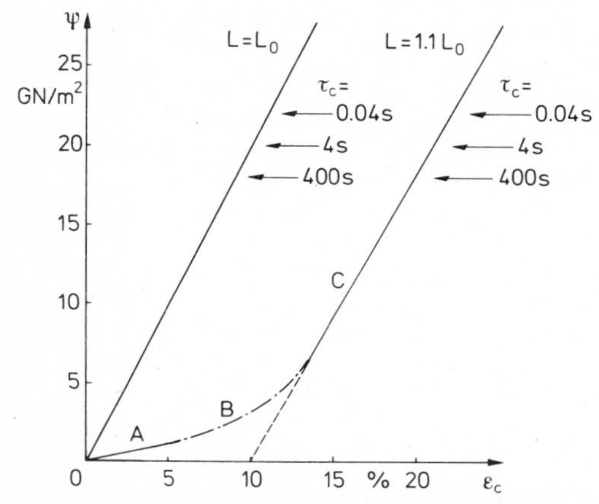

Fig. 5.11. Stress-strain diagram of polyamide 6 chain segments. L = contour length, 5.2 nm, 12 weak bonds, L_0 = distance between segment ends in the unstressed state, $\tau_c = 1/k_c$ = average lifetime according to Eq. (5.57), A = response due to hindered rotation of gauche bonds, B = intermediate region, C = energy-elastic response of extended chain.

References for Chapter 5

1. P. J. Flory: Principles of Polymer Chemistry, Ithaca, New York: Cornell University Press 1953.
2. L. R. G. Treloar: The Physics of Rubber Elasticity, 2nd Ed., London: Oxford University Press 1958, a) p. 61, b) p. 103ff, c) p. 108.
3. H. M. James, E. Guth: J. Chem. Phys. *11*, 455 (1943).
4. A. Abe, R. L. Jernigan, P. J. Flory: J. Am. Chem. Soc. *88*, 631 (1966).
5. J. P. Flory, A. D. Williams: J. Polymer Sci. A-2, *5*, 399 (1967).
6. P. R. Saunders: J. Polymer Sci. A-2, *2*, 3765 (1964).
7. W. Pechhold: Kolloid-Z. Z. Polymere *228*, 1 (1968).
8. H. G. Zachmann: Kolloid-Z. Z. Polymere *231*, 504 (1969).
9. J. A. Sauer, A. E. Woodward: Polymer Thermal Analysis, II., P. E. Slade, Jr., and L. T. Jenkins, Eds., New York: Dekker, 1970, p. 107.
10. P. B. Bowden: Polymer *9*, 449–454 (1968).
11. D. S. Boudreaux: J. Polymer Sci.: Polymer Phys. Ed. *11*, 1285 (1973).
12. W. J. Dulmage, L. E. Contois: J. Polymer Sci. *28*, 275 (1958).
13. T. R. Manley, C. G. Martin: Polymer *14*, 491–496 (1973) and ibid. 632–638 (1973).

14. H. Hahn, D. Richter: Colloid + Polymer Sci. *255*, 111–119 (1977).
15. E.g. J. Meixner: Handbuch der Physik, Vol. III/2, S. Flügge, Ed., Berlin: Springer-Verlag 1959, p. 413.
16. D. Langbein: Theory of Van der Waal's Attraction, Springer Tracts in Modern Physics, Vol. 72, Heidelberg: Springer-Verlag, 1974.
17. P. Lindenmeyer: Mechanical Behavior of Materials, Kyoto, Japan: Soc. Mat. Sci. 1972, Special Vol., p. 74.
18. H. Scherr: Dissertation Universität Ulm, Juli 1973.
19. H. H. Kausch: J. Polymer Sci. *C 32* (Polymer Symposia), 1 (1971).
20. A. D. Chevychelov: Polymer Science USSR *8*, 49 (1966).
21. H. H. Kausch, D. Langbein: J. Polymer Sci., Polymer Physics, Ed. *11*, 1201–1218 (1973).
22. H. H. Kausch, J. Becht: Deformation and Fracture of High Polymers, H. H. Kausch et al., Eds., New York: Plenum Press 1973, p. 317.
23. H. Baur, B. Wunderlich: Fortschr. Hochpolymeren Forsch. *7*, 388 (1966).
24. E. W. Fischer, G. Hinrichsen: Kolloid-Z. Z. Polymere *213*, 28 (1966).
25. J. D. Ferry: Viscoelastic Properties of Polymers, 2nd Edition, New York: J. Wiley and Sons, Inc. 1970.
26. A. Kadim: Thesis, Dept. of Chem. Engng., Technion, Israel Institute of Technology, Haifa, Israel 1968.
27. A. H. Abdel-Alim, A. E. Hamielec: J. Appl. Polymer Sci. *17*, 3769–3778 (1973).
28. J. W. Breitenbach, J. K. Rigler, B. A. Wolf: Makromol. Chemie *164*, 353–355 (1973).
29. A. M. Basedow, K. Ebert: Adv. Polymer Sci. (Fortschr. Hochpolymerenforsch.) *22*, 83–143 (1977).
30. E. M. Hagerman, C. C. Mentzer: J. Appl. Polymer Sci. *19*, 1507–1520 (1975).
31. Yu. Ya. Gotlib, A. A. Darinskii: Vysokomol. soyed. *A 16*, 2296–2302 (1974), Polymer Science USSR *16*, 2666 (1974).
32. S. N. Zhurkov, V. E. Korsukov: J. Polym. Sci., Polymer Physics Ed. *12*, 385–398 (1974).
33. E. A. Guggenheim: Handbuch der Physik, S. Flügge, Ed., Vol. III/2. Berlin: Springer-Verlag 1959.
34. S. Glasstone, K. J. Laidler, H. Eyring: The Theory of Rate Processes, New York: McGraw-Hill 1941.
35. H. H. Kausch: J. Macromol. Sci., Revs. Macromol. Chem., *C 4* (2), 243–280 (1970).
36. E. E. Tomashevskii: Soviet Physics – Solid State *12*, 2588–2592 (1971).
37. S. N. Zhurkov, V. I. Vettegren', V. E. Korsukov: 2nd Internat. Conf. Fracture, Brighton, England April 1969, paper 47.
38. H. I. Bernstein: Spectrochim. Acta *18*, 161 (1962).
39. P. J. Sheth, J. F. Johnson, R. S. Porter: Polymer *18*, 741–742 (1977), C. B. Wu, P. J. Sheth, J. F. Johnson: Polymer *18*, 822–824 (1977).
40. J. Skelton, W. D. Freeston, Jr., H. K. Ford: Appl. Polymer Symposia *12*, 111–135 (1969).
41. U. G. Gafurov, I. I. Novak: Polymer Mechanics *6*/1, 160–161 (1972).
42. S. N. Zhurkov, V. I. Vettegren', V. E. Korsukov, I. I. Novak: Soviet Physics – Solid State *11*/2, 233–237 (1969).
43. B. E. Read, D. A. Hughes, D. C. Barnes, F. W. M. Drury: Polymer *13*, 485–494 (1972).
44. P. Bouriot: Premier Colloque du Groupe Français de Physique des Polymères, Strasbourg, 21–22 mai 1970.
45. S. Sikka: PhD Thesis: Application of FTIR to study external stress effects on the PETP backbone, University of Utah, Salt Lake City, December 1976. S. Sikka, K. Knutson: to be published.
46. V. S. Kuksenko, V. A. Ovchinnikov, A. I. Slutkser: Vysokomol. soyed A11 *9*, 1953–1957 (1969).
47. H. Hahn: Personal communication.
48. S. H. Gibson, J. S. Holt, P. S. Hope: J. Polym. Phys. Ed. A2-*17*, 1375–1394 (1979).
49. D. H. Reneker: J. Polym. Sci. *59*, 39 (1962).
50. D. H. Reneker, J. Mazur: Polymer *24*, 1387–1400 (1983).

51. D. P. Pope, A. Keller: Colloid & Polym. Sci. *256*, 751 (1978).
52. A. Keller, J. A. Odell: The Extensibility of Macromolecules in Solution – A New Focus for Macromolecular Science, IUPAC Bucharest, 1983, Colloid + Polymer Sci. *263*, 181 (1985)
53. J. A. Odell, A. Keller, M. J. Miles: Polymer Comm. *24*, 7 (1984).
54. A. Keller: Bristol, personal communication 1984 (to be published) J. Polymer Sci., Polymer Phys. Ed. *24* (1986)).
55a. E. W. Merrill, P. Leopairat: Polym. Eng. Sci. *20*, 505 (1980).
55b. E. W. Merrill, A. F. Horn: Polymer Commun. *25*, 144 (1984).
56. B. Crist, Jr., J. Oddershede, J. R. Sabin, J. W. Perram, M. A. Ratner: J Polym. Sci.: Polym. Phys. Ed. *22*, 881 (1984)
57. A. I. Melker, A. V. Ivanov: Phys. Stat. Sol. (a) *84*, 417 (1984).
58. G. R. Strobl, R. Eckel: J. Polym. Sci.: Polym. Phys. Ed. *14*, 913 (1976).
M. Kobayashi, K. Sakagami, H. Tadokoro: J. Chem. Phys. *78*, 6391 (1983).
59. R. P. Wool, R. H. Boyd: J. Appl. Phys. *51*, 5116–5124 (1980).
The references used by Wool and Boyd to compile the diagram shown in Fig. 5.1 are the following:
T. Miyazawa: J. Polym. Sci. *55*, 215 (1961).
T. Shimanouchi, M. Asahina, and S. Enemoto: Polym. Sci. *59*, 93 (1962).
M. Asahina and S. Enemoto: J. Polym. Sci. *59*, 101 (1962).
I. Sakurada, U. Nukushina, and T. Ito: J. Polym. Sci. *57*, 651 (1962).
I. Sakurada, T. Ito, and K. Nakamae: J. Polym. Sci. *15*, 75 (1966).
R. N. Britton, R. Jakeways, and M. Ward: J. Mater. Sci. *11*, 2057 (1976).
J. Clements, R. Jakeways, and I. M. Ward: Polymer *19*, 639 (1978).
R. A. Feldkamp, G. Venkaterman, and J. S. King: In *Neutron Inelastic Scattering* (IAEA, Vienna, 1968), Vol. II, p. 159.
A. Odajima and T. Maeda: J. Polym. Sci. C *15*, 55 (1966).
J. F. Rabolt and B. Fanconi: J. Polym. Sci. B *15*, 12 (1977).
R. G. Schaufele and T. Shimanouchi: J. Chem. Phys. *42*, 3605 (1967).
P. J. Barham and A. Keller: J. Polym. Sci., Poly. Lett. Ed. *17*, 591 (1979).
D. S. Boudreaux: J. Polym. Sci. *11*, 1285 (1973).
B. Christ, M. A. Ratner, A. L. Brower, and J. R. Sabin: J. Appl. Phys. *50*, 6047 (1979).
G. Wobser and S. Blasenbrey: Kolloid, Z. Z. Polym. *241*, 985 (1970).
R. L. McCullough, A. J. Eisenstein, and D. F. Weikart: J. Polym. Sci. *15*, 1837 (1977).
G. Zerbi and L. Piseri: J. Chem. Phys. *49*, 3840 (1968).
60. M. Maroncelli, S. P. Qi, H. L. Strauss, R. G. Snyder: J. Amer. Chem. Soc. *104*, 6237 (1982).
61. P. J. Flory, D. Y. Yoon, K. A. Dill: Macromol. *17*, 862 (1984).
62. D. Y. Yoon, P. J. Flory: Macromol. *17*, 868 (1984).
63. M. Maroncelli, H. L. Strauss, R. G. Snyder: personal communication 1984 (to be published J. Chem. Phys.).
64. I. H. Hall, M. G. Pass: Polymer *17*, 807 (1976).
65. H. W. Siesler: Makromol. Chemie, *180*, 2261 (1979).
66. H. W. Siesler: J. Polymer Sci., Polymer Letters Ed., *17*, 453 (1979).
67. A. Garton, D. J. Carlsson, L. L. Holmes, D. M. Wiles: J. Appl. Polym. Sci., *25*, 1505 (1980).
68. K. Holland-Moritz, H. W. Siesler: Polymer Bulletin *4*, 165 (1981).
69. R. L. McCullough, A. J. Eisenstein, D. F. Weikart: J. Polym. Sci., Polym. Phys. Ed. *15*, 1805 and 1837 (1977).
70. D. B. Roitman, B. H. Zimm: J. Chem. Phys. *81*, 6333 and ibid. 6348 and 6356 (1984).
71. P. Pincus: Macromolecules *10*, 210 (1977).
Y. Rabin, J. W. Dash: ibid. *18*, 442 (1985).
72. T. Q. Nguyen and H. H. Kausch discuss in their recent papers (J. Colloid and Polymer Sci. *263*, 306 (1985) and ibid. *264*, 1 (1986) and Chimia *40*, 129 (1986)) firstly the effect of temperature on the mechanical degradation of diluted polystyrene solutions (in dioxan and dekalin) in elongational flow which was created by forcing the fluid through a steep contraction. Thermal activation had little influence on the yield for chain scission in this system due to the short residence time and transient strain rate inherent to this type of flow. Chain

scission in both solvents showed a weak negative activation energy which could only be partially accounted for by a change in solvent viscosity with temperature. Below the θ-temperature in dekalin solutions an abrupt decrease in chain scission was recorded after phase separation. Formation of interchain aggregates below the θ-point gave rise to a transient topological network. These physical crosslinks prevented the macromolecule from being fully stretched under flow and exerted a protective effect by lowering the local axial stress acting on the chain. Secondly the authors investigated the effect of the form of the flow field (laminar or turbulent regime) on the degradation. They found that chain scission in laminar flow was appreciable only for exceptionally large molecules ($M_w > 10^7$ g mol^{-1}) in a viscous solvent; otherwise it was provoked by the onset of turbulence in flow.

Chapter 6

Identification of ESR Spectra of Mechanically Formed Free Radicals

I. Formation

The action of axial tensile forces on a molecular bond R_1-R_2 results in a decrease of the apparent binding energy of the bond and thus in an increase of the probability for bond scission. If the reduction of the apparent binding energy is sizeable the mechanical action may be considered as the main cause of chain destruction. In as much as the scission of a chain molecule into organic radicals and the resulting appearance of unpaired free electrons is governed by mechanical forces a study of radical formation and of radical reactions will reveal information on the forces acting on a chain at a molecular level. The method of investigating free radicals by paramagnetic resonance techniques has been highly developed during the last thirty years [1, 2]. Since then it was successfully applied to elucidate the mechanism of the formation of free radicals in chemical reactions and under irradiation of visible and ultraviolet light, of x- and γ-rays, and of particles [1.3]. Also the value of the spectroscopic splitting factor g, the magnetic surrounding of the unpaired free electron spin, and the structure of the free radical have been studied. In all these cases the free electron spin acts as a probe, which, at least temporarily, is attached to a certain molecule, takes part in the motion of the molecule, and interacts with the surrounding magnetic field.

In investigations of the deformation and fracture of high polymer solids [4–73] by electron paramagnetic resonance techniques (EPR) similar experimental problems arise as during the above mentioned applications of EPR:
1. Interpretation of the form of the observed spectra, i.e. determination of the position of the free electron within the molecule.

2. Measurement of spectrum intensity, i.e. determination of the number of free electrons.
3. Observation of the changes displayed by the EPR signal and elucidation of the nature and kinetics of radical reactions.

From these primary data, information is sought to obtain with respect to the deformation behavior of chain molecules, their scission and their role in fracture initiation.

In the following a brief description of EPR spectroscopy will be given in order to provide an understanding of the main experimental technique which is used in these investigations. Subsequently reactions, identification, and concentration of mechanically formed free radicals will be discussed.

II. EPR Technique

A. Principles

The EPR spectroscopy of free electrons is based on the paramagnetism of free electron spins. For this reason one also speaks of electron spin resonance (ESR) spectroscopy. Electrons on fully occupied molecular orbitals generally do not give rise to a magnetic moment since – according to the Pauli principle – pairwise spins compensate each other. If, however, a bond is broken through homolytic scission, then free radicals with impaired electron spins are formed which are detectable. A free electron has the magnetic moment μ given by

$$\mu = g \beta s \tag{6.1}$$

where g is the spectroscopic splitting factor, β is the Bohr magneton, and s is the spin vector. In a magnetic field **H** the spin may assume one of two positions (spin quantum numbers m being either $+1/2$ or $-1/2$). The energy difference between these two states is

$$\Delta E = g \beta H. \tag{6.2}$$

The populations of the two energy levels at equilibrium are given by Boltzmann statistics which indicate that more of the free electrons will normally reside in the $-1/2$ "ground state". However, when electromagnetic energy of an appropriate (microwave) frequency ν is supplied, an electron can be "pumped" to the higher $+1/2$ state. The resonance condition is fulfilled when $h\nu$ is just equal to the energy difference between the two levels:

$$\Delta E = h \nu = g \beta H_{res} \tag{6.3}$$

where h is Planck's constant. In resonance, microwave energy is absorbed by the resonating sample if the upper energy level is constantly depopulated by thermal interaction between the spins and the molecular lattice. Generally the microwave frequency employed by commercially available ESR spectrometers is fixed (at about 9.5 GHz) and the magnetic field is swept through the resonance (at about 3200 Oe).

B. Hyperfine Structure of ESR Spectra

Since the "free" electrons are attached to molecules, they will sense not only the presence of the applied external magnetic field but will show magnetic interactions with nuclei of surrounding atoms which have a magnetic moment. These interactions give rise to the "splitting" that can be used to identify particular radicals. Equation (6.3) is accordingly transformed into

$$H_{res} = h\nu/g\beta - \sum_{i=1}^{n} (A_i + B_i)\, m_{Ii}. \tag{6.4}$$

Here m_{Ii} is the spin quantum number of the interacting nucleus i, A_i the isotropic part of the magnetic coupling between radical electron and nucleus i, and B_i the anisotropic part which depends on the orientation of the radical within the external magnetic field.

For the important class of hydrocarbon radicals, B is different from zero only for the α-protons, i.e. for the nuclei of hydrogen atoms directly bound to the radical C atom. In well ordered systems (highly oriented polymers, single crystals) these radicals do have a strongly orientation dependent hyperfine structure.

In the simplest case of no interaction (A = 0, B = 0) there will be one line with its center at H_{res}. If only one proton interacts isotropically (B = 0) with the radical electron then the spectrum will show two lines displaced from H_{res} by $\pm$A. If two equivalent nuclei interact with the electron then m_I can assume the values +1, 0, and −1 so that a triplet of lines appears with distances of −A, 0, and +A from the center respectively. Since the state with $m_I = 0$ appears twice as often as the other ones the individual resonance lines have intensities as 1:2:1. It may generally be said that n equivalent protons give rise to a spectrum with (n + 1) lines which show an intensity distribution given by the binomial coefficients of exponent n. For illustration Figure 6.1 shows the spectra of alkyl radicals $-\underset{\beta}{CH_2}-\dot{C}H-\underset{\beta}{CH_2}-$ in highly oriented polyethylene monofilaments which were produced by irradiation with fast electrons [37]. In one orientation (fiber axis parallel to H, upper spectrum) the total splitting parameter (A + B) of the α-protons is just equal to the isotropic parameter of the 4 equivalent β-protons. Therefore, a spectrum is observed which might be caused by 5 equivalent protons, i.e. a sextet with intensities 1 : 5 : 10 : 10 : 5 : 1 (some background including the small center line belongs to a second radical present within the sample). In the case of a perpendicular arrangement of fiber axis and magnetic field the spectrum shown in the lower portion of Figure 6.1 is observed. Now the α- and β-protons have different splitting parameters and a quintet of lines appears due to 4 equivalent β-protons with each quintet line split into a doublet by the α-proton. In all 10 lines result with intensity ratios of 1:1:4:4:6:6:4:4:1:1:. Again the small background of a superimposed second spectrum is observed. The orienta-

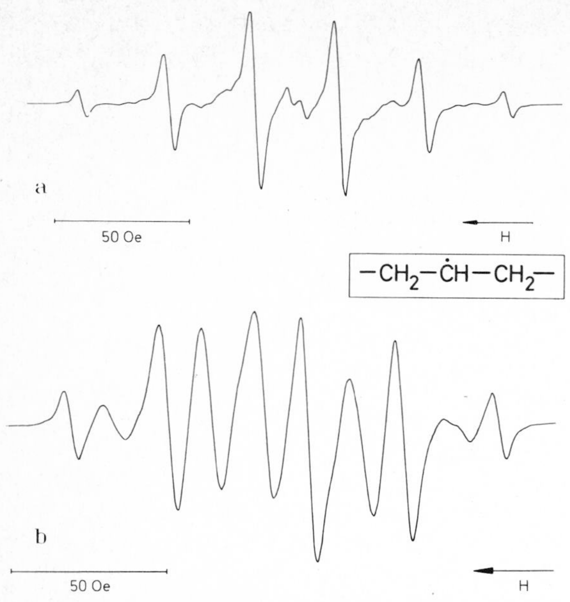

Fig. 6.1. Orientation dependency of the hyperfine structure of main chain (alkyl) free radicals: a. axis of highly oriented polyethylene fiber parallel to magnetic field H_0, b. axis of fiber perpendicular to H_0 (after [37, 38]).

tion dependency of the hyperfine structure of hydrocarbon radicals offers the opportunity to study the orientation of radical carrying chains.

C. Number of Spins

In most spectrometers lines are recorded which are the first derivative of the energy absorption peaks with respect to the magnetic field $\mathbf{H}_0$. It is the plot of these first derivates that is usually called the ESR spectrum.

The total microwave energy absorption can be obtained by double integration of the ESR spectrum. In the absence of saturation effects the total power absorbed, $I(\infty)$, depends upon the number of spins N_0, the intensity of the oscillating magnetic (microwave) field H_1, and the frequency ν according to

$$I(\infty) = 2\pi^2 \nu^2 H_1^2 \frac{1}{4kT} N_0 g^2 \beta^2 . \tag{6.5}$$

To calculate absolute intensities of spins usually comparison of the unknown intensity N_0 is made with a standard sample with a known concentration N_s of spins [1]. Frequently a solution of the stable free radical 1,1-diphenyl-2-picryl hydrazyl (DPPH) in benzene is used as standard radical. If the "unknown" spectrum is symmetrical and narrow and the experimental conditions (sample size and shape, conditions of spectrometer operation) are the same in both cases the relation

$$N_0 = N_s I(\infty) / I_s(\infty) \tag{6.6}$$

is valid where N_s and $I_s(\infty)$ refer to the standard sample. If lines of identical shape and width are compared it is sufficient to measure the peak height h of the first derivative. The numbers N_1 and N_2 of unpaired electrons causing these peak heights are then related by:

$$N_1/N_2 = h_1/h_2. \tag{6.7}$$

For most cases of stress induced free radical formation this relation has been used for evaluation of time-dependent intensities.

The number of free radicals necessary for detectable peak heights depends on the line shape and the effective volume of the spectrometer cavity. At the time covered by the referenced investigations (1960–1976) the lower limit of detectability of polymer radicals in the solid state can be given as 10^{13} spins/cm^3.

It may be mentioned that the number of free radicals present in a liquid or at the surface of a solid can also be determined by a non-spectroscopic method. This method exploits the color changes associated with the consumption of free radicals from a test solution e.g. of DPPH [39]. It should be emphasized again that great caution must be exercised in comparing absolute values of free radical concentrations obtained from differently shaped samples and/or in different cavities. According to general experience deviations of up to a factor of 2 must be expected.

III. Reactions and Means of Identification

The assignment of the specific peaks to specific types of free radicals is not always obvious; it sometimes involves highly intriguing techniques and the knowledge of related spectra and of chemical reactions to be expected [64, 67]. The main obstacles to be overcome are the broad line widths of resonance lines from solid specimens and the high rate of many radical reactions. It is evident that a large line width often prevents an effective resolution of nuclear hyperfine structure. The so-called "5 + 4 line" spectrum of mechanically destroyed methacrylic polymers [4] is an example of a spectrum which had only been identified after comparison with the 16-line-spectrum of an aqueous solution of the polymerization radical of methacrylic acid. Thus it was learned that the former spectrum is the unresolved form of the latter and has to be assigned to the same radical [40].

The difficulties in identifying an observed spectrum are well illustrated by the spectrum of 6 polyamide shown in Figure 6.2. It was obtained after stretching a fiber sample at room temperature in the ESR cavity by an automatic servo-control system to the point of radical detection, then scanning the magnetic field from 3071 to 3571 Oersted (mean, 3321 Oe) at 9.433 GHz and a time constant of 0.5 second.

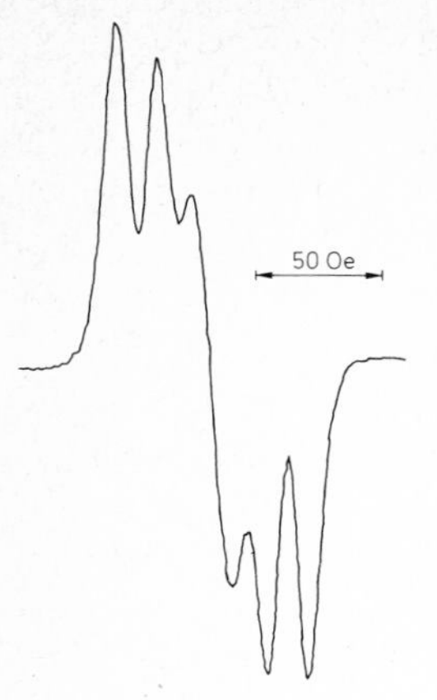

Fig. 6.2. Apparent 5-line spectrum as obtained after stretching of a bundle of 6 polyamide fibers at room temperature in an ESR cavity.

This apparent 5-line spectrum is obtained, under exclusion of oxygen, when highly stressed fibers or finely ground or milled material are investigated at room temperature [41]. A similar spectrum is observed after γ or electron irradiation of PA 6 [42, 43]. Two different interpretations of this 5-line spectrum have been given. Graves and Ormerod [42] suggested that an apparent quartet coming from the radical $-CH_2-CO-NH-\dot{C}H-CH_2-$ (radical I) is superimposed by a singlet which may be due to $-CH_2-CH\dot{O}-NH-CH_2-$ (II). Kashiwagi [43] deduced, after comparison with radicals from polyurea and succinamide, that the apparent 5-line spectrum from 6 polyamide and the 6-line spectra from 610, 66 and 57 polyamides are a triplet of doublets stemming from the radical I. It is, therefore, generally concluded that the radicals observed at room temperature in mechanically degraded polyamide material are mainly of type I, superimposed to a varying degree by a component of type II. The radicals I are secondary chain radicals which can only have been produced by the reaction of a hydrogen from a methylene group with an other – a primary – radical. Since the secondary radicals are reaction products they give no direct information on the nature and location of the primary process of mechanical degradation of the above solids.

The appearance of free radicals in the process of mechanical degradation reveals conclusively that under the action of local mechanical forces main chains have been broken and that the breakage of a main chain bond has led to the formation of a chain end free radical at either side of the broken bond. Knowledge of these facts can already be employed to study the relation between environmental parameters and the rate of chain scission. The kinetics of chain scission and its possible role in fracture initiation is investigated in detail in Chapters 7 and 8. Apart from the intensity which gives a quantitative measure of chain scission the form of free radical spectra has been used to determine the location of primary and secondary radicals within the molecule [64] and within morphological units (within crystals, on crystal surfaces, in amorphous regions).

If degradation and ESR investigation of a sample are carried out at liquid nitrogen temperature, the rate of radical reactions is slowed down sufficiently and a direct observation of the primary radicals produced by mechanical degradation becomes possible. In a thorough investigation Zakrevskii, Tomashevskii, and Baptizmanskii [10] have cleared up the scheme of radical reactions for 6 polyamide (caprolactam, capron). At 77 K they observed a complex spectrum showing the hyperfine structure of a sextet superimposed on a triplet. From the distance between the different lines of the sextet (splitting) and their intensity ratios the above authors established the presence of the radical $R{-}CH_2{-}\dot{C}H_2$ (III). This radical will form after rupture of any of the bonds 1 to 6 in the caprolactam unit:

$$\overset{7}{-}CO\overset{1}{-}CH_2\overset{2}{-}CH_2\overset{3}{-}CH_2\overset{4}{-}CH_2\overset{5}{-}CH_2\overset{6}{-}NH\overset{7}{-}CO-.$$

If the superimposed sextet is subtracted from the original spectrum a triplet remains which must be due to a scission product of the above molecule. The exclusive breakage of bonds 3 and 4 can be ruled out immediately since it would only lead to radicals of type III. Breakage of bonds 6 and 7 was eliminated after comparison with capron deuterated in the imino groups; breakage of bond 1 would not lead to a radical with a triplet spectrum. After inspection of all other possibilities, including secondary radicals, there remain as likely scission points only the bonds 2 and 5. After having studied α and ϵ methyl substituted caprolactam the authors [10] finally were able to state that in a stressed 6 polyamide molecule both the bonds 2 and 5 do break and with equal probability. The rupture of a PA 6 molecule, therefore, leads to three primary radicals:

$$-CH_2{-}\dot{C}H_2 \text{ (III, 50\%)}, -NH{-}\dot{C}H_2 \text{ (IV, 25\%) and } -CO{-}\dot{C}H_2 \text{ (V, 25\%)}.$$

Upon slowly rising the temperature from 77 K radical III begins to disappear leaving at 120 K only IV and V. At higher temperatures IV and V are gradually transformed into I which prevails above 240 K. Within the experimental limits mentioned above with respect to comparing the absolute numbers of radicals of different species there seems to be no loss of radicals during conversion of III, IV, and V into radical I.

In the following Sections and Tables the experimental conditions are reported which were employed during mechanical production and ESR investigation of free radicals. The assignments of primary and/or secondary radicals are given. For a discussion of the nature and kinetics of possible transfer reactions the reader is referred to the comprehensive reviews of Rånby et al. [2] and of Sohma et al. [64] and to the work of Bartoš and Tiňo [73]. Special problems in view of polymer morphology and of the reduction of chain strengths will be treated in Chapters 7 and 8.

IV. Assignment of Spectra

A. Free Radicals in Ground High Polymers

The appearance of macroradicals in mechanically destroyed polymers was first shown in 1959 [4–6]. Since then, fairly systematically natural and synthetic organic materials have been investigated with respect to formation of free radicals in mechanical degradation (see for instance the monograph by Ranby and Rabek [2] and the review papers by Butyagin et al. [7], Kausch [8], and Sohma et al. [64]). Owing to the originally limited sensitivity of ESR-spectrometers, the first experiments have been carried out with crushed polymers which have a high ratio of fracture surface to volume and, therefore, a comparatively large ESR signal.

The experimental equipment is described by various authors (referenced in 2). Grinding, milling, cutting or crushing of the polymer sample is done in vibration mills or mortars, under inert atmosphere, in vacuo, or in liquid nitrogen. Magnetic drills have been reported by which the shavings or grindings are produced within a sealed glass tube immediately at the ESR-cavity. Most equipment used in these experiments allows for a control of the gaseous environment and for temperature control from liquid nitrogen temperature (77 K) to above room temperature.

In Table 6.1 the presently available details of sample preparation (crushing technique, temperature, environment), the treatment of the crushed sample, the temperatures at which the ESR-spectra were taken, and the assignments of the observed spectra to primary and/or secondary free radicals are listed for 39 different *homopolymers.* As opposed to the numerous studies on homopolymers, little work has been done on *random or block copolymers:* Lazar and Szöcs [28] investigated random, block, and graft copolymers of methylmethacrylate (MMA) and styrene (S). They found in a degraded mixture of equal parts MMA and S homopolymers equal concentrations of MMA and S radicals. In a random copolymer (MMA: S = 1 : 1) the styrene radical prevailed clearly. DeVries et al. [33] had investigated a ground styrene-butadiene block copolymer. By comparison with the ESR signals of ground pure BR and PS they concluded that grinding of the block copolymer leads preferentially to rupture within the butadiene phase.

The common result of practically all of the reported ESR work [4–36] on comminuted polymers is that the mechanical action leads to the severance of a main chain bond and to the formation of chain end radicals as primary radicals. The only exception to this rule are the substituted polydimethylsiloxanes (No. 32 to 35) where there is a strong indication that the Si–O bond decomposes by an *ionic mechanism* rather than by homolytic bond scission [36]. In no case free radicals are produced by a mechanical stripping off of sidegroups or -atoms from the main chain. For this to occur stresses are required which simply cannot be transmitted onto the relatively small sidegroups present in the materials listed in Table 6.1. In fact the attempted degradation of low molecular weight compounds (paraffins, ethanol, benzene) with a molecular weight equal to or larger than that of those sidegroups failed, although mechanical means identical to those used successfully for macromolecules had been employed [13, 14, 62].

The recent claims of Sakaguchi et al. [74], to have traced *ionic fracture products* in solid PP through ion transfer to tetracyano ethylene (TCNE), seem as yet to be inconclusive.

B. Free Radicals in Tensile Specimens

Mechanical degradation which leads to formation of free radicals can be achieved in various ways but generally two types of samples are being used in experiment:

powders of finely ground polymers as discussed above or uniaxially stressed fibers, monofilaments, or strips of films. In samples of the first type all molecular fracture processes are completed before the ESR investigation begins. These samples can primarily be used to obtain information on the nature and absolute number of free radicals present within the degraded polymer.

Samples of the second type, tensile specimens, are in principle suitable for studying the time-dependent formation of free radicals and their effect on strength and on other macroscopic properties. It can be said a priori that the free radical intensities of tensile specimens – be it before or after rupture – will be very much lower than the intensities found in finely ground polymers with fracture surfaces of several thousand cm^2/g. Even with the increased sensitivity of presently available ESR spectrometers the investigation of stressed samples is limited to a few with favourable properties. The only experiments leading to a successfull investigation of time-dependent formation of free radicals were carried out with highly oriented, semi-crystalline fibers like 6 and 66 polyamide, polyoxamides, polyethylene, polyethylene terephthalate, natural silk Kevlar®, or with preoriented rubbers (natural rubber, polychloroprene) near liquid nitrogen temperatures.

In Table 6.2 experimental conditions and types of radicals observed are listed.

The experimental equipment necessary for stressing fibers within the resonant cavity has been extensively reviewed by Rånby et al. [2]. It is quite more elaborate than that for investigating powders, although the requirements for controlled temperature and atmosphere are nearly the same. Lever systems or servo-controlled electric or hydraulic loading systems have been reported [29, 37, 44, 46] through which programmed load or strain functions can be applied to bundles of fibers (or other tensile specimens) *in the resonant cavity.* The technique today applied throughout had been developed by Becht [46] in Darmstadt, the principle is shown in Figure 6.3. The stretching of the tensile specimens necessarily has to be carried out in that

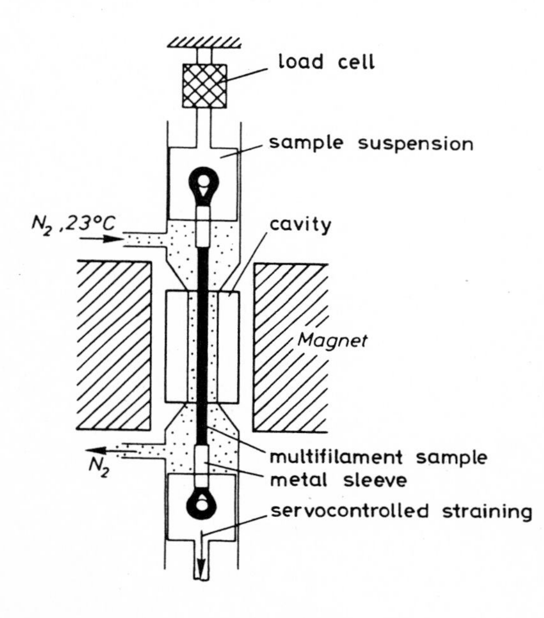

Fig. 6.3. Device of straining fiber bundles *within* the resonant cavity as originally developed by Becht [46]; the tensile specimens consist of several thousand filaments; within the metal sleeves these filaments are glued to each other and to the metal sleeve by an epoxy resin; after curing about 40% of the fibers outside the sleeves are removed in order to assure a homogeneous stress application to the remaining fibers which traverse the resonant cavity (courtesy Deutsches Kunststoffinstitut, Darmstadt)

Table 6.1. Condition of Degradation and Assignment of Resulting ESR Spectra of Polymers

Degraded Polymer		Sample preparation			(Heat) treatment time at temperature	
		process	temper-ature	environment	min	K
1. Polyethylene	PE	milling	77	liqu. N_2 + 10% O_2		
$-CH_2-CH_2-$		milling	240			
		milling	80–100	vacuum		
		milling	77	helium		
		sawing	77	liqu. N_2	none	
		milling	77	helium		
		sawing	77	liqu. N_2	5	132
		milling	77	helium		
		sawing	77	liqu. N_2	5	152
					5	233
		milling	77	helium	warming to 270	
2. Polypropylene	PP	milling	80–100	vacuum		
$-CH_2-(CH_3)CH-$						
		milling	80–100	vacuum	warming to 140	
		grinding	77			
		sawing	77	liqu. N_2		
		ball mill	77	vacuum		
		ball mill	77	vacuum	introducing air	
3. Polyisobutylene	PIB	milling di-		hexane, carbon		
$-CH_2-(CH_3)_2C-$		lute solution	77	tetrachloride		
		cutting	77	liqu. N_2 + 10% O_2		
		milling	77	liqu. N_2 + 10% O_2		
Poly(1-butene)	PB	vibrating mill	77	inert		
$-CH(CH_2-CH_3)-CH_2-$		vibrating mill	77	inert		200
4. Polybutadiene	BR					
$-CH_2-CH{=}CH-CH_2-$						
cis-rich		sawing	77	liqu. N_2		
54% trans		sawing	77	liqu. N_2		
		ball mill	77	vacuum		
5. Polyisoprene		milling	77	liqu. N_2 + 10% O_2		
$-CH_2-CH_3C{=}CH-CH_2-$						
6. Polystyrene	PS	milling di-		various solvents		
$-CH_2-(C_6H_5)CH-$		lute solution	77	plus monomers		
		milling	room	vacuum		

ESR spectra taken at K	Assignment of observed spectra to		Ref.
	primary radicals	secondary radicals	
1. 77			4, 9, 10
240			5
77		$-CH_2-\dot{C}H-CH_2-$	11
77	$-CH_2-\dot{C}H_2$	none	12
77		none	13, 14
135		$-CH_2-\dot{C}H-CH_3$	12
77		$-CH_2-\dot{C}H-CH_3$	13, 14
160	none	$-CH_2-\dot{C}H-CH_3$	12
77	none	$-CH_2-\dot{C}H-CH_3$ and	
		$-CH_2-HCO\dot{O}-CH_3$ or	
		$-CH_2-HCO\dot{O}-CH_2-$	13, 14
77	none	$-CH_2-HCO\dot{O}-CH_3$ or	
		$-CH_2-HCO\dot{O}-CH_2-$	13, 14
77	none	$-CH_2-CH=CH-\dot{C}H-CH_2-$	12
2. 77		$-CH_2-CH_3\dot{C}-CH_2-$(increasing	11
		with milling time)	11
140	$-CH_2-CH_3\dot{C}H$	$-CH_2-CH_3\dot{C}-CH_2-$(Sextet)	9
77		none	13, 14
77		$RO\dot{O}$	
77	$-CH_2-CH_3\dot{C}H$ and		
	$-CH_3CH-\dot{C}H_2$	none	15
77	$-CH_2-CH_3\dot{C}H$	$-CH_3CH-H_2CO\dot{O}$	15
3.			
77		$R_1-\dot{C}H-R_2$	16
77	$-CH_2-(CH_3)_2\dot{C}$	none	5
77		$RO\dot{O}$	4
77	$-CH_2-CH(CH_2-CH_3)-\dot{C}H_2$ and		65
77	$-CH_2-\dot{C}H(CH_2-CH_3)$	$-CH_2-\dot{C}(CH_2-CH_3)-CH_2-$	65
4.			
77	none	$RO\dot{O}$	13, 14
77	none	$RO\dot{O}$	13, 14
77	none	$R-\dot{C}H-CH=CH-CH_3$	63
5.	none	$RO\dot{O}$	4
6.			
77	$-CH_2-C_6H_5\dot{C}H$	reaction of $\dot{R}$ with monomer	17
room	none	$RO\dot{O}$	4

Table 6.1. (continued)

Degraded Polymer		Sample preparation			(Heat) treatment time at temperature	
		process	temper-ature	environment	min	K
6. (continued)		milling	77	liqu. $N_2 + O_2$		
		cutting	77	liqu. N_2 + 10% O_2		
		shaving	77	vacuum	none	
		shaving	77	vacuum	none	
		shaving	77	vacuum	heating in air to 273	
7. Poly α methylstyrene $-CH_2-(C_6H_5)(CH_3)C-$		milling dilute solution		various solvents		
			77	plus amines		
			77	toluene		
			77	toluene		
8. Polytetrafluorethylene $-CF_2-CF_2-$	PTFE	sawing	77	liqu. N_2	none	
		milling	77	liqu. N_2 + 10% O_2		
		ball mill	77	vacuum		
9. Polyvinylalcohol $-CH_2-OHCH-$	PVAL	grinding dilute solution	80	H_2O; D_2O		
		grinding	80	vacuum		
10. Polyvinylacetate $-CH_2-(OCOCH_3)CH-$	PVAC	shaving	77	vacuum		
		shaving	77	vacuum	heat. in vac. to 300	
		cutting	77?	liqu. N_2 + 10% O_2		
		milling dilute solution	77	various solvents and monomers		
11. Polymethacrylate $-CH_2-(CO_2CH_3)CH-$	PMA	cutting	77	liqu. N_2 + 10% O_2		
		milling dilute solution	77	various solvents and monomers		
12. Polymethylmethacrylate $-CH_2-(CO_2CH)(CH_3)C-$	PMMA	milling	77	liqu. N_2 + 10% O_2	heat. in air to 300	
		milling	77	liqu. N_2 + 10% O_2		
		sawing	77	liqu. N_2		
		shaving	77	vacuum	none	
		shaving	room	vacuum		
		shaving	77	liqu. N_2 + 10% O_2		
		grinding	77	vacuum		
		grinding	77	vacuum		
		grinding	77	vacuum		
		grinding dilute solution	77	benzene		
13. Polyacrylonitrile $-CH_2-CNCH-$	PAN	vibro mill	77			
			77			
		vibro mill		air		

ESR spectra taken at K	Assignment of observed spectra to: primary radicals	secondary radicals	Ref.
77	none	$RO\dot{O}$	4
77	none	$RO\dot{O}$	5
77	$-CH_2-C_6H_5\dot{C}H$		18, 19, 20
300	none	$-CH_2-\dot{C}_6H_4CH_2$	6
273	none	$RO\dot{O}$	19
7.			
77 to 300 }	$-CH_2-(CH_3)(C_6H_5)\dot{C}$	reaction of $\dot{R}$ with amines	21
77 }		$R_1-\dot{C}H-R_2$	16
177	none	$R_1-\dot{C}H-R_2$	16
8. 77	none	$-CF_2-CF_2O\dot{O}$	13, 14
77	none	$-CF_2-CF_2O\dot{O}$	4
77 and 243	$CF_2-\dot{C}F_2$	none	63
9. 80–170	$-CH_2-OH\dot{C}H$ and $-\dot{O}HCH-\dot{C}H_2$	none	22
80–350	$-CH_2-OH\dot{C}H$	$-CH_2-OH\dot{C}-CH_2-$	22, 23
10. 77	$-CH_2-(OCOCH_3)\dot{C}H$	none	19
room	none	none	19
77? }	$-CH_2-(OCOCH_3)\dot{C}H$		5
77 }		reaction of $\dot{R}$ with monomer	17, 21
11. 77 }	$-CH_2-(COOCH_3\dot{C}H$		5, 9
77 }		reaction of $\dot{R}$ with monomer	17
12. room }	$-CH_2-(COOCH_3)(CH_3)\dot{C}$	$RO\dot{O}$	4
77		$RO\dot{O}$	4
77		slight $RO\dot{O}$	13,14
77		$R_1-\dot{C}H-R_2$	19
room		none	19
300		none	5
77		$R_1-\dot{C}H-R_2$	16, 63
273		no change of $R_1-\dot{C}H-R_2$	16
283		$R_1-\dot{C}H-R_2$ disappears	16
77 }		$R_1-\dot{C}H-R_2$	16
13. 77 }	$-CH_2-CN\dot{C}H$		23
195 }			23
195		$RO\dot{O}$	24

Table 6.1. (continued)

Degraded Polymer	Sample preparation			(Heat) treatment time at temperature	
	process	temper-ature	environment	min	K
14. Poly(oxymethylene) POM $-CH_2-O-$	sawing grinding	77 80	liqu. N_2 vacuum		
15. Polyethyleneoxide PEO $-CH_2-CH_2-O$	grinding grinding dilute solution	80 80	vacuum $H_2O + CH_3COOH$		
16. Polypropyleneoxide $-CH_2-CH_3CH-O-$	grinding	80	vacuum		
17. Polyethyleneterephthalate PETP $-O-CH_2-CH_2-O-CO-C_6H_4-CO-$	vibro mill	77			
18. Polycarbonate PC $-O-C_6H_4-C(CH_3)_2-C_6H_4-O-CO-$		77			
19. Poly(2,6-dimethyl-p-phenylene oxide) PPO $-(CH_3)_2C_6H_2-O-$	 shaving 	77 77 room	liqu. $N_2 + O_2$ vacuum air		
20. Polycaprolactam 6-PA $-CO-(CH_2)_5-NH-$	cutting milling grinding milling milling milling milling	77 77 77 77 77 77	liqu. N_2 + 10% O_2 various (liqu. N_2, helium) vacuum helium vacuum helium	 heated to	 295
21. Poly α methylcaprolactam $-CO-CH(CH_3)-(CH_2)_4-NH-$ Poly ε methylcaprolactam $-CO-(CH_2)_4-CH(CH_3)-NH-$	milling milling	77 77	vacuum vacuum		
22. Poly(hexamethyleneadipamide) 66-PA $-NH-(CH_2)_6-NH-CO-(CH_2)_4-CO-$	grinding	77 77	liqu. N_2 liqu. N_2	 O_2 introduced	
23. Polyurethane PU (Solithane 113) $-O-CO-NH-(CH_2)_n-$	grinding grinding	77 room	liqu. N_2	 quenched in liqu. N_2	
24. Natural Silk	milling milling milling	77 77 77	helium helium helium		

ESR spectra taken at K	Assignment of observed spectra to		Ref.
	primary radicals	secondary radicals	
14. 77 88–350	$-CH_2\dot{O}$	$-CH_2O\dot{O}$ $-O-\dot{C}H-O-$ or $-O-\dot{C}H-OH$	13, 14 22
15. 80–230 80–200	$-O-\dot{C}H_2$ or → $-O-\dot{C}H_2$	$-O-\dot{C}H-CH_2-$	22 22
16. 80–200	$-O-CH_3\dot{C}H$ or →	$-O-CH_3\dot{C}-CH_2$	22
17. 77	$R-CO-O-\dot{C}H_2$		25
18. 77	$-O-\dot{C}_6H_4$ and $-OC-O-H_4\dot{C}_6$		26
19. 77 77 room	$-(CH_3)_2C_6H_2-\dot{O}$ (for all three)	$RO\dot{O}$ none none	27 27 27
20. 77 77 77 155 195 113 77	none $-CH_2-\dot{C}H_2$; $-CO-\dot{C}H_2$ $-NH-\dot{C}H_2$ $-CO-\dot{C}H_2$; $-NH-\dot{C}H_2$ (77, 155) none none	mostly $RO\dot{O}$ $-CO-NH-\dot{C}H-CH_2-$ $-CO-NH-\dot{C}H-CH_2-$ and singlet	5 9, 10, 12, 10, 66 12 22 9, 29 12
21. 77 77	$-CO-\dot{C}H(CH_3)$ $-NH-\dot{C}H(CH_3)$		10 10
22. 77 233	$-CO-\dot{C}H_2$; $-NH-\dot{C}H_2$	$-CO-NH-\dot{C}H-CH_2$ $RO\dot{O}$	30, 31, 32 30, 32
23. 77 77		singlet singlet	30, 33 33
24. 77 120 295	$-NH-\dot{C}H_2$; $-CO-\dot{C}H_2$ $-NH-\dot{C}H-CH_3$; $-CO-\dot{C}H-CH_3$ none	none $-NH-\dot{C}H-CO-$	12 12 12

Table 6.1. (continued)

Degraded Polymer	Sample preparation			(Heat) treatment time at temperature	
	process	temper-ature	environment	min	K
25. Polyethylene-Sulfide $-CH_2-CH_2-S_n-$	milling	77	liqu. N_2		
26. Polypropylene-Sulfide $-CH_2-CH_3CH-S_n-$	milling	77	liqu. N_2		
27. Ebonite, Thiokol, and other polymers containing sulfide bonds in the main chain	shaving compression shaving	room room 77	air air liqu. N_2 + 10% O_2		
28. Polydimethylsilmethylene $-Si(CH_3)_2-CH_2-$	drilling	77	liqu. N_2	none	
29. Polyoxymethylenedimethylsiloxane $-Si(CH_3)_2-CH_2-O-CH_2-O-$ $-CH_2-Si(CH_3)_2-O-$	drilling	77	liqu. N_2	none	
30. Polydimethylsilphenylene-dimethyl siloxane $-Si(CH_3)_2-C_6H_4-Si(CH_3)_2-O-$	drilling	77	liqu. N_2	none	
31. Polydimethylsildiphenylene-oxide dimethylsiloxane $-Si(CH_3)_2-C_6H_4-O-C_6H_4Si(CH_3)_2-O-$	drilling	77	liqu. N_2	none	
32. Polydimethylsiloxane $-Si(CH_3)_2-O-$	drilling	77	liqu. N_2	none	
33. Polydimethyl-(methylvinyl)-siloxane $-Si(CH_3)_2-O-$ 99.9%	drilling	77	liqu. N_2	none	
34. Polydimethyl-(methylethyl)-siloxane $-Si(CH_3)_2-O-$ 92 % $-Si(C_2H_5)(CH_3)-O-$ 8%	drilling	77	liqu. N_2	none	
35. Polydimethylsiloxane (crosslinked polymer) $-Si(CH_3)_2-O-$ 98.5% $-Si(O-)(CH_3)-O-$ 1.5%	drilling	77	liqu. N_2	none	

ESR spectra taken at K	Assignment of observed spectra to		Ref.
	primary radicals	secondary radicals	
25. 77	$-S-\dot{C}H_2$	$-CH_2-S-\dot{C}H-CH_2-$	34
26. (?)	no $R-\dot{S}$		34
27. room 77 77	$R-\dot{S}$		5, 34, 35
28. 77	traces of $\equiv Si-\dot{C}H_2$	$RO\dot{O}$	36
29. 77	traces of $\equiv Si-\dot{C}H_2$	$RO\dot{O}$	36
30. 77	none	$RO\dot{O}$ and traces of $\equiv Si-\dot{C}H_2$	36
31. 77	none	$RO\dot{O}$ and traces of $\equiv Si-\dot{C}H_2$	36
32. 77	none	none (ionic degradation)	36
33. 77	none	none (ionic degradation)	36
34. 77	none	none (ionic degradation)	36
35. 77	none	none (ionic degradation)	36

Table 6.1. (continued)

Degraded Polymer	Sample preparation			(Heat) treatment time at temperature	
	process	temper-ature	environment	min	K
13a. Poly(N-vinylene carbazole) PVK	drilling drilling	77 77	vacuum vacuum	heat. in air to 200	
19a. Poly[p-(2-hydroxy ethoxy) benzoic acid] PEOB	grinding	77	liqu. N_2	annealing at diff. temp.	
36. Cellulose $-C_6O_4H_{10}-O-$	vibro mill vibro mill vibro mill	77 77 room	vacuum liqu. air air		
37. Polyethylene glycol methacrylate PGMA (cross-linked 40%) $-CH_2(CH_3)(HOCH_2CH_2OOC)C-$	vibro mill	77	vacuum		

ESR spectra taken at K	Assignment of observed spectra to		Ref.
	primary radicals	secondary radicals	
13a. 100	$\sim CH_2-\dot{C}HR$ and $\dot{C}H_2CHR\sim$		68
200		$\sim CH_2-CH\dot{R} \sim$	68
19a. 223	$\dot{O}-C_6H_4-COO-$	peroxy, vanishing with incr. temp.	69
36. 77	$R_{cell}-C_4-\dot{O}$ and $R_{cell}-C_1$ and rupture of $-C_2-C_3-$		70, 76
77	$R_{cell}-C_4-\dot{O}$	peroxy rad.	70, 76
77	$R_{cell}-C_4-\dot{O}$	peroxy rad.	70, 76
37. 123	$-CH_2(CH_3)\dot{C}COOR$	no exact assignment poss.	71

Table 6.2. Form of Spectrum and Concentration of Free Radicals Formed in Tension

Polymer	Sample and treatment	Environment
1. Polyethylene	bundle of filaments	hexane at 243 K
	yarn, 300 filaments, 0.15 mm diam.	N_2
	moulded sheet, 3.5 mm thick	$N_2 + O_2$, 160 to 294 K
2. Polypropylene	yarn, 200 filaments, 0.20 mm diameter	N_2
3. Acrylonitrile-butadiene	cut sheets	2.8 mg O_3/l
	100% preoriented at 300 K	N_2 at 118 to 168 K
4. cis-Polyisoprene (natural rubber)	vulcanized sheet, 2 mm thick	
	prestretched; sulfur-cured;	Ar + N_2 at 93 K
	dicumyl-peroxide cured	Ar + N_2 at 93 K
5. Polychloroprene	dicumyl-peroxide cured, 100% preoriented at 300 K	Ar + N_2 at 93 K
6. Polystyrene	stack of strips	hexane at 243 K
	film, 1.3 mm thick	(air?)
7. Styrene copolymers	film, 1.3 mm thick	(air?)
8. Polymethylmethacrylate	bundle of filaments	hexane at 243 K
9. Polyethylene terephthalate	bundle of filaments	hexane at 243 K
	yarn, 7000 filaments, 0.02 diameter	N_2
	(mono filament?)	air at room temperature
10. Polycarbonate	bulk, drawn to necking	air at room temperature
11. Polycaprolactam	bundle of fibers	hexane at 243 K
	bundle of fibers	vacuum
	bulk material	vacuum
	yarn, 7000 filaments, 0.02 diameter	N_2
	(monofilament?)	air at room temperature
	monofilament	air at room temperature
12. Poly(hexamethylene adipamide) (66 polyamide)	preoriented film, 0.5 mm thick	air at room temperature
13. Poly(laurolactam) (12 polyamide)	yarn, 6800 filaments, 0.02 diameter	N_2
14. Polyoxamide	bundle of fibers	air at room temperature
15. Natural silk	bundle of fibers	hexane at 243 K
16. Silicone rubber	cut sheet	2.8 mg O_3/l
	cut sheet, 100% prestrain at 300 K	N_2 at 194 K
17. Poly(p[2 hydroxyethoxy] benzoic acid)	bundle of fibers	vacuum
	200 multifilaments	N_2 at 187 K
18. Poly(p-phenylene terephthalamide) (Kevlar® 49, PPTA)	50 strands of yarn (~13000 filaments)	–
	– unstressed	air or vacuum
	– stressed almost up to fracture	air or vacuum
	– following fracture (within cavity)	air
	– following fracture (within cavity)	high vacuum

ESR spectra taken at K		Observed radical	Concentration spins/cm^3	Ref.
1.	77	peroxy (R–O–$\dot{O}$)	$5 \cdot 10^{16}$	45
	291	singulet	5 to $50 \cdot 10^{14}$	37, 46
	103	peroxy	1.6 to $48 \cdot 10^{15}$	47
2.	291	none	10^{14}	37, 46
3.	room	asymmetric triplet	up to 10^{18}	48
	118 to 168	asymmetric triplet	up to $2 \cdot 10^{16}$	49
4.		asymmetric quintet		
	113	($-CH_3C{=}CH-\dot{C}H_2$?)	4 to $11 \cdot 10^{17}$	49, 50
	113	peroxy	well observable	50
5.	113	peroxy		51
6.	77	none	10^{14}	45
	(room)	none	10^{11}*	52
7.	(room)	none	10^{13}	52
8.	77	none	10^{14}	45
9.	77	peroxy	$8 \cdot 10^{15}$	45
	291	none	10^{14}	37, 46, 53
	113	peroxy	up to $15 \cdot 10^{16}$	53
10.	room	ill resolved quartet	up to $1.6 \cdot 10^{16}$	54
11.	77	quintet ($-CO-NH-\dot{C}H-CH_2-$)	$5 \cdot 10^{17}$	44, 45
	298	quintet	$1 \cdot 10^{17}$	55
	298	none		55
	291	quintet	$1 \cdot 10^{18}$	32, 37, 46
	113	peroxy	up to $7 \cdot 10^{16}$	53
	77	quintet	$5 \cdot 10^{16}$	56
12.	300	triplet (unresolved quintet?)		57
13.	291	quintet	$5 \cdot 10^{14}$ to $5 \cdot 10^{15}$	37, 46
14.	300	quintet	$1 \cdot 10^{16}$	58
15.	77	doublet	$7 \cdot 10^{17}$	44, 45
16.	room	multiplet	up to $2 \cdot 10^{16}$	48
	194	multiplet	up to $2 \cdot 10^{16}$	59
17.	room	slightly asymmetric triplet	well observable	60
	187	phenoxy + peroxy	well observable	69
18.	–			
	293	background from Mn^{+2}, Cu^{+2}	–	72
	293	(and possibly Fe^{+2}, Fe^{+3},	–	72
	293	Cr^{+3}, Ti^{+3})	–	72
	293	weak singlet or doublet superimposed on background	$2 \cdot 10^{10}$/filament ($2.5 \cdot 10^{16}$ spins/cm^3)	72

* (indirectly estimated)

temperature regime within which free radical spectra can be well observed. For thermoplastic fibers this meant 200 to 320 K; for preoriented rubbers ambient temperatures of 93 to 123 K were necessary. At these temperatures the primary free radicals are sufficiently mobile to react rapidly with atom groups of the same or other chain molecules, with absorbed gases, or with additives or impurities acting as radical traps. For this reason the observed free radicals listed in Table 6.2 are secondary radicals – with the only exception of the highly stable primary radical of poly-p-(hydroxyethoxy) benzoic acid (PEOB) studied by Nagamura and Takayanagi [60]. The secondary radicals – if sufficiently resolved – have been assigned to either peroxy radicals ($R{-}O{-}\dot{O}$), main chain radicals ($R{-}\dot{C}H{-}R$), stabilizers or unknown impurities [55, 62], or deliberately added radical traps as, for instance chloranyl [61, 75].

References for Chapter 6

1. P. Ayscough: Electron Spin Resonance in Chemistry: Methuen, London 1967.
2. B. Rånby, J. Rabek: ESR Spectroscopy in Polymer Research. Heidelberg: Springer-Verlag 1977.
3. H. Fischer: Magnetische Eigenschaften freier Radikale: Landolt-Börnstein-Tabellen, Neue Serie II/1, K.-H. Hellwege, A. M. Hellwege, Eds., Berlin: Springer-Verlag 1965.
4. S. E. Bresler, S. N. Zhurkov, E. N. Kazbekov, E. M. Saminskii, E. E. Tomashevskii: Zh. tekhn. fiz. *29,* 358 (1959), Soviet Physics – Techn. Physics *4,* 321–325 (1959).
5. S. E. Bresler, E. N. Kazbekov, E. M. Saminskii: Polymer Sci. USSR *1,* 540 (1959).
6. P. Yu. Butyagin, A. A. Berlin, A. E. Kalmanson, L. A. Blumenfeld: Vysokomol. Soyed. *1,* 865 (1959).
7. P. Yu. Butyagin, A. M. Dubinskaya, V. A. Radtsig: Uspekhi khimii *38,* 593 (1969). Russian Chem. Rev. *38,* 290–305 (1969).
8. H. H. Kausch: Rev. Macromol. Chem. *5(2),* 97 (1970).
9. K. L. DeVries, D. K. Roylance, M. L. Williams: Report from the College of Engng., Univ. of Utah, Salt Lake City, UTEC DO 68–056, July 1968.
   D. K. Roylance, K. L. De Vries, M. L. Williams: P. II. Internat. Conf. Fracture, Brighton, April 13–18, 1969.
10. V. A. Zakrevskii, E. E. Tomashevskii, V. V. Baptizmanskii: Fizika Tverdogo Tela *9,* 1434–1439 (1967); Soviet Physics – Solid State *9,* 1118–1122 (1967).
11. V. A. Radtsig, P. Yu. Butyagin: Vysokomol. Soyed. *A 9,* 2549 (1967); Polymer Sci. USSR *9,* 2883 (1967).
12. V. A. Zakrevskii, V. V. Baptizmanskii, E. E. Tomashevskii: Fizika Tverdogo Tela *10,* 1699 (1968), Soviet Physics – Solid State *10,* 1341–1345 (1968).
13. J. Sohma, T. Kawashima, S. Shimada, H. Kashiwabara, M. Sakaguchi: Nobel Symposium 22, Almquist & Wicksell, Stockholm, 225–233 (1972).
14. T. Kawashima, S. Shimada, H. Kawashibara, J. Sohma: Polym. J. (Japan) 5, 135–143 (1973).
15. H. Yamakawa, M. Sakaguchi, J. Sohma: Rep. Progr. Polymer Physics, Japan, XVI, 543 (1973). M. Sakaguchi and J. Sohma: Rep. Progr. Polymer Physics, Japan, XVI, 547 (1973). M. Sakaguchi, H. Yamakawa and J. Sohma: Polymer Letters *12,* 193 (1974).
16. P. Yu. Butyagin, I. V. Kolbanev, V. A. Radtsig: Fizika Tverdogo Tela *5,* 2257–2260, Soviet Physics – Solid State *5,* 1642–1644 (1964).
17. A. M. Dubinskaya, P. Yu. Butyagin: Vysokomol Soyed. *A 10,* 240–244 (1968), Polymer Sci. USSR *10,* 283–288 (1968).
18. A. M. Dubinskaya, P. Yu. Butyagin: Vysokomol. Soyed. *B 9,* 525 (1967).

19. S. N. Zhurkov, E. E. Tomashevskii, V. A. Zakrevskii: Fizika Tverdogo Tela *3,* 2841–2847 (1961), Soviet Physics – Solid State *3,* 2074 (1962).
20. J. Tino, M. Capla, F. Szöcs: European Polymer J. *6,* 397 (1970).
21. A. M. Dubinskaya, P. Yu. Butyagin, R. R. Odintsova, A. A. Berlin: Vysokomol. Soyed. *A 10,* 410–415 (1968), Polymer Sci. USSR *10,* 478–485 (1968).
22. V. A. Radtsig and P. Yu. Butyagin: Vysokomol. Soyed. *A 7,* 922 (1965), Polymer Sci. USSR *7,* 1018 (1965).
23. P. Yu. Butyagin: Dokl. Akad. Nauk *140,* 145 (1961).
24. P. Yu. Butyagin, I. V. Kolbanev, A. M. Dubinskaya, M. Yu. Kisljuk: Vysokomol. Soyed. *A 10,* 2265–2277 (1968).
25. S. A. Kommissarov, N. K. Baramboim: Vysokomol. Soyed. *A 11,* 1050–1058 (1969), Polymer Sci. USSR *11,* 1189–1199 (1969).
26. V. A. Zakrevskii, E. E. Tomashevskii: Vysokomol. Soyed. *B 10,* 193 (1968).
27. T. Nagamura, M. Takayanagi: J. Polymer Sci., Polymer Phys. Ed. *13,* 567 (1975).
28. M. Lazar, F. Szöcs: J. of Polym. Sci. Part C *16,* 461 (1967).
29. D. K. Roylance: Ph. D. Dissertation, Dept. of Mech. Engng., Univ. of Utah, Aug. 1968.
30. D. K. Backman: Ph. D. Thesis, Univ. of Utah, May 1969.
31. D. K. Backman, K. L. De Vries: J. Polymer Sci. A-1 *7,* 2125 (1969).
32. M. L. Williams, K. L. De Vries: 14th Sagamore Army Materials Res. Conf. Aug. 23 (1967).
33. K. L. De Vries, D. K. Roylance, M. L. Williams: J. Polymer Sci. A-2, *10,* 599 (1972).
34. V. A. Zakrevskii, E. E. Tomashevskii: Polymer Sci. USSR *8,* 1424 (1966).
35. S. N. Zhurkov, V. A. Zakrevskii, E. E. Tomashevskii: Fizika Tverdogo Tela *6,* 1912–1914 (1964), Soviet Physics – Solid State *6,* 1508–1510 (1964).
36. K. A. Akhmed-Zade, V. V. Baptizmanskii, V. A. Zakrevskii, S. Misrov and E. E. Tomashevskii: Vysokomol. Soyed. *A 14,* 1360–1364 (1972), Polymer Sci. USSR *14,* 1524–1530 (1972).
37. J. Becht: Dissertation, Technische Hochschule, Darmstadt (1971).
38. H. H. Kausch, J. Becht: Kolloid-Z. u. Z. Polymere, *250,* 1048 (1972).
39. T. Pazonyi, F. Tudos, M. Dimitrov: Angew. Makrom. Chemie *10,* 75–82 (1970).
40. H. Fischer: Polymer Letters *2,* 529 (1964).
41. J. B. Park: Ph. D. Thesis, Univ. of Utah, June 1972.
42. C. T. Graves, M. G. Ormerod: Polymer *4,* 81 (1963).
43. M. Kashiwagi: J. Polymer Sci. *A 1,* 189 (1963).
44. S. N. Zhurkov, A. Ya. Savostin, E. E. Tomashevskii: Dokl. Akad. Nauk. SSR, *159,* 303 (1964), Soviet Physics – Doklady *9,* 986 (1964).
45. S. N. Zhurkov, E. E. Tomashevskii: Proc. Conf. Physical Basis of Yield and Fracture, Oxford 1966, p. 200.
46. J. Becht, H. Fischer: Kolloid-Z. u. Z. Polymere *240,* 766 (1970).
47. L. A. Davis, C. A. Pampillo, T. C. Chiang: Bull. Amer. Phys. Soc. *18,* 433 (1973), J. Polymer Sci., Polymer Phys. Ed. *11,* 841–854 (1973).
48. K. L. De Vries, E. R. Simonson, M. L. Williams: J. Macromol. Sci. – Phys. *B-3,* 671 (1970).
49. R. T. Brown, K. L. De Vries, M. L. Williams: Polymer Preprints *11/2,* 428 (1970).
50. R. Natarajan, P. E. Reed: J. Polymer Sci. A-2, *10,* 585 (1972).
51. E. H. Andrews, P. E. Reed: In: Deformation and Fracture of High Polymers: H. H. Kausch, J. A. Hassell and R. I. Jaffee, Eds., Plenum Press, New York 1973, p. 259.
52. L. E. Nielsen, D. J. Dahm, P. A. Berger, V. S. Murty, J. L. Kardos: J. Polymer Sci., Polymer Phys. Ed. *12,* 1239 (1974).
53. T. C. Chiang, J. P. Sibilia: J. Polymer Sci., Polymer Phys. Ed. *10,* 2249 (1972).
54. B. Yu. Zaks, M. L. Lebedinskaya and V. N. Chalidze: Vysokomol. Soyed. *A 12,* 2669 (1970), Polymer Sci. USSR *12,* 3025 (1971).
55. D. Campbell, A. Peterlin: Polymer Letters *6,* 481 (1968).
56. B. Crist, A. Peterlin: Makromol. Chem. *171,* 211–227 (1973).
57. G. S. P. Verma, A. Peterlin: Polymer Preprints *10,* 1051 (1969).
58. P. Matthies, J. Schlag, E. Schwartz: Angew. Chem. *77,* 323 (1965).
59. R. T. Brown, K. L. De Vries, M. L. Williams: Polymer Letters *10,* 327 (1972).

60. T. Nagamura, M. Takayanagi: J. Polymer Sci., Polymer Phys. Ed. *12,* 2019–2034 (1974).
61. J. Ham, M. K. Davis, J. H. Song: J. Polymer Sci., Polymer Phys. Ed. *11,* 217 (1973).
62. V. A. Lishnevskii: Doklady Akad. Nauk. SSSR *182*/3, 596 (1968).
V. A. Lishnevskii: Vysokomol. Soed. *B11*/1, 44 (1969), translated in „Sowjetische Beiträge zur Faserforschung und Textiltechnik" *6*/5, 225 (1969).
63. M. Sakaguchi, S. Kodama, O. Edlund, J. Sohma: J. Polymer Sci., Polymer Letters Ed. *12,* 609 (1974).
M. Sakaguchi, J. Sohma: J. Polymer Sci., Polymer Phys. Ed. *13*, 1233 (1975).
64. J. Sohma, M. Sakaguchi, in: Adv. Polymer Sci., Vol. 20, Berlin, Heidelberg: Springer-Verlag 1976.
65. V. A. Radtzig: Vysokomol. Soyed *A 17,* 154–162 (1975). Polymer Sci. USSR *17,* 179–190 (1975).
66. J. Tino, J. Placek, F. Szöcs: European Polymer J. *11*, 609–611 (1975).
67. D. Campbell: J. Polymer Sci, Part D, Macromol. Rev. *4,* 91–181 (1971).
68. J. Tino, F. Szöcs, H. Hlouskovà: Polymer *23,* 1443 (1982).
69. T. Nagamura, K. L. DeVries: Polymer *22,* 1267 (1981).
70. D. N.-S. Hon: J. Appl. Polymer Sci. *23,* 1487 (1979).
71. J. Pilar, K. Ulbert: Polymer *16,* 730 (1975).
72. I. M. Brown, T. C. Sandreczki, R. J. Morgan: Polymer *25,* 759 (1984).
73. J. Bartos, J. Tino: Polymer *25,* 274 (1984).
74. M. Sakaguchi, H. Kinpara, Y. Hori, S. Shimada, H. Kashiwabara: Polymer *25,* 944 (1984).
75. R. E. Florin: J. Polym. Sci., Polym. Phys. Ed. *16,* 1877 (1978).
76. G. V. Abagyan, P. Yu. Butyagin: Vysokomol. soyed. A *26*, 1311 (1984), Polymer Science USSR *26*, 1466 (1984).

Chapter 7

Phenomenology of Free Radical Formation and of Relevant Radical Reactions (Dependence on Strain, Time, and Sample Treatment)

I. Radical Formation in Thermoplastics

A. Constant Rate and Stepwise Loading of Fibers

Historically, the first ESR experiments with mechanically produced free radicals have been made in the Ioffe-Institute in Leningrad in 1959 [1] with ground or milled polymers, the samples being investigated after completion of the fracture process. To elucidate the effect of structural or environmental parameters on the kinetics of stress-induced free radical formation it is necessary to study – by ESR – highly stressed chains *during* the loading process. As outlined in Chapter 5 a notable energy elastic straining of a chain can only be achieved if the chain cannot relieve stresses internally by a change of conformation or externally by slippage against the field exerting the uniaxial forces. *Vice versa the mechanical scission of a chain must be taken as an indication that at the time of chain rupture axial stresses* ψ *equal to the chain strength* ψ_c *had not only been reached but also maintained during the average lifetime period* τ_c *of the chain segment.*

In view of the fact that the slippage of chains, microfibrils, or fibrils reduces or prevents the mechanical scission of chains, highly oriented thermoplastic fibers with little potential for plastic deformation are the most suitable objects for a study of chain scission kinetics. ESR investigations using fibers of polyamide 6, 66, and 12, PE, PP, PETP and of other materials have been carried out in a limited number of laboratories, at first in the USSR, later in the USA, Germany, Great Britain, and Japan (see Table 6.2). Practically all researchers worked with highly oriented monofilaments, stacks of strips, or mostly with commercial yarns made up of a few hundred filaments each having a diameter of about 20 μm. As discussed in Chapter 2 the filaments consist of fibrils and these in turn of microfibrils with a diameter of 20 to 40 nm. The experimental set-up allowing uniaxial straining of these samples under controlled atmosphere and directly within the ESR cavity has been described and referenced in Chapter 6, Section IV B. It has also been seen that the primary radicals formed in tensile straining can – with one exception – be assigned to the breakage of main chains. In the following discussion, concentrations of secondary – rather than of primary – radicals are usually analyzed. It can be understood that the conversion factor from the number of primary radicals to the number of secondary radicals is constant. Whether it is close to unity will have to be discussed later in more detail.

If a semicrystalline microfibril is subjected to stresses, the resulting deformation will be non-homogeneous at the molecular level. The bulk of the deformation will be born by the amorphous regions. As discussed in Chapter 5 the largest stresses are transferred upon extended chain segments which share the strain imparted onto an amorphous region. The stress induced scission of chains must, therefore, be expected to occur within the amorphous regions.

Becht and Fischer [2] have demonstrated directly that free radicals are formed in the amorphous regions. These authors observed that polycaprolactam samples swollen in methacrylic acid did not show in subsequent stressing the usual ESR spectrum of the polyamide radical but the polymerization radical of the methacrylic acid. Based on the logical assumption that only amorphous regions had been swollen it was proved, therefore, that free radicals had formed in the amorphous regions only. Verma et al. [3] reached at the same repeatedly confirmed conclusion studying irradiation-produced radicals in semicrystalline polymers. Those radicals had been created under γ-irradiation within the whole volume of a PA 66 film i.e. as well within amorphous as within crystalline regions. At room temperature Verma observed – depending on sample orientation within the ESR cavity – a three, four, or six line spectrum. He assigned the pronounced anisotropy of the spectrum to the fact that the majority of the remaining radicals is located in the well oriented crystalline blocks. If the free radicals were introduced into the same material by stretching, no conspicuous anisotropy of the ESR spectrum could be observed. In this case the radicals were obviously located in sites showing rather small local chain orientation. In terms of fiber morphology the chain scission events must have taken place, therefore, in the amorphous regions between the crystal lamellae within the microfibrils, at microfibril surfaces, or in interfibrillar regions (Fig. 2.6). The question of the location of chain scission points will be touched upon repeatedly within this chapter.

At first the phenomenology of free radical production in preoriented yarns will be discussed taking 6 and 66 polyamide as examples. In constant load rate tests a very steep increase in concentration of fairly stable secondary radicals is observed in a strain regime of between 8% and rupture strain (16 to 25%) which corresponds to a stress regime of between 500 and 900 MN/m^2 (Fig. 7.1). Load rate tests are a suitable means to demonstrate the effect of mechanical action on radical formation and accumulation, but they are not sufficient to discriminate between the simultaneous effects of stress, strain, and time on the rate of radical formation. From constant load tests it was learned that, under constant load, the rate of radical formation decreases quite rapidly whereas the amount of radicals formed in virgin samples reaches a constant equilibrium value (Fig. 7.2). If a sample is strained repeatedly to a strain level ϵ_1 then free radicals will be formed, i.e. chains will break, only during the first straining cycle. If, however, in subsequent straining the level ϵ_1 is exceeded, more chains will be seen to break. In a step-strain test the radical population increases stepwise (Figs. 7.3 and 7.4).

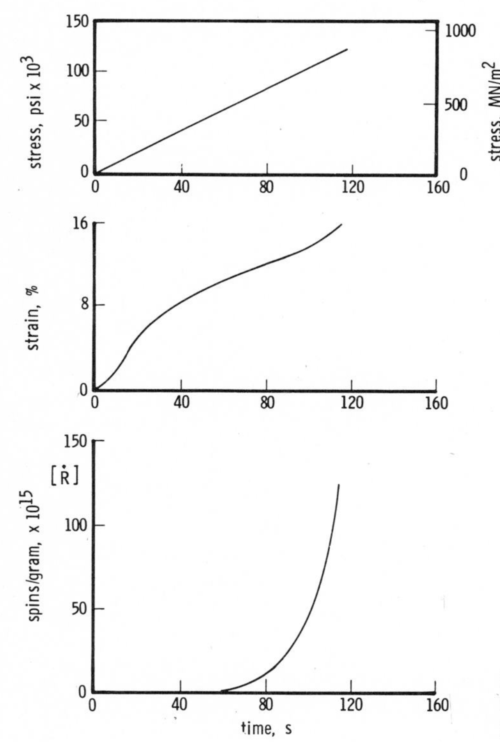

Fig. 7.1. Typical stress, strain, and free radical formation curves in constant load-rate straining of 6 polyamide fibers [4].

In the first paper on this subject [7] Kausch and Becht gave a detailed discussion of the notable fact that within each strain step the radical concentration increases concurrently with a decrease of the macroscopic tensile stress. The conclusion was drawn that the local molecular stresses are unevenly distributed. They are not simply a constant multiple of the macroscopic stresses. The chain segments breaking towards the end of a strain interval quite obviously do not participate in the macro-

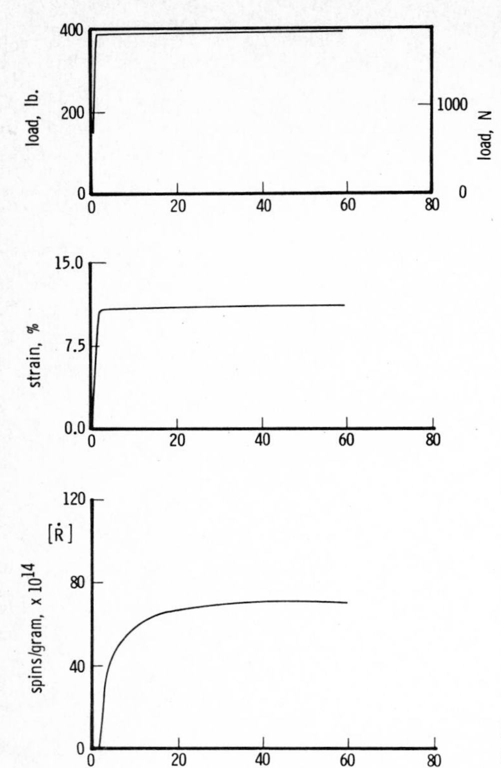

Fig. 7.2. Load, strain, and free radical production in a constant-load test of 6 polyamide fibers [4].

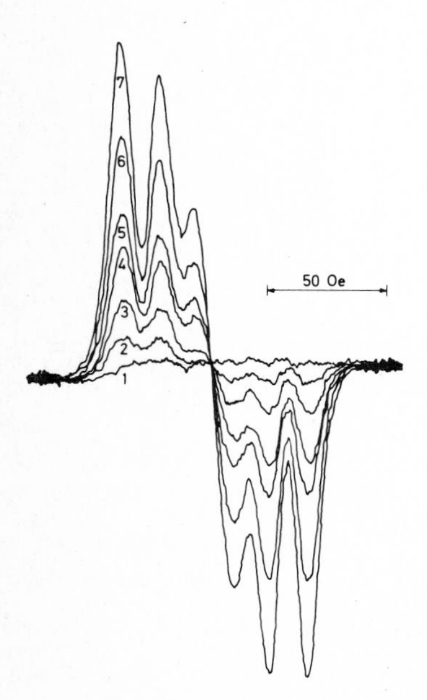

Fig. 7.3. ESR spectra of secondary radicals produced after stepwise loading of 6 polyamide fibers. The numbers correspond to stress levels of 0, 690, 760, 794, 828, 876, and 897 MN/m^2 respectively [5].

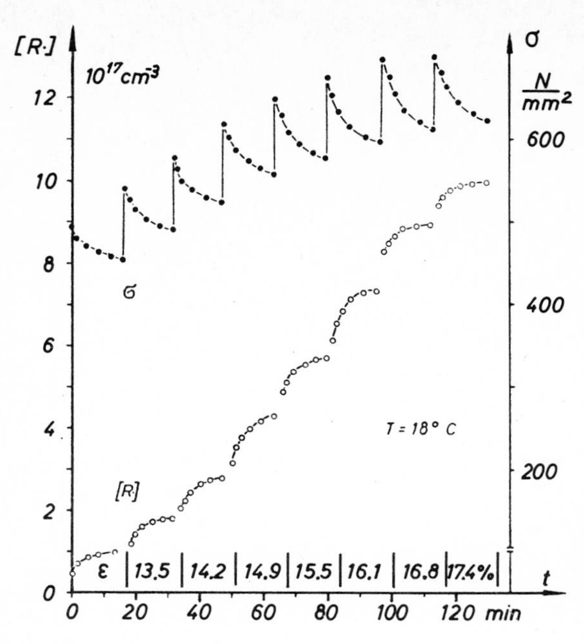

Fig. 7.4. Concentration of free radicals formed in stepwise straining of 6 polyamide fibers and measured uniaxial stress as a function of strain and time [6].

scopic stress relaxation (a decrease of the axial chain stress to 0.9 ψ_c increases the lifetime by two orders of magnitude and virtually stops the chain scission process). It was furthermore concluded that the microfibrils do not unload through slippage and that the lateral rigidity of the crystal blocks within the sandwich-structured microfibrils must be large enough so as to permit the build-up of large stress concentrations. If the lateral rigidity of the crystal blocks were insufficient the fibril would break up into submicrofibrils. This situation is schematically shown by Figure 7.5. The

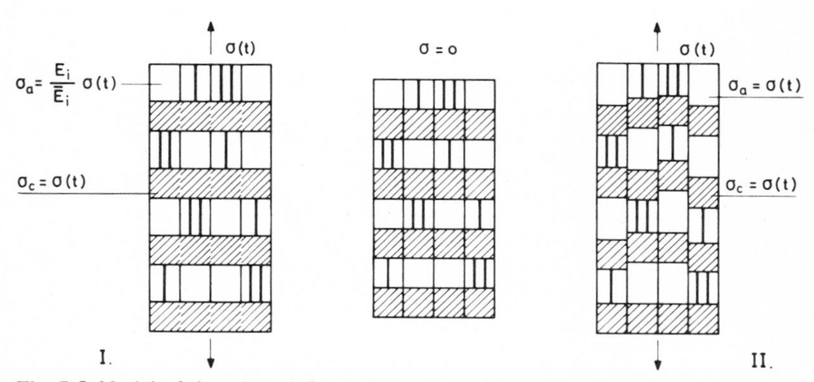

Fig. 7.5. Model of the micro-deformation and resulting local stresses within different amorphous (σ_a) and crystalline (σ_c) regions of semicrystalline fibers; case I. large, case II. small lateral rigidity of crystalline phases; E_i: Young's moduli of amorphous subregions, supposedly depending on the concentrations of highly extended tie molecules which are schematically indicated by the number of heavy lines between crystalline regions.

central diagram represents the alledged submicrofibrillar structure within an unstressed microfibril. If the crystal blocks break up while being strained the tensile stresses in every part of the structure will equilibrate and be equal to the external stress (right hand side). The external stresses are, as one knows from Chapters 1 and 5 always much smaller than the chain strength. If, however, the layered structure is also preserved at large strains stresses proportional to local strain and tensile modulus E_i of the amorphous subregions can be exerted (left hand side of Fig. 7.5). This means that within a subregion with large E_i correspondingly large – although locally well confined – stresses $\epsilon_a E_i$ – can be reached. For this reason the left hand side model is proposed to describe the mechanical response of an axially strained microfibril.

The observed fact that chains break even after (20 minutes of) stress relaxation not only demands the integrity of the crystal blocks but also *an intimate and persistent lateral cohesion between microfibrils* within a fibril and between fibrils within a filament.

That such an intimate cohesion does exist had been concluded by Gezalow et al. [8] from the uniformity of stress induced long period changes. In his recent PhD thesis Frank [55] convincingly demonstrated that almost up to fracture the macroscopic tensile strains (between 16 and 52%) applied to PA 6 and 66 filaments always corresponded within experimental error to the microfibrillar tensile strains calculated from long period changes. Thus, it can be taken for granted that at room temperature the stress-strain behavior of PA 6 and 66 fibers described in this Section reflects exactly the stress-strain behavior of the microfibrils. The stress-transfer lengths of the microfibrils (cf. Chapter 5) obviously must be short compared to their length. It must also be concluded that stress relaxation at constant fiber elongation takes place within the microfibrils. The continued rupture of chains indicates that the axial *strains* of the microfibrils are maintained during such stress relaxation. Those strains, however, can only be maintained if large scale slip of microfibrils or fibrils does not occur.

In the described experiments with PA 6 fibers the cause of stress relaxation must obviously be sought within the microfibrils. Following the model shown on the left hand side of Figure 7.5 decrease of σ (t) at constant microfibril elongation can be caused.

- either by the gradual and fairly general decrease of the relaxation moduli $\overline{E}_i$ of *practically all* amorphous regions within the microfibril
- or by the preferential and large decrease of the $\overline{E}_i$ of just a few amorphous regions, accompanied by their large deformation and by the retraction of others.

The previously discussed intimate lateral cohesion and the X-ray results [8, 55] indicate that stress relaxation occurs through a more or less homogeneous decrease of all relaxation moduli. Evidently, the situation will be entirely different if the loading is done at elevated temperatures, at say 100 to 150 °C, where microfibrillar slip is well discernible [56].

The following statements will serve, therefore, as the basis for a mathematical analysis of the (static) chain scission experiments such as step-strain or small loading rate tests at sufficiently low temperatures:

- stress relaxation or creep is not caused by slippage of fibrils or microfibrils;
- chain scission occurs in tie-segments, i.e. in those segments of a chain molecule

which connect two different crystal blocks traversing in a fairly extended conformation the amorphous region in-between (e.g. Fig. 5.2); most frequently tie-segments interconnect adjacent crystal blocks within a microfibril, but also more distant or interfibrillar ties have to be considered;

- eventually breaking tie-segments are *solidly held,* most probably by the crystal blocks; although the *clamping* extends over 2 to 5 nm the tie-segments are treated as if they had a well defined end-to-end distance L_0 and contour length L;
- chain scission occurs if the axial chain stress ψ reaches chain strength ψ_c;
- the large value of molecular stress concentration (ratio of ψ to σ of the order of 50) follows from the fact that individual (highly rigid and nearly extended) chains are subjected to the comparatively large strains of the intra- or interfibrillar amorphous regions they span.

At this point only the results of the mathematical analysis with respect to the kinetics of the formation of free radicals are presented. A discussion of the possible role of chain scission in fracture will be resumed in Chapter 8.

The analysis concerns the general case of fairly extended tie-segments of contour length L_i, end-to-end distance L_0, and elastic modulus E_k. Each segment spans an arbitrary number n_{ai} of amorphous regions (width L_a) and $n_{ci} = n_a - 1$ of crystalline regions (width L_c). If it is assumed that within a strained amorphous region a chain is at first fully extended and only thereafter strained elastically then energy elastic chain stresses ψ_i will only be built up in "taut" chains whose contour length L_i is smaller than $L_a (1 + \epsilon_a)$:

$$\psi_i = E_k \left[\frac{L_a}{L_i} (1 + \epsilon_a) - 1 + \frac{L_0 - L_a}{L_i} (1 + \epsilon) \right]. \tag{7.1}$$

At a given average axial microfibril strain ϵ the stresses ψ_i are largest for tie-segments interconnecting adjacent crystal blocks ($L_0 = L_a$). The third term in Eq. (7.1) then disappears. If the tie-segments extend over several sandwich layers and crystal blocks of the same or different microfibrils the first term in Eq. (7.1) becomes small. In that case the average *tensile* strain ϵ determines the stress level ψ_i. It is important to point out again that in this analysis no loading of interfibrillar tie-segments through *shear strains* (slippage between microfibrils) will be considered. At a degree of crystallinity of 50% ϵ is by a factor of $(2\text{-}E/E_c)$ smaller than ϵ_a, where E and E_c designate the relaxation modulus of the microfibril as a whole and that of the crystalline regions.

The mathematical analysis of the free radical production curves aimes at a determination of the distribution $N_0(L_i/L_0)$ and of the kinetic parameters U_0, ω_0, and β. It can be carried out taking into consideration only tie-segments between adjacent crystal blocks ($L_0 = L_a$). If only long interfibrillar tie-segments were present the resulting distribution $N_0' (L_i/L_0)$ would be *narrower* and approximately obtainable through the transformation $N_0'\left(\frac{L_i}{L_0}\right) = N_0 \left(\frac{[1 + \epsilon_a]\, L_i}{[1 + \epsilon]\, L_0}\right)$. The following analysis also assumes a homogeneous distribution of amorphous strains within a microfibril. Consideration of a non-homogeneous distribution of strains ϵ_{ai} would again result in *narrowing* $N_0(L_i/L_0)$. From Eqs. (5.57) and (7.1) one obtains straight forwardly the expectation value of the number of broken chains or formed free radicals:

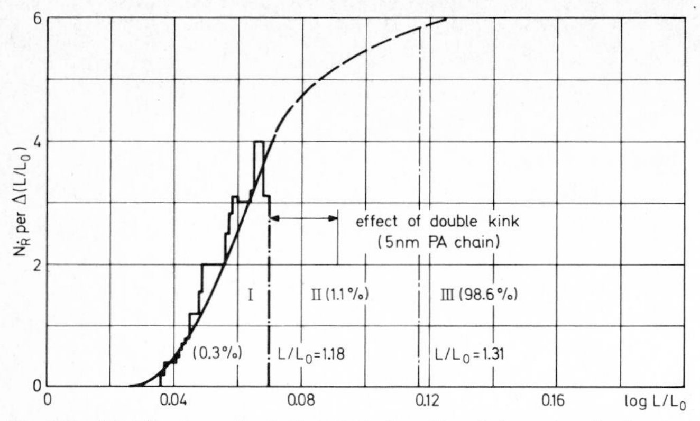

Fig. 7.6. Distribution of relative chain lengths of tie molecules in the amorphous regions of 6 polyamide fibers as obtained from numerical evaluation of step-strain ESR data [4, 7]. I chains broken until fracture, II chains fully extended and contributing with chain modulus E_k to sample modulus E, III chains in non-extended conformation.

$$[R\cdot] = 2 \sum_i N_0 (L_i/L_0) \left\{ 1 - \exp\left[- \int_0^t \bar{\omega}_b \exp(\beta \psi_i/RT) \, d\tau \right] \right\} \quad (7.2)$$

where $\bar{\omega}_b = \omega_0 \exp(-U_0/RT)$.

Through a numerical iteration procedure, a distribution of L_i/L_0 can be determined [7] in such a way that Eq. (7.2) describes the experimental points (e.g. those radical concentrations shown in Fig. 7.4). The length distribution of the intrafibrillar tie-segments obtained in that case is shown in Figure 7.6, it ranges from $L/L_0 = 1.06$ to 1.18. If the same experimental radical concentration points were to derive from the rupture of fairly long interfibrillar tie-segments then their relative lengths must have a distribution of L/L_0 between 0.97 (!) and 1.05. Since a given distribution of segment lengths is essentially reflected by the step heights of radical equilibrium concentrations a histogram of these step heights serves to characterize the apparent segment length distribution without introducing any assumptions on the nature of the tie-segments. In Figure 7.7 histograms from step strain tests carried out at different temperatures are reproduced.

The distribution of relative tie-segment lengths has been determined by several authors [3–7, 10, 11, 13, 27, 55–60], for PA 6 they all find relative lengths between 0.96 and 1.4 with a maximum around 1.16–1.18. Especially noteworthy in this respect are the observations of Zhizhenkov et al. [60] since they obtain their distribution curves from an analysis of the NMR signal of strained PA 6 fibers. The authors observe that the area C_m under the narrow component of their NMR spectra, which corresponds to the "mobile fraction" of amorphous chains, increases with temperature and decreases with fiber strain. They state that the transition from a partly to a fully extended segment (cf. Fig. 5.1) is equivalent to the transition from a mobile to an immobile state. Thus, they derive from the variation of C_m with strain the original tie-chain length distribution function for their PA 6 fibers [60]. Interestingly enough this distribution function shows a maximum at $L/L_0 = 1.16$. The total

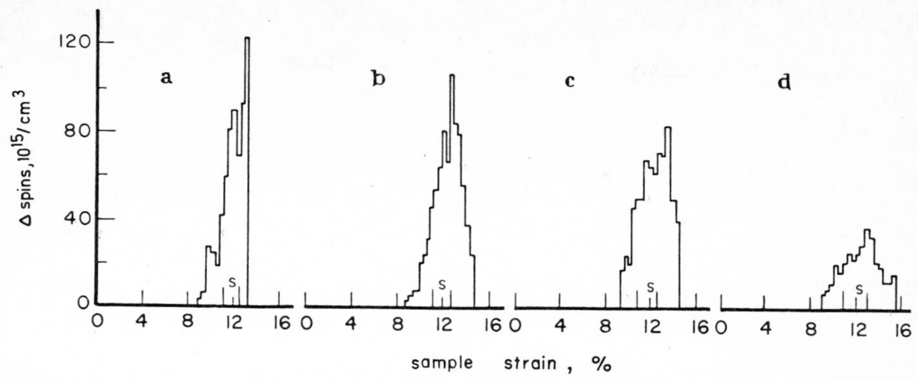

Fig. 7.7. Distributions of relative chain lengths shown as histograms of free radical concentrations obtained in step-straining of 6 polyamide fibers at (a) –25 °C, (b) room temperature, (c) + 50 °C, and (d) +100 °C [4, 5].

number of overloaded tie-chains within an amorphous region is indicated by Zhizhenkov and Egorov as being (as a function of fiber treatment) between 3 and 10% of the number of stems that theoretically could intersect the same region [60].

The association of the observed free radical distributions with apparent chain length distributions requires the smallest number of additional assumptions. If it were to be assumed that the microfibrils contained a monodisperse distribution of relative segment lengths (e.g. $L/L_0 = 1.0$) it were also necessary to assume that

- the amorphous strains are unevenly distributed throughout the fiber
- the molecular extensibility is comparable to the amorphous strains, i.e. reaches 25 to 35%
- there is a rapid strain hardening of an amorphous region after breakage of the tie molecules (otherwise collapse of that region would lead to complete stress relaxation in the whole microfibril).

Particularly the latter two conditions seemed to be unrealistic. So the early proponents [3–7] were led to their interpretation that a fiber contains tie chains of different conformations ("tautness").

A third interpretation of step-strain data was advanced by Crist et al. [9]. They suggested that there is an uneven distribution of strains due to differences in the lengths of the many thousand filaments strained simultaneously in any of the above experiments. The uneven-length effect undoubtedly broadens the observed distributions of relative chain lengths. But the premature breakages of individual filaments, and the formation of their fracture surfaces, cannot account for the number of free radicals formed. To further disprove this contention Hassell and DeVries examined the free radicals formed in straining of highly oriented bulk nylon 66 banding material [10]. They obtained quite similar and even slightly wider histograms in straining bulk strip material than bundles of nylon 66 fibers. It is only at a level of deformation close to the rupture strain that a difference between the straining of a single fiber and of a bundle thereof becomes noticeable. This has recently been shown directly by Frank and Wendorff [61]; their stress-strain curves of single fibers and of multi-filaments of PA 6 coincided up to amorphous strains of 25%, i.e. well beyond the strain of 20% at which radicals appeared for the first time (see Fig. 8.15).

Convincing evidence against the assumption that segment length distributions are macroscopic artefacts was also derived from step-temperature tests carried out by Johnsen and Klinkenberg [11] which will be discussed in Section C.

B. Effect of Strain Rate on Radical Production

Local molecular stresses can partly be relieved by slip or uncoiling of molecules. In thermoplastics the relaxation times governing those viscoelastic deformations at room temperature are of the order of milliseconds to minutes, i.e. smaller or comparable to the duration of the mechanical excitation. Rapid loading, then, can bring more nonextended chain segments to high stresses and eventually to chain scission than slow loading during the same strain interval. But their number should be small. On the other hand, the higher rigidity of a rapidly loaded viscoelastic body leads to a decrease of the strain at fracture and, therefore, also to decrease of the largest strain, ϵ_a, imposed on the amorphous regions.

The effect of strain rate on chain scission of rigidly clamped segments follows from Eq. (7.2). The higher the loading rate the smaller the time t necessary to reach a certain strain, the shorter also the time interval a stressed chain spends at the stress level $\sigma(t)$ and the smaller the probability for chain scission at that level [7]. In Figure 7.8 the decay curves of polyamide 6 segments of *monodisperse* contour lengths $L = 1.09\ L_0$ have been plotted. At 20% *amorphous* strain ϵ_a a segment of length L will assumedly have a fully extended conformation and experience a *chain* strain of 10%. At this strain level chain scission becomes noticeable at small strain rates ($0.1 \cdot 10^{-6}\ s^{-1}$). At larger strain rates an equivalent decay is observed later. For the given system an increase of strain rate by a factor of 10 will shift the decay curve to higher strain values by $\Delta\epsilon_a = 0.9\%$.

The effect of a change of strain rate in the course of a tensile experiment can be predicted from Figure 7.8. An instantaneous increase in strain rate is equivalent

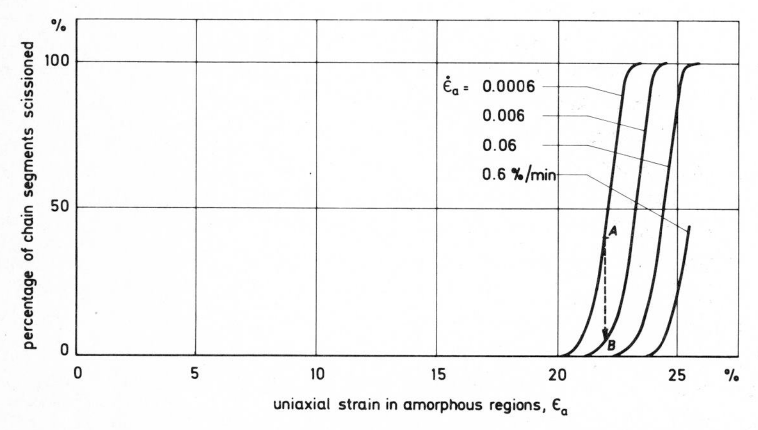

Fig. 7.8. Effect of rate of strain on the scission of polyamide 6 segments of contour length $L = 1.09\ L_0$ (after 12.7).

to changing from a certain level of depletion (e.g. 40% at point A) to a level where the depletion at that rate should have been much lower (6% at B). In other words a fraction of 34% of those chains which are supposed to scission beyond B are already gone. Consequently the rate of chain scission following an increase in strain rate should be reduced. This has been quantitatively verified by Klinkenberg [13].

Gaur [52] found that [R·] was a unique function of ϵ_a for strain rates up to $20 \cdot 10^{-3}\ s^{-1}$; at $80 \cdot 10^{-3}\ s^{-1}$, however, a decrease in [R·] was observed which corresponded to a strain retardation of 1 to 2%. These findings agree within experimental error with the careful analysis of Klinkenberg [13]. Only a few other experiments are known where the concentration of free radicals as a function of strain rate has been investigated [14–16]. In Figure 7.9 the total number of *spins at fracture* of a 6 polyamide fiber sample is shown to increase in the range of clamp displacement rates between 1 and $10 \cdot 10^{-3}$ cm/s. Above $20 \cdot 10^{-3}$ cm/s (which corresponds to a strain rate of $1.43 \cdot 10^{-3}\ s^{-1}$) a drop in free radical production is indicated [14]. The total rate effect seems to be not very large, however. For the present experiment, it did not exceed 50%. Similar values have been reported [16] from an investigation of the cold drawing of polycarbonate [15]. Here the total concentration of free radicals obtained increased by 140% with strain rates varying from $0.017\ s^{-1}$ to $1.7\ s^{-1}$ (Fig. 7.10). The effect of strain rate on the formation of free radicals in polychloroprene will be discussed in Section II A.

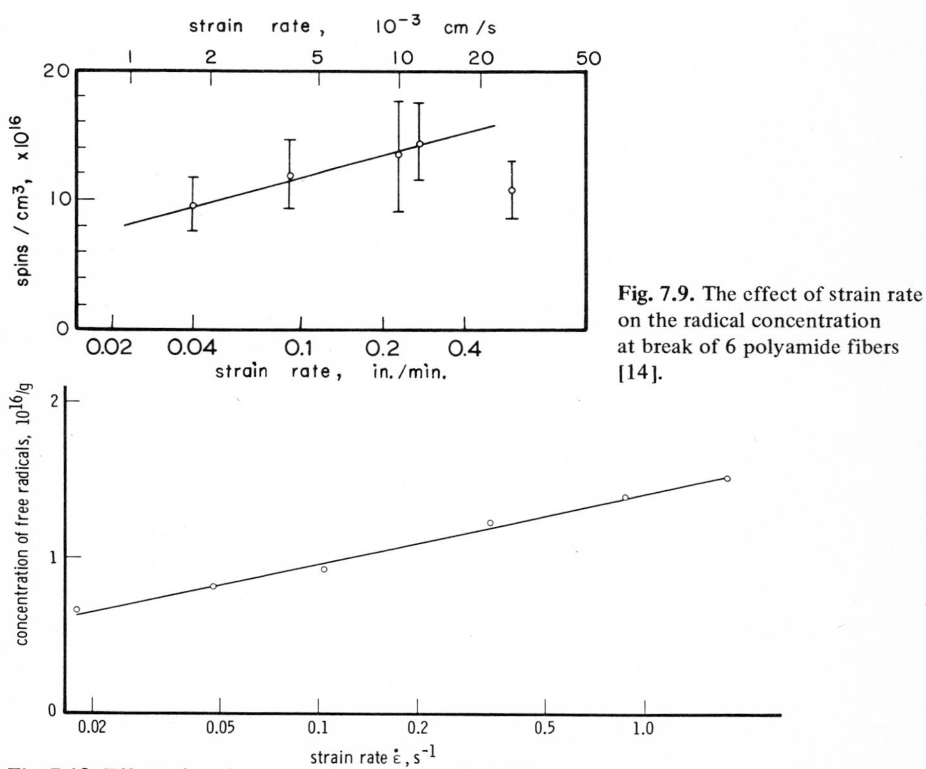

Fig. 7.9. The effect of strain rate on the radical concentration at break of 6 polyamide fibers [14].

Fig. 7.10. Effect of strain rate on the concentration of free radicals after necking of polycarbonate [15, 16].

In all of these cases the variation of free radical concentration seems to reflect more the changes of material response than the kinetics of decay of stressed chains. The effect of strain rate on radical production in comminution (grinding, milling) would be very difficult to determine and no results have been reported.

C. Effect of Temperature

1. Apparent Energy of Bond Scission

It is presumed that the stress-induced scission of chain molecules is achieved through the cooperative action of mechanical forces (lowering the bond's binding potential barrier) and of statistically fluctuating thermal vibrations supplying the remaining increment of energy which is necessary to split the loaded bond. It is also believed that Eq. (5.57) provides an adequate description of how mechanical and thermal energy affect the rate k_c of the thermo-mechanical chain scission process. If this presumption is correct, the lack of thermal vibrational energy should increase the stability of a stressed bond. Vice versa an increase in thermal energy should make previously stable bonds reach the critical level of excitation and should break them. It is of interest to quantitatively analyze this aspect of interaction of thermal and mechanical energy terms in the scission kinetics of 6-polyamide chains.

The largest energy available to one $RCONHCH_2{-}CH_2R'$ bond within a time period of $1/k_b$ is (Eq. 5.57 for $n_c = 1$):

$$U(T) = \beta\psi + RT \,\ell n\, \omega_0/k_b . \tag{7.3}$$

Using as before a value of ω_0 equal to $10^{12}\ s^{-1}$, a value of β of $5.53 \cdot 10^{-6}$ m^3/mol, and a rate $k_b = 1/\tau_b = 1/48\ s^{-1}$ the thermal term in Eq. (7.3) has, at a temperature of 20 °C, a value of 77.2 kJ/mol. The stress term $\beta\psi$ must reach, therefore, a value of 110.8 kJ/mol before U(T) is larger than the U_0 of PA 6 and *bond* rupture at room temperature can occur. At 200 K the thermal term has decreased to 52.7 kJ/mol. In order to provide a total energy of 188 kJ/mol, the mechanical term must now reach 135.3 before bond breakage will occur within 48 s. For a *chain segment* containing n_c weak bonds the rate of scission, $1/\tau_c$, employed in the above calculations will be $1/\tau_c = n_c/\tau_b$.

If one assumes that neither β nor U_0 depend on temperature, one has to conclude that at 200 K it is possible to raise the molecular stress to a level which is 17% higher than the largest molecular stress attainable at room temperature. In other words, a polyamide fiber highly strained at 200 K will quite stable contain many bonds with $\beta\psi$ values between 110 and 135 kJ/mol. Those bonds will break, of course, if the temperature is raised to room temperature provided that no other means of stress relaxation (slip, uncoiling) become available to the strained chain segment.

The search for chain scission events during an upward temperature scan can be called a critical experiment for the validity of both the kinetic equation (Equation 5.57) and the morphological model (Fig. 7.5). These investigations have been carried out in the Deutsches Kunststoff-Institut in Darmstadt. Following earlier experiments by Becht [5] Johnsen and Klinkenberg [11, 13] step-strained 6 polyamide fibers at 206, 227, 248, 269 and 290 K. These tests showed the same stepwise increase of radical concentration and a relaxing stress as reported earlier by them [6, 7, 11, 17] and other authors [3, 4, 10, 14] and as described in Section A. The result of a step-temperature experiment is shown in Figure 7.11. Whereas the stress relaxation is much more pronounced here than it is in a step-strain test, the radical accumulation

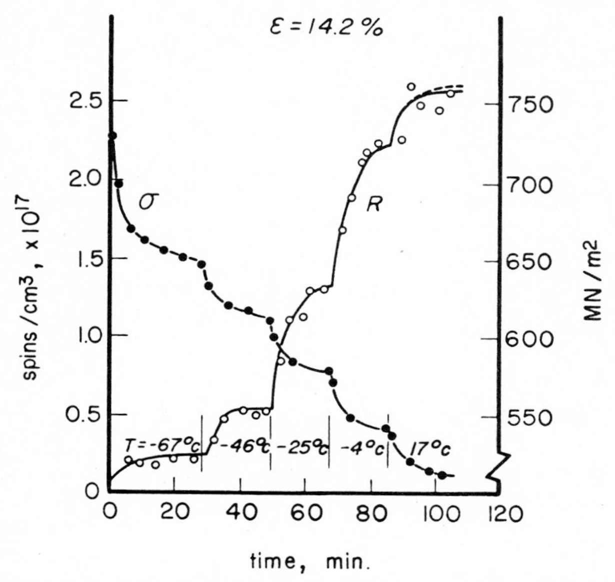

Fig. 7.11. Concentration of free radicals and uniaxial stress in step-temperature test as a function of temperature and time for 6 polyamide fibers [11].

appears to be very similar in both experiments. The authors [11] established that the final radical concentration $[\dot{R}(T, \epsilon)]$ practically did not depend on the way T and ϵ were attained. Their experiment has proven the contention made earlier about the kinetics of thermomechanical chain scission. It has also proven that the highly loaded chains do show in the region $T < 260$ K essentially no slip or anelastic deformation during the time they are held at the high load, despite the considerable amount of overall stress relaxation. A slight loss of stressed chains becomes apparent at $T > 270$ K.

The equivalence of thermal and mechanical action on a given network of chains is further illustrated by calculating the relative length L/L_0 of those intrafibrillar chains which are just about to break. One considers that condition to be fulfilled if the $\psi(L)$, calculated from Eq. (7.1) for $L_a = L_0$ and introduced into Eq. (7.3), yields U_0:

$$\left(\frac{L}{L_0}\right) = \frac{(1 - xE/E_c)\epsilon/(1 - x) + 1}{1 + (U_0 - RT\ell n\omega_0/k_c)/\beta E_k} \,. \tag{7.4}$$

E and E_c are respectively the (temperature-dependent) Young's moduli of the sample as a whole and of the crystal lamellae in chain axis direction, x is the degree of sample crystallinity.

Equation (7.4) relates sample strain ϵ, temperature T, and relative length L/L_0 of critically loaded chain segments, but subject to the condition that no chain slip occurs simultaneously. Under this condition the following equivalence between incremental changes of sample strain and temperature is obtained from Eq. (7.4):

$$\Delta\epsilon = -\frac{(1-x)\,\Delta T}{(1-xE/E_c)}\left[\frac{L\,R\,\ell n\,\omega_0/k_c}{L_0\,\beta\,E_k} - \left\{\left(1+\frac{U_0 - RT\,\ell n\,\omega_0/k_c}{\beta\,E_k}\right)\frac{L}{L_0} - 1\right\} \cdot \frac{d(E/E_c)/dT}{1-xE/E_c}\right]. \tag{7.5}$$

The two terms in the square brackets refer respectively to the change of chain strength and of strain distribution with temperature. Using the kinetic parameters employed in evaluating Eq. (7.3) for PA 6 chains, values of $E_c = 25\ GN/m^2$ and of $E_k = 200\ GN/m^2$ as previously, and an estimated temperature coefficient $d(E/E_c)dT$ of $-0.8 \cdot 10^{-3}\ K^{-1}$ one obtains at a temperature of $-20\ °C$ the following quantitative evaluation of Eq. (7.5):

$$\Delta\epsilon = -\ [0.12 \cdot 10^{-3}\ L/L_0 + 0.48 \cdot 10^{-3}\ (1.1\ L/L_0 - 1)]\ \Delta T. \tag{7.6}$$

This means that for a critically loaded PA 6 chain segment with $L/L_0 = 1.1$ a temperature increase of 10 K can be compensated by a decrease of strain of only 0.23%. The result of this calculation agrees fairly well with the average value of 0.3%/10 K derivable from the experiments of Johnsen and Klinkenberg [11]. For preoriented PA 6 fibers they observed that at $-60\ °C$ a sample strain of 14.6% led to the same number of (10^{17}) free radicals as a strain of 12.2% at $+20\ °C$.

The loading of possibly present long interfibrillar tie-segments is essentially determined by $1 + \epsilon$ rather than $1 + \epsilon_a$. Accordingly in Eq. (7.5) the first term in square brackets and a front factor equal to ΔT must be retained. For PA 6 this leads to a $\Delta\epsilon$ equal to $-0.24 \cdot 10^{-3}\ \Delta T\ L/L_0$, i.e. to 0.26%/10 K at $L/L_0 = 1.1$. This comparison does not permit, therefore, to rule out the existence of interfibrillar tie-segments.

2. Rate of Bond Scission

A *rate of bond scission* can be defined in terms of time as $d[\dot{R}]/dt$ or in terms of strain as $d[\dot{R}]/d\epsilon$. The relation between the two rate expressions is defined by the identity

$$\left(\frac{d[\dot{R}]}{dt}\right)_{\epsilon_1,\,\dot{\epsilon},\,T} = \dot{\epsilon}\left(\frac{d[\dot{R}]}{d\epsilon}\right)_{\epsilon_1,\,\dot{\epsilon},\,T} \tag{7.7}$$

which has to be taken at constant $\dot{\epsilon}$ and T at corresponding sample strains ϵ_1.

In the absence of plastic flow (chain scission at strains of 10 to 30%) the rate can be derived numerically from Eq. (7.2); since the rate will depend on the shape of the tie-segment length distribution, $N_0(L_i)$, no general analytical function can be given for $d[\dot{R}]/d\epsilon$. The effect, however, of a change of $\dot{\epsilon}$ on $d[\dot{R}]/d\epsilon$ is the same as that illustrated by Figure 7.8. The effect of the strain ϵ_1 at which the rate of bond scission is determined, depends on the shape of the segment length distribution, i.e. on the number $\Delta N(L/L_0)$ of chains having a relative length L/L_0 suitable to be broken by an increment $\Delta\epsilon$ of sample strain:

$$\left(\frac{d[\dot{R}]}{d\epsilon}\right)_{\epsilon_1, T} = \frac{2\Delta(L/L_0)}{\Delta\epsilon}\left(\frac{dN}{d\,L/L_0}\right)_{\epsilon_1, T}. \tag{7.8}$$

In Figures 7.6 and 7.7 typical segment length distributions are given.

A change of temperature will affect the rate of radical production in three ways. Firstly the number N of highly extended chains will decrease if thermally activated slip becomes possible. Secondly the activation energy U(T) decreases with increasing temperature. Thirdly the sample modulus E and the crystalline modulus E_c are affected by temperature.

If in a region of sufficiently low temperatures chain slip can be excluded i.e. if $\frac{dN}{d\,L/L_0}$ is constant with T the temperature affects solely the width $\Delta(L/L_0)$ of the chain segment population which is critically loaded by a strain increase:

$$\frac{\Delta(L/L_0)}{\Delta\epsilon} = \frac{(1 - xE/E_c)/(1 - x)}{[1 + (U_0 - RT\ln\omega_0/k_c)/\beta E_k]}. \tag{7.9}$$

Employing the same values as used in the evaluation of Eq. (7.3) one arrives at $\Delta(L/L_0) = 1.7\,\Delta\epsilon$ at 20 °C and at $\Delta(L/L_0) = 1.6\,\Delta\epsilon$ at –60 °C. This means for example that a given distribution of segments with relative lengths L/L_0 between 1.1 and 1.3 break at room temperature within the strain interval from 12 to 13.5%. At a temperature of –60 °C sample strains of between 13.5 and 15.1% are required.

A completely different chain loading mechanism prevails during the plastic deformation of polymers at strains of between 30 and several hundred per cent. In that case a chain will rupture as a consequence of the frictional forces encountered by the chain itself or by other morphological units in the dynamic shearing (cf. Chapter 5 II E). The obtainable axial chain stresses are proportional to the molecular or fibrillar friction coefficients and to the strain rate $\dot{\epsilon}$. The number of critically loaded chains will reflect, therefore, the strong increase of the friction coefficient with decreasing temperature. Davis et al. [19] strained sheets of high molecular weight polyethylene in air and observed the formation of peroxy radicals. At a level of true strain $\ln \ell/\ell_0$ of e.g. 1.1 – which corresponds to a conventional strain of 200% – the concentration of the peroxy radicals increased from $5 \cdot 10^{14}$ cm^{-3} at 294 K to 10^{16} cm^{-3} at 160 K. The rate of radical accumulation, $d[\dot{R}]/d\ln(\ell/\ell_0)$, revealed two transition regions, one between 180 and 200 K and another beginning at 250 K.

The low temperature transition was assigned to the transition of amorphous PE to the glassy state, the other to slippage of chains within crystal lamellae [19].

The formation of free radicals during plastic deformation has also been reported for PP, PETP and PA 6 [62], PC [15], PVDC [63] and PVC [64]. Detection in the latter two materials was made by indirect methods: by respectively a radical scavenger method based on chloranyl [63] and through the initiation of free radical polymerization of vinyl monomers in contact with deformed PVC [64].

3. Concentration at Break

The increase of temperature has a fourfold effect on the number of free radicals observed at the point of macroscopic fracture. Firstly as seen previously the bond strength is decreased and thus chain scission at a given molecular stress is facilitated. Secondly the decrease in intermolecular attraction and the increase in molecular mobility lead to more rapid relaxation of molecular stresses. For the same reason – thirdly – the density of stored elastic energy at a given strain level is decreased which will in turn affect crack stability and propagation. Fourthly, the increased reactivity of the free radicals may lead to an increased discrepancy between the concentrations of radicals formed and of radicals observed at the point of failure.

Whenever possible the fourth point has been taken into consideration and radical concentrations have been corrected for losses due to recombination or other decay reactions of the (secondary) radicals under consideration. In the case of 6 polyamide fibers [5, 11, 17, 18] the radical decay is a second order recombination reaction with the rate constant depending on type, location, and mobility of the radicals (see Section III below).

In Figure 7.12 the concentration of free radicals at fracture of 6 polyamide fibers is shown as observed and also corrected for radical decay during and after sample fracture [18]. The indicated decrease of the radical concentration at break towards

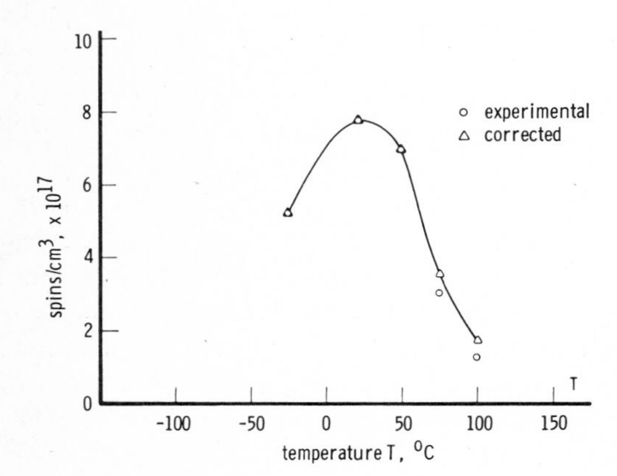

Fig. 7.12. Concentration of free radicals at break of 6 polyamide fibers as a function of breaking temperature [4, 5], ○ observed concentrations, △ concentration corrected for decay of free radicals.

lower temperatures was well confirmed by Johnsen and Klinkenberg [11] who extended their measurements to –67 °C. At that temperature they obtained a concentration which was by a factor of 8 smaller than that at room temperature.

The first three of the initially mentioned temperature effects can be better understood if one makes use of the segment length distribution model. If, in fact, at low temperatures the changing bond strength is the only temperature dependent parameter then a chain length distribution derived at about −25 °C should exactly include the (narrower) distributions at still lower temperatures. Vice versa from one measured step-strain free-radical curve all narrower ones must be predictable. Johnsen and Klinkenberg [11, 13] have confirmed that the chain length distributions obtained at temperatures between −67 and −25 °C are derived from one and the same "master" distribution which is skewed to a different extent at the right hand side. At higher temperatures (−4 and 17 °C) discrepancies between the apparent distributions and the master distribution are observed. They are interpreted as indication of structural changes leading to a relaxation of some overstressed chains. These chains may, however, break at a later stage of the loading process [11, 13, 65]. The relaxation and slip effect is stronger at higher temperatures and at 50 °C it clearly outweighs the effect of diminished bond strength, i.e. the total number of broken bonds at fracture decreases. Beyond 50 °C most of the chain segments belonging to the master distribution escape chain scission because the crystal blocks are no longer able to exert and maintain large enough molecular stresses. At this point it may remain open which mechanisms are responsible for the stress relieving structural changes.

In the above sections were discussed the scission of tie chains in highly oriented fibers. Judging from the absence of any scale effect the molecular fracture events are evenly distributed throughout the amorphous volume of the stressed sample. Those "volume" concentrations of 10^{15} to 10^{18} spins/cm^3 accounted for almost the total radical intensity whereas the final fracture surfaces (with estimated radical populations of 10^{12} to 10^{13} spins/cm^2) did not noticeably contribute towards the total number of radicals observed in tensile straining of the fiber samples. This is different if the mechanical degradation is achieved through milling, grinding and/or cutting. In all these techniques comminution is achieved through failure of virgin material resulting from the application of multiaxial stress particularly involving compressive components. In impact loading of platelike or granular samples compressive waves may be generated. Upon reflection on boundaries or discontinuities, compressive waves become dilational waves giving rise to tensile fracture. All forces combine to overcome the cohesive forces and to create new surface area. Those chains which break in this process are either intersecting weak cross-sections and/or are contained in shear zones along a suitable direction. It must be assumed, therefore, that chains are broken preferentially – but not exclusively – in the zones of new surface.

The surface concentration of free radicals in ground polymers has been determined by different investigators [20–23]. As to be expected the mobility of the chains within the degrading material strongly effects the obtainable radical concentrations. Backman and DeVries [21, 22] sliced 66 PA, PP, and PE in a nitrogen atmosphere and transferred the slices within less than 0.5 s into liquid nitrogen and determined the number of free radicals by ESR. The degradation temperature was varied between −20 °C and 90 °C. The authors observed a rapid decrease of surface concentrations at the respective glass transitions (Fig. 7.13). It should be noted that the rate of decay of the radicals formed also increases beyond T_g, but that this loss

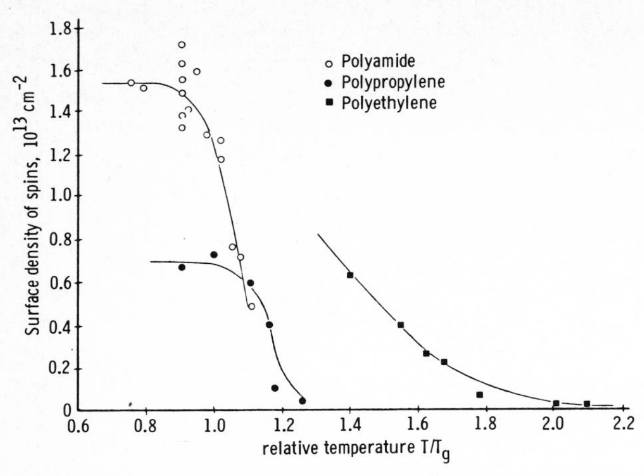

Fig. 7.13. Surface densities of spins observed on comminuted polymers as a function of the degradation temperature T plotted as fraction of the corresponding glass transition temperature [20].

of radicals amounts to less than 30% [18]. From the level of the low temperature radical concentrations it can be concluded that the average fracture plane breaks between 5 and 10% of all chains intersecting the plane. Pazonyi et al. [22] investigated polyvinylchloride with differing contents of plasticizer (30, 40, 50, and 60% respectively of dioctylphthalate, DOP). They estimated that cutting at 30 °C led to the scission of (128, 105, 55 and 34% respectively of all) chains intersecting an ideally sharp fracture plane. They also investigated low and high density polyethylene and found 1.39 and $1.17 \cdot 10^{15}$ spins/cm^2 which means they had to cut 120 and 104% respectively of the intersecting chains. If one realizes that even a "sharp fracture plane" is in reality a strongly deformed surface zone of a depth of 1 μm or more the above "cutting ratios" are absolutely reasonable. Pazonyi et al. did their cutting in an ethanolic solution of diphenyl picryl hydrazyl the color change of which was used as a measure for the number of free radicals formed. The unusually large molecular cutting ratios given by Pazonyi et al. and their discrepancy with the values cited earlier are noted and discussed by Salloum and Eckert [23]. They point out that reactions of DPPH with ethanol and with polymer surfaces not containing free radicals may take place thus accounting for the excessively high radical numbers determined by Pazonyi. Salloum and Eckert drilled PE, Nylon, and PP in a solution of DPPH in benzene. They obtained 0.1–2 (PE), 0.5–4 (6 PA) and $7\text{–}9 \cdot 10^{13}$ spins/cm^2 (PP) respectively. The first two values are in agreement with the results of Backman and DeVries, the latter is not.

Both laboratories [20–22] conclude that free radicals are formed within a surface layer of finite thickness. From a study of oxygen diffusion rates and of the conversion of the original ESR signal from secondary 6 nylon radicals to that of peroxy radicals Backman and DeVries determine that 50% of all "damage" occurred within a layer of less than 0.6 μm from the surface. The remaining 50% of chain radicals were found in a depth of up to 3 μm from the surface. Considering the morphology of the degraded polymers, the mechanics of the comminution process, and the mobility of the primary formed free radicals the spatial distribution of the secondary radicals can be understood. In this case the pull out and breakage of *individual PA chains* appears unlikely in view of the crystal strength. As we have discussed in Chapter 5, a 6 PA chain embedded within a crystallite with more than 1.7 nm of its length will rather break than be pulled away from the crystallite. The pull out of *whole microfibrils* from a fracture surface very probably will occur and will be accompanied by breakage of interfibrillar tie chains thus introducing damage within a surface layer of up to 1 μm depth. Even more important will be the heavy plastic deformation of the material ahead of a growing fracture surface. The propagation of free radicals certainly contributes to deepen the layer of apparent damage. The layer depths observed in these experiments do coincide, however, very well with the lower limits of particle sizes obtainable in mechanical degradation [24] indicating, that the damage may indeed be introduced mechanically up to the observed depths.

This phenomenon, namely the chain breaking efficiency of a plane fracture event, has been further investigated by Wool et al. [66]. They microtomed atactic polystyrene in 1- and 2-μm slices and measured the ensuing decrease in viscosity average molecular weight. From their results they concluded that each molecular coil with M_n larger than the critical entanglement molecular weight M_{cr} which intersects the (ideally sharp) fracture plane is cut ξ times. From their model they obtained average ξ-values of between 5 at + 80 °C, 14.45 at room temperature, and 22 at − 100 °C. They assume that a PS molecule having an M_n of 450,000 occupies a sphere of diameter D = 60 nm. Any given area A_0 will be intersected by chain segments belonging to N_D such spheres. The number of these chain segments is $N_{seg} = A_0/q$. It is easily verified (cf. Tab. 9.2) that $N_{seg} = 16\ N_D$ (450,000). It follows that at room temperature the total number of chain cuts $(\xi \cdot N_D)$ is comparable to the total number of chain segments intersecting an ideal plane. At lower temperatures it will be larger. In view of the fact that the chain scission events do occur within a material volume several 10 to 100 coil diameters deep, it is quite conceivable that the total number of chain scission events is 5 to 20 times larger than the number of coils intersecting an ideal reference plane.

D. Effect of Sample Treatment

The effect of heat treatment on structure, fracture strength and chain scission has been very extensively investigated by Statton, Park and DeVries [25–27] and by Lloyd [5]. With respect to our understanding of chain breakages in annealed fibers the following morphological changes seem to be particularly noteworthy. Relaxed heat treatments result in [25]:

1. Increases of the degree of crystallinity, perfection of crystallites, density, chain folding, Young's modulus, and intensity of the fluid-like component of the NMR-signal.
2. Decreases of sample length (shrinkage), breaking strength, sonic modulus, and T_g.

The changes are the more intense the higher the annealing temperature. Tension during annealing tends to modify or reduce the above listed effects. Relaxed and tension annealing do, however, increase the long period.

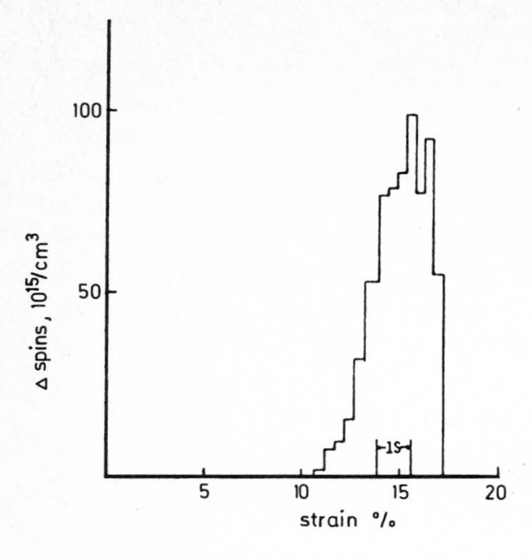

Fig. 7.14. Histogram of radical concentration from 6 polyamide fibers slack annealed at 199 °C and subsequently step-strained at room temperature [4. 5].

The morphological changes certainly have a bearing on the molecular stress distribution and chain breakage. In Figure 7.14 the histogram from an ESR step-strain test is reproduced which was obtained from a tension-free annealed polyamide sample at 199 °C [27]. The histogram is broader, indicating a less homogeneous distribution of molecular strains, and it is shifted towards higher strains by about two to four per cent of strain. This may be seen from Figure 7.15 where the chain length distribution of the annealed sample is compared with that of an unannealed reference sample. The used probability plot had been calculated under the assumption that the histogram in Figure 7.14 represents a normal distribution of chain lengths skewed at the right hand side.

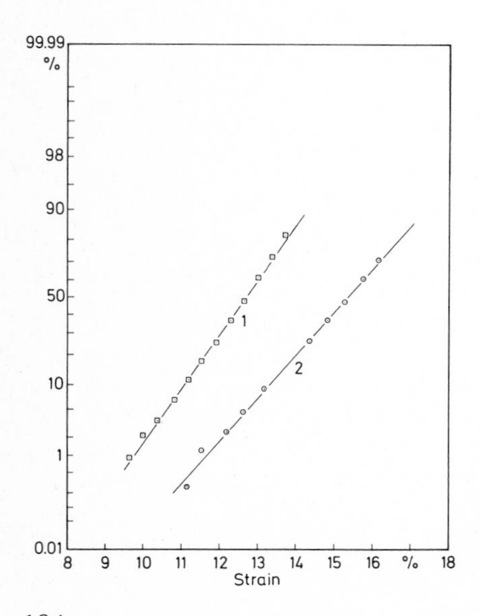

Fig. 7.15. Probability plot of histogram of radical concentration of 1. reference sample and 2. sample slack annealed at 199 °C [4, 5].

The observed conformational changes in slack annealing have been interpreted [25–27] as an increase in relative length of (the formerly) extended tie chain segments due to defect migration out of the crystal blocks. The number of regular chain folds also increases. Shrinkage of the yarn appears to be governed by the number of folds introduced. The structural changes occurring during annealing are mechanically stable and not simply reverted under tensile stress. In Figure 7.16 a model representation of the conformational changes in annealing are given [4, 5]. Because of defect migration one obtains in tension annealing a relaxation of local molecular strains and thus a larger number of extended chain segments. On the basis of this experimental evidence Statton has constructed a string model of slack annealed, tension annealed, and reference fibers which is shown in Figure 7.17 [28]. The effect of the annealing process on the total number of free radicals formed in subsequent straining at room temperature and on fracture strength is shown in Figures 7.18 and 7.19. The annealing at maximum tension (11.7% elongation) does lead to a smaller number of spins at fracture and may be explained by the breakage of chains during tension annealing. Intermediate positive annealing strains (not shown in Fig. 7.18)

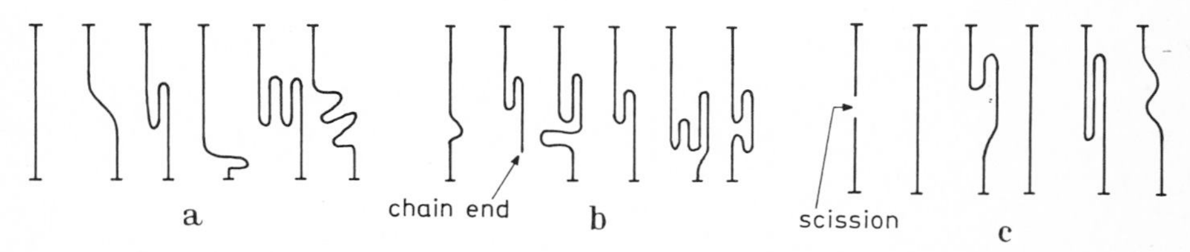

Fig. 7.16. Model of tie chain conformation in (a) reference sample, (b) fiber slack annealed, and (c) fiber annealed under tension [4, 5].

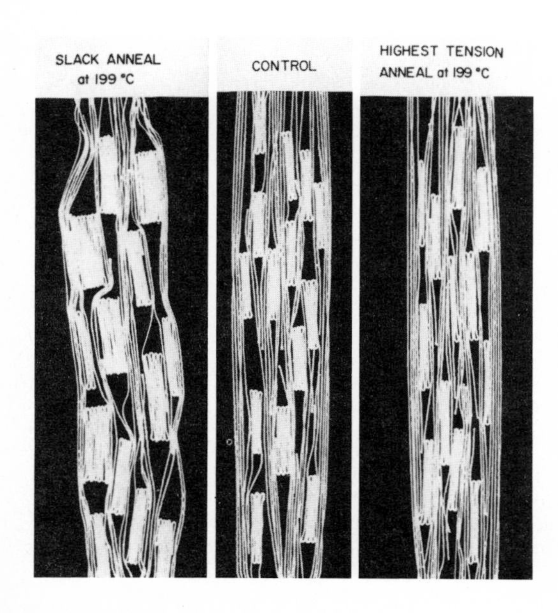

Fig. 7.17. String-model by Statton [27, 28] representing effect of annealing on fiber morphology and chain conformation.

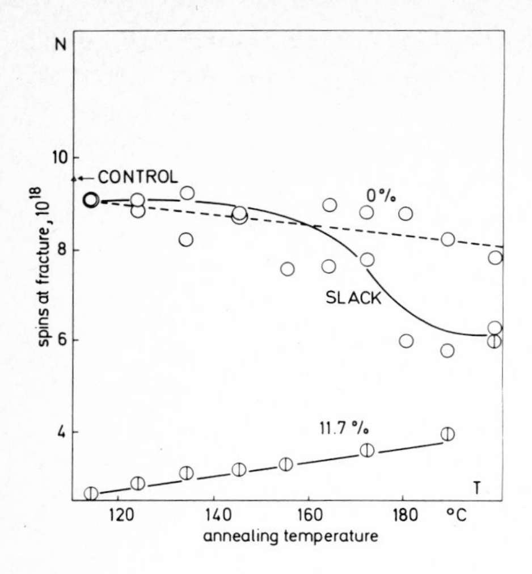

Fig. 7.18. Concentration of free radicals at fracture (at room temperature) as function of annealing temperature [27].

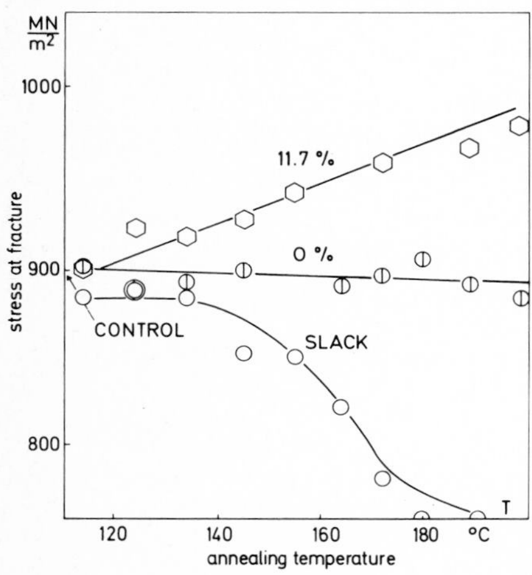

Fig. 7.19. Effect of annealing temperature and annealing condition on room temperature fracture stress of 6 polyamide fibers [27].

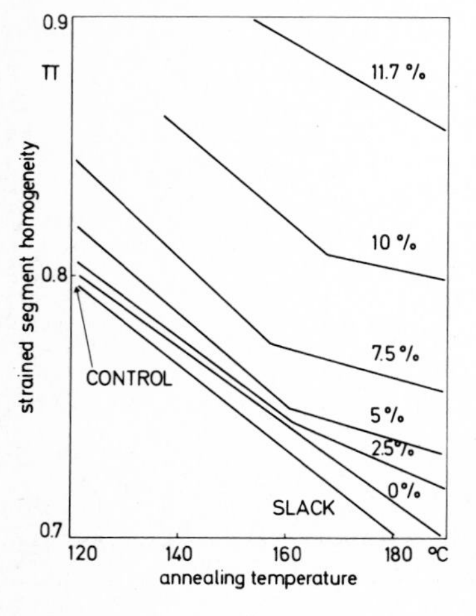

Fig. 7.20. Effect of annealing temperature and constraint during annealing on the strained segment homogeneity π which is defined as the ratio of that strain where the first radicals are observed to the strain at break [27].

do slightly increase the number of spins. The opposite effects of annealing temperature and of constraint of the sample during annealing on the homogeneity of the chain length distribution are shown in Figure 7.20. These effects are discussed in more detail in Chapter 8.

II. Free Radicals in Stressed Rubbers

A. Preorientation, Ductility, and Chain Scission

In Section I we have discussed the stress-induced chain scission of thermoplastic polymers. As we have seen chain scission could occur whenever the intermolecular forces acting on tightly embedded sections of extended (tie) molecules were strong enough to resist segment slip during deformation in such a manner as to give rise to axial forces equal to or larger than the mechanical strength of the molecular chain. Exactly the same considerations apply to chain scission in elastomers which are in a temperature region below T_g. Andrews, Reed, and Natarajan [29–31] have studied under these aspects the deformation behavior of sulphur cured natural rubber (cis-polyisoprene), dicumyl peroxide-cured natural rubber, and polychloroprene. They observed that (see Fig. 7.21) in the temperature region I from glass transition (200 K) to 160 K the deformation behavior of natural rubber was very similar to that of a thermoplastic polymer, with a clearly discernible yield point and a yield stress which

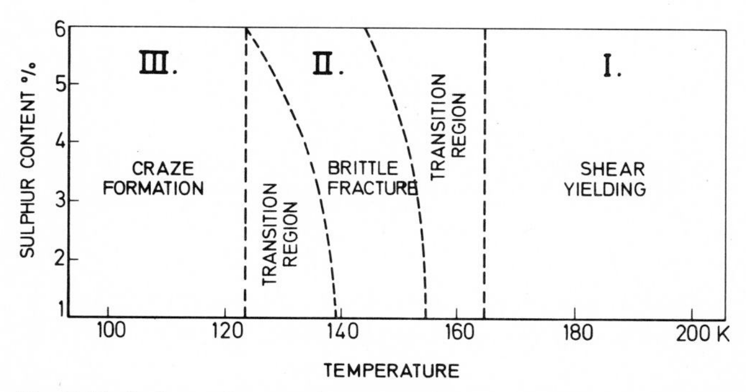

Fig. 7.21. Deformation and fracture behavior of preoriented polyisoprenes and polychloroprene (after Andrews and Reed, [31]).

increased with decreasing temperature [31, 67]. Following a transition region there was a narrow temperature band (II) between 150 and 130 K within which brittle fracture occurred at strains of less than 5%. For still lower temperatures – and only for these temperatures – the authors [29–31] made the very interesting observation that the deformation behavior was determined by the state or orientation of the stressed network. Unoriented rubber specimens behaved brittle just as they do within the temperature band above. A second group of samples was prepared by first pre-orienting them at room temperature by uniaxial straining to 100 to 200%, and then cooling them down to 90 to 130 K (region III). These samples showed a ductile deformation behavior (Fig. 7.21). In this case the ductile low temperature deformation was accompanied by microvoiding in narrow striations, the formation of detectable amounts of free radicals, and by absorption of environmental gases within the microvoids. These gases expanded upon warming of the deformed specimens and gave rise to a peculiar foaming effect. The observed free radical spectra of sulphur-cured polyisoprene have been assigned to allyl main chain radicals stemming from the scission of the (weakest) bond located between the two α-methylene groups. These radicals appeared together with a contribution of possibly $R\dot{S}$ radicals. The spectra of dicumyl peroxide-cured polyisoprene and polychloroprene were assigned to secondary peroxy radicals $RO\dot{O}$. The concentration of free radicals (in polychloroprene) varied with strain rate and yielded a maximum of $6 \cdot 10^{16}$ spins/g at a strain rate of $0.01\ s^{-1}$ [31].

Taking all these observations together, the following model of the deformation behavior of elastomers below their glass transition temperature is proposed: In region I intermolecular attraction is sufficiently strong to permit energy elastic deformation of chain segments. Gradually at first, and exclusively beyond the yield point, slip and reorientation of chain segments occur. Chain scission is negligible because chains will rather slip than break. In temperature region II where brittle fracture occurs irrespectively of the state of preorientation the intermolecular attraction seems to be just large enough to permit the axial loading of chain segments up to their breaking stress. In the absence of local strain hardening the largest crack initiated within the specimen during deformation to the 55% level will rapidly expand and prevent the growth of any other crack nuclei. As discussed for thermoplastics, the formation of essentially one plane of fracture is hardly sufficient to produce a detectable amount of free radicals. Unoriented rubber specimens (samples frozen in the relaxed state) show the same brittle behavior also in temperature region III (90 to 130 K). Specimen preoriented by extensions of 100 to 200%, however, obviously have gained the property of local strain hardening. The nature of this effect can only be speculated on and probably is related to the build up of partially oriented microfibrils within the preoriented elastomers. The result of local strain hardening is that the flaws and defects which are still present develop into a system of interconnected microvoids and extended fibrils forming the large and clearly visible striations. Environmental gases are adsorbed to the newly formed surfaces. Since the microfibrils are strong enough to carry the load transferred upon them the strained sample will not break but more and more defects will develop into crazes and striations. The largest number of striations appears at medium extension rates ($0.01\ s^{-1}$ in polychloroprene). The large local deformation (up to more than 100% strain) which is found throughout the total volume of the sample together with the strong intermolecular attraction between different chain segments leads to a large axial excitation – and breakage – of those chain segments which play the role of tie molecules. As the discussed results show, the spin concentration at break of elastomers in temperature region III is comparable to that obtained from 6 polyamide. In case of a *large* preorientation of 300% (in polychloroprene) a ductile deformation behavior and the formation of free radicals are observed [31]. The deformation is macroscopically homogeneous. No striations or crazes are visible indicating that the strain hardening is more effective than at lesser preorientations. The growth of microvoids obviously is stopped before they coalesce and form crazes. The absence of large scale gas adsorption and of foaming upon subsequent heating together with differences in the observed ESR spectrum support this view [31].

Brown, DeVries, and Williams duplicated these experiments and extended them to Hycar 1043, an acrylonitrile-butadiene rubber [32] and to Silastic E RTV, a silicone elastomer [33]. They confirmed for these polymers the above described effect of prestrain on the stress-strain behavior at low temperatures (118 to 193 K) the formation of free radicals with increasing sample strain, and the effect of strain rate on radical concentration with a maximum at a rate of $9 \cdot 10^{-4}\ s^{-1}$. For these polymers they did not observe, however, the three temperature zones mentioned above which are characterized by distinctly different types of mechanical response. They also propose that during preorientation numerous regions of higher strength and degree of orientation are created which later serve to arrest developing microcracks before those can attain a critical size. The fact that for the latter two polymers [32, 33] only one region of ductile and one region of brittle fracture were observed may be attributed to different physical and chemical properties of those networks (configuration, cross-link density, glass transition temperature). Some of these structural parameters will be studied in the following section.

B. Cross-Link Density, Impurities, Fillers

We have seen above that in low temperature ductile deformation of rubbers a large number of chain segments break which indicates that the local axial stresses exceed the strengths of the chain segments. We have shown earlier that the stresses required for bond breakage are more than two orders of magnitude larger than those applied macroscopically to deform the rubber specimens. The appearance of free radicals, therefore, indicates the highly inhomogeneous distribution of local stresses. The radical intensity gives an account of the number N of segments which have been stressed beyond their strength; N will depend, therefore, on all parameters which raise the local stresses like cross-link density or content of reinforcing fillers, which decrease the bond strength like presence of hetero-groups in the main chain, or which affect the number of observed radicals through radical reactions.

Andrews and Reed [31] observed an increase in radical intensity with increasing cross-link density of sulphur cured natural rubber (Fig. 7.22) using the preorientation-technique described above. This observation is well in keeping with the fact that tensile stresses of identically strained rubbery specimens increase with increasing cross-link density i.e. decreasing chain length between cross-links. The effect of impurities on the concentration of formed free radicals although clearly demonstrated by the data of Figure 7.22 is not fully understood. In the absence of conspicuous radical reactions or a direct effect on bond strengths it is believed that the impurities constitute stress raising inhomogeneities [31].

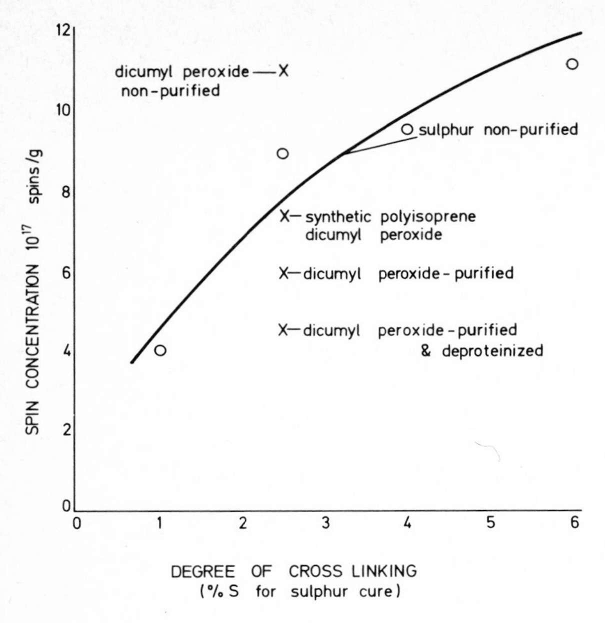

Fig. 7.22. Effect of crosslink density and crosslinking agent on the concentration of free radicals at fracture at 100 to 120 K (after Andrews and Reed [31]).

It is not always necessary to prestrain rubbers to induce the development of a large number of microcracks distributed throughout the bulk of the rubber sample. Wilde et al. [34, 35] have shown that certain granular fillers can promote bond rupture during sample fracture. Again the studies were by necessity conducted at temperatures below 210 K where the radicals were sufficiently stable. Fillers in elastomers are commonly used for a variety of reasons. In any case the presence of such fillers increases the complexity of the micromechanics of deformation and fracture. The presence of the filler results in very non-homogeneous localized stress fields ranging from levels of high strain concentration to complete separation between the matrix and filler (dewetting). The filler particles may serve as crack arrestors and/or initiators and thereby alter the fracture strength, resilience, and/or toughness. Wilde studied four systems all filled to approximately 50% by weight. These were:

1. 29 μm glass beads in EPDM,
2. silane treated 15 μm glass bead filled polyisoprene,
3. NaCl filled polyisoprene and
4. HiSil 233 (silica) filled polyisoprene.

The first system provided very weak interaction between filler and matrix, the last very strong interaction, and the other two demonstrated intermediate filler-matrix interaction. All systems were tested in temperature region I (170 to 200 K), a region within which an unfilled rubber without preorientation produces no free radicals in tensile straining. Three of the four filled systems, however, behaved differently. It was found that the untreated glass filled elastomer fractured with no detectable production of free radicals, fracture in the rubbers filled with silane treated glass and NaCl produced easily detectable concentrations of radicals ($3.21 \cdot 10^{14}$ spins/cm^3). The HiSil system produced at some temperatures and strain rates, up to an order of magnitude more radicals than NaCl or treated glass ($7.86 \cdot 10^{14}$ spins/cm^3). In his PhD thesis Wilde [35] details a rather comprehensive comparison of scanning electron microscope photographs with the ESR results. The photographs indicate that at room temperature dewetting started to occur in the untreated glass system at strains of less than 10 to 20%, in the treated glass beads and NaCl systems it occurred at strains of between 50 and 100%, and in the HiSil system at strains greater than 200%. Scanning electron micrographs of the fracture surfaces at all temperatures between 150 and 300 K revealed the same findings and showed 1. that untreated glass beads were sitting essentially loose in the "smooth" vacuoles or voids on the fracture surface, 2. that treated glass surfaces and NaCl behaved similar to untreated glass except that here the vacuoles were not so smooth and that there was material adhering to the filler particles, and 3. that the HiSil particles were embedded in the matrix at the fracture surface with the rubber adhering completely to the particle surfaces.

A slightly different two-phase system with strong bonds at phase boundaries is that obtained from triblock copolymers such as styrene-butadiene. As discussed in Chapter 2 such a copolymer molecule is composed of rigid (styrene) end blocks joined by elastomeric (butadiene) center blocks. The styrene blocks aggregate to form small domains which act as cross-links leading to a *rubberlike elasticity* of the block copolymer at *ambient temperatures* but to the possibility of *plastic deformation* at *high temperatures.* For an understanding of the fracture behavior of these systems it would be useful to identify in which of the phases molecular rupture most often occurs. The direct means of accomplishing this goal would be to fracture the material and analyze the hyperfine structure of the resulting ESR spectra. In the temperature range between liquid nitrogen and room temperature, however, tensile straining does not lead to sufficiently observable accumulations of free radicals. As a consequence, De Vries, Roylance, and Williams [36] adapted the less conclusive but more straight-forward method of comparing the spectrum of styrene-butadiene block copolymers (SBS) with those of styrene and butadiene alone. These studies were done on grindings to accumulate large amounts of fracture surface and at liquid nitrogen temperatures. The low temperature made the radicals more stable and hopefully "froze" the primary radicals. Comparison of the spectra of the three materials showed that the spectrum in the SBS contained all the lines of the butadiene radical but not of the styrene radical. The radical of the SBS system was, therefore, assigned to the butadiene phase. Unfortunately the study leaves open the questions as to whether or not the radical formed by grinding at low temperatures is the same as that presumably formed under normal conditions at room temperature and whether the observed radical is the primary radical or a secondary one.

Taking together the experimentally observed effects of the five parameters *impurity content, degree of cross-linking, ease of crystallite formation, strength of filler-matrix adhesion,* and *presence of different phases* on the nature and intensity of free radicals formed we derive at the following conclusions. All five parameters tend to increase the apparent cross-link density and decrease the extensibility of chain segments between cross-links. They thus enhance the forces acting at a given strain and also the ultimate forces that can be transmitted onto chain segments. Both conditions are necessary before chain stresses can exceed chain strengths. Furthermore all five parameters increase the number of actual or potential defect sites and cause a better spatial distribution of chain scission events. If chains fail they will always be located in the amorphous or more flexible regions of the (hetero-phase) elastomers.

III. Mechanically Relevant Radical Reactions

In Sections I and II of this Chapter primary and secondary free radicals have been treated as microprobes which characterize occurrence and molecular environment of chain breakages. As shown in Chapter 6 the primary mechano-radicals are always chain end radicals which are mostly unstable. At a rate depending on temperature these radicals will transfer the free electrons and thus "convert" to secondary radicals. This reaction and also subsequent conversion and decay reactions including recombination are relevant with respect to an interpretation of the fracture process in two ways. Firstly these reactions interfere with a determination of the concentration and molecular environment of the original chain scission points. Secondly they change the physical properties of other network chains through the introduction of unpaired electrons and the formation of crosslinks [67]. For a discussion of *spectroscopic details* and of the *stability* and *conformation* of the free radicals the reader is referred to the comprehensive monograph of Rånby and Rabek [37] and to the review articles of Campbell [38] and of Sohma et al. [39]. In the following the *mechanically* relevant aspects of radical transfer and decay are reported.

A. Transfer Reactions

At this point the geometrical aspect of transfer reactions is to be investigated. Through breakage of an elastically strained chain the primary radicals are formed at either side of the breakage site. Because of the axial chain stresses the new chain end radicals rapidly retract from each other. They are thus being prevented from recombination. The primary radicals then convert to secondary ones mostly by transfer of a hydrogen atom (see Table 6.1). If hydrogen transfer occurs *along* one and the same chain no effect on the physical network properties is to be foreseen. If, however, a hydrogen is abstracted from a *neighboring* chain the radical is transferred to a mid-chain position at a hitherto uneffected chain segment. There is strong evidence for the occurrence of both processes, the radical migration *along* a chain [39] and the hydrogen hopping *accross* chains [40]. The hydrogen hopping process was demonstrated by Dole et al. [40] who had molecular deuterium gas, D_2, diffuse into irradiated polyethylene; D_2 was then converted to HD by the following reactions:

$$R + D_2 \longrightarrow R_1D + \dot{D}$$
$$\dot{D} + R'H \longrightarrow \dot{R}' + HD.$$

Although in this experiment a gaseous tracer had been used it is believed that hydrogen abstraction from a hydrocarbon chain proceeds in the same manner. In view of the mobility of the chain end radical the hydrogen abstraction seems to be a logical vehicle for the transfer of a free electron to a central position within another chain segment.

The radical migration along a chain was found by Sohma et al. [39] to give the only plausible explanation for his observation that in PE and PMMA chain end radicals convert to a near-end radical:

$$\begin{array}{cccccccccccc}
 & H & & H & & & H & & H & & H & \\
- & C & - & C^{\cdot} & \longrightarrow & - & C & - & C & - & C & - H \\
 & H & & H & & & H & & \cdot & & H & \\
 & \multicolumn{3}{l}{(77\ K)} & & & \multicolumn{5}{l}{(5\ min.\ at\ 152\ K)} &
\end{array}$$

$$\begin{array}{ccccccccc}
CH_3 & H & CH_3 & H & & CH_3 & H & CH_3 & H \\
-C & -C- & C & -C^{\cdot} & \longrightarrow & -C & -C- & C & -CH. \\
R & H & R & H & & R & \cdot & R & H
\end{array}$$

(after short time milling at 77 K) (after 0.5 to 24 h. milling at 77 K)

Sohma indicates that also in these polymers radical migration along the chain and successive hydrogen abstraction are both important mechanisms of radical migration.

Transfer reactions are, of course, influenced by the presence of impurities or additives which are active as radical acceptors or radical scavengers. Their mechanical relevance lies in the fact that they reduce or prevent transfer of free radicals to central chain positions, radical recombination, and/or cross-linking. The latter aspect is discussed in Chapter 8. For a review of radical reactions with H_2, D_2, O_2, SO_2, NO and NO_2, NH_3, H_2S, and various halogens the reader is referred to the works of Campbell [38] and of Rånby and Rabek [37], the latter being especially concerned with the reactions of oxygen in its different forms (triplet and singlet molecular oxygen, atomic oxygen).

B. Recombination and Decay

The *normal decay* of macroradicals by recombination in the solid phase is a second order reaction [5, 11, 17–18, 37–39, 41–43]

$$[R(t)] = [R_0]/(1 + k[R_0]t). \tag{7.10}$$

The decay rate term kR_0 depends – as expected – on temperature: for unloaded fibers under He atmosphere values of $0.0014\ s^{-1}$ (75 °C) and $0.0110\ s^{-1}$ (100 °C) have been determined [18]. This value ties in with the $0.001\ s^{-1}$ measured at 50 and 60 °C under N_2 [11, 41]. It should be noted that the recombination reaction is slowed down by the application of tensile forces as for instance during the pre-fracture loading period: under near maximum tensile stress the decay rate terms were only 0.00035 s^{-1} (75 °C) and 0.00075 s^{-1} (100 °C) respectively [18]. The kinetics of decay is also affected by conversion reactions (e.g. of alkyl into allyl radicals) or catalytic effects [37].

The effect of the mobility of the radical site on the recombination rate has been clearly shown by the pressure [44–46] and temperature dependence [41–43, 47–49] of radical decay in e.g. PE, PP, PA 6, 66, and 12, and PPO. The decay curves of free radical populations in (irradiated) polymers as a function of increasing temperature reveal a number of plateaus each corresponding to radicals trapped in a particular morphological site. A transition from one plateau to the next indicates that one *species* of the trapped free radicals has become sufficiently mobile for recombination (chemically identical radicals in crystalline and amorphous regions may differ in steric configuration which also effects the rate of recombination [37, 42, 47]). The transitions are hence associated with the corresponding mechanical γ, β, and α dispersions [41–43].

As already indicated earlier apparent steps in those radical concentration curves which are obtained by straining fibers by equal increments at different temperatures [11, 19], are not solely related to dispersion regions of the modulus but reflect details of the relative segment length distribution as well.

For chemically different radical species generally different decay constants with different activation energies are observed. Tino et al. [42] determined for the central polyamide radical $-NH-\dot{C}H-CH_2-$ in the temperature region of 15 to 50 °C an activation energy for decay of 10 kcal/mol, for the chain end radical $-CONH-\dot{C}H_2$ a value of 5.2 kcal/mol.

Studies of the effect of hydrostatic pressure on free radical decay have the aim of elucidating the mobility and the distance of migration. A series of experiments has been carried out at the Polymer Institute of the Slovak Academy of Sciences in Bratislava. The rate constants of free radical decay decrease exponentially with increasing pressure [44–46]. At low temperatures the rate constant varies only slightly with pressure. The retarding effect becomes stronger at higher temperatures; the closer the decay temperature to the glass transition temperature T_g the more pronounced the stabilizing effect of pressure. Of course the pressure effect shows a saturation once the pressure is so large as to inhibit the molecular motion in question, 8 kb for the α relaxation in PE and PVAC at 80–110 °C [44], 15 kb for the α relaxation in 6 PA [44].

Another mechanical effect, the formation of additional cross-links in rubber vulcanizates stressed at low temperatures, has been investigated by Kusano et al. [67]. They point out that free radicals are formed in NR stressed at –195 °C; these radicals begin to react at $T > -40$ °C and apparently prefer a cross-linking reaction to a reaction with oxygen. An interesting technique to study the inverse phenomenon, the weakening of a network by breakage of cross-links, has been indicated by Huang and Aklonis [68] who used the photo-induced scission of disulfide linkages.

C. Anomalous Decay

Sohma et al. [39] investigated the thermal decay of mechano-radicals. They observed for PE, PP, and PTFE an increase in the concentration of free radicals while increasing the temperature from 77 to 170 K. They termed this behavior which is not found for irradiation-produced radicals, anomalous. The anomalous increase is enhanced by excess triboelectric charges of the sawed samples and by the presence of oxygen [39]. On the basis of their extensive investigations the authors advance a mechanism for the creation of free radicals through heat treatment at the rather low temperatures of 77 to 170 K. They postulate the formation of (spectroscopically neutral) carbanions during the degradation process at 77 K, here illustrated for PP:

$$-CH_2-(CH_3)\dot{C}H + e^- \rightarrow -CH_2-(CH_3)\dot{C}H^-$$

$$\uparrow$$

(carbanions, no ESR spectrum)

$$\downarrow$$

$$-(CH_3)CH-\dot{C}H_2 + e^- \rightarrow -(CH_3)CH-\dot{C}H_2^-.$$

During the "heat" treatment (100–173 K) the carbanions react with O_2 to form spectroscopically active peroxy radicals. The released electrons leak to the ground:

$$-CH_2-(CH_3)\dot{C}H^- + O_2 \rightarrow -CH_2-(CH_3)HCO\dot{O} + e^-$$

$$-(CH_3)CH-\dot{C}H_2^- + O_2 \rightarrow -(CH_3)CH-H_2CO\dot{O} + e^-.$$

D. Radical and Electron Trapping Sites

Mechanically relevant information has been obtained from an analysis of radical trapping sites. Thus it has been concluded in previous sections of this chapter that it is the amorphous regions of a semicrystalline polymer where mechano-radicals are formed. Apart from that changes in sample morphology during mechanical working have been investigated by ESR technique. Kusumoto, Takayanagi et al. [50–51] studied PE and PP films by successively removing the amorphous material through nitric acid etching. They then analyzed the ESR spectra obtained by γ irradiation of the so treated films. Thus they were able to correlate the octet observed in PP with radicals trapped at defect sites within the crystallites, the nonet with radicals in the amorphous fold surfaces. The latter are especially effective radical trapping sites. The authors analyzed the effect of quenching, annealing, and cold-drawing on the mosaic-block structure of their films.

From a change in line shape parameter of the PE sextet spectrum and of the temperature dependence of radical recombination rates the authors determined the increase in unoriented material content during mechanical fatigue [51]. For the study of the surfaces of as-grown and annealed PE single crystals Kusumoto et al. [52] also employed the spin-probe method.

The ESR technique has permitted to identify initial rupture points in mechanically degraded copolymers. As already discussed in Chapter 6 Lazar and Szöcs [53] investigated random, block and graft copolymers of methylmethacrylate and styrene (see 6 IV A).

By comparing the reactions of mechano-radicals in ball-milled PP with those of "γ-radicals" Kurokawa et al. [54] concluded that the mechano-radicals were produced and trapped *on the fresh surfaces* created by the breaking-up of solid polypropylene.

A quite interesting mechanism influencing polymer strength has recently been advanced by Zakrevskii et al. [69]. They propose the tunneling of an electron from a highly strained chain to an electron-trap. The chain thus becomes ionized and its rupture (into a radical and a radical ion) is, therefore, facilitated. The electronic properties of polymers and their influence on mechanical degradation still deserve more attention. Some aspects will be discussed in 9 III E.

References for Chapter 7

1. S. E. Bresler, S. N. Zhurkov, E. N. Kazbekov, E. M. Saminskii, E. E. Tomashevskii: Zh. tekhn. fiz. *29,* 358 (1959); Sov. Phys.-Techn. Physics *4,* 321–325 (1959).
2. J. Becht, H. Fischer: Kolloid-Z. u. Z. Polymere *229,* 167 (1969).
3. G. S. P. Verma, A. Peterlin: Polymer Preprints *10,* 1051 (1969).
4. H. H. Kausch, K. L. De Vries: Int. Fracture *11,* 727–759 (1975).
5. B. A. Lloyd: Ph. D. Thesis, Department Mechanical Engng. Univers. of Utah (June 1972); B. A. Lloyd, K. L. DeVries, M. L. Williams: J. Polymer Sci. A-2, *10,* 1415–1445 (1972).
6. J. Becht: Dissertation, Technische Hochschule, Darmstadt (1971).
7. H. H. Kausch, J. Becht: Rheologica Acta *9,* 137 (1970).
8. M. A. Gezalov, V. S. Kuksenko, A. I. Slutsker: Mechanica Polymerov *8,* 51 (1972); Polymer Mechanics USSR *8,* 41 (1972).
9. B. Crist, A. Peterlin: Makromol. Chem. *171,* 211–227 (1973).
10. W. H. Hassell: M. S. Thesis, Department Mechanical Engng. Univers. of Utah (June 1973).
11. U. Johnsen, D. Klinkenberg: Kolloid-Z. u. Z. Polymere *251,* 843 (1973).
12. H. H. Kausch: J. Polymer Sci. C *32,* 1–44 (1971).
13. D. Klinkenberg: 38. Dtsch. Physikertagung, Nürnberg, 23.–27. Sept. 1974. D. Klinkenberg, Dissertation, Technische Hochschule, Darmstadt 1978.
    D. Klinkenberg: Progr. Coll. & Polymer Sci. *66,* 341 (1979), and Coll. & Polymer Sci. *257,* 351 (1979).
14. M. Williams, K. L. De Vries: Proc. Fifth Internat. Congr. Rheology, S. Onogy, Ed., Vol. 3, University of Tokyo Press, Tokyo 1970, 139.
15. B. Yu. Zaks, M. L. Lebedinskaya, V. N. Chalide: Vysokomol. soyed *A 12,* 2669 (1970); Polymer Sci. USSR *12,* 3025 (1971).
16. H. H. Kausch, J. Becht: Kolloid-Z. u. Z. Polymere *250,* 1048 (1972).
17. J. Becht, H. Fischer: Kolloid-Z. u. Z. Polymere *240,* 766 (1970).
18. K. L. DeVries, B. A. Lloyd: Univers. of Utah, personal communication.
19. L. A. Davis, C. A. Pampillo: J. Polymer Sci., Polymer Phys. Ed. *11,* 841–854 (1973).
20. D. K. Backman, K. L. De Vries: Department Mechanical Engng. Univers. of Utah, June (1969).
21. D. K. Backman, K. L. DeVries: J. Polymer Sci. A-1, *7,* 2125 (1969).
22. T. Pazonyi, F. Tudos, M. Dimitrov: Angew. Makrom. Chem. *10,* 75–82 (1970).
23. R. J. Salloum, R. E. Eckert: J. Appl. Polymer Sci. *17,* 509–526 (1973).
24. T. Pazonyi, F. Tudos, M. Dimitrov: Plaste u. Kautschuk *16,* 577–581 (1969).
25. W. O. Statton: J. Polymer Sci., (Polymer Symposia) *32,* 219 (1971).

26. J. B. Park: Ph. D. Thesis, Department Mechanical Engng. Univers. of Utah (June 1972).
27. J. B. Park, K. L. DeVries, W. O. Statton: J. Macromol. Sci. – Phys. (USA) *B15*. 205 (1978); ibid. 409 (1978).
28. W. O. Statton: Univers. of Utah, personal communication.
29. E. H. Andrews, P. E. Reed: J. Polymer Sci. B-*5*, 317 (1967).
30. R. Natarajan, P. E. Reed: J. Polymer Sci. A-2, *10*, 585 (1972).
31. E. H. Andrews, P. E. Reed in: Deformation and Fracture of High Polymers, H. H. Kausch, J. A. Hassell, and R. I. Jaffee, Eds. New York: Plenum Press 1973, 259.
32. R. Brown, K. L. DeVries, M. L. Williams: Polymer Preprints *11*/*2*, 428 (1970).
33. R. T. Brown, K. L. DeVries, M. L. Williams: Polymer Letters *10*, 327 (1972).
34. K. L. DeVries, T. B. Wilde, M. L. Williams: J. Macromol. Sci-Phys. B7 (*4*), 633 (1973).
35. T. B. Wilde: Ph. D. Dissertation, Department of Mechanical Engng. Univers. of Utah (1970).
36. K. L. DeVries, D. K. Roylance, M. L. Williams: J. Polymer Sci. A-2, *10*, 599 (1972).
37. B. Rånby, J. F. Rabek: ESR-Spectroscopy in Polymer Research, Berlin, Heidelberg, New York: Springer 1977.
38. D. Campbell: J. Polymer Sci., Part D (Macromol. Rev., Vol. 4) 91–181 (1970).
39. J. Sohma, M. Sakaguchi: Advances in Polymer Sci. *20*, 109–158 (1976).
40. M. Dole, D. R. Johnson, S. Kubo, W. Y. Wen: Polymer Preprints *15*/1, 495 (1974)
41. F. Szöcs, J. Becht, H. Fischer: Europ. Polymer J. *7*, 173–179 (1971).
42. J. Tino, J. Placek, F. Szöcs: Europ. Polymer J. *11*, 609–611 (1975).
43. T. Nagamura, N. Kusumoto, M. Takayanagi: J. Polymer Sci. (Polymer Phys. Ed.) *11*, 2357–2369 (1973).
44. F. Szöcs, J. Tino, J. Placek: Europ. Polymer J. *9*, 251–255 (1973).
45. F. Szöcs, O. Rostasova, J. Tino, J. Placek: Europ. Polymer J. *10*, 725–727 (1974).
46. F. Szöcs, O. Rostasova: J. Appl. Polymer Sci. *18*, 2529–2532 (1974).
47. H. Kashiwabara, S. Shimada, J. Sohma: ESR Applications to Polymer Research (Nobel Symposium 22), Stockholm: Almqvist & Wiksell 1973.
48. T. Nagamura, M. Takayanagi: J. Polymer Sci. (Polymer Phys. Ed.) *13*, 567–578 (1975).
49. J. Tino, J. Placek, P. Slama, E. Borsig: Polymer *19*/2, 212–214 (1977).
50. N. Kusumoto, K. Matsumoto, M. Takayanagi: J. of Appl. Polymer Sci. A-1, *7*, 1773–1787 (1969).
51. T. Nagamura, N. Kusumoto, M. Takayanagi: Intern. Conf. on Mech. Behavior of Materials, Kyoto, Japan, Aug. 1971.
52. N. Kusumoto, M. Yonezawa, Y. Motozato: Polymer *15*/12, 793–798 (1974).
53. M. Lazar, F. Szöcs: J. of Polymer Sci. Part C *16*, 461–468 (1967).
54. N. Kurokawa, M. Sakaguchi, J. Sohma: Polymer J. *10*/1, 93–98 (1978).
55. O. Frank: Dissertation, Technische Hochschule, Darmstadt, 1984.
56. V. A. Marikhin, L. P. Myasnikova: J. Polymer Sci. – Polym. Symp. *58*, 97 (1977).
57. H. A. Gaur: Coll. & Polymer Sci. *256*/10, 64 (1978).
58. D. K. Roylance, K. L. DeVries: Research Report R76-9, MIT, August 1976.
59. D. Roylance: Int. J. Fract. *21*, 107 (1983).
60. V. V. Zhizhenkov, E. A. Egorov: J. Polymer Sci. – Polym. Phys. Ed. *22*, 117 (1984).
61. O. Frank, J. H. Wendorff: Coll. & Polymer Sci. *259*/11, 70 (1981).
62. V. A. Lishnevskii: Doklady Chem. *182*/1–3, 834 (1968).
63. R. E. Florin: J. Polymer Sci. – Polymer Phys. Ed. *18*, 1877 (1978).
64. C. Vasiliu Oprea, M. Popa: Colloid & Polymer Sci. *258*, 371 (1980) and ibid. *263*, 25 and 738 (1985).
65. H. H. Kausch: Polymer Eng. & Sci. *19*/2, 140 (1979).
66. R. P. Wool, A. T. Rockhill: J. Macromol. Sci. – Phys. *B20*/1, 83 (1981).
67. T. Kusano, K. Kobayashi, K. Murakami: Rubber Chem. & Techn. *52*, 773 (1979).
68. W. N. Huang, J. J. Aklonis: J. Macromol. Sci. – Phys. *15*/1, 45 (1978).
69. V. A. Zakrevskii, V. A. Pakhotin: Fiz. Tverd. Tela *20*, 371 (1978) (Soviet Phys. Solid State *20*, 214 (1978)) and Vysokomol. soyed A 23, 658 (1981) (Polymer Sci. USSR *23*, 741 (1981)).

Chapter 8

The Role of Chain Scission in Homogeneous Deformation and Fracture

B. Loading of Chains before Scission

To date two methods have been employed for the determination of the distribution $N(\psi)$ of (large) axial chain stresses:

- the indirect method of observing the formation of free radicals after a lowering of chain strengths through increasing temperatures
- the direct method of evaluating the stress-induced shift of infrared (IR) absorption bands.

The indirect method of obtaining stress-distributions is based on the fact that the strength of a chain segment decreases with increasing temperature. During an upward temperature scan, therefore, those chain segments will be broken successively whose strength $\psi_b(T)$ becomes smaller than their axial load. The method has been discussed in Section IC1 of Chapter 7 for groups of chain segments with different relative lengths L/L_0. The ensuing relative length distribution $N(L/L_0)$ can easily be transformed into a stress distribution $N[\psi(L/L_0)]$ with the aid of Eq. (7.1).

This indirect method has a rather restricted applicability. It is limited to a temperature region $T < T_1$ where interfibrillar slip is still negligible and to molecular stresses $\psi > \psi_b(T_1)$. Furthermore the method does not permit the solution of a problem inherent to all ESR measurements; namely to determine whether the absolute concentration of broken bonds is equal to one half the observed concentration of free radicals or whether it is (much) larger.

To overcome this difficulty Zhurkov and his collaborators began in 1965 to study by an independant method, the IR technique, the effect of axial stress on molecular chains [4–16]. Subsequently similar [35, 37–40] infra-red studies were carried out in the USA. All these investigations are based on the fact that frequency and intensity of skeletal or backbone vibrations of chain segments respond to a superimposed deformation of the vibrating skeleton. In polypropylene for example C–C stretch vibration modes have been assigned [16, 36, 41, 42] to the parallel dichroic bands (Fig. 8.1) at 1326, 1168, 1045, 975, 842, 456, and 398 cm^{-1}. All of these bands have been found to be stress sensitive [36]. The stress effect manifests itself as a shift of the frequency of maximum absorption and in the occurrence of a "tail" at longer wavelengths (Fig. 8.2).

The mechanisms of the distortion of IR absorption bands of stressed polymers have been discussed especially by Gubanov [7–9], Kosobukin [13], Vettegren and Novak [15] and Wool [36]. There is a general agreement that the distorted IR absorption band profile $D(\nu)$ can be related to a large number of separate oscillators with strongly overlapping absorption bands whose maxima are shifted by different amounts. The possible causes of the frequency shift of an individual stressed oscillator have been considered to be the quasielastic deformation of the harmonic oscillator (reduction of force constant under stress), the elastic increase of bond angles, changes of segments conformation, and the creation of defects. For small deformations one predicts [4–16, 36], that for the first three mechanisms there is a quite linear change of frequency with *molecular* stress ψ:

$$\Delta\nu' = \nu_\psi - \nu_0 = \alpha'\psi. \quad (8.1)$$

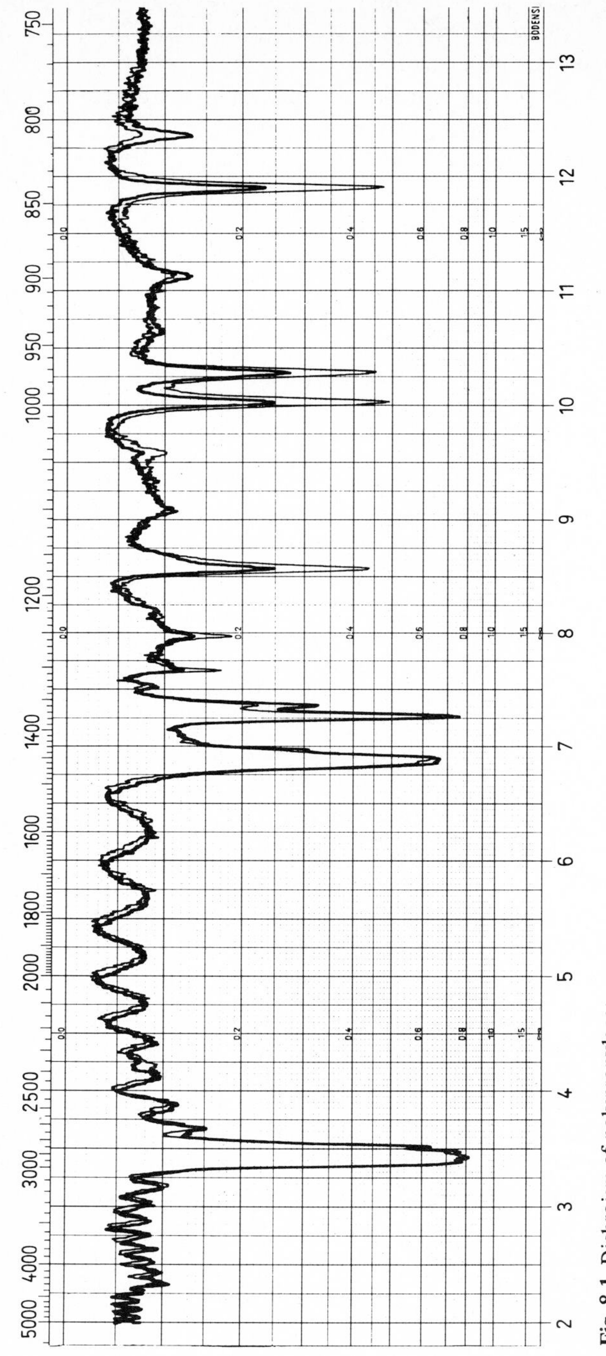

Fig. 8.1. Dichroism of polypropylene;
parallel dichroic vibrations at 842, 975, 998, 1045, 1168, 1254, 1304 cm^{-1}
perpendicular dichroic vibrations at 809, 899, 941, 1330 cm^{-1}.

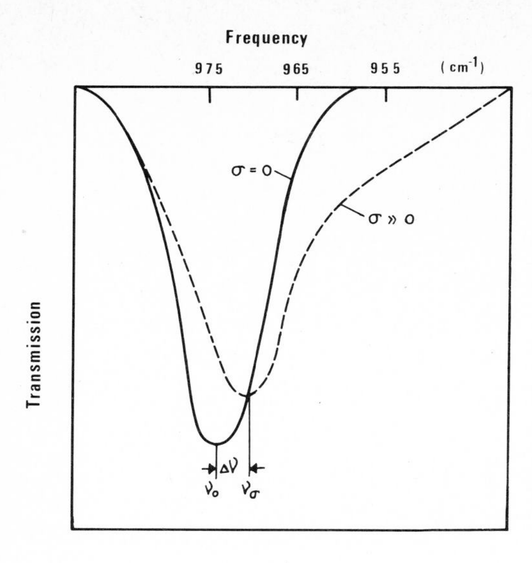

Fig. 8.2. Schematic effect of stress on a skeletal vibration of a semicrystalline, highly oriented polymer after ([5, 36]).

Wool and Boyd [221] have calculated values of α' for different vibrations of trimethyl octane (TMO) which can be considered as a model substance for a helical isotactic PP molecule. They conclude that application of an axial force on the terminal C atoms results in considerable torsion, bending, and bond stretching, the average ratios of $\Delta\phi/\phi : \Delta\theta/\theta : \Delta R/R$ being 9 : 2.4 : 1. Only anharmonic oscillators gave the negative frequency shift which is experimentally observed. The effect of stress and temperature on the nature of the frequency shifting (in PP) was further studied by Bretzlaff and Wool [222a]. They found that the only way to interpret the *asymmetric* wing on a stressed IR band was through the assumption of a distribution of "overstresses". These results were also supported by FTIR and Raman Studies of stressed ultraoriented polyethylene [222d]. They found that the Raman active C—C-stretching modes at 1030 cm^{-1} and 1160 cm^{-1} shifted with calculated frequency shifts using the anharmonic Morse potential function. The observed magnitudes of the frequency shifts for the CH_2 rocking, bending, wagging and stretching modes were also in agreement with their calculations. From the potential energy function of the polyethylene chain E_c was determined as 267 GPa (which is in reasonably good agreement with the value of 290 GPa used in this monograph). Wool also showed [222e] that the fracture stress of PE should be 45.5 GPa with an activation energy of about 112 kJ/mol.

As discussed previously molecular stresses are a function of molecular strains which depend on sample morphology and chain orientation. Any real sample, therefore, contains a variety of differently stressed oscillators. The profile of the deformed band, $D(\nu)$, representing such a system of oscillators can be expressed by a convolution integral:

$$D(\nu) = \int_{-\infty}^{\infty} F(n)\, U(\nu - n)\, dn\,. \tag{8.2}$$

Here F (n) is the distribution of stressed oscillators, U (ν) the shape of the normalized undeformed band, and n an integration variable. The distribution F (ν) can be obtained by deconvolution and transformed into a molecular stress distribution by Eq. (8.1).

As already indicated and shown in Figure 8.2 the distribution of molecular stresses in semicrystalline polymers contains a *peak of apparently homogeneously stressed chain segments* – related to the crystalline regions – and a *tail of very differently stressed chains* – related to highly stressed tie segments. For a *regular* sandwich structure of crystalline regions having a modulus E_c and a chain axis orientation described by cos θ one observes a frequency shift $\Delta\nu$ of the *crystalline peak:*

$$\Delta\nu = \alpha' \frac{E_k}{E_c} \sigma \cos^2 \theta = \alpha \sigma. \tag{8.3}$$

Thus, the experimental frequency shifting coefficient, a, depends on orientation, temperature, morphology, deformation rate and incident beam polarization [222c]. As shown in Fig. 8.3 quite different a values have been observed for identical bands in different polypropylene samples [36].

The 1168 cm^{-1} band of the same polymer has an even larger stress sensitivity of 10 cm^{-1} per GPa [15]. For the 930 cm^{-1} band of PA 6 α values of 2.2 [15, 16] and 3.7 cm^{-1}/GPa [223] have been reported; for the 974 cm^{-1} band of PETP α ranged from 3.6–6 [15, 16] to 8.4 cm^{-1}/GPa [223]. For the latter band depending on sample treatment Mocherla and Statton find stress sensitivity values of 12 to 20 cm^{-1} per GN/m^2 [40]. More recently both authors [16, 223] have obtained a stress sensitivity value of 8 cm^{-1} per GN/m^2 at 1350 cm^{-1} for polyacrylonitrile.

In the tail region of the "stressed 975 cm^{-1} band" of PP frequency shifts of up to 35 cm^{-1} are discernible which correspond to molecular stresses up to 12 GN/m^2. Although the significance and dynamics of "stressed-IR" investigations are not yet understood in full detail a number of interesting observations have been made:

- the straining of preoriented (PE, PP, PETP) samples initially leads to the expected decrease of gauche-isomers [178, 223–225];
- overstressed segments begin to appear (in PETP) at a stage of further sample deformation where the rate of gauche-trans transitions diminishes considerably [223];

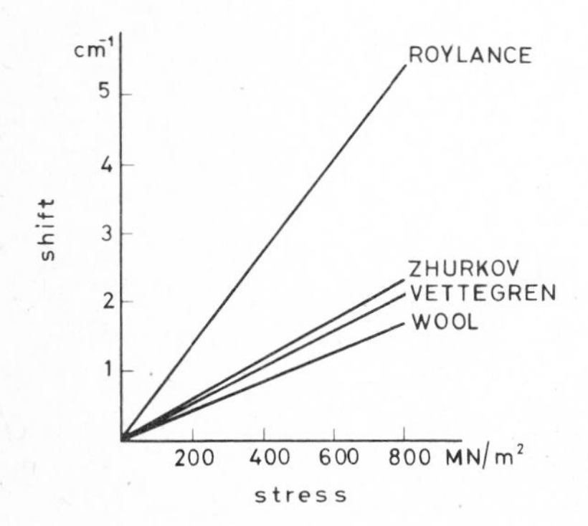

Fig. 8.3. Stress-induced shift of the peak of the 975 cm^{-1} absorption band of polypropylene as found by different investigators [6, 15, 34, 36].

- the integrated area of a deformed band (975 cm^{-1} PP) remained constant with σ increasing from 0 to 700 MN/m^2 [5] indicating that no new oscillators had been created under stress;
- the area fraction in that band related to overstressed segments increased with stress to 15% at $\sigma = 400$ MN/m^2 and 18% at 490 MN/m^2 [12]; for the 1170 cm^{-1} band an area fraction of 6% was reported [16] together with 4% [223] and 15% [16] for PETP (976 cm^{-1}), and 7% for PA 6 (930 cm^{-1}), all at a stress of 500 MN/m^2;
- it is interesting to note that the segment density 1/q L for PP segments of 5 nm length amounts to $56 \cdot 10^{19}$ segments per cm^3 of which 15% or $8.5 \cdot 10^{19}$ segments/cm^3 would be overstressed at $\sigma = 400$ MN/m^2;
- at a constant stress level this tail area is directly proportional to the content of non-crystalline material, i.e. the overloaded segments seem to be contained in the amorphous regions [5];
- the tail regions of the deformed bands (930 cm^{-1} PA 6; 970 cm^{-1} PETP; 975 cm^{-1} PP) show fairly well defined limits at the lower frequency end corresponding to *fairly well defined limits of sustained molecular stress* ψ; these limits are determined [6, 15, 16] to be 11–13 GN/m^2 in PP, 15–20 GN/m^2 in PETP (lavsan), 8 GN/m^2 in PAN and 16 to 20 GN/m^2 in PA 6;
- with some samples [12, 24, 25] these limits are already reached at moderate macroscopic stresses smaller than 50% of sample strength; in other samples the limits are only attained at a stress level close to the breaking stresses [16];
- more recently Vettegren et al. [384,385] advanced an extremely interesting *dilaton* theory: the overstressed bonds are found in those regions which are temporarily expanded by fluctuating thermal energy.

Figure 8.4 reproduces the results of Vettegren et al. [16] on the maximum stresses on molecular segments for three morphologically different samples each of PP and PETP; the – non-specified – morphological differences do not seem to effect the limiting stress value; in samples a) the largest molecular stress concentration

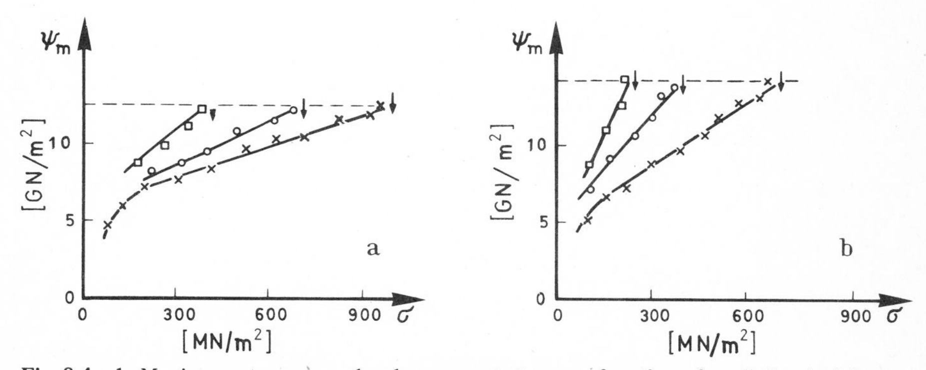

Fig. 8.4 a, b. Maximum stress on molecular segment, ψ_m, as a function of applied uniaxial stress σ for PP **(a)** and PETP **(b)** for samples of different strengths σ_b; σ_b indicated by ↓ (after [16]). T = 26 °C.

(ψ_m/σ) is larger than in b) or c) but in all three samples it is apparently extremely homogeneous as indicated by the fact that the maximum value of ψ_m is only reached at the maximum of σ; this behavior ties in with the observation [5] that the deformation of a band (975 cm^{-1} PP) of a highly drawn sample ($\lambda = 10$) is much less than that of a sample drawn to $\lambda = 5$;

- the observed molecular stress limit is also a function of temperature; in PETP Zhurkov et al. [6] determined a decrease from 22 GN/m^2 at 100 K to 15 GN/m^2 at 400 K (on the basis that the observed molecular stress limits correspond to the chain strengths ψ_b and using $\omega_0/k_b = 48 \cdot 10^{12}\ s^{-1}$ as for PA 6 segments one derives an activation volume of $11.7 \cdot 10^{-6}\ m^3/mol$ and an activation energy $U_0 = 316$ kJ/mol for the PETP segments); from later experiments Vettegren et al. [16] obtained activation volums of $13.2 \cdot 10^{-6}\ m^3/mol$ for PETP, $15.0 \cdot 10^{-6}\ m^3/mol$ for PA 6 and $15.6 \cdot 10^{-6}\ m^3/mol$ for PP;
- the dichroism of a band permits the determination of the average degree of orientation ($\overline{\cos^2\theta}$) of loaded chain segments; Zhurkov et al. [5] found for a PETP with $\overline{\cos^2\theta} = 0.75$ that the most highly stressed segments approached $\overline{\cos^2\theta} = 1$, i.e. complete uniaxial orientation (Fig. 8.5);
- the distribution of molecular stresses in PP varies during deformation (creep) and stress relaxation as indicated by the different time dependent intensity changes of fixed frequencies in the deformed part of the spectrum (dynamic polarized IR studies [6, 35])
- creep deformation tends to increase the highly loaded chain area of the 975 cm^{-1} band in PP at the expense of intermediately stressed segments [6, 35] further supporting the hypothesis that the band deformation is indeed due to a *shift of frequencies* and not to a *creation of new oscillators*

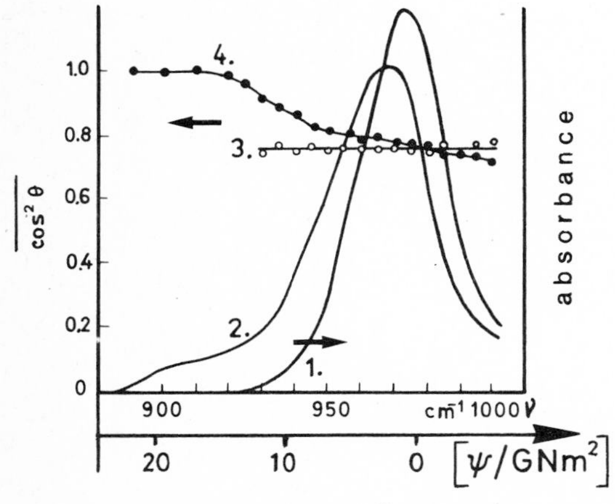

Fig. 8.5. Average orientation ($\overline{\cos^2\theta}$) of highly stressed PETP segments as a function of axial chain stress ψ; as abscissa the frequency shifts of the 974 cm^{-1} band and the corresponding axial chain stresses ψ (using a stress sensitivity factor $\alpha = 3.6\ cm^{-1}$ per GN/m^2) have been employed (after [5]). 1. unstressed IR band, 2. IR band in sample stressed at $\sigma = 500\ MN/m^2$, 3. average chain orientation as a function of frequency; $\sigma = 0$, 4. average chain orientation as a function of frequency at $\sigma = 500\ MN/m^2$.

- stress relaxation after shock loading of highly oriented PP showed an initial 10 to 70 s period of fast decay at practically constant degree of crystal orientation [35] as evidenced by the absorption of the 899 cm^{-1} band which is a highly dichroic, crystalline or "helix" band [43]; at later times a "steady state" relaxation takes place which is accompanied by a decrease of absorption of the 899 cm^{-1} band indicating a decrease of crystallinity or of helical regularity [35];
- in shock loading PP (e.g. up to a strain of 10.5% in less than 0.1 s) the largest absorption at 955 cm^{-1} is observed after t = 69 s when considerable stress relaxation has occurred whereas in ramp loading at a rate of 10%/min the largest absorption is attained at the maximum of stress at 10.5% strain; the largest increase in intensity at 955 cm^{-1} is (3.2 times) larger in shock loading than in ramp loading [38]; the molecular stress transfer in this highly oriented PP, therefore, is a viscoelastic process involving the deformation of the amorphous regions and the resistance to uncoiling of helices; Wool [39] has carried out a detailed experimental and computational analysis of stress relaxation, dynamic IR behavior and bond rupture; he concluded that different stress sensitivity values for crystalline regions (2.1 cm^{-1} per GN/m^2) and individual chains (8 cm^{-1} per GN/cm^2) should be employed, that the most highly stressed chains (952 cm^{-1}) relax first thus contributing to the increase in intensity at higher frequencies (e.g. 955 and 960 cm^{-1}), and that bond scission will be negligible if its activation energy U_0 is equal to or larger than 121 kJ/mol (29 kcal/mol); with U_0 = 105 kJ/mol very little chain scission occurs (affecting IR tail intensities by less than 0.3%), with 84 kJ/mol the majority and with U_0 = 63 kJ/mol (15 kcal/mol) all of the overloaded chains are expected to break; in his PP material chain scission did not seem to contribute measurably to the relaxation phenomena, this argument was backed by the observation that the absorption bands characteristic for chain-end groups remained constant within 0.5% [39]; on the other hand in the PP material of Vettegren et al. [16] stress relaxation at 100 °C was accompanied by the measurable appearance of chain-end groups.

The dynamic IR behavior of PETP has been investigated by Mocherla et al. [40]. Apart from the already discussed stress effects the authors observed the existence of an initial region of small stresses (up to 70 MN/m^2) within which stress effects (on the 973 cm^{-1} band of PETP) were absent. They concluded from extended studies of differently heat-treated films that this initial region coincides with the range of elastic response of the sample; obviously small elastic stresses are predominantly transmitted by secondary forces not giving rise to axial distortion of the chains.

Summarizing the above results from this important method one may say that in (highly) oriented samples a fraction of some 4 to 18% of the amorphous chain segments supports chain stresses ψ which are up to twenty times larger than the average sample stress σ. There are upper limits of molecular stress which in the case of polyamide 6 seem to be given by the chain strength (21 GN/m^2). On the other hand the upper limit of molecular stresses in the highly, uniaxially oriented polypropylene material of Wool is due to the inception of a distortion of helical chain conformations and of crystal lamellae. In that case no bond breakages could be traced.

In loading their oriented PP films Zhurkov, Vettegren et al. [6–16] observed, however, that an initially present concentration of $1-10 \cdot 10^{18}$ carbonyl groups per cm^3 increased [24]. They found this increase and the formation of other end groups (see below) related to the decrease of the number of overloaded chains [16]. The Russian authors concluded from the equality of the activation energies of thermal and mechanical destruction of their PP films (29 kcal/mol) with that of end-group accumulation (30 kcal/mol) that the kinetics of all three processes is controlled by the same molecular process, the thermomechanical scission of chain segments. According to the calculations of Wool, chain scission should practically never occur in PP if U_0 is equal to 29 kcal/mol. Those calculations, however, most probably refer to a constant temperature in the network of loaded and overloaded chains. Consideration of the large strain energy release at chain breakage would lead to the adjustment that local temperatures T are higher than the ambient temperature T_0 and that the effective chain strengths, $\psi_b(T)$, are smaller than $\psi_b(T_0)$. It seems likely, therefore, that the phenomenological observations do not contradict each other especially taking into account possible structural differences of the PP samples of the laboratories.

The upper limits of molecular stresses deduced for PETP ($15-20$ GN/m^2) and PAN (8 GN/m^2) were also related to chain breakage by the Russian authors [16]. With these materials, however, no comparable results on scission products (end groups, free radicals) have been reported. It must be concluded, therefore, that in these cases as with the PP material of Wool the upper stress limit seems to derive from the limited resistance of the crystal lamellae against distortion and break-up.

The latter conclusion is supported by the investigations of Becht and Kausch [44–48] concerning the deformation of highly oriented semicrystalline fibers. In a regular sandwich structure critical axial forces can be exerted onto tie segments if, and only if, the crystal lamellae can withstand stresses up to the chain strength. Otherwise crystal disintegration would precede chain scission. Using the spin probe technique, calorimetry and molecular weight measurements Becht [44–47] demonstrated the strain effect on crystal integrity. He irradiated samples of highly oriented 6 PA, 12 PA, PP, PETP, and PE by 1-Mev electrons at liquid nitrogen temperature. Subsequently, all samples were heated to their glass-transition-temperature (or above) for at least 5 minutes; thus, all radicals in the amorphous phase disappeared, and radicals remained within the crystallites only. These samples were then strained within the cavity of an ESR spectrometer at room temperature.

The rate-determining steps of the radical-decay reactions in these cases are transfer reactions and oxygen diffusion. They are both very strongly correlated with the mobility of the radical-carrying molecular segments with respect to their surroundings, i.e., the crystal lattice. An increase in these rate constants, therefore, is a measure of an increase in mobility, which may be due to a decrease in intermolecular attraction and/or to the introduction of new crystal defects.

In Figure 8.6, the concentration of free radicals is plotted versus a reduced time $K_0 \Delta t$, where K_0 is the rate of radical decay at zero strain and Δt is the time elapsed from the inception of sample straining. It should be noted that in PE, PP, PETP, and 12 PA, the application of sample strain has a twofold result: an immediate and irreversible decrease of the (crystalline) radical population and a stress-dependent re-

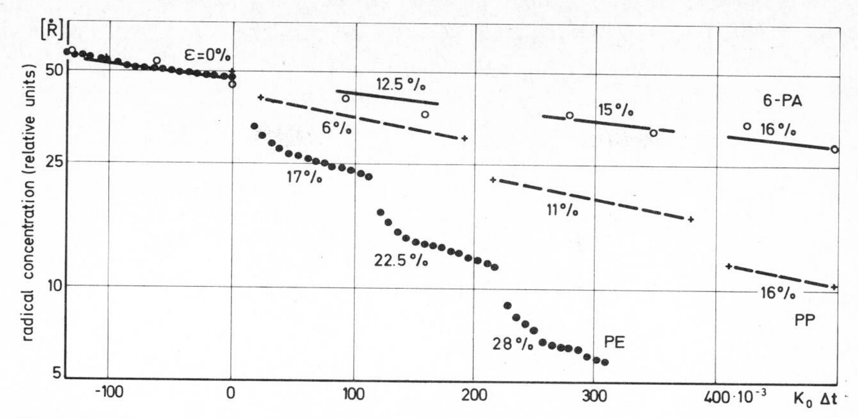

Fig. 8.6. Effect of sample strain ϵ on crystal distortion as evidenced by the rate of radical decay in irradiated and annealed highly oriented fibers (after [48]).

versible increase in the rate of decay. In 6 PA neither of these two effects can be observed. It was concluded from this experiment that in strained 6 PA, the tie molecules fail before axial chain stresses are reached which are sufficient to distort the crystal lamellae. In PE, PP, PETP, and 12 PA, however, forces can be transmitted which clearly affect the crystal lamellae. It is logical to assume that these forces are primarily transmitted by the tie segments. One would have to conclude that the axial load bearing capability of microfibrils of PE, PP, PETP and 12 PA is limited through the beginning crystal distortion. Thus chain strengths are not fully utilized in these polymers.

Still another combination of chain and crystal properties seems to exist in the PE films used by Vettegren et al. [15–18] for their end-group studies. Under stress vinyl, methyl, and oxygen-containing end groups formed readily and in large concentrations while no stress effects on IR bands were reported.

C. Spatially Homogeneously Distributed Chain Scissions

1. As Analyzed by IR Technique

The scission of molecular chains generally leads to highly reactive chain end radicals and eventually to the formation of new end groups (cf. Chapters 6 and 7). Unless oxygen is excluded the following end groups are formed most frequently: methyl, aldehyde, ester, carbonyl, carboxyl, or vinyl groups. All of these groups have characteristic IR absorption bands (Table 8.1). Based on this consideration Zhurkov and his collaborators quantitatively investigated the build-up of new groups in loaded polymer films [16–18, 22–26]. They employed a double beam technique the principle of which is illustrated in Figure 8.7. The optical densities $D = \ln I_0/I$ of a stressed (σ) and an unstressed reference film (r) are compared:

$$\Delta D = D_\sigma - D_r = k(C_\sigma d_\sigma - C_r d_r). \tag{8.4}$$

Table 8.1. IR Absorption Bands Suitable for End-Group Analysis [17, 18, 24, 39].

(End) Group	Structure	Absorption Band ν (cm^{-1})	Absorption Coefficient [24] k (10^{-19} cm^2)
Vinyl	$-CH_2-CH=CH_2$	910	2.52
Vinylene	$-CH_2-CH=CH-CH_2-$	965	1.46
Aldehyde	$-CH_2-CHO$	1070	2.52
Methyl	$-CH_2-CH_3$	1379	1.02
Vinyl	$-CH_2-CH=CH_2$	1645	0.68
Carboxyl	$-CH_2-COOH$	1710	8.92
Aldehyde	$-CH_2-CHO$	1735	2.47
Ester	$-CH_2-COO-CH_2-$	1745	6.36

Here k is the absorption coefficient, d the thickness of the film, and C the concentration of end-groups.

With the advent of computer aided IR spectroscopy the *difference spectroscopy* (be it in transmission, reflexion, or dispersion) has gained considerably through the ease of manipulation and the increased sensitivity [226–230]. Nevertheless, in interpreting the lines appearing positively or negatively in a difference spectrum some particularities of method or sample must be taken into consideration. Thus, the limited signal/noise ratio and differences in sample thickness and homogeneity, and in the physical state (orientation, phase separation) cause and influence the obtained difference spectrum even if the chemical composition of the compared samples is identical [227–230]. Figure 8.8 shows one of the first difference spectra recorded by Zhurkov et al. [17, 18] from a polyethylene film ruptured at room temperature. The spectra have been evaluated with respect to the kinetics of accumulation of new groups. It must be underlined that Zhurkov and his collaborators interpreted the appearance

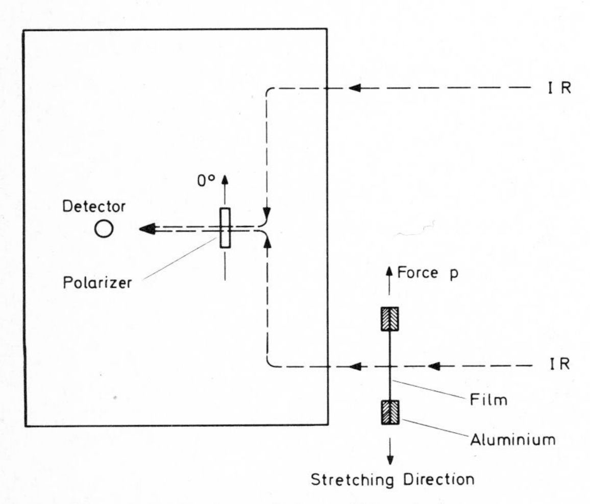

Fig. 8.7. Principle of double-beam IR technique.

of these new groups to be the direct result of chain scission, each chain scission event giving initially rise to two free radicals and subsequently to two new "end" groups. However, since some of these groups are formed by a chain of radical reactions with oxygen no quantitative transformation is guaranteed. A chain-end radical which transforms into a peroxy radical could easily react with a neighboring segment:

$R\,\dot{C}HX + O_2 \rightarrow R\,CHXO\dot{O}$ (formation of a polymer hydroperoxide)

$R\,CHXO\dot{O} + R'CH_2R'' \rightarrow R\,CHXOOH + R'\dot{C}H\,R$ (propagation reaction)

It is well documented [208 b] that the latter free radical continues the reaction chain multiplying the number of newly formed, oxygen-containing groups which are not indicative of a primary chain scission event. The same is true for some stress-activated ozone reactions which could give rise to the formation of oxygen-containing groups and which *do not involve* primary chain scission. This mechanism has recently been studied by Zaikov et al. [356], it will be discussed in Section 8 III B 1. It follows from these observations that the kinetics of accumulation and the concentration of "IR-groups" will, to a certain extent, be influenced by the nature and kinetics of the multiplication reaction. This should be kept in mind in the following discussion; it may explain some of the discrepancies which exist with respect to the concentrations of defects at fracture determined by different methods (see Section C 2 and Table 8.3).

Subjecting uniaxially oriented films to uniaxial loads Zhurkov et al. [16–18, 22–26] measured by such a technique the accumulation of aldehyde groups in high pressure and low pressure polyethylene and in polypropylene after predetermined loading intervals. Typical results are reproduced in Figure 8.9. The concentration C(t) of aldehyde-end groups increased with loading time according to a first-order kinetic equation:

$$C(t) = C^* (1\text{-}\exp\text{-}Kt) \qquad (8.5)$$

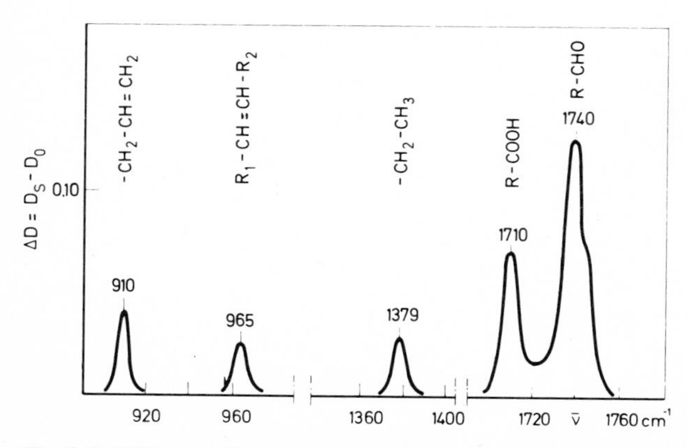

Fig. 8.8. Difference IR spectrum of unstressed and stressed oriented polyethylene (after [16–18]).

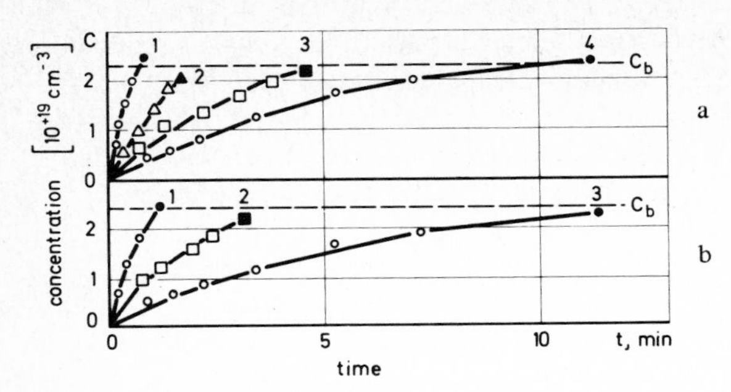

Fig. 8.9 a, b. Accumulation of aldehyde-end groups formed in high density polyethylene (after [16–18]). **(a)** T = 20 °C; σ (MN/m²): 1. 360, 2. 350, 3. 330, 4. 280; **(b)** σ = MN/m²; T (°C): 1. 67, 2. 46, 3. 20.

with rate constant K and apparent concentration C* of mechanically excitable – and thus breakable – bonds. The authors found that the concentration C* depended only slightly on temperature and stress (changes from 1 to $4 \cdot 10^{19}$ cm^{-3} for stresses between 180 and 500 MN/m² at temperatures between –70 °C and +20 °C) whereas the rate "constants" K varied by several orders of magnitude (Fig. 8.10). The authors noted that the rate constant K for accumulation of end groups was governed by an effective, stress-dependent activation energy E_* which is plotted in Figure 8.11. Using the apparent linear relation between E_* and σ one obtains:

$$K = K_0 \exp[-(E_A - \alpha_* \sigma)/RT]. \tag{8.6}$$

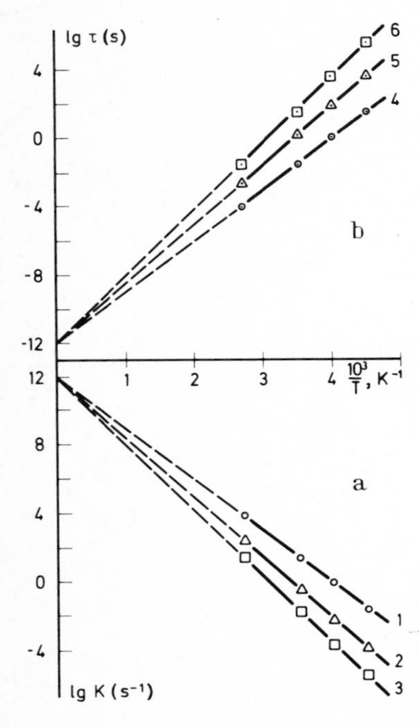

Fig. 8.10. Rate constants K for end-group accumulation and times to fracture τ for high density polyethylene (after [16–18]); σ(MN/m): 1. 400, 2. 300, 3. 200, 4. 400, 5. 300, 6. 200.

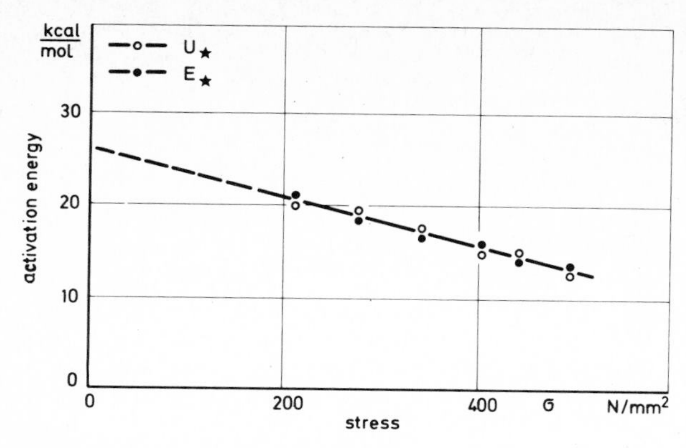

Fig. 8.11. Apparent activation energy. U_* for macroscopic breakage under uniaxial stress σ; E_* for accumulation of end groups under stress σ (after [18]).

The activation energies E_A and activation volumes α_* for the three polymer films investigated are listed in Table 8.2 together with the preexponential factors K_0.

In these experiments the authors also observed that the concentration of end groups at break for a series of specimens did not depend on stress, temperature, or type of loading (Fig. 8.12). The apparent "critical concentration" of *all* newly formed end groups in PE-HD amounted to $3.9 \cdot 10^{+19}$ cm^{-3} or to $\simeq 2 \cdot 10^{+19}$ bond ruptures per cm^3. Vettegren et al. [70, 383] estimate that in a *surface layer* of 1 μm thickness the concentration of end groups may even be higher by a factor of 20 to 60. For polypropylene they derived concentrations of $6 \cdot 10^{19}$ cm^{-3} broken bonds in the surface layer – as compared to $0.25 - 0.40 \cdot 10^{19}$ cm^{-3} reported by Zhurkov et al. [17, 18] for the interior of the film. The latter value should be compared with the number of overloaded 5 nm segments. In Section I B a concentration of $8.5 \cdot 10^{19}$ cm^{-3} overloaded segments had been determined from the area of the tail region of the "stressed IR band". Such a comparison indicates that of $N_0 = 56 \cdot 10^{19}$ cm^{-3} segments present in the interior material of the PP film $0.15\ N_0$ are initially overloaded and $0.0045\ N_0$ will have been broken at the point of mac-

Table 8.2. Kinetic Parameters of End Group Formation in Certain Polymer Films (after Zhurkov, [18]).

Polymer	Preexponential factor K_0 (s^{-1})	Activation energy E_A (kcal/mol)	Activation volume α	
			$\frac{\text{kcal mm}^2}{\text{N mol}}$	$\frac{10^{-6}\text{m}^3}{\text{mol}}$
Polyethylene (low-pressure)	10^{11}	27	0.026	110
Polyethylene (high-pressure)	10^{12}	28	0.059	250
Polypropylene	10^{12}	30	0.051	210

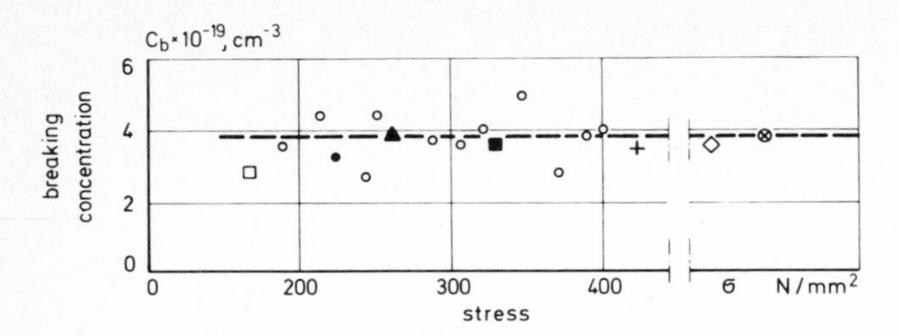

Fig. 8.12. Concentration C_b of end groups just prior to macroscopic fracture of high density polyethylene (after [18]); conditions of loading: samples broken in creep at constant stress at (+) 200 K; (■) 273 K; (○) 293 K; (□) 373 K; (●) 333 K; (▲) 323 K; (◇) samples broken under linearly increasing deformation with time; (⊕) samples broken under stress increasing linearly with time.

roscopic fracture. In another study on PP [25] a concentration of broken bonds of $3 \cdot 10^{19}$ cm^{-3} is reported, i.e. a fraction of 35% of the overloaded bonds is broken throughout the total volume of the sample. In the latter case wholesale destruction of the polymer must have occurred since on the average *each molecule* in a (monodisperse) M_n = 50000 g/mol sample *would have been broken 2.5 times.* In the former case the effect is ten times smaller, but M_n is still reduced to 40000 g/mol and M_w to 43750 g/mol. The *fracture criterion* derived for the three polymer films on the basis of the kinetics of thermal and mechanical destruction and of end-group formation was that of a *critical concentration of newly broken chains* [16–18]. In later publications this model has been varied and enlarged. Thus it was proposed [384–388] that the overstressed bonds are due to density fluctuations (*dilatons*) extending over hundreds and thousands of atoms. When these regions disintegrate they should give rise to "fracture-nucleating" submicrocracks.

The IR-technique has also been applied to polymer stress effects by other laboratories particularly in the USA [231–237] and in Germany [228]. Whereas it was generally qualitatively confirmed that in stressed films new groups, especially carbonyl and vinyl groups, do appear, none of the laboratories was able to repeat the detailed kinetic analysis. In view of the fact that the detection of an intensity difference requires for most bands about 10^{19} new groups [238–240], one realizes that for stress induced defects the IR technique operates at the limits of detectability – especially if chain scission is accompanied by structural changes. In the following the absolute number of chain scission events and its effect on fracture progress will have to be discussed.

2. Final Concentration of Chain Scission Events as Determined by Different Techniques

The breakage of a highly stressed chain can potentially be traced in several ways:

- through the formed free radicals
- after radical reaction through new end-groups
- through the ensuing molecular weight reduction
- through the energy irreversibly dispersed in (repeated) mechanical loading.

If the number of breaking chain segments is as large as indicated by the IR *group analysis* and if each segment breaks at the limiting stress value derived from the *analysis of deformed-IR bands* then the accumulated molecular stresses would be comparable within an order of magnitude to the applied macroscopic stress. In that case it must be assumed that apart from conformational rearrangements and chain slippage chain scission events noticeably influence observed stress-strain curves. So far the PP material from Leningrad is the only polymer which has lent itself successfully to both of the above mentioned IR analysis. In the cited literature [4–33] there is no reference to an attempt to explain stress-strain or stress-time data of this PP material on the basis of its end group-formation kinetics.

As discussed frequently within this text *stress-relaxation* data can not and must not be related solely to chain scission [2–52]. Attempts have been forwarded, however, to explain the *stress-strain* curves of polyamide-6 [49–51] and PEOB fibers [52] on the basis of free radical production kinetics. In these models the concept was employed that tie segments show a distribution of relative lengths. The *(tautness)* distribution, $N(L/L_0)$, was assumed to remain unchanged in a wide time-temperature regime. As detailed in Chapter 5, a tie-chain segment will respond to a strain ϵ_a predominantly rubber-elastically if $\epsilon_a < (L-L_0)/L_0$, it will break if $\epsilon_a > (1 + \psi_b/E_k)L/L_0 - 1$, and it will respond energy-elastically in the intermediate region where it happens to be fully extended and subcritically loaded. From these considerations a homogeneous four-phase model can be derived (Fig. 8.13). It is based on the assumption that a highly oriented semi-crystalline fiber can be represented by crystalline and amorphous regions in series with the latter consisting of three fractions, i.e. of the fully extended segments, of the non-extended tie-chains, and of the remaining matrix (folds, ciliae, broken chains). Once the strain-dependent widths of the fractions are obtained from a relative segment length distribution and appropriate moduli are assigned to the individual phases a *stress-strain* curve can be constructed. Such a procedure was proposed in 1971 by DeVries and Williams [49]. It has been executed by Lloyd [50] and by Klinkenberg [51] for PA 6 fibers and by Nagamura et al. [52]

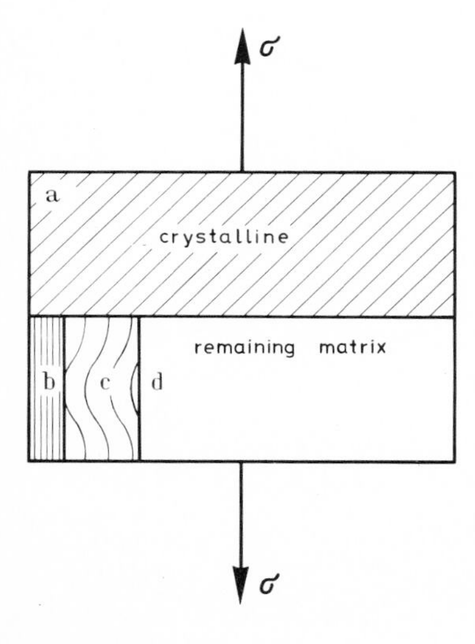

Fig. 8.13. Four-phase model of highly oriented semicrystalline fiber. **(a)** crystalline lamellae, **(b)** fully extended tie segments, **(c)** non-extended tie segments, **(d)** matrix (folds, ciliae, broken chains).

for PEOB multi-filaments. The calculated stress-strain curves can be made to fit the stress-strain response of virgin material; they contain as possible sources of uncertainty the *choice of the "elastic" moduli* and of the *number and distribution of tie-segment lengths,* i.e. of $N(L/L_0)$. The choice of the form of $N(L/L_0)$ determines which number of segment lengths is assigned to the fraction of non-extended chain segments by way of extrapolation. Lloyd [50] and Nagamura et al. [52] have chosen a normal distribution to represent $N(L/L_0)$.

A fit of observed and calculated stress-strain curves of first and second stretching cycles was obtained if the calculated number of radicals formed (or of chains broken) was taken as considerably larger than the observed number. Nagamura et al. [52] report a factor f_c of 40 for PEOB fibers, Klinkenberg [51] of 20 for PA 6 fibers. If one wishes to explain such a discrepancy while maintaining the basic model one is forced to make additional assumptions:

- either the number of broken chains is systematically larger by f_c than the number of observed radicals, e.g. due to a Zakrewskii-mechanism [20] to be discussed later;
- or the breakage of N_1 chains in a particular volume element leads to the preferential unloading of f_c N_1 extended chains; this amounts to the consideration of a microfibrillar substructure;

In recent years this question has been further discussed on the basis of more extended comparative studies also including molecular weight measurements [220, 231–238, 241–246]. Thus, Gaur [241] and Frank and Wendorff [243] analyzed the stress-strain behavior of PA 6 filaments (Fig. 8.14). They determined the amount of energy W_{irr} irreversibly dissipated by defect formation in a first load cycle ① in Fig. 8.14), the viscous energy dissipation ② and the elastically stored energy ③. They observed that almost up to fracture there was a linear correlation between W_{irr} and

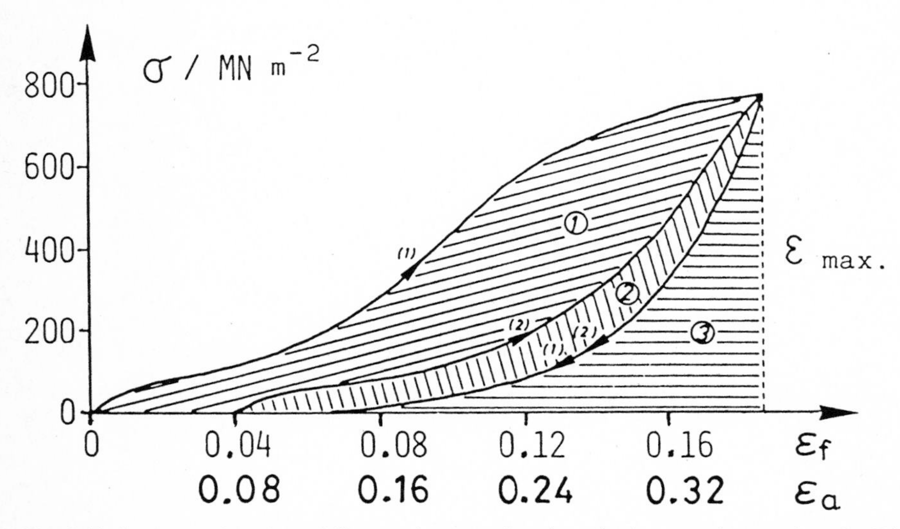

Fig. 8.14. Stress as a function of fiber strain ϵ_f or of strain ϵ_a in the amorphous regions for PA 6 fibers (from [243]). (1) First loading cycle, (2) Second (and further) loading cycle(s), ①Irreversibly dissipated energy, ②Viscous energy dissipation, ③Elastically stored energy.

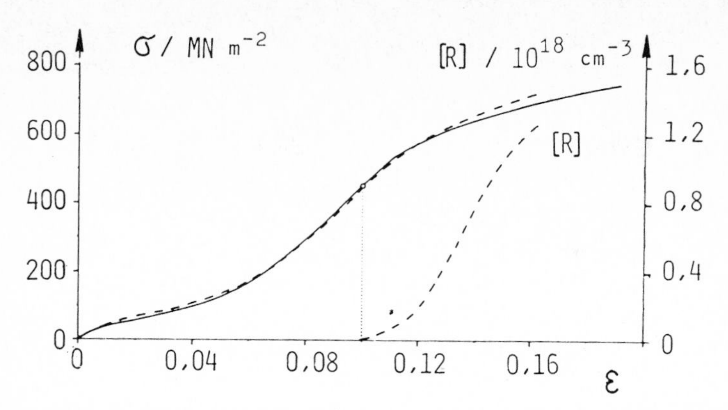

Fig. 8.15. Stress-strain diagram of a PA 6 single filament (–) and of a fiber (- -); and concentration [R] of formed free radicals (from Frank and Wendorff [243]).

free radical concentration [R·] with a ratio of $15 \cdot 10^{-18}$ J spin^{-1}. They also noted that the inflection point of the stress-strain curve coincided with the beginning of the free radical production (Fig. 8.15). For this material and some other PA fibers Frank [238] also gave the decrease in molecular weight provoked by chain scission during straining up to ~20% (Table 8.3).

In the following it will be attempted to reconcile some of the above, contradictory observations. To do so the model developped in Chapter 7 will be used. Thus, it will be assumed that the above defects are essentially located within the microfibrils. As indicated in Chapters 2 and 7 the microfibrils can be modelled as a series arrangement of crystalline and amorphous regions, where the latter are traversed by highly extended tie molecules of various degrees of tautness. The fraction of taut tie molecules in PA 6 has been estimated by Friedland et al. [247] from analysis of the stress effect on IR absorption bands to vary between 0.03 and 0.05; more recently Zhizhenkov and Egorov [248] proposed, on the basis of NMR results, values between 0.04 and 0.10 depending on sample preparation. The mechanical estimates of Mishra et al. [249], who deduced from a Takayanagi model tie-chain fractions of between 0.05 (at draw ratio 2.5) and 0.20 (at 4.5) seem to be somewhat high (see also [397]).

The careful X-ray studies of Frank [238] have again confirmed that (at room temperature) the macroscopic strains ϵ_f also apply to the change of the long period L of the microfibrils. In view of the large longitudinal stiffness of the crystalline layers (E_c = 183 GN m^{-2}) and at a degree of crystallinity of 50% one easily calculates the strain ϵ_a of the amorphous regions to be:

$$\epsilon_a = \frac{\Delta L_a}{L_a} = \frac{\Delta L - \Delta L_c}{L_a} = \frac{\epsilon_f L - 0.5\epsilon_c L}{0.5L} = 2\epsilon_f(1 - 0.5\ \epsilon_c/\epsilon_f) \sim 2\epsilon_f; \qquad (8.7)$$

ϵ_a is also indicated in the abszissa of Figure 8.14. It should be kept in mind that the assumption of a homogeneous strain within the molecularly quite heterogeneous amorphous regions is restricted to distances of the order of 5 nm and is essentially based on the large lateral stiffness of the crystalline regions. It becomes clear from Figure 8.14 that there are two regions of mechanical response of a microfibril: the small strain region with $\epsilon_a < 0.08$ and the large-strain region with $\epsilon_a > 0.15$. Frank carefully investigated this response. For fibers held at 50% relative humidity, he measured in a first load cycle an initial modulus E_1 of 4.6 GN m^{-2}; taking into consideration the existence of an intermicrofibrillar amorphous fraction of 0.4 he deduced in his PhD thesis for the microfibrils a modulus of the amorphous regions of $E_{1a} = 2.88$ GN m^{-2} In a second load cycle he obtained $E_2 = 1.80$ GN m^{-2} and $E_{2a} = 1.13$ GN m^{-2}. He also determined the moduli of humidity saturated amorphous regions as $E_{1ah} =$ 1.00 GN m^{-2} and $E_{2ah} = 0.75$ GN m^{-2}. The decrease of the elastic modulus after a first load cycle is a common observation and has correctly been explained by the "destruction of some structure" [see e.g. 238, 249].

The following molecular interpretation can be given to the observed small-strain behavior. In this region stress is essentially transmitted in a continuous manner [391] through the hydrogen-bond coupled segments; the elastic modulus decreases with strain [238, 249] due to the gradual destruction of hydrogen bonds with segment deformation and orientation. If some *completely* extended segments should be present within the amorphous regions at zero strain, their number must be extremely small so as not to change the strain softening behavior.

In a second region, ($0.08 < \epsilon_a < 0.20$), E_a increases and this behavior is determined by the combined entropy- *and energy*-elastic loading of fairly extended tie chains [238, 242, 247–251]. The stressing of a partly extended chain of contour length L_i provokes to some extent conformational transitions towards an all-trans conformation; evidently, it depends on the nature and intensity of the *inter*molecular interactions whether entropy- or energy-elastic behavior prevails in this region.

If it is assumed that within a strained amorphous region a chain is at first fully extended and only thereafter strained elastically then energy elastic chain stresses ψ_i will only be built up in "taut" chains whose contour length L_i is smaller than $L_a(1+\epsilon_a)$:

$$\psi_i = E_k \quad \frac{L_a}{L_i} (1+\epsilon_a) - 1 \quad . \tag{8.8}$$

Given an amorphous strain ϵ_a of 20% and a chain stiffness E_k of 200 GN m^{-2}, one arrives at the result that ψ_i of chains with $L_i \leqslant 1.086\ L_a$ exceeds the chain strength of 21 GN m^{-2}, they must break. As has been shown in detail in Chapter 5 the condition $L_i \leqslant 1.086\ L_a$ is equivalent to demanding that there are 5 or less $gt\bar{g}t$ kinks in an otherwise all-trans planar segment. Such highly extended chain segments do exist in PA fibers and they start breaking at about these strains (ϵ_a>20%). In their publications Frank and Wendorff give values of $1.5 \cdot 10^{-17}$ [243] and $4.4 \cdot 10^{-17}$ J/spin [238] for the ratio of irreversibly dissipated energy and free radical concentration which corresponds to an energy dissipation of 18 and 60 mJ per mol of chain scission events. These values are 20 to 69 times larger than the value of 870 kJ mol^{-1} to be calculated in Section E for the scission of highly stressed amorphous chains of 5 nm length solidly anchored in two opposite crystal lamellae. A small portion of this difference can be ascribed to the fact that the chain segments breaking towards the end of the loading are longer than 5 nm; they will, therefore, contain (up to 15%) more elastic energy. The major portion of the energy difference, however, will have to be sought elsewhere. Frank et al. propose that the free radical concentration *determined by ESR* is not a correct measure of the number of chain scission events since rapid radical recombination or chain scission without radical formation may occur. In that case, one chain scission event would not give rise to two free radicals seen by ESR but to a smaller number. Their hypothesis is based on the evaluation of the (slight) decrease in molecular weight M_n of their samples after rupture (Table 8.3).

This may be perfectly true. But even then it is claimed that the principal source of energy dissipation is the stress-induced conformational transition of partly extended chains containing a few double kinks. The annihilation of a double kink would increase the end-to-end distance of a chain segment by $\Delta L = 0.26$ nm. If the segment had been subjected to 5 GN m^{-2} then an energy dissipation of $0.25 \cdot 10^{-18}$ J/segment would be observed (Friedland et al. even estimate that the *average* stress on the highly stressed bonds in PA 6 is 10 GN/m^2 which would correspond to an energy dissipation of $1 \cdot 10^{-18}$ J per relaxing segment).

The above data may now be correlated with each other. In the amorphous regions at 32% strain one determines by ESR that $4 \cdot 10^{17}$ (out of 10^{21}) chains/cm^3 are broken. If each detected chain event would trigger between 120 and ~400 conformational transitions then the overall dissipated energy would exactly correspond to that measured by Frank. It seems to be quite realistic to expect such a number of rearrangements. From 10000 amorphous chain segments of 5 nm length 300 to about 1000 would classify as taut chains (IR [247], NMR [248], and mechanical [249] data); only four of these taut chains would be "seen" by ESR as being broken up to sample fracture; but even if Frank's hypothesis is correct that 10 times more chains are broken than are registered by ESR, the dominant role of viscoelastic processes in the deformation of highly extended chains is still maintained [242]. In this case the loading of a hydrogen bonded chain segment causes in this order the break-up of hydrogen bonds, the entropy and energy-elastic stressing of the segment in a highly oriented microfibril leading to stress-induced transitions towards a higher extended conformation and, if the stress level exceeds chain strength, chain scission.

On the basis of the above model some comments may be made with respect to the other defect concentrations indicated in Table 8.3. The comparative IR measurements by DeVries et al. [233] seem to confirm that ESR radical concentrations are systematically lower than IR group concentrations. For PE those authors determined that [R·] was by a factor of 2 to 5 smaller. Molecular weight analysis shows that macroscopic fracture leads to roughly a 10% decrease in molecular weight M_n. In random scission this corresponds to the breakage of every tenth molecule or of 0.3% of all segments of 5 nm length. These values are compatible with the mechanical model given above. However, as indicated by Nemzek et al. [235], the form of the molecular weight distribution may influence M_η values by up to 20%, so that ΔM_η values may have to be corrected. Undoubtedly further progress is to be expected in the domain of molecular weight determinations so that more precise data will become available.

At this time the best interpretation of the sometimes contradictory data on defect concentration is offered by an analysis of the experiments of DeVries et al. [233, 235]. It is noted that the discrepancies between ESR, IR, and M_η data are small, if the defects are produced rapidly (grinding) and at low temperatures. Under those conditions a modest loss of ESR traced free radicals occurs (perhaps of the order of one half) and only a modest multiplication of IR groups takes place (by a factor of about 1.3). However, if more time is allowed for defect production and, consequently,

Table 8.3. Defect Concentrations as Determined by Different Methods

Material:	Method: / Treatment:	ESR $[R\cdot]$* 10^{17} spins cm^{-3}	IR new groups 10^{17} cm^{-3}	molec. weight analysis [kg/mol] N_{rupt} 10^{17} cm^{-3}	visc. $\Delta M_\eta/M_\eta$	GPC M_n	GPC M_w	Ref.
HDPE	fracture	0.05	100					Zhurkov [17]
	stressing at	0.035	3.2					DeVries [233b]
	500 MPa	0.045		3.9	–4/58			DeVries [246]
	fracture							
	grinding at 77 K	0.10–0.12	4.9					DeVries [233b]
	irrad. 3 Mrad	3.9	21					DeVries [233a]
	irrad. 50 Mrad	48	110					DeVries [233a]
UHMW	grinding at 77 K	~ 1	11					DeVries [233b]
PS	grinding at 77 K	0.45–0.63	1.5–2.5	1.1–1.7	–10/200 to –166/916			DeVries [235]
PETP	11% strain			19	–0.8/28			Stoeckl [220]
PA 6	reference	5	not traceable, < 100					
	18% strain					34.2	74.7	Frank [238]
						31.2	70.6	Frank [238]
	17% strain	5.4		33	–3.1/36.4			Stoeckl [220]
	reference					30.5	69.2	Roylance [231]
	fracture			31		26.7	60.6	Roylance [231]
	fracture	1.1		11	–3.5/38.8			DeVries [245]
PA 66 St.	reference					25.5	54.8	Frank [238]
	20% strain	1.1				27.3**	54.8	Frank [238]

* It is interesting to note that the experiments of Narzullaev et al. [389b] predict that the presence of 10^{17} radicals/cm^3 will diminish the strength of **PE** by 1.5 kPa, of **PP** by 4 kPa and of **PA6** by 5 kPa (see also Chapt. 4 II B 3)

** The apparent small increase in M_n after fracture of the **PA 66 St** material is of the order of the experimental error.

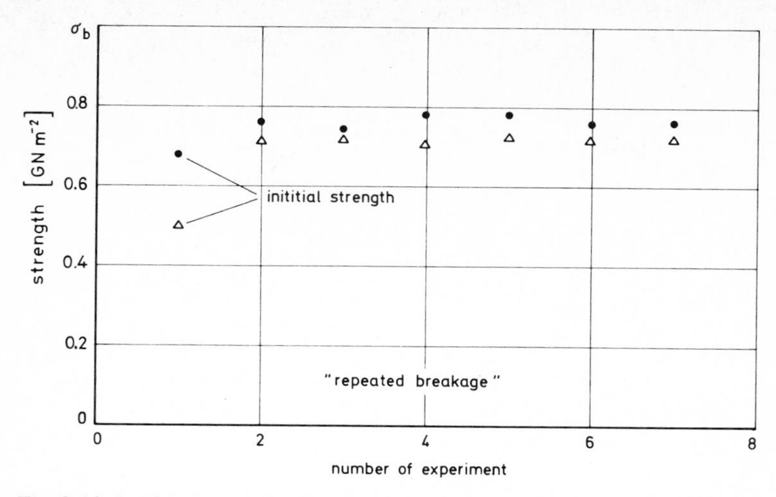

Fig. 8.16. Load-to-break of PA-6 fiber segments strained repeatedly (after [53]).

for radical reactions the discrepancies between different data increase considerably (Table 8.3 and recent observations by Szöcs [423]).

It should be mentioned that the absolute number of chain breakages at fracture is small. Thus, it is not surprising that the load bearing capability of the fiber material outside of the immediate fracture zone does not necessarily suffer. Prevorsek [53] showed that the strengths of fiber sections subjected repeatedly to breaking loads did not decrease. The two ruptured portions obtained in a first loading exhibited a larger strength than the original fiber and maintained this strength in subsequent loadings (Fig. 8.16).

3. Significance of Chain Length Distributions

One word of caution must be said with respect to the extrapolation of the segment length distribution $N(L/L_0)$ into the unknown regions of L/L_0. For this purpose DeVries et al. [49, 50] and Nagamura et al. [51] used the assumption that the "visible" part of the segment length distribution (Fig. 7.14) constitutes the skewed part of a normal distribution. A plot of the skewed distribution on probability coordinates (Fig. 7.15) apparently does not disprove the contention made. They had derived the parameters of the normal distribution (mean, variance) using the existence and location of an inflection point in the – always only partially known – cumulative distribution. It cannot be excluded that such inflection points result from the increased unloading of microfibril ends at larger strains since the stress-transfer length L_ϵ of a micro-fibril (with modulus E_f, diameter d) increases in case of perfect interfibrillar adhesion (of strength τ) with $L_\epsilon = E_f d_f \epsilon / 2\tau$. The eventual error of an extrapolation of $N(L/L_0)$ into the range of not fully – or even slightly – loaded segments is of not too much importance in calculating accumulated molecular stresses. It is of importance, however, if the distribution $N(L/L_0)$ is used to derive a fracture criterion. *A fracture criterion* [252] *in which macroscopic failure coincides with breakage of the last tie segment of the total distribution must be ruled out.* Against such a fracture criterion the above mentioned argument holds that at the point of macroscopic failure 99.7% of all tie-segments are still unbroken (see also Fig. 7.6, Sect. I C 2 and refs. [395, 397]).

Although DeVries, Lloyd, and Williams [49, 50] work with a normal distribution of relative chain lengths they do not put it to a full and questionable use. Their assumptions may be reformulated on the basis of the above considerations. It is assumed 1. that the segment length distribution $N(L/L_0)$ – derived from spatially homogeneous chain scission events of virgin material – be representative for the homogeneity of molecular stress distribution in any average amorphous region of the fiber, 2. that an initially narrow distribution continues to rise sharply with segment length and 3. that the *complete* break-down of tie segments within microfibrillar amorphous regions be restricted to a comparatively small fracture zone which does not contribute measurably to the free radical population. They use the widths of their chain length distributions $N(L/L_0)$ and correlate it with macroscopic strength (Fig. 8.17). In other words they correlate the non-homogeneity of the molecular stress distribution with macroscopic strength – and find a negative correlation.

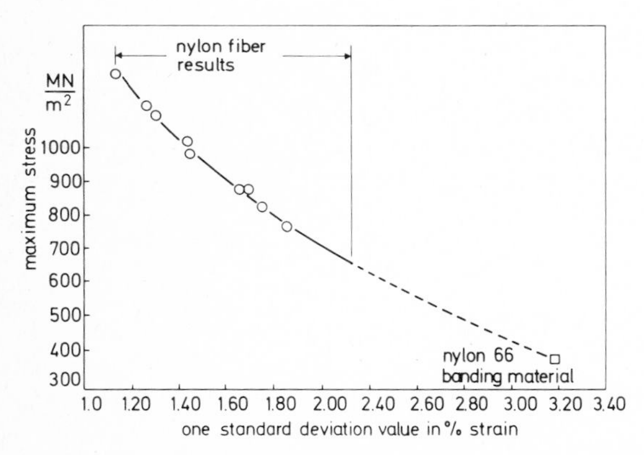

Fig. 8.17. Correlation of the homogeneity of segment lengths (narrow relative length distribution) with uniaxial strength σ_b [49, 50, 54].

It has been stated frequently in this work that the tie segments which eventually break are not the main source of strength of highly oriented fibers. The total number of radicals formed at macroscopic fracture does not *measure,* therefore, the strength of the sample. The wide differences in radical concentrations $N(\dot{R})$ at break of comparably strong fibers (cf. Table 6.2) illustrate this statement. Again these concentrations should be taken only as an indication of the *homogeneity* of *stress distribution between* different microfibrils and of *stress response within* different microfibrils. The later a possible defect zone starts its accelerated growth the longer the phase of spatially homogeneously distributed chain scission and the larger the number of chains which are loaded up to breakage. If one compares, under this aspect, members from one family of samples strength-related conclusions seem to be justified.

In Chapter 7 the morphological changes in PA 6 fibers due to annealing and their effect on chain scission have been discussed. In Figures 7.16 and 7.17 it had been indicated that slack annealing is accompanied by a relative lengthening of tie-segments and by a widening of the length distribution. The indicated losses of homogeneity lead to accelerated flaw growth already at stresses lower than those exerted on the control sample, i.e. to a loss of strength (Fig. 7.18; "slack"). Annealing at fixed fiber ends leads to a certain loss in homogeneity while maintaining the average relative segment length. The loss in strength (Fig. 7.19) and in the number N of radicals at break (Fig. 7.18) as compared to the control sample is small. This result confirms the findings of Levin et al. [21] with highly oriented tensile Kapron annealed at 200 °C. Annealing under a 11.7% prestrain increases the relative segment length (Figs. 7.16 and 7.17), segment homogeneity (Fig. 7.20), and sample strength (Fig. 7.19).

The *opposing effects* of increasing annealing temperature and increasing tensile prestrain on segment homogeneity are clearly indicated on Figure 7.20. The concentration $N_{\dot{R}}$ of radicals at break formed after annealing initially increases with pre-

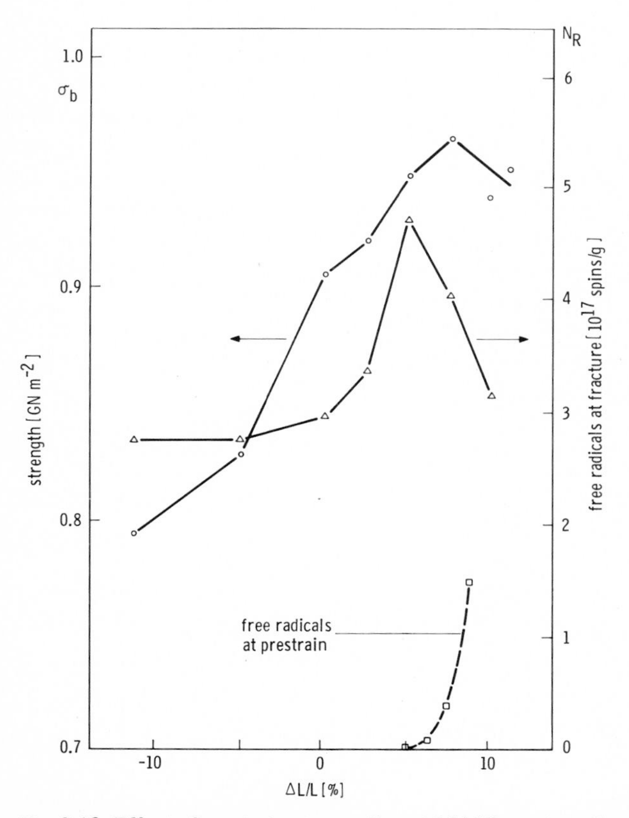

Fig. 8.18. Effect of prestrain at annealing at 164 °C on strength and number of broken bonds of 6 polyamide fibers (after [55–57]).

strain (Fig. 8.18), showing a maximum at a 5% prestrain. At that strain radical formation in prestraining (N_p) starts. The total number of radicals ($N_p + N_{\dot{R}}$) is approximately constant in that region. The sample strength increases by 20% up to a maximum at 7.5% prestrain and decreases slightly thereafter (Fig. 8.18).

D. Formation of Microcracks

All of the foregoing considerations have been concerned with a distribution of chain scission events homogeneous on a macroscopic scale. The accelerating effects of stress concentrations at microstructural inhomogeneities and of mutual interaction of scission points have so far been neglected. Such effects were not apparent in the discussed IR and ESR investigations of chain rupture; in any event they did not prevent tremendous concentrations of scission points being accumulated before final fracture. This fact can certainly be taken as an indication that during a considerable period of straining, chain scission points remain isolated "defects" and do not give rise to unstable cracks.

On the other hand the appearance of *submicrocracks* has been claimed for many years by the Leningrad school [17, 18, 27, 28, 254, 384–387] which employed X-ray scattering techniques to search for them. Such submicrocracks were found in loaded PE, PP, PVC, PVB, PMMA, and PA 6. The authors noted two important regularities of the submicrocrack formation [28]. First of all, the submicroscopic cracks appear at finite sizes with their transverse dimensions practically independent of load duration, stress value, and temperature (Table 8.4). Secondly their transverse size is determined by the polymer structure. For oriented crystalline polymers the transverse size coincides with the microfibril diameter; for non-oriented, amorphous polymers having a globular structure it coincides with the diameter of the globules [28]. It was claimed that these cracks nucleate fracture [384–387].

Two explanations have been given for the formation of these submicrocracks: Zakrewskii [20] proposes a radical chain reaction and Peterlin [58] suggests that the ends of microfibrils are the crack nuclei.

Table 8.4. Characteristics of Incipient Cracks in Polymers (after Zhurkov Kuksenko et al., [27, 29])

Polymer	Micro crack diameter longitudinal nm	Micro crack diameter transverse nm	N_{cr} cm^{-3}	Density change $\Delta\rho/\rho$ calculated	Density change $\Delta\rho/\rho$ measured
Polyethylene	15	17	$6 \cdot 10^{15}$	–	–
Polypropylene	20	32–35	$7 \cdot 10^{14}$	–	–
Polyvinyl chloride	–	60	$8 \cdot 10^{14}$	–	–
Polyvinyl butyral	–	50	$3 \cdot 10^{14}$	–	–
Polymethylmethacrylate	80	170	$4 \cdot 10^{12}$	$0.4 \cdot 10^{-2}$	$0.3 \cdot 10^{-2}$
Polycaproamide	5	9–25	$5 \cdot 10^{16}$	$1{,}8 \cdot 10^{-2}$	$2.0 \cdot 10^{-2}$

The Zakrewskii-mechanism was proposed by the Russian school following their studies of the formation of free radicals and of end-groups. They had noted that the concentration of free radicals was either by a factor of 4000 (PE-HD) smaller than the concentration of newly formed end-groups or even unmeasurably small (PP). This led them to search for a reaction by which chains might break without increasing the concentration of free radicals.

The mechanism of the proposed mechano-chemical reaction chain is illustrated in Figure 8.19. The thermo-mechanical scission of a regular covalent bond (*a*) leads to the formation of two chain-end radicals (*b*). As amply discussed in Chapter 6 the latter radicals are highly reactive. They will transform by hydrogen abstraction into stable end groups at the same time creating main-chain radicals within neighboring chains (*c*). It is then assumed by Zakrevskii and Zhurkov [17, 20] that the decreased energy of activation for the thermal scission reaction of radicalized chains (cf. Chapter 4) also leads to a lower tensile strength of these chains causing them to rupture (*d*). Repetition of the steps *c* and *d* would not increase the number of free radicals present but would multiply the number of chains broken. In Figure 4.3 the energy levels of the proposed steps of the reaction chain are indicated. It should be noted, however, that the lowering of the activation energy for scission of a radicalized chain is effective only where thermal equilibrium can be established (cf. Chapter 4 II B).

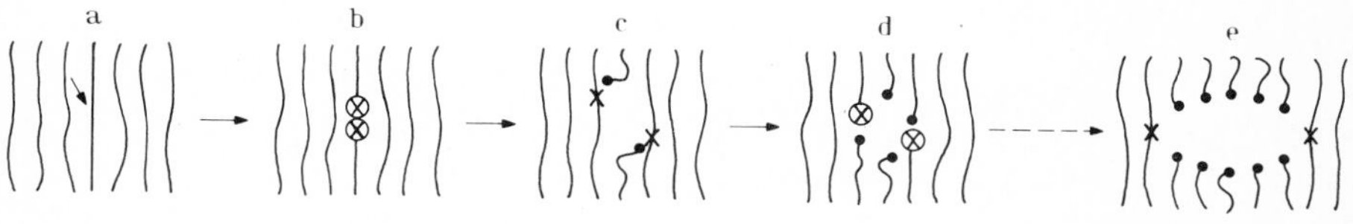

a b c d e

$R_1R_2 \rightarrow \dot{R}_1 + \dot{R}_2 \rightarrow R_1H + R_2H \rightarrow R_1H + R_2H \dashrightarrow R_1H + R_2H$

$2(R\text{-}\dot{C}H\text{-}CH_2\text{-}R')$ $\quad R\text{-}CH{=}CH_2 + 2\dot{R}'$ $\quad n(R\text{-}CH{=}CH_2) + 2\dot{R}'$

Fig. 8.19. Mechano-chemical reaction chain after Zakrevskii and Zhurkov [17, 20, 55]: (a) stress-induced chain scission, (b) formation of chain-end radicals, (c) radical reaction leading to main-chain radicals, (d) scission of radicalized chains, (e) formation of a submicrocrack by repetition of the steps (c) and (d); ⊕ chain-end radical e.g. $-CH_2-\dot{C}H_2$, x main-chain radical e.g. $-CH_2-\dot{C}H-CH_2-$, • stable end groups e.g. $-CH_2-CH_3$.

The authors suggest [17, 18] that the 3300 new end groups formed per submicrocrack in PE-HD (5000 in PP) are created by such a mechanism. Their data indicate also that there is less than a single pair of free radicals available per submicrocrack (0.4 in HDPE, 0.3 in PA 6 and an immeasurable small fraction in PP). These discrepancies, the unfavorable conditions for thermal equilibrium in the splitting reaction, and the absence of a well defined termination of the reaction chain raise considerable concern. They make it difficult to believe that such a reaction is the major reason for the formation of submicrocracks which remain fairly stable during extended and repeated loading periods [220].

Peterlin [58] has based his proposal on the origin of submicrocracks on morphological considerations and analysis of the previously mentioned X-ray data

[17–21, 27]. He suggests that the *ends of microfibrils*, preferentially situated on the outer surface of the *fibrils*, retract under stress (Fig. 8.20). In low strength PA 6 they thus open up about 10^{16} oblate spheroids per cm^3 having 6 nm diameter in fiber axis direction and 10 nm in the perpendicular direction. He estimates that the number and size of cracks in high strength polyamide is considerably smaller and is in agreement with the much higher draw ratio employed in producing those fibers. For highly oriented films with microfibrillar substructure similar considerations on the origin of the voids should hold.

Considering the two important regularities of submicrocrack formation [28], the inconsistencies of the Zakrewskii-mechanism [17], and the morphological aspects [58] the retraction of microfibril ends seems to be a logical interpretation of the opening-up of fixed-geometry voids. This means at the same time that the *submicrocrack formation is a process inherently independent of chain scission* or end-group formation [220, 421]. The voids constitute structural irregularities and contribute as such toward the general non-homogeneity of the distribution of stresses. Their direct influence as individual stress concentrator is weak and ineffective with regard to accelerating chain scission. This conclusion is based on the following facts:

- obviously the existing voids do not prevent the further build-up of large elastic strains in the fibers or films – as evidenced by the subsequent scissioning of chains and the appearance of deformed-IR bands,
- if chain scission would occur preferably adjacent to microfibril ends further lateral widening of the there existing microcracks would then invariably accelerate chain scission.

The above statement on the independence of chain scission and microcrack formation does not preclude, of course, the sudden-death failure of a stressed fiber by coalescence of microcracks as proposed by Morgan et al. [421] for Kevlar®.

Theses conclusions have found a certain confirmation by the studies of spherulitic POM by Wendorff [255]. After straining of different melt-crystallized or quenched POM samples he observes defects (of an average diameter of between 8 and 11 nm in lateral and ~8 nm in axial direction depending on annealing temperature). The size does not change with sample strain, the concentration, however, increases exponentially. He interprets the defects as local yielding sites which disappear upon annealing. He also

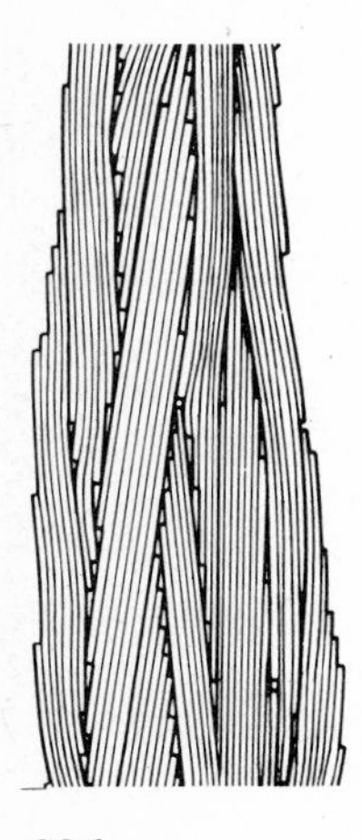

Fig. 8.20. Fibrillar model of fibrous structure with practically all ends of microfibrils concentrated on the outer boundary of fibrils (after [58]).

indicates that the non-linearity of the stress-strain curve of POM is related to the appearance of these elliptical void-like defects.

E. Energy Release in Chain Scission

A molecular chain segment stressed almost to its load-bearing capacity constitutes an extremely powerful source of elastically stored energy. In thermo-mechanically activated bond scission only a minor fraction of this stored energy is needed to break the chemical bond, namely the mechanical contribution $\beta \psi_b$ to the activation energy U_0. The remaining, major part of the energy is available for mechanical interaction with surrounding chains or is dissipated as heat. The dissipated heat has a twofold effect through the ensuing local temperature rise: it increases the mobility of other chain segments and decreases their rupture strength ψ_b (T). Both effects tend to facilitate the further degradation of the stressed polymer.

The total elastic energy stored in a segment of elastic modulus E_k, length L, and cross-section q amounts to

$$W_k = qL\psi^2/2E_k. \tag{8.9}$$

For a polyamide 6 segment of 5 nm length with $q = 0.189\ nm^2$, $\psi_b = 21\ GN/m^2$ and $E_k = 200\ GN/m^2$ one obtains a value of $1 \cdot 10^{-18}$ J per segment or 600 kJ/mol for the elastically stored energy. To break a C—C bond at room temperature a *mechanical* energy of 110 kJ/mol is needed (cf. Chapter 7 I C 1). In setting up an energy balance the contribution of the elastic forces must not be forgotten which hold the highly stressed *tie-segment ends within the crystal lamellae* [48]. At the rupture stress this energy amounts to about 190 kJ/mol for each segment end (cf. Table 5.4). One thus arrives at an energy of 870 kJ/mol which is liberated at the moment of chain scission. If this energy were to be confined within the volume of the segment and of its ends ($L_{total} \simeq 10$ nm) it would constitute an energy density of $W_{total}/q \cdot L_{total}$ = 764 MJ/m^3. Chain scission events, therefore, have been justly termed *micro-explosion* by Zhurkov and his colleagues.

On the other hand at a (large) concentration of $(0.6 \cdot 10^{+24})$ bond ruptures per m^3 (i.e. of 0.83 mol/m^3) the total energy having been stored in the retracting chain ends amounts to an average of 722 kJ per m^3 which will be dissipated as heat Q_b. The chemical energy U_b of this number of broken bonds amounts to 156 kJ/m^3. These average energy values should also be compared against the density of elastically stored energy, i.e. against $\sigma^2/2E$. This term reaches about 125 MJ/m^3 for very high strength PA 6 fibers, which in turn is only one sixth of the cohesive energy density of that material. The energy dissipation due to ruptured bonds is slightly smaller, therefore, than the pure hysteresis *losses* encountered in loading and unloading a polyamide fiber (for PA-6 at 10 to 30 Hz $\tan \delta$ is of the order of $5 \cdot 10^{-3}$).

The important problem of how the external work W done on a polymer sample during a loading-unloading cycle serves to increase the internal energy U, to change the entropy S, or to be irreversibly dissipated as heat $Q_{ir} = Td_iS$ was attacked by Müller in 1956 in his classical experiments [59–62]. According to Eqs. 5.2 and 5.3 one can write

$$dW = dU - TdS + Q_{ir}. \tag{8.10}$$

Müller and his co-workers measured several deformation-related thermodynamic quantities: the reversible and irreversible parts of heat production during the plastic deformation of PE, PVC, PETP, PA 6 [59–61], PC [63], PS [64], and of various elastomers [61, 65, 66], the ensuing temperature rises [67], the change of internal energy with the time of storage [68] and its effect on the energy of dissolution [69]. They noted that the entropy of thermoplastics during cold flow is decreased and that the internal energy is increased. They also measured the balance of energy in stressing (s) and during retraction (r) of PIB in a sequence of stretching cycles. The change of internal energy in an i th cycle can be written as:

$$\delta U_i = dW_s - dW_r + Td_sS - Td_rS + Q_{irs} - Q_{irr} \tag{8.11}$$

of which the entropy terms are equal by definition. The authors observed that δU_i was different from zero only in the first cycle. The increase of internal energy during extension of elastomers and thermoplastics was ascribed by them to a reduction of close range order and to chain scission and formation of radicals as further sinks of external energy (see also 8 I C 2 and refs. [238, 241–245]).

The latter hypothesis was also tested by Godovskii et al. [31]. PA-6 fibers drawn to $\lambda = 5.5$ at 210 °C were repeatedly stretched at room-temperature. The authors found for PA 6 the same characteristic difference between the first and the following loading cycles as did Müller for PIB: δU was essentially different from zero only in the first loading cycle. The ratio of $\delta W_1/\delta U_1$ was found, independent of the macroscopic stress, to amount to 7.0. This constancy of $\delta W_1/\delta U_1$ is in fact remarkable. It indicates that processes are responsible for the increase of internal energy which occur, independent of σ, if locally a critical excitation is surpassed. Godovskii et al. assume that these processes are chain ruptures. They derive from δU_1 the number N_1 of chain scission events each of which contributes an increment of internal energy of $1.7 \cdot 10^{-19}$ J (100 kJ/mol). The dissipated energy per chain scission event, $\delta W_1/N_1$, amounted to 700 kJ/mol. These figures are only slightly smaller than the values found above on the basis of elasticity considerations for the mechanical contribution to chain scission (110 kJ/mol) and for the energy dissipated by the retracting segments (870 kJ/mol). This striking coincidence does not prove, however, the above hypothesis that δU_1 might be interpreted solely as an increase in chemical bond energy due to chain scission.

Müller [62] and Kausch and Becht [55] have indicated that slip processes and conformational changes may respond energetically similar to bond ruptures if they are occurring as a consequence of stress induced distortion of the polymer network or as a result of local heating. The increase in internal energy is then related to a decrease of close range order or the decrease of the number of hydrogen bonds or to internal stresses between chains and within crystallites. The existence of intermolecular forces of considerable magnitude in oriented polymers can be inferred from a number of optical, spectroscopical and mechanical experiments. Particularly noteworthy are the following observations. Vettegren et al. [70] noted that in perfectly annealed PETP film the maximum of the characteristic backbone vibration band lies at 975 cm^{-1}, whereas in oriented PETP film it lies at 972 cm^{-1}. Lunn and Yannas [71] found that the asymmetric stretch vibration band of methyl groups with a maximum at 2971 cm^{-1} in isotropic PC film splits into two components 1 cm^{-1} apart in a PC film drawn to about 85%. They assigned this frequency shift to the action of *intermolecular* rather than intramolecular forces. With the drawn material they also studied chain-segment orientation by determining the dichroic ratio of the 1364 cm^{-1} band. They obtained an interesting result that chain backbone motion (with an ensuing change of the degree of orientation) in drawn PC films was clearly observable more than 60 K below the glass transition temperature of 149 °C. The largest deviations from an equilibrium state (highest intermolecular tensions) existed in those films drawn at the lowest temperature (23 °C) where annealing at 81 °C resulted in measurable changes of the orientation distribution. Müller [62] determined that in the cold drawing of non-oriented polymers the fraction $\delta U/\delta W$ of internal energy rise to external work amounted to 0.2 (PE), 0.3 to 0.45 (PA 6, PETP) and 0.3 to 0.5 (PVC).

The question is still open, therefore, as to the exact amount contributed to the internal energy rise by the different mechanisms mentioned above and to the extent the local rearrangements are triggered by chain scission.

F. Fatigue Fracture of Fibers

The preceeding Sections have dealt exclusively with the response of chains and microfibrils to constant or monotonically increasing stress or strain. In service, however, fibers are frequently subjected to intermittent, alternating, or cyclic loads. The behavior of fibers, therefore, to repetitive cyclic loading has been studied for many years (e.g. 72–87). Following an extensive review of Hearle et al. [76] it can be said that in cumulative extension cycling a fiber fails by virtue of having reached its breaking extension. Under such a condition of constantly increasing maximum extension, fatigue can be predicted from an adequate knowledge of the anelastic deformation and the time-dependent breakage conditions of the fiber. As yet no *specific fatigue effects have been identified* in cumulative extension cycling [76].

In one of the earliest reports on fatigue of fibers and fabrics Busse et al. [72] pointed out that the vibration life of nylon, cotton, and rayon cords is inversely proportional to the rate of thermally activated flow steps.

A similar conclusion has been drawn by Regel et al. [74] and Tamush [75] for cyclic stressing. These authors assume the validity of the principle of damage accumulation ("Miner's rule", see also Chapt. 3 IV D):

$$\int_0^{2\pi N_F/\omega} dt \,/\, \tau(\sigma[t]) = 1 \tag{8.12}$$

with N_F the number of cycles to failure and ω the loading frequency. If τ, the expectation value of the time-to-break under *constant* stress, is taken to be the common Arrhenius expression one obtains

$$\int_0^{2\pi N_F/\omega} dt \,/\, \tau_0 \exp\left[(U_0 - \gamma\,\sigma[t])/RT\right] = 1. \tag{8.13}$$

Regel et al. [74] report that such a damage accumulation law did apply to PAN fibers loaded at 24 Hz up to $1.5 \cdot 10^7$ cycles. For PMMA films, viscose fibers, and capron fibers (PA 6) correspondance of experimental data with Eq. (8.13) could be obtained through air cooling of the fatigued samples, after preliminary drawing, or at elevated temperatures. The authors concluded that Eq. (8.13) would describe fatigue fracture within the kinetic concept of fracture if (ambient) temperature T and activation volume γ would be replaced by values T^* and γ^* which would depend on the parameters of the fatigue experiment (frequency, form of stress or strain pulse).

The fatigue failure of fibers has also been extensively investigated by Prevorsek, Lyons, and co-workers. Their many pertinent papers are referenced in [78]. The authors interpret the fatigue life as the time to void nucleation by rearrangement of molecular segments. They arrive at a kinetic equation relating the number of cycles to failure with various mechanical and molecular parameters [78].

Fatigue testing PA 66 and PETP fibers between a lower (σ_{min}) and an upper stress level (σ_{max}) Bunsell, Hearle, Oudet et al. [77, 79, 84] observed a distinctive fatigue fracture morphology (see Fig. 9.53); they also obtained the apparently unexpected result that fatigue failure was accelerated if σ_{min} *was decreased* (zero minimum load constituting the most severe fatigue condition). The authors ascertained that in addition to creep damage accumulation (Eq. 8.12) a distinctive fatigue failure mechanism exists. Such a mechanism may very well be the *progress in disentanglement of chain segments* proposed independently by Hertzberg [28] and Kausch (in the first edition of this book).

No special fatigue mechanisms were assumed by DeVries et al. [49] in explaining free radical formation in cyclic loading and by Kenney et al. [86] in testing filaments, yarns, and ropes of PA 66 at a fixed value of $\sigma_{min}/\sigma_{max} = 0.1$. In the latter samples abrasion and wear could be of importance at larger (bending) strain amplitudes [87].

Summarizing the above observations one has to conclude that initiation and propagation of fatigue cracks will be favored if chains and fibrils are given a chance to relax stresses, to degrade possible strain hardening effects, and to reorient. Characteristic fatigue mechanisms are also well observable in non-oriented polymers. They will be treated in Section II C of this and in the next chapter.

G. Ultra-high Strength Fibers

In the previous sections (A to F) the rupture mechanisms of conventional fibers have been discussed. At draw ratios λ of between 3 and 6 the amorphous regions within the microfibrils still contained more than 90% of *non*-extended chain segments which support most of the load [256]. This is different in ultra-highly oriented fibers with draw ratios λ of between 10 and more than 50. In linear PE for instance the number of non-extended, randomly oriented segments, of loops and chain folds decreases with increasing λ [257–266]. The microfibrillar structure with alternating amorphous and crystalline regions (see Fig. 2.6) is initially still observed (depending on molecular weight and sample preparation technique up to draw ratios of about 20 to 30 [263]).

Samples of a higher draw ratio, however, contain broader crystalline size distributions with a small component of long crystals (100 to 800 nm). Moreover the strong coherence between crystallites seems to indicate the existence of intercrystalline bridges [259, 262, 263]. The Ward-model (Fig. 8.21) gives a representation of these structural features [259]. The Ruland-model [276c] considers in addition the imperfections caused by *trapped loops*.

As underlined by the model calculations of Kanamoto and Porter [271 a] very high draw ratios ($\lambda > 200$) should be obtainable from the semicrystalline state by pure tearing away or unraveling of the crystal lamellae. In fact, single crystal mats of ultra-high molecular weight polyethylene (UHMWPE of $M_w = 2 \cdot 10^6$) have been drawn by them [264] to a λ of 247. At this extreme draw ratio the fiber elastic modulus approached a value of 220 GPa [264 a] which is close to the theoretical modulus of a perfect PE crystal (240–290 GPa). With UHMWPP films Kanamoto et al. [264 b] achieve an elastic modulus of 30 GPa at $\lambda = 58$; as they indicate the lattice modulus of a perfect PP crystal is only slightly larger (35–42 GPa).

According to all observations it has to be expected that the ultimate properties of ultra-high oriented fibers depend on the properties of the crystal blocks and on the way these blocks are linked together. Starting from this idea a simple *shear lag theory* of fiber composites has been proposed [258–261] to calculate the elastic moduli and the yield behavior of ultra-oriented fibers. In these investigations and more clearly in the creep studies of Ward and Wilding [267] two thermally activated mechanisms were identified. A first one which has a small activation volume (~0.1 nm^3) and is primarily

Fig. 8.21. A schematic representation of the structure of the crystalline phase of linear polyethylene (from [259]) highly oriented

affected by the draw ratio. This mechanism seems to be related to the α-relaxation process in the crystalline regions which reduces the effectiveness of the crystalline links between the crystal blocks; the other mechanism, which is not affected by changes in structure, has a larger activation volume (0.3–0.6 nm^3) and is tentatively associated with the "molecular network" [267]. Creep of aromatic polyamide fibers was explained [398] by crystal rotation and crystal boundary slip. Chain scission [399] seems to be a less probable explanation. The deformation behavior of polyethylene Shish-Kebabs produced by stirring induced crystallization is characterized by a transformation of lamellar into fibrillar material [266, 268].

Novel processing techniques of ultra highly oriented fibers are reviewed by Ohta [269] and in [403–409]. Their strength has been widely discussed [e.g. 257, 258, 269–273, 393, 395, 399]. A definite and positive correlation between the total ratio, the elastic modulus and the fracture strength is observed (see [257b] for an extended discussion). The creep sensitivity of PE gives rise to a pronounced rate dependency of strength; thus, at a temperature of -10 °C the stress at yield or break of ultra-oriented PE of comparatively low molecular weight ($M_w \sim 60–100 \cdot 10^3$) varies from ~0.6 GPa at 10^{-5} s^{-1} to 1.4 GPa at 0.1 s^{-1}; at 20 °C the stress values are about 0.2 GPa smaller [270]. With the above samples final fracture is initiated by slippage of chains. As proposed by Woods et al. [267c] this tendency for creep can be greatly reduced by controlled crosslinking by electron irradiation. In Figures 8.22a and b stress-strain curves of electron irradiated (A) and unirradiated fibers (B) at 23 °C are compared. The marked reduction in ductility, especially at the smaller loading rate is clearly visible.

Very high draw ratios could also be obtained with ultra-high molecular weight samples (UHMWPE, $M_w = 1.5–2 \cdot 10^6$) prepared by spinning from dilute solution [266, 272–275]. In a dilute solution the number of entanglements between chain molecules evidently decreases with decreasing volume fraction ϕ of the polymer. By quenching and solvent evaporation thin films can be prepared from the solution which maintain the small entanglement concentration [274]. If these films are uniaxially drawn the molecular segments between "entanglement points" are gradually straightened. The maximum draw ratio λ_{max} corresponds to the complete extension of such a segment (of average length). According to the classical theories of rubber elasticity

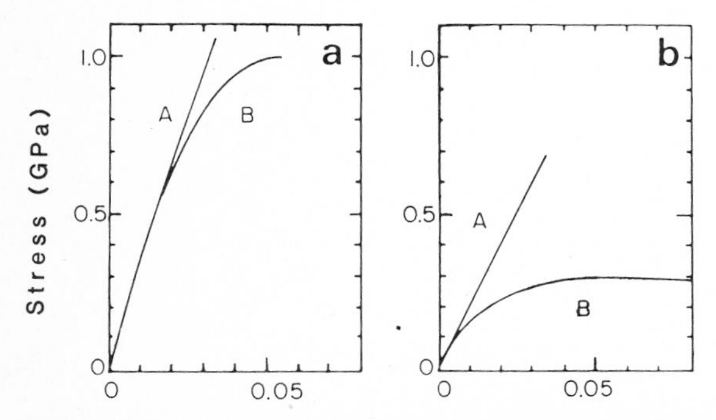

Fig. 8.22a, b. Stress strain curves for ultra-high modulus PE electron irradiated (A) or unirradiated (B) at strain rates of $8.3 \cdot 10^{-3}$ s^{-1} (a) or $8.3 \cdot 10^{-5}$ s^{-1} (b) at 23 °C (from Woods et al. [267 c])

the maximum draw ratio varies with the number N_c of statistical chain segments between entanglement points as:

$$\lambda_{max} = (N_c)^{1/2}. \quad (8.14)$$

In dilute solution this number increases with $1/\phi$. The experiments of Smith, Lemstra and Booij gave an excellent verification of this relation and of the entanglement concept [274]. Their results are shown in Fig. 8.23. In analyzing these results it should be kept in mind that a minimal number of entanglements will be required in order to assure the formation of a coherent physical network which can be oriented by drawing. This point has also been stressed by Chuah and Porter [271 b], Ward [257 b], and by Ruland et al. [276] for pressure crystallized, hot-drawn PE. Together with the draw ratio the strength of these ultra-oriented fibers increased. Marichin et al. [273] report strength values of about 3 GPa of as-spun ultra-oriented fibers and values of 4.6 to 5.6 GPa for fibers which they had further oriented by drawing at very high temperatures (130 to 150 °C). Savitsky et al. [272] even obtain tensile strengths of up to 7 GPa. This is much less than the (theoretical) tensile strength of an individual PE chain which has a value of between 20 and 40 GPa [48, 222e, 242b]. It coincides however exactly with the pull-out strength of a PE molecule from a crystal, calculated by Kausch and Becht [48] to be 1.37 nN per chain crossection of 0.19 nm^2 (see also Chapter 5 II B and C). All theoretical and experimental evidence points to slip processes as the origin of the rupture of ultra-oriented PE fibers.

High strength fibers containing extended aromatic chains behave somewhat differently. Figure 8.24 gives a model representation of the distribution of chain ends in Kevlar® 49 fibers [277]. It is believed that the relatively stiff and straight chains

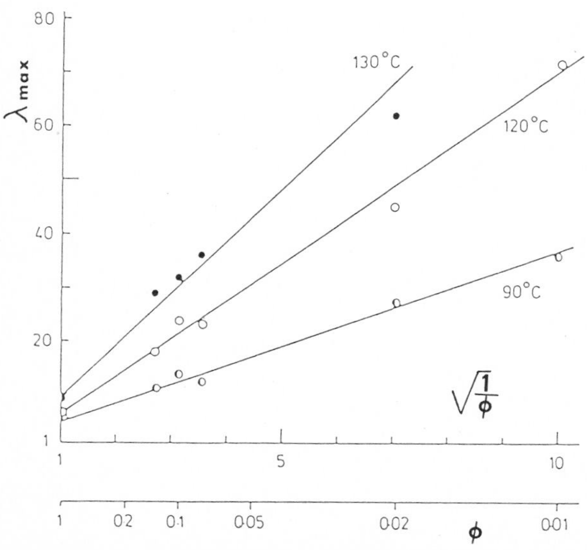

Fig. 8.23. Maximum draw ratio of UHMWPE vs. (initial polymer volume fraction)$^{-1/2}$ at the indicated draw temperatures (from Smith et al. [274])

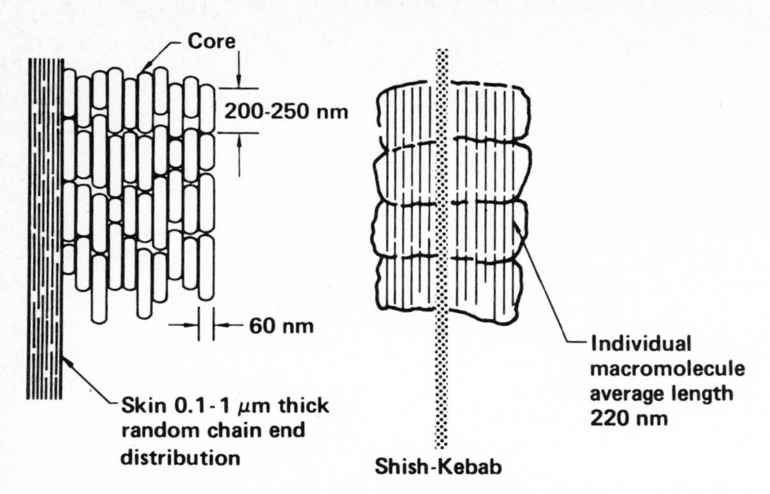

Fig. 8.24. A model of the chain end distribution in Kevlar 49 fibers (from [277]). The interlamellar regions should not be considered as gaps but as planes of maximum chain end concentration which are still traversed by a large percentage of extended chains [277–280]

are arranged during spinning from liquid crystalline dopes in the form of a highly oriented fibrillar skin enclosing a crystalline core of row lamellae [277]. Their impurities [400], static and dynamic properties [278, 401], and molecular weight [402] have been studied. Using a new technique by bonding these fibers to a bending beam Deteresa et al. [279] apply compressive strains of up to 3%. They show that regularly-spaced helical kink bands appear at about 0.53% strain. The tensile strength (3.4 GPa) was measured to be 5 times larger than the compressive strength and 17 times larger than the shear strength (0.18 GPa). From their extended electron microscopical observations Morgan et al. [277, 421] conclude in accord with the earlier observations of Konopasek and Hearle [148] that failure of such a structure is initiated by shear crack propagation through the voided fibrillar skin followed by transverse crack propagation along interlamellar planes combined with axial shear splitting (Fig. 8.25).

The failure mechanisms of Kevlar® 49 fibers have also been investigated by ESR techniques [281, 282]. Brown and Hodgeman [281] studied the thermal degradation; they noted the generally observed slightly asymmetic ESR signal of untreated roving material later assigned by Brown, Sandreczki and Morgan [282] to transition metal ions like Mn^{+2}, Cu^{+2} and possibly Fe^{+2}, Fe^{+3}, Cr^{+3} and Ti^{+3}. Upon heating to beyond 350 °C (in air) or 450 °C (in vacuum) free radicals appeared in large numbers which were considered to result from homolytic bond scission followed by crosslinking reactions leading to the formation of conjugated structures [281]. No stresses were applied in these studies.

Morgan et al. [282] searched for the stress-induced formation of free radicals. They concluded from their observations that the amount of chain scission *before* final fracture is below the detectable limit, i.e. less than 1/20 of that observed *after* fracture or less than 10^9 chain scissions per filament. However, the final highly fibrillar fracture event led to a measurable intensity of (probably secondary) free rad-

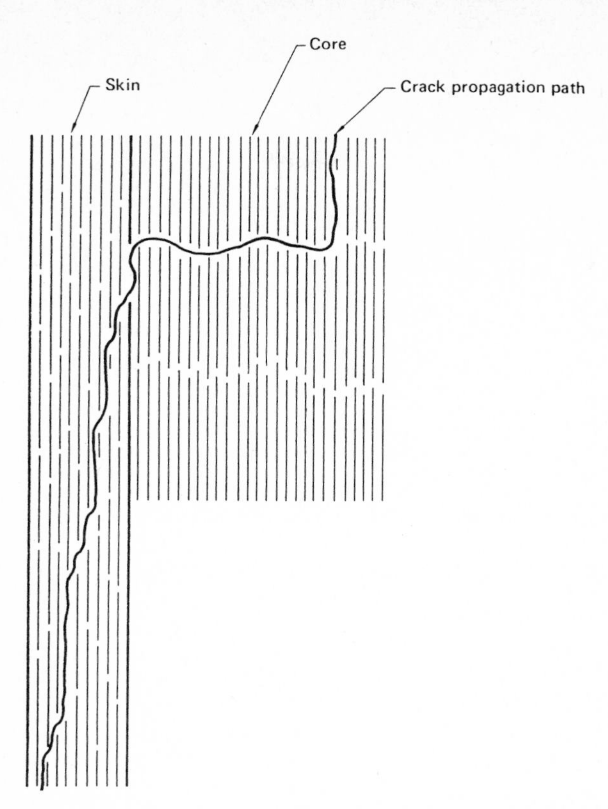

Fig. 8.25. Planar model of crack propagation path through macromolecular chain ends in fiber skin and core in Kevlar 49 fiber (from [277]). Strong indications that the interlamellar planes are the critical sites were also obtained by Avakian et al. [280] who observed steps of a minimum height of 250 nm in the fracture planes of HCl etched fibers; in this case the crack presumably followed the defects created along the advantageous diffusion paths by chemical attack

icals of $2 \cdot 10^{10}$ spins/filament. This amount corresponded to 50 times as many chains as are crossing one transverse fracture plane. Since the axial splitting extended over 2500 weak planes an average fraction of 2% of all tie molecules within this region had been broken during the final crack propagation. These observations strongly confirm that chain scission in stressed Kevlar® 49 fibers is the consequence rather than the origin of mechanical breakdown.

The question of the strength of other aromatic or rigid rod polymer fibers has also been dealt with experimentally and theoretically [283–288] not giving rise to any different conclusions concerning the role of chain scission. In the case of the highly regular polydiacetylene single-crystal fibers, however, practical strength values (of up to 1.5 GPa) have been obtained [289], which are very close to the chain strength (of 3 GPa).

II. Deformation, Creep, and Fatigue of Unoriented Polymers

A. Fracture at Small Deformations (Brittle Fracture)

1. Impact Loading

Polymeric materials are viscoelastic solids. Their propensity to anelastic and plastic deformation is reduced when they are tested at high load rates and/or at low temperatures. The reduced deformability causes a formerly tough or highly elastic polymer to respond in a brittle manner. The impact fracture of natural rubber at liquid nitrogen temperature is convincing evidence of this fact.

The resistance of polymeric materials to impact loading is of considerable technical importance as illustrated by the following arbitrary examples:
- aircraft windows have to withstand impinging particles or rain
- protective coverings of high voltage switches must resist the impact of metal chips
- water pipes or plastic floor tiles should not be damaged if hit accidentally
- car bumpers or packaging materials must absorb as much of the energy of a collision as possible
- household items are not expected to break the first time they fall to the ground
- gears should transmit even rapidly increasing forces
- safety belts should yield but not fracture.

In view of the technical relevance the impact resistance of polymers has been intensively investigated. The cited references [88–103] may serve as an introduction into the large body of literature. Vincent [88] and Bucknall et al. [89] give a general survey of the impact testing of polymers. Other references concern the molecular aspects [88–96], instrumentation [97–100], and particle impact [101–103]. The extensive literature on the fracture mechanical analysis of impact bending tests will be referenced in Chapter 9.

The fact that the impact resistance depends strongly on the presence and shape of stress concentrators, on the sample geometry, and on the testing conditions has made it very difficult to define and measure *a unique material property impact resistance.* Since the impact behavior under one condition (e.g. impact bending) cannot very well be predicted from the behavior under different conditions (e.g. falling weight) quite a number of impact tests have been devised. The four best known ones are the three point bending (Charpy unnotched and notched), two point bending (Izod), tensile-impact, and falling-weight tests which have been standardized (DIN 53453, 53448, 53373, 53443E; ASTM D 256, 1822, 2444, 3029).

In the context of this chapter mainly the three-point bending test (Charpy) will be discussed. In this test the energy loss A_n is measured which a pendulum incurs in striking and breaking a prismatic specimen (of thickness D and width B). The observed energy loss comprises basically four different terms:
- the energy W_e to bend the sample up to the point of crack initiation,
- the energy G_cBD to propagate the crack through the specimen,
- the energy W_{kin} required to accelerate the sample from zero to pendulum velocity,
- vibrational or otherwise dissipated energy.

The simple excess energy measurement of the pendulum does not permit discrimination between these different energy sinks. The separation of the first three terms becomes possible, however, if the bending force P acting between pendulum head and specimen is measured (so-called instrumented impact test). A further improvement of the analysis of impact phenomena is obtained if not only the pendulum head but also the impacted specimen is instrumented. A relatively simple and straightforward method to determine the time scales of local deformation and crack initiation in a shock-loaded specimen is provided by a sprayed-on graphite gauge as described by Stalder et al. [290]. Figure 8.26 shows a Charpy specimen prepared this way and the corresponding measuring circuit. The inverse gauge resistance, $R_0/R(t)$ and the force P(t) exerted on the hammer nose during a doubly instrumented Charpy test of a PC specimen are plotted in Fig. 8.27. The time scale and the nature of the different phases of the impact event become already clear from this diagram.

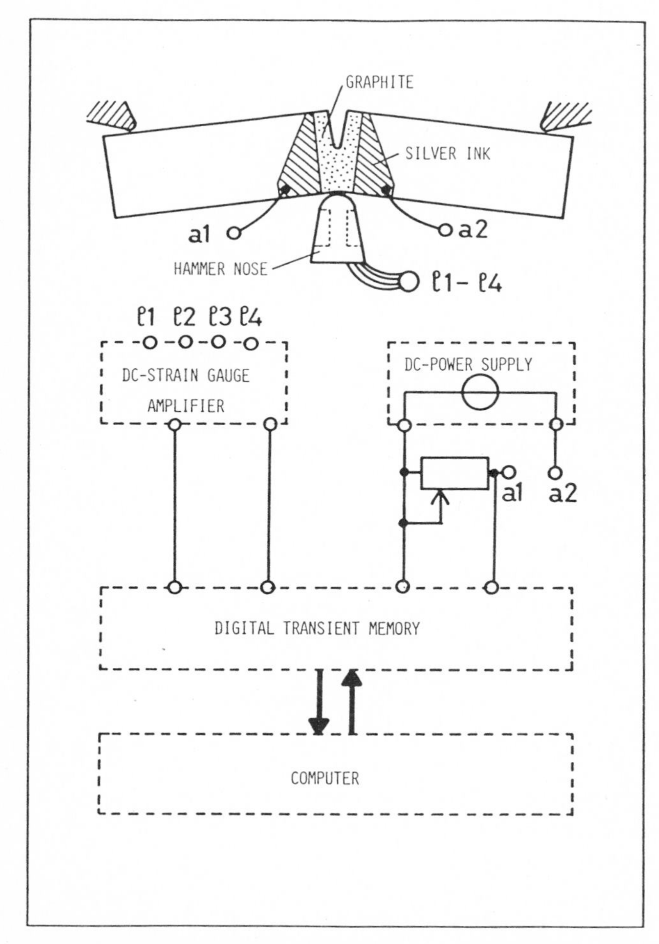

Fig. 8.26. Charpy specimen with sprayed-on graphite strain gauge and measuring circuit according to Stalder [290]

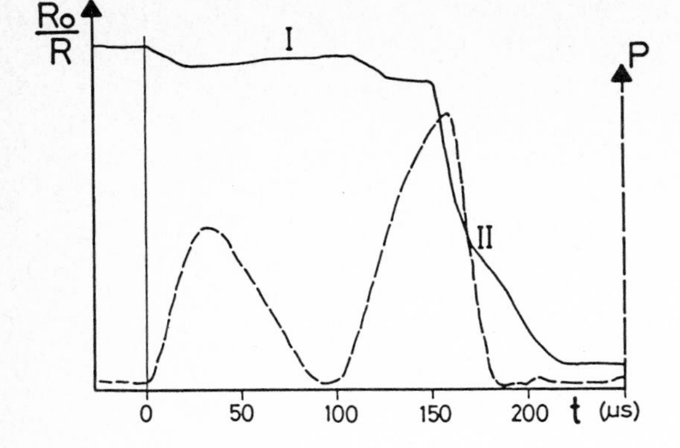

Fig. 8.27. Charpy test of a notched PC specimen

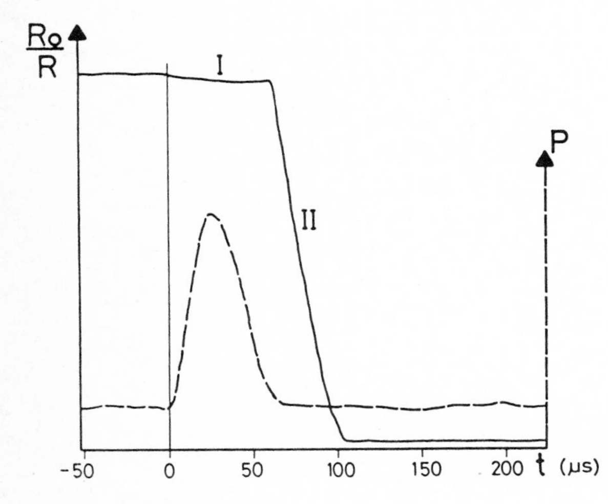

Fig. 8.28. Charpy test of a precracked PMMA specimen

Two phases in the time sequence of the event are evident:

Phase I begins as soon as the striker touches the specimen and ends when the crack begins to propagate. During this time, which is called the time-to-fracture initiation, the specimen center is accelerated up to the striker velocity. This acceleration may well occur in several bounces. Through the displacement of the specimen center bending stresses are imparted onto the specimen; in the given case of a notched sample the stresses have been large enough as to initiate and propagate a crack. In this phase (I) the graphite gauge acts as a strain gauge and records the elastic and plastic deformation of the sample. The variation of R_0/R is a function of the deformation of the specimen and of the plastic zone at the crack tip. Knowledge of the strain gauge factor would permit this deformation to be quantified.

Phase II during this time the gauge records in real time the relative position of the crack front, even if the velocity of the crack is high.

In the shown example of a notched PC specimen phase I reveals a notable deformation of the specimen and a relatively long time-to-fracture initiation. The first load peak is the inertial peak (acceleration of the specimen), the second provides the critical stress at the notch tip necessary to initiate and propagate a crack. In phase II, the graphite gauge recording shows that the crack starts to propagate with a high velocity (~200 m/s) which decreases rapidly.

The time delay which appears in Fig. 8.27 between crack initiation and maximum load is due to the fact that the shock waves generated at the moment of the second contact take a finite time to propagate from the *front* of the striker to the conventional strain gauge recording the load which was mounted at the *rear* of the striker. At time t = 0 (first contact) this delay does not appear because the shock waves must propagate from the point of impact to both detectors, i.e. through the sample to the graphite gauge and through the striker to the strain gauge. Both waves seem to reach the respective gauges simultaneously.

For the static elastic bending of a prismatic beam between supports at a distance L_s the following expressions for energy and strain are obtained.

The total bending energy W_e, as a function of beam deflection δ, is

$$W_e = P\delta/2 = \frac{2E\,\delta^2 BD^3}{L_s^3}, \qquad (8.15)$$

the largest density of stored elastic energy in the tensile fiber:

$$w_m = \sigma_m^2/2E = \frac{18E\,\delta^2 D^2}{L_s^4}, \qquad (8.16)$$

and the largest strain can be expressed as

$$\epsilon_m = \sigma_m/E = 6\delta\, D/L_s^2. \qquad (8.17)$$

From the foregoing considerations it can be concluded that fracture in impact bending of a given material will be delayed if the stored elastic energy w_m *available* to initiate and propagate a crack remains small and/or if the fracture resistance R of that material is large. The latter property will be examined in detail in Chapter 9. The influence of the former shall be further discussed at this point.

Apart from specimen geometry the stored elastic energy of a bent specimen depends on its elastic modulus E (Eq. 8.16). A "soft" specimen evidently will not break but fold. The correlation between elastic modulus and "impact strength" has been discussed repeatedly (see for instance [88–97]). In the classical Charpy test impact strength (Schlagzähigkeit, résistance au choc) has been defined as the ratio a_n of the measured pendulum energy loss A_n to the sample cross-section B · D.

Such a notation creates the allusion that a_n is a fracture surface specific material property. As pointed out above this is not the case. Neither W_e nor W_{kin} are proportional to the sample cross-section. A comparison of a_n values should only be made, therefore, if all values are obtained from one type of test, preferably even from specimens of identical geometry. Impact strength values a_n of Charpy un-notched specimens (DIN 53453) at 20 °C range from 3.5–12 kJ/m^2 for filled phenol melamine and urea resins, 4–22 kJ/m^2 for various filled epoxy and polyester resins, 12–20 kJ/

m^2 for PMMA, PS, and SAN, and 50–90 kJ/m^2 for ethyl cellulose, CA, styrene-butadiene copolymers, and POM. For many thermoplastics (ABS, CAB, PE, PP, PTFE, PC, PVC, PA) no fracture is observed under these test conditions [104].

The above impact strength values designate brittle behavior if $a_n < 40\ kJ/m^2$. The materials with a_n between 50 to 90 kJ/m^2 will generally be brittle if bluntly notched. Of those polymers not breaking in the Charpy un-notched test some will respond in a brittle manner if they are sharply notched, whereas others remain tough even then. Vincent [96] and Bucknall et al. [89] propose, therefore, the following qualitative impact resistance rating: *brittle* ($a_n < 40\ kJ/m^2$), *brittle if bluntly notched, brittle if sharply notched, tough but crack propagating,* and *very tough and crack arresting.*

Studying the impact behavior of 18 different thermoplastics Vincent [96] found that for two thirds of his data there existed a good negative correlation between impact strength rating and dynamic modulus.

The characteristic impact behavior of brittle polymers can conveniently be outlined using PMMA as an example. The impact of a specimen initially leads to its continuous straining and to the build-up of stresses. These stresses provoke the growth of defects such as crazes [105]. Final breakdown of the impacted specimen will start from such a defect (or from purposely introduced notches). In the case of brittle polymers final breakdown occurs through rapid propagation of a crack. The energy absorbed in the impact test is essentially the energy transferred onto the specimen up to the moment where rapid crack propagation set in. As shown by the diagramme in Fig. 8.28 this energy may completely be determined by the "intertia peak" the specimen being fractured "in flight" while being no longer in contact with the hammer. In

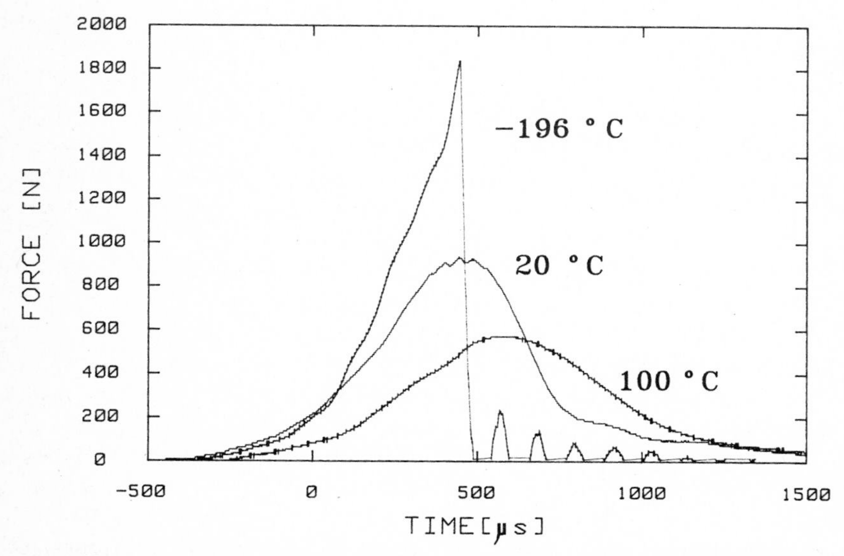

Fig. 8.29. Impact fracture of HDPE at different temperatures; for gradual force build-up the hammer tup had been equipped with a 1 mm-layer of plasticine; the large increase in energy for material separation with increasing temperature is readily visible; also visible are the oszillations of the hammer tup after fracture [290]

this case the measured energy loss A_n has no relation to the resistance of the material against fracture. In fact an already broken specimen will cause the same energy losses to the pendulum. With notched polycarbonate (Fig. 8.27) one measures more reliably the forces necessary to bend the specimen up to the point where crack propagation is initiated. In this case the bending energy W_e accounts for most of the energy loss A_n. With unnotched PE or ABS, however, a notable amount of energy, $W_s = G_0 BD$ will be needed to make the crack propagate. A quantitative discussion of W_s, which must make use of the mechanics of fracture, will be given in Chapter 9.

Both terms, W_e and W_s depend on temperature, as is readily seen from Fig. 8.29. The ductile crack propagation observed in HDPE at +100 °C and +20 °C changes into a brittle one at liquid nitrogen temperature.

Whatever has been said about the phenomenology of impact *bending* fracture is also valid for *tensile* impact: The impact resistance of a material is increased by all molecular properties or processes which assist the distribution and dissipation of the mechanical energy and permit the attainment of a large deformation before inception of rapid crack propagation. In particular the importance of the relaxation behavior [91, 93–96], of high molecular weight [106], of partial orientation [108, 303], and of the unoccupied volume [92] have been emphasized.

A *maximum fractional unoccupied volume* ($\bar{f}$) was defined by Litt et al. [92] through the densities of the amorphous (ρ_a) and the crystalline state (ρ_c) as:

$$\bar{f} = (\rho_c - \rho_a)/\rho_c. \tag{8.18}$$

The $\bar{f}$ values of common polymers at room temperature vary between zero and 0.12. Litt found a good – although not perfect – correlation between $\bar{f}$ and impact strength. He indicates that polymers with high impact strength all have $\bar{f} > 0.07$.

Considerable qualitative evidence has been cited in the literature (e.g. 88–105, 291, 292) that the intensities of the low-temperature relaxation peaks deriving from main-chain motion (β and γ peaks) are related to impact toughness. A positive correlation has to be expected since an increased molecular mobility

- reduces the level of stored elastic energy at a given sample deformation and load rate and retards crack initration
- prolongs the phase of thermal crack growth, i.e. retards the inception of an unstable crack,
- and increases the energy expenditure during crack propagation.

This correlation has been clearly demonstrated by Ramsteiner who determined for a dozen of the most important thermoplastics the brittle-ductile transition temperatures and the logarithmic decrement in torsion [292]. His results are reproduced in Fig. 8.30. Wherever a pronounced low temperature (β) relaxation peak is observed in torsion it coincides with the brittle-ductile transition temperature in impact bending. In the absence of such peaks (PIB, PMMA, PS) the transition is shifted all the way up to the beginning of the α relaxation. Ramsteiner also discusses the influences of thermal treatment (shock cooling of PC increases its molecular mobility and its toughness in a temperature range between –10 and +70 °C), water and rubber modification [292].

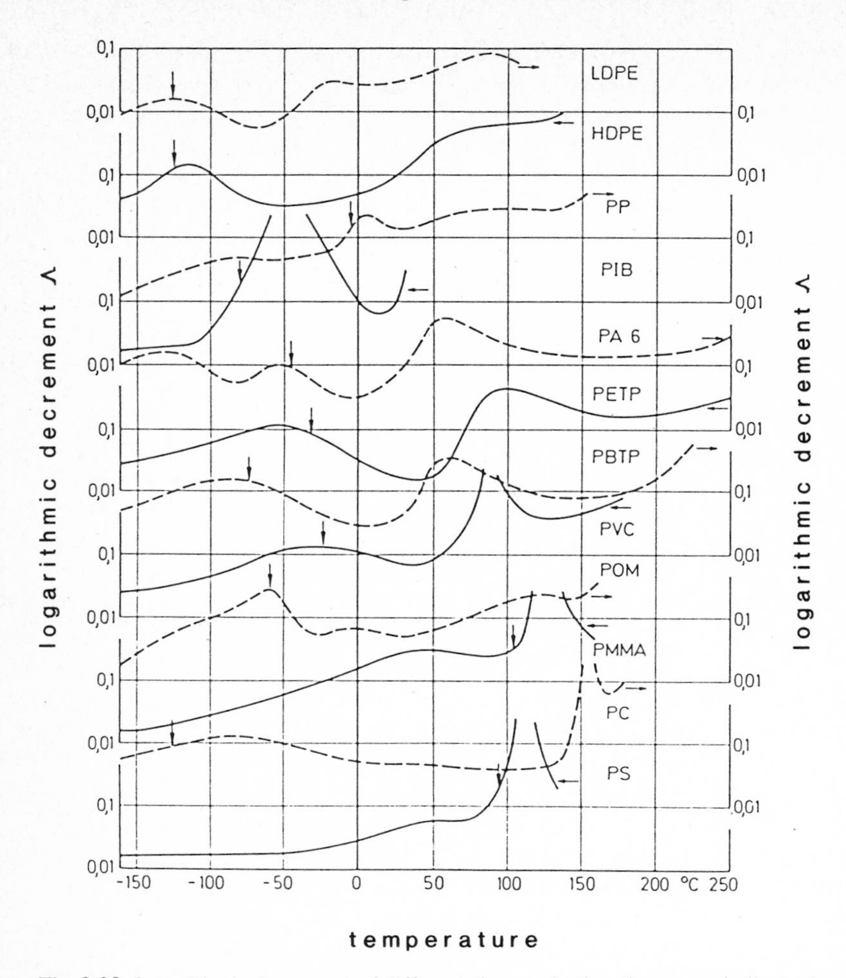

Fig. 8.30. Logarithmic decrement of different thermoplastics; the arrows indicate the temperature of the ductile-brittle transition in impact bending (from Ramsteiner [292])

A quantitative correlation will be limited by the fact that the relaxation behavior is determined in the linear anelastic range of stresses and at small strain rates. These conditions are quite different from those of an impact test. A quantitative discussion also has to be based on a fracture mechanical analysis of crack initiation and propagation (cf. Chapter 9).

The impact resistance of a polymer is reduced by all factors which lead to a general or local increase of stored elastic energy at a given deformation *but are not accompanied by an overproportional increase in strength.* Thus the *stress concentrating effect* of notches, flaws, or inclusions in an otherwise unchanged polymer considerably lowers its impact resistance. Increasing the *degree of cross-linking* beyond the point which insures the loading of all chains only leads to the formation of short, tightly anchored chain segments. Such segments are the first to be overloaded and scissioned upon straining. The small impact resistance of fully cured thermosetting resins underlines this point. Reinforcing thermoplastics by *short* random fibers increases their rigidity more effectively than their strength which leads to a net decrease of impact resistance.

A considerable toughening is to be expected, if strength and deformation can be made to increase simultaneously. Such an effect is realized through the partial orientation of un-oriented polymers [108, 303]. For PS drawn to $\lambda = 3.4$ Retting [108] reports increases in tensile strength from 47 to 80 MN/m^2 and of strain at break from 7 to 22%. The IUPAC Working Party "Structure and Properties of Commercial Polymers" studied systematically the effect of orientation of various polystyrene samples (homopolymers and rubber-modified polystyrenes) on optical and mechanical properties [109, 110]. It was observed that the impact strength a_n of unnotched homopolymer specimen increased with draw ratio from about 3 at $\lambda = 1$ to more than 50 kJ/m^2 at $\lambda = 6$ if the molecular orientation was parallel to the main specimens dimension. In a perpendicular direction a_n decreased from 3 to 0.3 kJ/m^2 for the homopolymer and from about 9 to 2 kJ/m^2 for a commercially available rubber-modified polystyrene [109, 110]. A similar increase in low temperature ductility has been observed by Daniels [293] for PA 66 if the preorientation draw ratio exceeded 1.8. The maximum impact *load* of PETP can be tripled [303].

The principal effect of chain backbone properties on impact strength has become clear. A chain does not so much contribute to impact strength through the level of stress ψ_b it supports at the moment of chain scission or disentanglement but rather through the energy dissipated before ψ_b is reached. Chains are loaded in shear which necessitates the displacement of chains with respect to each other. A maximum of energy dissipation is obtained, therefore, if the intermolecular shear stresses are *large but insufficient to break a chain* (cf. Eq. 5.41) and if the chains do not easily disentangle so as to insure chain slip over large volume elements.

2. Fracture at Cryogenic Temperatures

As to be expected fracture at cryogenic temperatures has much in common with impact fracture since the major relaxation mechanisms (α, β) are frozen in at liquid nitrogen temperature. Fracture of isotropic specimens will generally be brittle in nature and characterized by the same sequence of events as impact fracture. In a tensile or bending experiment one has:

- (more or less) steadily increasing load
- at an elevated load level growth of defects through e.g. scission of tie-molecules and/or void formation (as proposed for PE [294]) or through crazing
- onset of rapid crack propagation once defect size a and instantaneous stress σ fulfill the instability criterion to be discussed in Chapter 9.

The fracture surface of a HDPE specimen bent at liquid nitrogen temperature shows the growth region of a defect (Fig. 8.31, upper right hand corner). The so-called mirror grew during load application at a temperature of -196 °C from an unresolved, small material inhomogeneity to its critical size of 2–3 nm, at which unstable

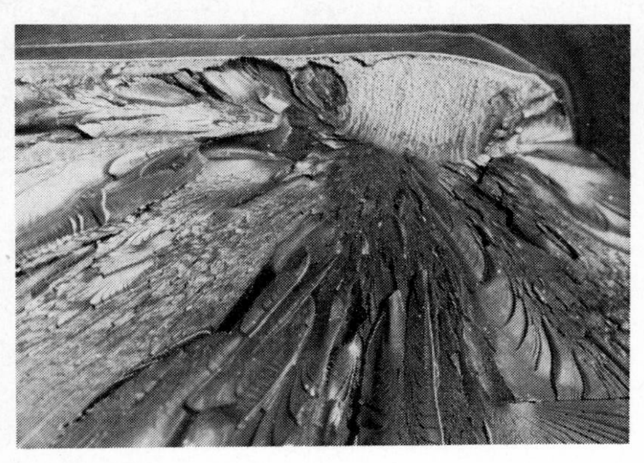

Fig. 8.31. Bending fracture of notched HDPE at liquid nitrogen temperature (enlarged 8 : 1, taken from [106])

crack propagation set in. Closer analysis of the fracture surface reveals that even at this temperature and at high crack velocity plastic deformation had occurred (see Fig. 9.18). Brown [295] recently estimated that plastic deformation in most of the common thermoplastics – except in PE and PP – occurs in the absence of thermal activation if the ratio of shear stress and shear modulus approaches 0.064 to 0.092. For PE and PP this ratio is of the order of 0.03. He introduces a model by which the plastic deformation is explained as a combination of three molecular motions which he calls *shearon* (intermolecular shear), *roton* (intramolecular shear), and *tubon* (motion parallel to the covalent bond).

In the same way as the impact behavior the low temperature deformation is influenced by molecular and structural features (molecular weight, thermal treatment, preorientation). For example rapidly quenched, "high" molecular weight (M_w = $2.2 \cdot 10^5$) HDPE samples break in a brittle manner if T < 95 K [294] whereas slowly cooled "low" molecular weight ($M_w = 1 \cdot 10^5$) samples show a ductile-brittle transition already at ~ 130–166 K (Fig. 8.32). Whereas these low temperature ductile-

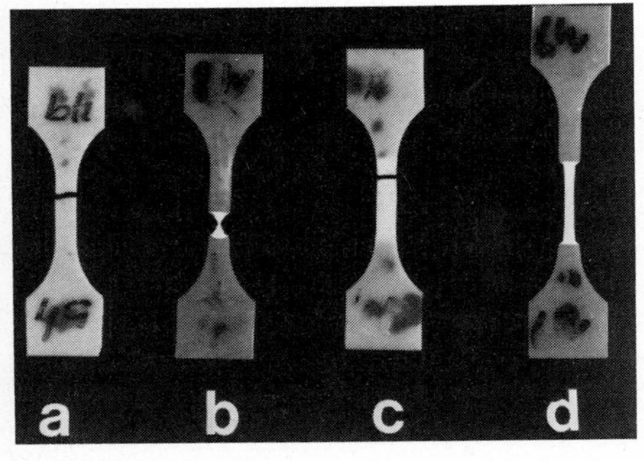

Fig. 8.32a–d. Photographs of different tensile failure processes. **a)** Brittle fracture; **b)** Necking/rupture; **c)** Homogeneous yielding; **d)** Necking/drawing (Courtesy Prof. I. M. Ward, [294])

brittle transitions are not very rate sensitive (in the range of 0.11 to $11 \cdot 10^{-3}$ s^{-1}) this will be different for the conventional transition from necking/rupture to necking/drawing situated at 243–273 K [294].

The low temperature fracture events generally start from very small structural irregularities. According to the severity of these irregularities fracture stresses can show, therefore, a very large variation. This is well represented by the fracture data of unplasticized PVC (Fig. 8.33).

In partially oriented polymers more and longer extended chain segments are pointing in the direction of orientation. If a sample is loaded in this direction the molecular stresses will be more homogeneously distributed and brittle rupture delayed (to larger values of macroscopic stress or strain). This was observed by Retting [108–110] for impacted preoriented PS (see previous section) and by Daniels [293] for the tensile behavior of partially oriented PA 66 at –196 °C. Up to draw ratios smaller than 1.8 the PA 66 broke at strains smaller than 20% (corresponding to a total elongation at break λ_b of less than 2). For $\lambda \geqslant 1.8$ the strains at break went up to 75% giving a λ_b of more than 3; λ_b increased further with increasing draw ratio λ [293].

An entirely similar behavior was found by Andrews and Reed for elastomers at cryogenic temperatures (see e.g. [297, 298] and Section 7 II). At these temperatures

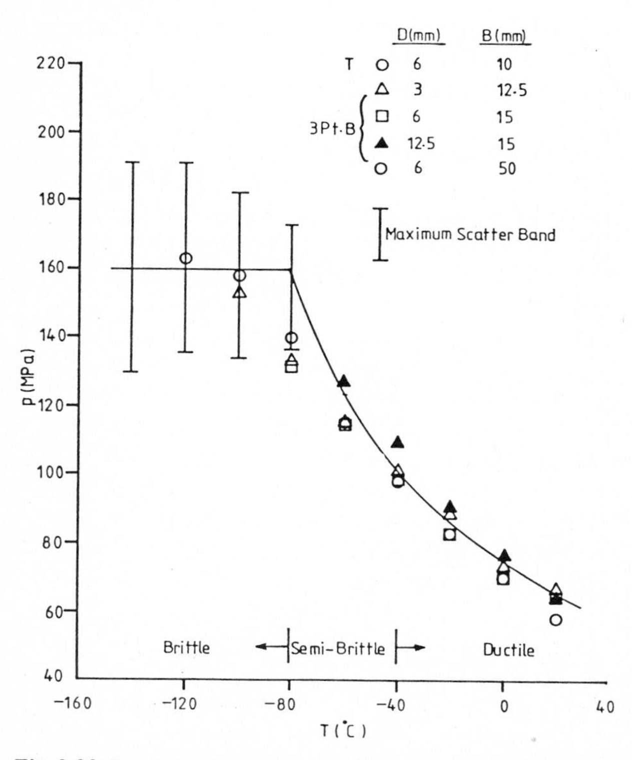

Fig. 8.33. Fracture stress in tension (T) or bending (3 Point Bending) of unplasticized PVC (from [296]); D and B are specimen dimensions

elastomers behave very much like thermoplastic materials. One particularity may be noted in the case of cross-linked rubbers where chain scission will be provoked at large extensions [297, 298]. The free radicals formed in this process are highly reactive; since they give rise to cross-linking reactions the cross-link density of a forcibly extended rubber vulcanisate may even increase [298].

B. Failure Under Constant Load (Creep and Long-Term Strength)

1. General Considerations

Subjecting a specimen to a constant load evidently can have one of three results:
- the specimen breaks upon application of the load
- the specimen fails after some time
- the specimen carries the load for an indefinite period of time.

The first case, failure *during* the application of the load, will be treated in other sections (8 II A if it occurs in a brittle manner, 8 II D for short-time ductile failure). In this section the technologically interesting long-term strength will be discussed. Owing to its fundamental importance the time-dependent strength of load-bearing polymers has been extensively investigated with temperature as the principal parameter. In Figures 1.4, 1.5 and 3.7 stress-time-temperature diagrams of various thermoplastics are represented. The phenomenon of a delay between load application and final failure has found numerous interpretations. Among these interpretations one group is based on *purely statistical considerations.* There the time-to-failure t_b is inversely related to the probability that a certain detrimental event takes place in an *otherwise unchanged material.* Such events are e.g. the formation of a void of sufficient size or the coincidence of local fracture events. The statistical aspects of fracture have been discussed in Section II of Chapter 3.

A second group of interpretations is based on *reaction rate theory* and on the *gradual exhaustion* of the polymer load carrying capability due to the breakage and/or displacement of elements and/or to the formation of defects (cf. Section IV of Chapter 3). Depending on the load-time-temperature regime apparently different failure patterns are observed. As an example the creep fracture morphologies of PVC tubes subjected to different stress levels have been reproduced in Chapter 1. At higher stresses (σ_v = 50 MN/m^2) PVC tubes fail in a brittle manner after little creep deformation, and one speaks of *delayed brittle fracture* (Fig. 1.1). At medium stresses (42 MN/m^2) and after prolonged times the tubes show large scale plastic deformation, i.e. *delayed yielding* (Fig. 1.2). At lower stresses ($\sigma_v < 20$ MN/m^2) failure does either not occur at all within experimental time scales or by a competing mechanism, the *formation of a creep crack* (Fig. 1.3).

In order to predict lifetimes of stressed specimens three different approaches have been followed: Treatment of the whole fracture event as a single *thermally activated process*; analysis of the *rate of creep strain* or calculation of the *rate of flaw growth*. These three approaches will be discussed in the following.

2. Kinetic Theories

According to the kinetic fracture theories it is admissible to describe the whole complex fracture phenomenon by the kinetics of its slowest mechanism. Generally measurable experimental parameters such as rate and extent of creep deformation or flaw size do not enter the constitutive equation as variables. (The theory of Hsiao-Kausch, developed for solids not showing large scale anelastic deformation introduces a time-dependent strain which derives, however, as a *consequence* of the gradual network degradation).

To a large extent kinetic theories are founded upon the material behavior that is represented by the shown stress-time temperature diagrams (Figs. 1.4, 1.5, 3.7, and 3.11). The essential feature certainly is that within a large interval of time the logarithm of the time-to-break t_b decreases linearly with increasing uniaxial or principal stress σ_0. Any one of these straight lines can be expressed by

$$\ell n\, t_b/t_0 = A - \alpha\sigma_0 \tag{8.19}$$

where t_0, A, and α are constant with respect to time and stress. As discussed in Section IV of Chapter 3 various authors have proposed that these "constants" have a more general significance. Most approaches work with an Arrhenius equation:

$$\ell n\, t_b/t_0 = (U_0 - \gamma\sigma_0)/RT. \tag{8.20}$$

In this form U_0, γ, and t_0 are supposedly independent of temperature. The experimental data of Savitskii et al. [119] plotted in Figure 3.7 verify this point for a cellulose nitrate film and the polyamide 6 monofilaments investigated. For each material only one set of constants is necessary to describe all data points. For cellulose nitrate the authors obtained $U_0 = 159$ kJ/mol, $\gamma = 644 \cdot 10^{-6}$ m^3/mol, $t_0 = 10^{-13}$ s and for the oriented polyamide 6 (Capron) $U_0 = 184$ kJ/mol, $\gamma = 113 \cdot 10^{-6}$ m^3/mol, $t_0 = 10^{-13}$ s.

In the case of PVC tubes one obtains from an evaluation of Figure 1.4 according to Eq. (8.20) values of $U_0 = 397$ kJ/mol, $\gamma = 1740 \cdot 10^{-6}$ m^3/mol and (a purely formal value of) $t_0 = 1.7 \cdot 10^{-52}$ s. It should be noted that this set of parameters describes the PVC fracture times notwithstanding the fact that these data belong to three different fracture modes. In the stress-time curves of non-oriented semicrystalline polymers (PE, PP) one observes a drop of strength at larger values of t_b, the well-known bend (Fig. 1.5). The data of the flat portions (relating predominantly to ductile failure) can be represented by

$U_0 = 307$ kJ/mol, $\gamma = 4390 \cdot 10^{-6}$ m^3/mol, and $t_0 = 3 \cdot 10^{-40}$ s

and the data of the steep portions (creep crack failure) by

$U_0 = 181$ kJ/mol, $\gamma = 3610 \cdot 10^{-6}$ m^3/mol, and $t_0 = 8 \cdot 10^{-20}$ s.

For oriented semicrystalline polymers Zhurkov et al. [18] report:

HDPE $U_0 = 113$ kJ/mol, $\gamma = 109 \cdot 10^{-6}$ m³/mol, $t_0 = 10^{-11}$ s
LDPe $U_0 = 117$ kJ/mol, $\gamma = 247 \cdot 10^{-6}$ m³/mol, $t_0 = 10^{-12}$ s
PP $U_0 = 125$ kJ/mol, $\gamma = 213 \cdot 10^{-6}$ m³/mol, $t_0 = 10^{-12}$ s.

These formal evaluations need discussion. The good correspondence between U_0 and the dissociation energy of the weakest main chain bonds for many of the (mostly oriented) polymeric materials investigated by him has led Zhurkov [120, 18] to establish his kinetic theory of fracture. This theory assumes that all important steps contributing to the fracture of a material are not only related to but even controlled by thermal density fluctuations (dilatons). As important steps have been considered chain scission, microcrack formation, and crack nucleation [3, 384–388]. The effect on fracture behavior of structure, orientation, dimensionality of the state of stress, morphology, and homogeneity of chemically "identical" polymers must be expressed, therefore, through the single parameter γ.

The parameter γ has the dimension m³/mol and is termed activation volume. With reference to Chapter 3, Section IV, it should designate the thermally activated volume if the stress σ_0 indeed acts on this volume directly and without intermediates. This is approximately the case in flow but not in chain scission. The scission of a chain necessitates molecular stresses ψ_b which are clearly much larger than the applied external stresses σ_0. For PA 6 molecular stress concentration factors ψ_b/σ_0 of between 10 and 40 are commonly determined (cf. Chapters 7 and 8 above). Taking $12 \cdot 10^{-6}$ m³/mol for the activation volume β of the chain scission process a parameter $\gamma = \beta_\psi/\sigma_0$ of 120 to $480 \cdot 10^{-6}$m³/mol should be predicted. The activation volume of $12 \cdot 10^{-6}$ m³/mol amounts to (0.27 nm)³ or to a stretching of the rupturing bond by 0.1 nm.

The effect of the structural parameter *chain orientation* on γ becomes apparent from Figure 3.11. There the stress-lifetime data of PAN fibers [74] are plotted with draw ratio λ as a parameter. The highest draw ratio ($\lambda = 17.3$) corresponds to a γ of $248 \cdot 10^{-6}$ m³/mol, $\lambda = 4$ to 590, $\lambda = 2.62$ to 841 and $\lambda = 1$ to $\gamma = 1200 \cdot 10^{-6}$ m³/mol. The γ values are the smaller the more homogeneously the macroscopic stress is distributed over the molecular chains.

The foregoing discussion shows that the γ values of PMMA, PVC, and PE can hardly be explained on the basis of chain scission unless extremely large stress concentration values are postulated. In the case of PVC the γ corresponds to (1.7 nm)³ which points to the involvement of larger domains in the activation of the creep process. The fact that the PVC fracture data of all modes are described by one equation seems to indicate that in the cited experiment [111] all modes depend on the creep deformation. Once a certain structural weakening has resulted from creep the final failure is accomplished by one of the three modes, e.g. by brittle fracture at large stresses, ductile failure, or creep crack development. Since the final period accounted only for a small fraction of t_b no effect on the slope of $\ln t_b/t_0$ was noted.

Comments are necessary concerning the significance of t_0 and σ_0 in Eq. (8.20). Values of 10^{-12} to 10^{-14} s have been frequently determined for t_0; they have been assigned to longitudinal vibrational modes of the backbone chain. For PVC and PE extremely small values have been obtained in this section which certainly cannot be interpreted in the same manner. With flow processes it must be recognized that the activation energy is not temperature independent but decreases with temperature:

$$U_0 (T + \Delta T) = U_{00} + \frac{\partial U_0}{\partial T} \Delta T. \tag{8.21}$$

In this case one has

$$\ell n\, t_b/t_0 = \frac{\partial U_0}{\partial T} \frac{\Delta T}{RT} + \frac{(U_{00} - \gamma\sigma_0)}{RT}. \tag{8.22}$$

A value t_0^d formally determined from experimental data according to Eq. (8.20) contains, therefore, the two terms t_0 and the temperature-independent part of $\frac{\partial U_0}{\partial T} \frac{\Delta T}{RT}$. The latter term seems to be responsible for t_0^d values of 10^{-40} s and less.

3. Creep Failure

The presence of long and flexible chains causes a polymeric material to show a monotonic, partly anelastic deformation under constant load, viz. a creep deformation. If not accompanied by strain hardening the creep deformation of a loaded specimen will invariably lead to a reduction in cross-section, to an increase in stresses and to the eventual (ductile) failure. This phenomenon is well known for practically all materials. Especially for metals the analysis of creep deformation is highly advanced (see for instance [300–302]).

As indicated above the continued reduction of sample cross-section due to the creep deformation will lead to an increase in stress. Sample thinning and possibly structural weakening associated with the continued creep deformation set the stage

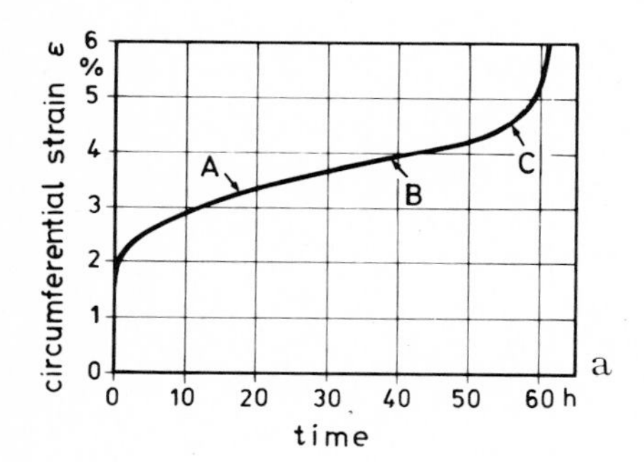

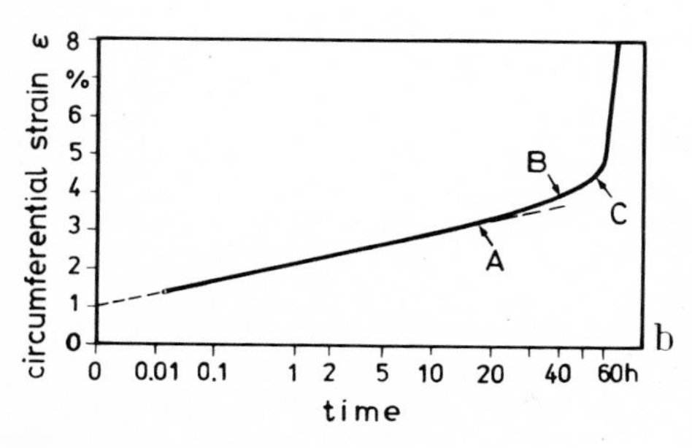

Fig. 8.34. Circumferential creep strain ϵ of a PVC pipe at $\sigma_v = 42$ MN/m^2 and T = 20 °C (after [109]). (a) linear time scale, (b) $\epsilon - \epsilon_0 \simeq (t/h)^{0.23}$. A: limit of primary creep phase, B: inflection point, C: appearance of stress whitening.

for the eventual local yielding. As an example the circumferential creep strain of the afore mentioned PVC pipe at $\sigma_v = 42 \text{ MN/m}^2$ is plotted in Figure 8.34. The curve clearly represents the characteristic parts of a creep curve: the instantaneous (elastic) deformation ϵ_0, a primary phase with a decreasing strain rate, a secondary phase where the strain rate is constant and a third phase of accelerated creep. During the latter phase creep rates are large, the material approaches the yield conditions. Such a con-

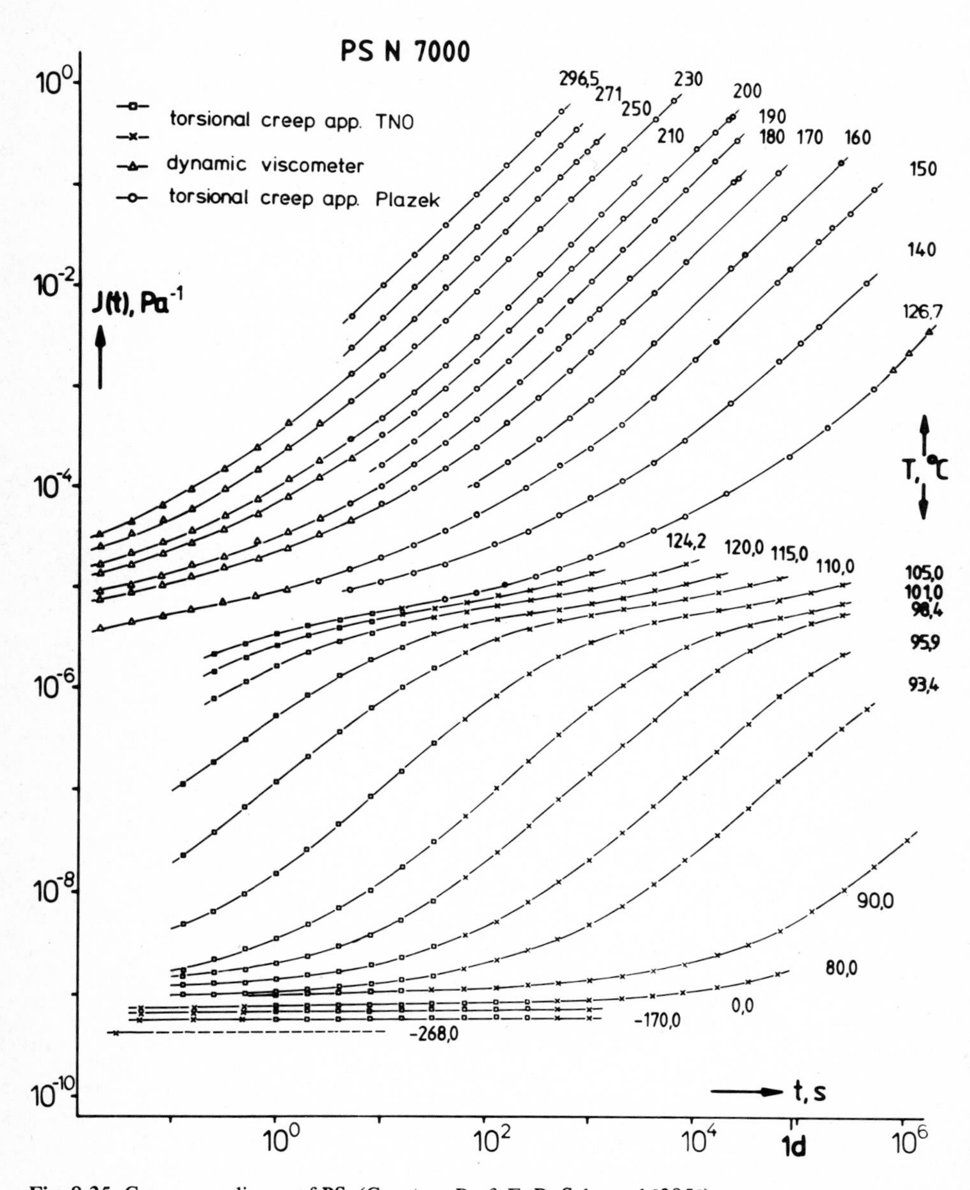

Fig. 8.35. Creep compliance of PS. (Courtesy Prof. F. R. Schwarzl [305])

dition is generally reached first within the most highly stressed material, i.e. at the smallest sample cross-section.

In order to predict the creep behavior and possibly the ensuing failure a number of approaches have been proposed. These are based respectively on the theory of viscoelasticity – including the concept of free volume – or on empirical representations of ϵ (t) or of the creep modulus $E(t) = \sigma_0/\epsilon(t)$. For polymers in the linear viscoelastic range the creep compliance $J(t,T) = \epsilon(t,T)/\sigma_0$ shows the expected inverse behavior of the relaxation modulus $E(t) = \sigma(t,T)/\epsilon_0$. The creep behavior has been studied intensively by Schwarzl et al. [304–305] within experimental time scales spanning more than 8 decades. At extremely short times and low temperatures the response is elastic with compliances J_0 of the order of 0.3 to $0.5 \cdot 10^{-9}$ Pa^{-1}. At slightly larger times the local motion leads to an increase of J to values of the order of $1 \cdot 10^{-9}$ Pa^{-1}. In the transition range (see Fig. 8.35) J increases rapidly. For many applications the approximation of J (t) in this region by:

$$\log J(t)/\mathrm{GPa} \sim m \log t/s + c \text{ with } 0.5 < m < 0.9 \tag{8.23}$$

has been useful [304].

At longer times and beyond the transition zone an uncrosslinked polymer flows:

$$J = J_e + t/\eta_0 \tag{8.24}$$

with compliances of the order of 1 Pa^{-1}.

The framework of the linear theory of viscoelasticity also permits the calculation of viscoelastic moduli from relaxation time spectra and their interconversion. The reduction of stresses and time periods according to the time-temperature superposition principle frequently allows establishment of master-curves and thus the extrapolation to larger values of t (cf. Chapter 2). The strain levels presently utilized in load bearing polymers, however, are generally in the non-linear range of viscoelasticity. This restricts the use of otherwise known relaxation time spectra or viscoelastic moduli in the derivation of ϵ (t) or E (t).

The limits of linearity are discussed in a review article by Yannas [112]. He concludes that for practically all polymers at $T\text{-}T_g < -20$ K a strain smaller than 1% forms the limit of linear response. For semicrystalline polymers (e.g. PP, PAN, PETP, PA 66) a limit of 0.1 to 0.4% seems to be valid also above the glass transition temperature T_g (even at $T - T_g > 100$ K).

For amorphous polymers (PC, PIB, NR) the limit increases around T_g from 1 to about 50% of strain [112]. There are some special approaches to a calculation of creep and recovery based on non-linear theories of viscoelasticity [112–114, 306, 307]. At present it is not possible, however, to predict with sufficient generality the onset of accelerated creep and thus of delayed yielding from non-linear viscoelastic theories. Promising results have been obtained, however, using the free volume concept to account for the shift of relaxation times with strain and thermal history and to predict changes of failure stress and failure mechanism [123, 124].

Empirical approaches to represent the creep curves of materials by mathematical functions have been known for more than 60 years. One of the first is the Andrade equation

$$\epsilon(t) - \epsilon_0 = \beta (t/t_0)^{1/3} + Kt \qquad (8.25)$$

where $t_0^{-1/3}$ and K are adaptable constants [115]. This equation – originally derived for metals – describes the *uniaxial* creep of different materials such as Pb, Cu, Fe, Cd, PMMA, PA, PVC within certain time periods very well [111, 115, 116]. Various empirical modifications of this equation have been reported [116, 308, 309, 319]. Thus the *circumferential* creep is described by a similar potential law

$$\epsilon(t) - \epsilon_0 = \beta(\sigma_0, T)(t/t_0)^n. \qquad (8.26)$$

For PVC tubes the exponent n was found by several investigators to be about 0.23 [111]. For the dependency of β on σ_0 a hyperbolic sine function has been frequently proposed in the literature (e.g. 111, 116). For PVC tubes at 20 °C β assumes the form [111]:

$$\beta = 0.019 \sinh(\sigma_0/8.85 \text{ MN m}^{-2}). \qquad (8.27)$$

Equations (8.26) and (8.27) permit the calculation of $\epsilon(t, \sigma_0)$ at 20 °C. They thus also permit one to determine at which stress a specified strain value ϵ_1 is being reached after very long times. One finds that a strain value $\epsilon_1 = 3\%$ – at which the creep rate of moderately stressed PVC tubes ($\sigma_v = 35$ to 48 MN/m^2) changes from phase I to phase II – will be reached at $\sigma_0 = 22$ MN/m^2 only after 50 years. An eventual fracture should take place even later than that.

A still simpler and more accessible approach to predict the creep strain of engineering components is based on a logarithmicly linear approximation of the creep modulus, E(t):

$$\log[E(t)/E(t_0)] = \frac{1}{3} \log(t/t_0) \log[E(1000\text{ h})/E(1\text{ h})]. \qquad (8.28)$$

Values of the 1-h and 1000-h moduli are generally given in the technical literature (e.g. 104). At 20 °C and at small stresses (5 to 20 MN/m^2) the ratios of E (1000 h) E(1 h) of common polymers are of the order of 0.96 (SB), 0.92 to 0.93 (PC, PETP, fiber reinforced thermoplastics), 0.88 to 0.90 (PS, PVC, PMMA, POM), and 0.72 to 0.79 (HDPE, PP, ABS). Whereas extrapolation methods are frequently the only means available they are subject to all the limitations introduced by the gradual development of structural weakening (demonstrated by the change from a decelerating creep rate to an accelerating one). Also the action of competing processes (creep crack development) is generally not taken into account by this extrapolation method.

In this section different approaches to predict creep amplitudes have been discussed. In order to calculate the time of creep failure knowledge of the "critical failure conditions" is necessary. In ductile failure the *yield criteria* will generally serve this purpose. For strain-hardening, fiber-forming polymers Section I may be consulted. Concerning the latter case the observation of Takaku [310] is also interesting in that the true ultimate creep strain $\lambda_p \cdot \lambda_b$ attained by partially oriented polypropylene fibers after prolonged loading was independent of the extent of preorientation λ_p.

These fibers seemend to fail once a predetermined, stress- and temperature-dependent state of orientation had been achieved.

A final remark concerns the effect of the state of stress on failure mode and time. The term σ_0 in Eq. (8.20) has been taken to represent the macroscopic uniaxial stress σ_0 in the event of uniaxial stretching – or the largest principal stress component σ_v in the event of multiaxial states of stress. This does not mean that the times to fracture of tubular and uniaxial specimens are always identical if only σ_v is equal σ_0. As discussed above the creep functions in these two cases are described by different potential laws, namely by Eqs. (8.25) and (8.26) respectively.

A systematic study of the effect of the state of stress on the lifetime of PVC tubes has been carried out by Smotrin et al. [151]. They found that at short times to fracture (at stresses of 50 MN m^{-2}) the simple Rankine criterion $\sigma < \sigma^*$ represented their failure data in two-dimensional stress space. With increasing times to fracture, however, the von Mises criterion gave the better representation. Gotham [152] studied the uniaxial creep failure of 15 different polymer materials at 20 °C. Within a time range of up to 10^7 s he observed brittle failure with injection molding grade PMMA, PS, SAN, and glass-fiber reinforced PA 66, and ductile failure for PP, bulk cast PMMA, PC, PSU, PVC, ABS, POM, PA 66, and P4MP.

4. Creep Crack Development

The development of a creep crack is the dominating failure mechanism of non-oriented thermoplastics at smaller stresses (see diagrams Figs. 1.4, 1.5 and 8.39). The fracture surface morphology of a typical creep crack is reproduced in Figs. 1.7 and 8.36. The final failure of these samples had been caused by crack growth over the whole wall thickness.

The pattern of creep-crack failure shows the three well-known phases: a period t_i of apparently homogeneous deformation up to the point of crack initiation, a period t_a of thermally activated slow crack growth, and a final period t_r of unstable

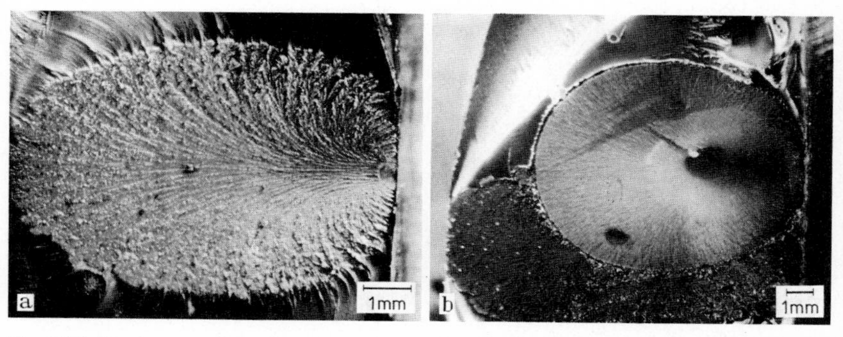

Fig. 8.36. Fracture surface of creep cracks in LDPE tubes (Courtesy of P. Stockmayer, [118]). (a) crack initiation at inside boundary of tube wall, $\sigma_v = 2$ MN m^{-2}, T = 80 °C, $t_b = 140$ h, (b) crack initiation within tube wall, $\sigma_v = 2$ MN m^{-2}, T = 80 °C, $t_b = 7082$ h.

crack growth. An inspection of loaded samples during and after the creep experiment reveals a number of morphological features:

- creep cracks initiate at flaws or material inhomogeneities, more often in the surface zones (Figs. 1.7, 8.36a) but also in the bulk material (Fig. 8.36b) [111, 117, 118, 311, 312];
- the initiated crack grows through thermal activation during a period t_a under apparently constant conditions; this is evidenced by the appearance of the "mirror zone", the frequently ideally circular or semicircular, smooth interior fracture surface (Figs. 1.7, 8.36b); depending on the stress level the mirror zone may extend over almost the complete wall thickness of tubular specimens;
- at higher magnifications of the mirror zones it can be seen that the surfaces are not smooth on a molecular level; in the case of amorphous PVC an irregular, rough, and scaly surface is observed and in the case of semi-crystalline PE a cellular, locally highly drawn surface morphology becomes apparent (Fig. 1.8);
- inspection of some transparent tubular samples under load and of non-transparent samples after an interruption of the creep test showed that in those PVC and PE pipes creep crack development started rather late in the experiment, i.e. t_i was larger than 0.90 t_b [111, 117]; in PP tubes the first creep cracks could be detected at t_i being about 0.5 t_b [117]. In other cases, especially in particle initiated fractures, the thermal growth period t_a can be much larger than the initiation period t_i. Johansson [312] reports this for high quality PVC pipes where he found up to 100 simultaneously growing cracks.

Morphological effects on the initiation of creep cracks were also studied by Stockmayer [118]. Using an elaborate peeling technique he transformed whole tube sections into thin films of up to 0.06 mm thickness (Fig. 8.37). The walls of LDPE tubes, transformed into films of 0.9 m^2 area, then were scanned under a microscope for flaws or material inhomogeneities (Fig. 8.37). Inhomogeneities could be detected in every tube and have been related to an imperfect blending of the LDPE with carbon black [118]. Stockmayer also found two interesting correlations:

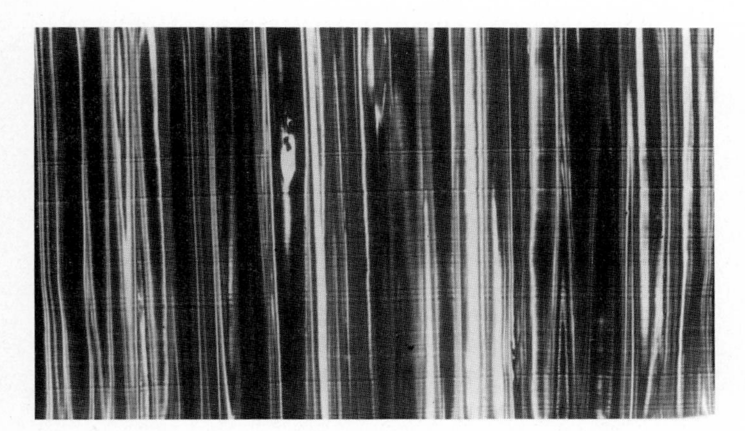

Fig. 8.37a. Part of a film cut from LDPE tube wall showing inhomogeneous blending in extrusion (Courtesy P. Stockmayer, [118]). Thickness 0.06 mm; width 50 mm

Fig. 8.37b. Section of a film cut from LDPE tube wall showing spots which are free from carbon black. (Courtesy P. Stockmayer, [118])

- creep cracks were initiated preferentially in those zones which also contained (the most) inhomogeneities (Fig. 8.36 shows such creep cracks)
- tube lifetimes were greatly reduced if the creep cracks had been initiated in the outside or inside zones of the wall; from an ensemble of tubes tested at 80 °C and having an average lifetime $\overline{t_b}$ of 2340 h tubes with inhomogeneities (and crack initiation) in the boundary zones failed after a $\overline{t_b}$ of 740 h; tubes with crack initiation in the central region of a wall lasted to $\overline{t_b}$ = 7400 h, i.e. a factor of 10 longer [118, 149]. This correlation is clearly brought out by the plot shown in Fig. 8.38. Although the differences in stress distribution within the pipe wall due to frozen-in tensions may have influenced crack initiation, it seems to be more probable that network defects are mostly responsible for this observation. As discussed in Section III B2 the inner walls of extruded pipes are especially exposed to thermal degradation. It was precisely in this region that Stockmayer and Wintergerst found the origins of the most rapid failures [118].

To test the influence of the observed irregularities further the authors performed tensile drawing and static loading experiments on samples cut from the peelings. In these experiments, there was practically no correlation between yield stress and time to failure on the one hand and kind and concentration of flaws and inclusions on the other. However, the effective elongation of films containing defects was much smaller since the necking samples generally broke whenever the neck had reached the defect. This behavior confirms the statements made before namely that in these cases creep craze nuclei must be considered as network defects which are not detectable in short time loading tests unless the defective zone is locally subjected to large deformations.

The term "network defect" needs some deliberation. As indicated before many of those defects are created during sample fabrication. If they are due to particulate inclusions, imperfect blending, or oxidation during processing, they exist from the very beginning of sample life; in these cases a sizeable portion of the sample life may in fact be used for defect growth ($t_a \sim t_b$). However creep cracking is also observed

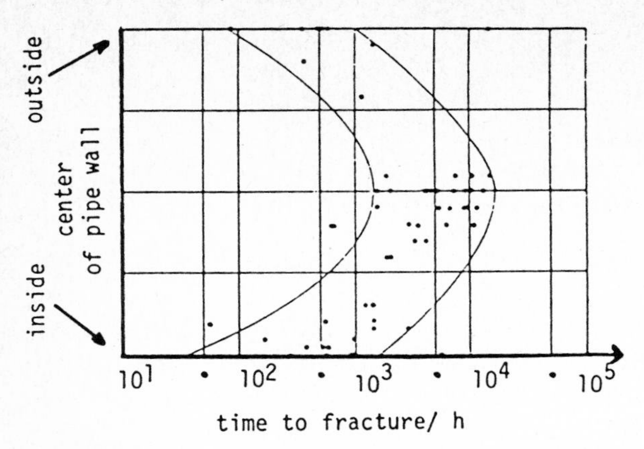

Fig. 8.38. Position of creep craze initiation within wall of low density PE pipe vs. time to fracture (after Stockmayer [118b])

in the apparent absence of preexisting defects. In these cases creep cracks are initiated during the service life of a sample following the (local) physical and/or chemical degradation of the network (see also [410]).

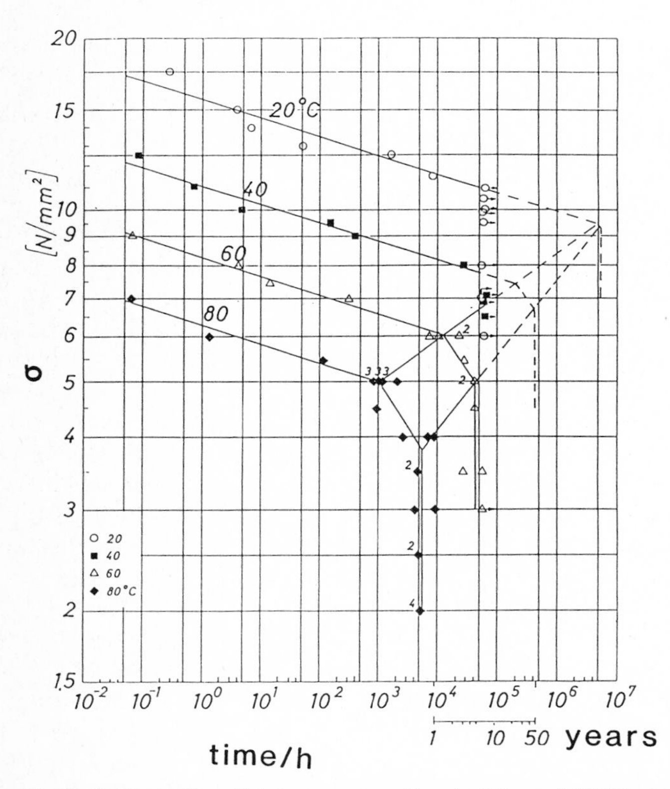

Fig. 8.39. Circumferential stress σ versus time to failure of HDPE pipes (after Gaube et al. [117b]); the small numbers indicate number of specimens with almost identical times to failure

In Fig. 8.39 long term test results with HDPE pipes are plotted [117b]. In this case the σ-t curves reveal three regions, the well known, slightly inclined portion on the left-hand side (ductile failure), the somewhat steeper creep-crack region, and a vertical portion. The studied material (Hostalen GM 5010 T2 of Hoechst AG) resists better to creep cracking, consequently the slightly inclined "ductile portions" of the curves extend to longer times than those for the HDPE material represented in Fig. 1.7.

Kagan et al. [121] have studied the effect of supramolecular structure on the time-dependent strength of HDPE. They found a good correlation between the 1-h (plastic) creep strength and the yield strength for materials of different crystallinity, density ($0.945 < \rho < 0.960$ g cm^{-3}) crystallite size, and spherulite diameter. These parameters hardly affected the activation volume γ but slightly the activation energy of the plastic deformation process (parallel displacement of the *ductile branch* of the σ-log t/t_0 curve). On the other hand they found a clear tendency to higher *long-term brittle strength* (resistance against creep crack formation) with an increase in density and size of crystallites and with a reduction of spherulite diameter. Gaube et al. [117] also report that an increase in crystallinity (i.e. density) of PE, PP, PETP, POM, PA leads to a higher yield strength – at lower strain values – and to a higher ductile strength. They point out, however, that the decrease of brittle strength in the steep region begins sooner, the higher the crystallinity and the lower the molecular weight. The ultrahigh molecular weight polyethylene GUR does not show creep crack formation at all. Judging from these observations the process of creep crack formation seems to be related to the *gradual disentanglement of chains* and the ease of void opening in the inter-crystalline and/or inter-spherulitic regions. A similar concept has recently been proposed by Lustiger et al. [313] who assume that the interlamellar disentanglement of chains causes the creep cracking of PE pipes.

The vertical drop of the σ-t diagram (Fig. 8.39) so far has only been observed at higher temperatures. After extended studies of this phenomenon Koppelmann et al. [363] and Gaube et al. [117b] have assigned this complete loss of mechanical strength to the increasing thermal degradation after exhaustion of the thermal stability of the material (as evidenced by the coincidence of the reduction of the oxidation induction time at 200 °C to zero and the rapid drop in strength).

To summarize the observations with failure under constant load: three mechanisms are of importance which generally have quite different kinetics. Creep prevails at higher stresses, the corresponding times to failure depend strongly on the applied stress; creep crack formation is observed at smaller stresses, in semi-crystalline materials possibly caused by disentanglement or pull-out of chains leading to a branch of the σ-t curve with a somewhat steeper slope; it is noteworthy that in the presence of an active environment this steeper branch is very often shifted, mostly towards smaller times to failure [107, see also Section III B4]; the third mechanism, auto-accelerated oxidation, is only observed at elevated temperatures *and* after exhaustion of the thermal stability; after initiation oxidation occurs at increasing rates so that on a logarithmic timescale it appears equivalent to instantaneous degradation and loss of strength [107, 117b, 363].

Concluding this section one may say that there is very little evidence that the loading and breaking of chains controls the creep of unoriented polymers. The softening mechanism – common to all linear polymers [122, 128, 154] – responsible for the change from decelerated to accelerated creep is generally related to segmental motion and changes in intermolecular attraction but not to chain degradation. A search for free radical formation in creep of non-oriented polymers carried out by Jansson et al. [125] is as yet also inconclusive. These authors had included radical scavengers in PVC which led to a considerable reduction in creep rate (by a factor of up to about 10). They state, however, that it has not been possible to separate the stabilisation effect from the effect of mechanical reinforcement [125].

C. Homogeneous Fatigue

1. Phenomenology and Experimental Parameters

Polymeric engineering materials subjected to strong mechanical and environmental excitation show – as any other material – a gradual degradation of their performance including eventual failure. If the property changes are mostly due to *chemical reactions* one speaks of *corrosion,* or of *radiative degradation.* The term *fatigue (Zerrüttung)* is used if a deterioration of material properties is caused by the repeated cyclic or random application of mechanical stresses. The synergistic interaction of mechanical and environmental attack leads to the phenomenon of environmental stress corrosion (ESC). The role of molecular chains in ESC, and in corrosive, thermal, and radiative degradation will be discussed in the final Section of this Chapter. The fatigue of polymeric fibers has already been treated in Section 1 F of this Chapter. The present Section, therefore, will deal mainly with the principal aspects of fatigue and with isotropic thermoplastics. The fatigue of polymeric engineering materials has received a growing attention in the last thirty years. Review articles on this complex subject have appeared from time to time; the reader is referred to the ones authored by Andrews [126], Manson and Hertzberg [127–128], Oberbach [129–130], Schultz [131], and, more recently, by Sauer et al. and Takemori [314, 315]. A comprehensive monograph on this subject has been written by Hertzberg [316]. In addition, a large number of research papers have been published of which an incomplete listing is given in this Section [116, 132–147, 153, 317–324].

It is the general observation that failure in repeated loading occurs at load levels which are lower than the stresses sustained under the conditions of static loading (creep) or of monotonously increasing deformation (drawing). The lower the stress level to which a material is subjected the larger the number N of load cycles which are sustained. The total time t_f, however, which a fatigued sample spends under load is generally much smaller than the time to failure under static loading conditions. Load reversals or load interruptions have an accelerating effect, therefore, on the loss of load bearing capability; load reversals or interruptions constitute the elements of fatigue. It may already be stated at this point that the accelerating effect of load reversals has to be related to two characteristic material properties:

- the anelastic response which leads within each loading cycle to a dissipation of mechanical energy as heat and

- the occurrence of molecular rearrangements which may lead to the disentanglement and slip of chain segments, to local reorientation, and/or to the formation of voids.

It has to be expected, and must be recognized, therefore, that three principal mechanisms may contribute to fatigue failure: *thermal softening, excessive creep or flow,* and/or the *initiation and propagation of fatigue cracks.*

The experimental parameters of a fatigue test have been aptly defined by Andrews [126]:

- a periodically varying stress system having a characteristic stress amplitude, $\Delta\sigma$;
- a corresponding periodic strain amplitude, $\Delta\epsilon$;
- a mean stress level, σ_m;
- a mean deformation, ϵ_m;
- a frequency, ν; $\omega = 2\pi\nu$;
- a characteristic wave form (sinusoidal, square, etc.) for both the stress and strain;
- ambient and internal temperatures, which in general will not be precisely identical;
- a given specimen geometry, including notches, if any.

More recently the (computer-programmed) random loading has also found growing application in testing structural components [132]. Of course, the above parameters can be combined in different ways and be changed within a test. They partially depend upon each other and on environmental conditions.

The experimental equipment will not be described here, the reader is referred to the cited review articles or pertinent research papers (e.g. 132–138). In practice five different types of stress-strain control are employed which are classified by Andrews [126] and Manson et al. [127] as:

- periodic loading (between fixed stress limits, in tension or compression); [133–135]
- periodic loading (between fixed strain limits, in tension or compression); [133–135]
- reversed bending stress (implicit in flexing a sheet in one dimension); [136]
- reversed bending stresses in two dimensions (by rotary deflection of a cylindrical specimen); [137]
- reversed shear stresses, e.g. obtained by torsional deformation. [138]

2. Thermal Fatigue Failure

In cyclic loading of a viscoelastic solid an amount of energy ΔW is dissipated per cycle. For sinusoidal loading in the anelastic range, one obtains

$$\sigma(t) = \sigma_0 \sin \omega t \tag{8.29}$$

$$\Delta W = \pi \sigma_0 \epsilon_0 \sin \delta . \tag{8.30}$$

This hysteresis heating leads to a noticeable temperature rise. At low test frequencies and low stress levels the temperature increase of the test specimen generally approaches a finite value. A PA 6 sample for instance fatigued at 50 Hz, at a constant stress amplitude of 8 MN m^{-2}, and at an ambient temperature of 21 °C, assumed a stable temperature of 27 °C after some 10^4 cycles [139–140]. At this temperature mechanical energy input and thermal energy loss were in equilibrium.

At higher frequencies ω, higher ambient temperatures, or higher stress amplitudes, however, the energy input into a sample becomes larger than the heat which can be transferred out of the sample through conduction or radiation. Depending on the experimental setup, failure of the sample then occurs through thermal softening and/or plastic flow. The stress and frequency range, where this type of thermal failure has to be expected, can be calculated from the energy input, the geometry, and the thermal properties of a sample. Thermal equilibrium in a cylindrical sample of radius R is reached once

$$\frac{\partial T(r,t)}{\partial t} = \frac{\omega \sigma_0^2 \sin\delta}{2\rho c_p E'} + \frac{\lambda}{\rho c_p}\left(\frac{\partial^2 T}{\partial r^2} + \frac{1}{r}\frac{\partial T}{\partial r}\right) \tag{8.31}$$

approaches zero. The highest temperature, T_e, observable at equilibrium in the center of the cylinder then amounts to

$$T_e = T_a + \left[\frac{R}{2h} + \frac{R^2}{4\lambda}\right]\frac{\omega \sigma_0^2 \sin\delta}{2E'} \, . \tag{8.32}$$

T_a designates the ambient temperature, h the coefficient of convection at the sample surface; and the constants ρ, c_p, and λ have their usual significance as density, specific heat, and heat conductivity of a material. The term $\lambda/\rho c_p$ is also known as temperature conductivity number. For conventional polymeric materials this number assumes values between $1.0 \cdot 10^{-7}$ m^2/s (PTFE) and about $2.1 \cdot 10^{-7}$ m^2/s (HDPE, POM).

It should be mentioned at this point that the hysteresis heating ΔW in a fatigue test with *constant-strain amplitude* decreases with increasing temperature, since ΔW is proportional to $\sigma_0 \sin\delta$, i.e. to E''. Under this condition thermal equilibrium can generally be established. The same effect, a decrease of E'', can of course be reached if a sample is plasticized. The plasticization of an engineering component subjected to fatigue at constant-strain amplitude can be, therefore, a suitable way to prolongue the fatigue life of the sample [152]. Machyulis et al. [152] point out that the thermo-stabilization, the fatigue-stabilization, and the plastization effects of various additives on polyamide are not related to each other. For optimal results on fracture energy and cycles to failure Kuchinskas et al. recommend diffusion stabilization from a 5% ethanolic quinhydrone solution at 343 K for 1 h [152].

The conditions of thermal failure are especially discussed by Oberbach [129–130] for a variety of polymers, by Zilvar [139–140] for PA 6, and by Alf [141] for PMMA and unsaturated polyester resins.

It is only through their effect on the above constants and on the complex modulus E* that the molecular chain structure influences thermal failure. In the context of this monograph this type of failure will, therefore, not be discussed any further.

3. Wöhler Curves

The common observation that the number of cycles to failure, N_F, depends on the stress or strain amplitude is generally represented in the form of Wöhler curves (Wöhlerkurven, S – N curves). Such a representation accounts for the fact that stress (or strain) amplitude and number of load cycles are the most important of the many (and closely interacting) parameters which affect the fatigue life.

The fatigue data represented by Figure 8.40 were obtained from fully reversed, uniaxial tension-compression tests at 0.1 Hz with temperature rises T_e-T_a below 2 K [142]. The three different regions of the S – N curve of polystyrene are fairly typical for the fatigue-life curves of other glassy polymers as well. *In region I* crazes form during the first tensile quarter cycle. The fatigue behavior is essentially that of a pre-crazed sample. Failure occurs through the initial slow growth of a crack through craze material followed by catastrophic crack propagation. The number of cycles to failure N_F is strongly influenced by the stress amplitude, presumably through the strong dependency of the number of crazes on the stress level.

Region-II fatigue fracture is characterized by the fact that a period of craze nucleation precedes craze growth and the formation, slow growth, and catastrophic propagation of a crack. This type of fatigue fracture is observed at stress levels just below the immediate craze initiation stress σ_i. The dependency of N_F on σ is considerably reduced. This accounts for the delay in craze initiation as well as for the reduced rate of slow crack growth at lower stress levels. The slope ($\simeq$ 14 MN m^{-2} per decade of N_F) seems to be characteristic for a variety of polymers [142, 153].

The long-life *region III* essentially forms the endurance limit of the material. The dependency of σ on N_F is extremely small. Crack initiation has a very long incubation period. Analysis of the fracture surfaces of fatigue failures in this region reveals furthermore that the slow crack growth mode has two phases. The first shows little evidence of crazing, the second resembles the region-II failure mode.

The above considerations concern polystyrene fatigued by completely reversed stress ($\sigma_m = 0$, $\Delta\sigma = 2\sigma$). In principle similar observations are also made with other polymers and at other conditions of stress, e.g. under constant maximum stress at various stress amplitudes [127, 153]. At this point it should be emphasized that the "fatigue process" under conditions II and III comprises at least two phases: an initial phase of crack or craze initiation with an apparently *homogeneous* response of the material and a second phase of *heterogeneous* fatigue with localized and growing

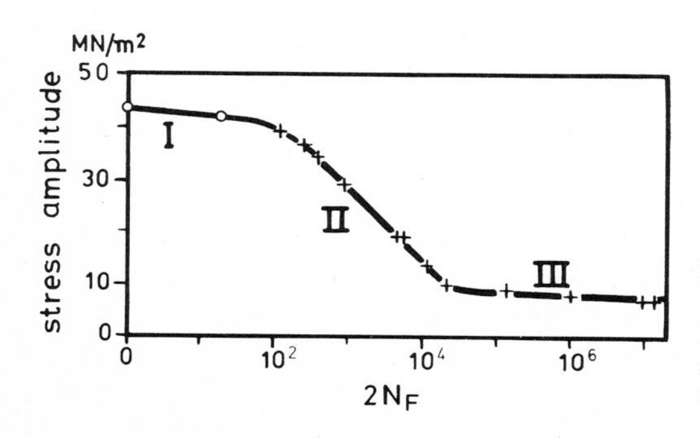

Fig. 8.40. Wöhler curve of polystyrene subjected to fully reversed uniaxial tension-compression at 0.1 Hz at ambient temperature (after [142]). N_F number of cycles to failure.

material discontinuities (fatigue cracks). The second phase in turn comprises several modes of crack propagation (thermal and athermal growth, catastrophic failure). This will be discussed in more detail in Chapter 9.

The discrimination between homogeneous and heterogeneous is certainly useful although not necessarily always significant. Most of the *kinetic theories of failure* treated in Chapter 3 encompass the two phases of crack initiation and slow crack growth. Thus they implicitly assume that the entity of molecular rearrangements between the application of load and the end of the slow crack growth period can be described by one constitutive equation.

4. Molecular Interpretations of Polymer Fatigue

Different experimental observations permit the conclusion that the prolonged periods of craze or crack initiation at lower stresses are not simply due to the reduced probability for crack nucleation in an otherwise unchanged material. The nature of the changes occurring on a molecular level during sample fatigue have been investigated by several authors [e.g. 138, 143–147, 153]. Thus molecular mobility, interaction, and damping have been studied by the torsion pendulum [138, 134–144] and by infra-red technique [138], molecular packing and defect structure by density measurements and X-ray scattering technique [144–146] and by the use of samples of different molecular weight [153], and morphological changes by the recombination kinetics of trapped free radicals [146]. The results obtained from these different experimental methods characterize the molecular rearrangements, but they do not yet permit the establishment of quantitative fatigue limits.

Through simultaneous application of several of these methods Sikka [144] has been able to characterize some molecular rearrangements in homogeneous fatigue. He fatigued in cyclic extension thin films (0.075 mm thick) of polystyrene (Tricite) and polycarbonate. Subsequently he investigated these films by Fourier transform infrared (FTIR) and mechanical spectroscopy and by X-ray diffraction technique. The fatigued PS samples were scanned under an electron microscope to detect any crazes that might have developed due to fatigue. No crazes were observed in the samples fatigued up to 2500 cycles.

The FTIR results obtained on a PS film after 2500 cycles are shown in Figures 8.41 and 8.42. In the *superimposed* FTIR spectra slight distortions of IR bands are noticeable in the frequency range of 900 to 1700 cm^{-1} (Fig. 8.41). To detect minute changes caused by the fatigue process Sikka also recorded *computer-subtracted* FTIR spectra. One such spectrum (fatigued-90% unfatigued) is plotted in Figure 8.42, the changes observed in this spectrum are listed in Table 8.5. However, as indicated in Section I C 1, great care must be taken in interpreting such difference spectra.

From X-ray diffraction data Sikka obtained a mean Bragg distance d_{Bragg} of about 0.48 nm for the unfatigued PS films which showed a decrease of $\simeq$ 0.01 nm in the fatigued sample (2500 cycles). This negative shift in d_{Bragg} was believed to be conclusive. It was related to the decrease in interphenyl and intraphenyl distances. The fatigue to 2500 cycles also resulted in a change of dynamic mechanical losses [144, 317]. Figure 8.42 shows a plot of tan δ versus temperature obtained at a fre-

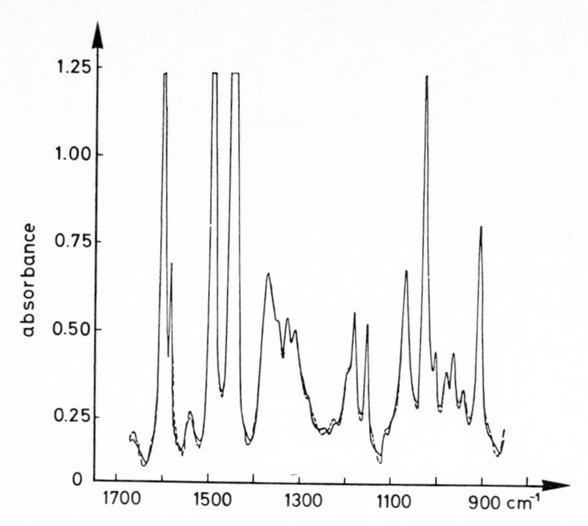

Fig. 8.41. Superimposed FTIR spectra I_n of unfatigued (——) and I_f of fatigued (– – –) polystyrene film (from [144]).

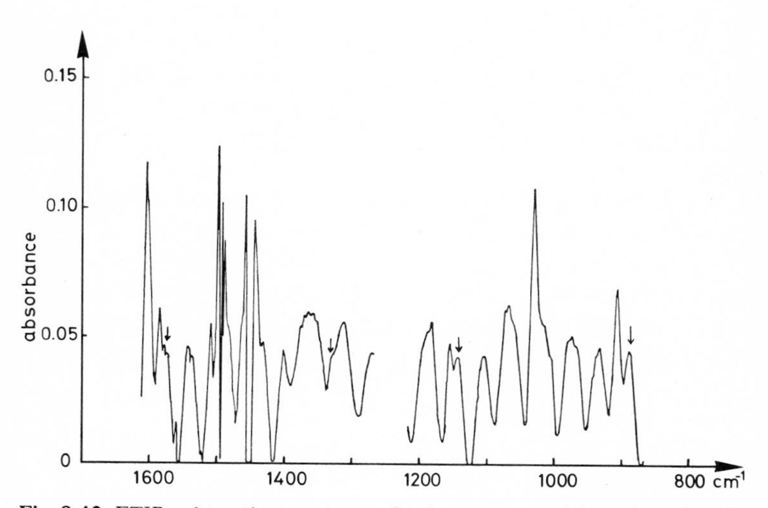

Fig. 8.42. FTIR subtraction spectrum of polystyrene, $I_f - 0.9\ I_n$ (from [144]).

quency of 3.5 Hz. For plasticized PVC Takahara et al. [318] made a similar observation; they noted that tan δ exhibited a minimum on approaching the point of failure.

The FTIR results seem to indicate that the fatigue process has caused changes in the absorption frequencies of various vibrational modes of the *phenyl side groups.* Also the IR bands associated with bending and stretching modes of CH_2 groups and bending modes of CH groups show frequency shifts and band distortions. These results suggest that fatigue has modified the internal and external environment (intrachain and interchain interactions) of PS chains.

Table 8.5. Fatigue-affected IR Bands of Polystyrene (from Sikka [144]).

IR Band Frequency (cm^{-1})	Observation
540	Multiple splittings
906	Part of the band shifts to lower frequency at 888 cm^{-1}. Shift amount is -18 cm^{-1}.
1154	Part of the band shifts to lower frequency at 1145 cm^{-1}. Shift amount is -9 cm^{-1}.
1310	Part of the band is contributed by fatigue process.
1376	Part of the band appears at a higher frequency of 1400 cm^{-1}. Shift amounts to 24 cm^{-1}. Rest of the band is resolved into bands at 1352, 1358, 1365, 1371, 1376 cm^{-1}.
1448	Shows strong splitting.
1478	Two additional bands appear at 1489 and 1509 cm^{-1}.
2850	Weak splitting observed.
3062	Medium intensity splitting observed.
3083	Medium intensity splitting observed.

On the basis of these assignments and of his X-ray studies Sikka proposes that during fatigue weaker Van der Waal's and possibly some backbone bonds are broken. This has a twofold effect. The loosening of the structure permits on one hand the PS chains to locally rearrange themselves in a more perfect manner and to reduce the average intraphenyl and interchain distances. The local rearrangements are, on the other hand, accompanied by an increase in free volume in interstitial areas between the domains of improved order.

In his experiments with polycarbonate films Sikka [144] used cyclic tensile stresses between 50 and 91% of the yield stress. In the upper stress range the stress-strain curve was clearly non-linear. He employed the same experimental techniques as mentioned before to study molecular fatigue effects. Sikka found some indication

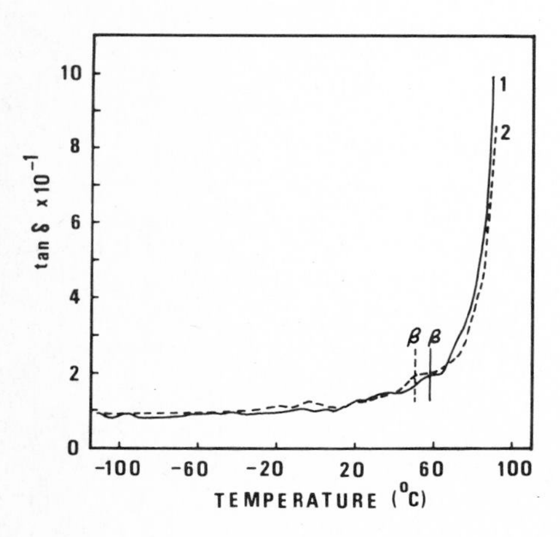

Fig. 8.43. Tensile mechanical loss tangent of polystyrene at 3.5 Hz (from [144]). 1. unfatigued sample; 2. sample fatigued to 2500 cycles.

of a stretching of the backbone chains. X-ray diffraction again revealed a change of the mean interchain distance. It amounted to a decrease of 1.25% after 5800 cycles under a tensile stress reaching 83% of the yield stress σ_y. No effects of fatigue on orientation could be detected in such samples by birefringence measurements. The densities decreased by a fraction of up to 0.083% to 1.1966 g/cm^3.

Some dynamical mechanical loss data are represented in Figure (8.44). The α and β relaxation peaks in fatigued samples appear at lower temperatures in comparison to the corresponding peaks in unfatigued samples. In all cases the samples fatigued to large numbers of cycles (> 2000 cycles at stress levels $\sigma_{max}/\sigma_y > 0.56$) fractured during the dynamical mechanical tests.

Also in this case one has to conclude that under fatigue stresses chains are stretched and locally packed more closely whereas the overall free volume increases.

The observations discussed in detail above seem to be representative for the findings of other authors as well. Bouda et al. [138] report a *decrease of density* with cycle number for PA 6 and PMMA which they ascribe to the *formation of microvoids.* The *decrease of heat capacity* in PMMA in the temperature range 350 to 400 K, the decrease of tan δ at 93 K (1 Hz) and an *increase in shear modulus* G′ in the range 100 to 250 K (where no crazes had formed) seemed to indicate a *nonuniform local volume contraction* during the fatigue process.

Schrager [134] investigated epoxy resins of different degrees of cure. A fully cured sample showed a monotonous decrease of G′ and an increase in tan δ. Notable is a complete and persistant recovery of G′ and tan δ within 20 h. An incompletely cured sample responded to fatigue loading by an initial rise of G′ (indicating completion of chemical cure simultaneously to chain scission in other parts of the matrix). After some $20 \cdot 10^3$ cycles chain scission and a decrease of G′ dominated.

Wendorff [255] studied fatigue-induced density fluctuations in polyoxymethylene (Hostaform T 1020) by X-ray scattering. He reports several interesting structural details of these defects:

- a constant concentration of about 10^{12} cm^{-3}
- a size growing with stress amplitude and cycle number
- a shape resembling that of a rotation ellipsoid (with axes 2a = 2 to 6 nm, 2b = 9 to 50 nm)

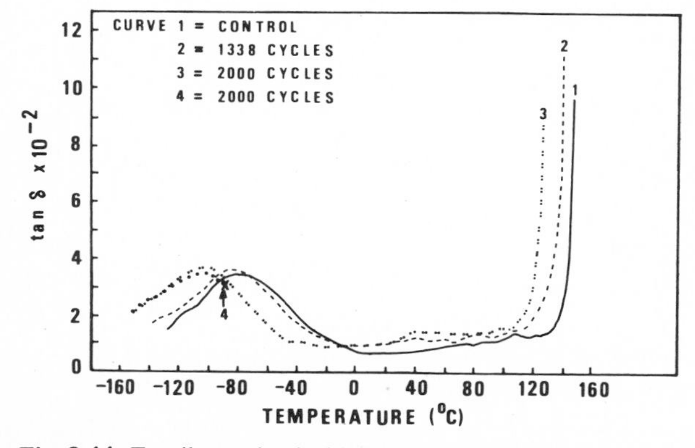

Fig. 8.44. Tensile mechanical loss tangent of polycarbonate at 3.5 Hz (from [144].

- the defects have a compliance some 3 to 5 times larger than that of the bulk material
- the defects show viscoelastic behavior and a size distribution
- annealing close to the melting temperature leads to partial healing; the defects reappear if cycling is resumed
- form and size of the defects depend on crystallization conditions.

These observations indicate that the defects are microvoids appearing preferentially at interlamellar boundaries oriented perpendicular to the loading direction. Similar defects are observed in homogeneous deformation during *static* loading at those strains where the stress-strain curve begins to deviate from linearity [145].

Whereas Wendorff states that fatigue does not affect the crystalline regions of POM Nagamura et al. [146] report a change of the crystalline mosaic block structure of HDPE. They arrived at this conclusion by analysis of the trapping and decay behavior of γ-ray induced free radicals.

Measurable macroscopic parameters which have been related to the progress of material fatigue are creep strain and strain rate [72, 116, 122, 123, 147, 319]. Mindel et al. [122] studied the creep rate as a function of strain in pure compression of polycarbonate. They found that fatigue loading becomes effective through an increase of strain rate after each interruption of loading. Since the strain level after which accelerated creep began remained constant (at 8.8%), fatigue lives were reduced. Tensile creep has frequently been held responsible for the fatigue failure of polymers. In 1942 Busse et al. [72] suggested this mechanism for polyamide, cotton and rayon. Brüller et al. [147] stated cyclic creep strains were predictable from the Boltzmann superposition principle. Trantima has predicted cycles to fracture of PP from creep rupture data [319].

Sauer et al. [153] studied particulary the effect of molecular weight on fatigue of PS and PE under different alternating stresses. They observed that an increase in molecular weight dramatically increases the fatigue life of PS (an increase of the number of cycles by a factor of 10 for an increase of M by a factor of 5). This effect is partly assigned to the retardation in craze initiation because of the fewer number of chain ends and the larger number of entanglements in high molecular weight samples. The major contribution derives, however, from the increased resistance of the crazed material to rupture (cf. Chapter 9.II).

Comparing the fatigue behavior of PE and PS Sauer et al. [153] conclude:

- in both crystalline polymers, like polyethylene, and amorphous polymers, like polystyrene, increases of molecular weight lead to increased fatigue lifetimes, for a given imposed alternating stress, and to increased endurance limits,
- the ratio of the estimated fatigue endurance limit strength to the nominal static fracture strength for specimens prepared from narrow molecular weight polystyrene standards rises appreciably with increasing molecular weight, from a value of 0.19 for a molecular weight of 160,000 to a value of 0.30 for a molecular weight of $2 \cdot 10^6$.
- if fatigue failure due to excessive thermal heating is avoided, the resistance of polyethylene of medium or high molecular weight to alternating loading is considerably greater than that of polystyrene, based on static stress values to produce either yield or fracture. Thus, at comparable molecular weights of about $2 \cdot 10^6$,

polyethylene gives an estimated ratio of endurance limit strength to yield stress of about 0.90 as compared to the 0.30 ratio for polystyrene,

- the greater resistance of crystalline PE to fatigue type failure as compared to amorphous PS is attributed primarily to two characteristics. The first concerns the phase of homogeneous fatigue. Because PE has a much lower amorphous-phase T_g and consequent greater molecular and chain mobility of its amorphous chains, it is much less sensitive to surface or volume flaws; and secondly, because of its heterogeneous nature, with crystallites imbedded in an amorphous matrix, it is also much more resistant to fatigue *crack propagation* (FCP). The latter phenomenon which also strongly depends on molecular weight will be discussed in Chapter 9 (Sections IIc and IIIc).

5. Fatigue-life Enhancement Through Surface Treatment

Mechanical fatigue is an aging process which initially leads to *localized* volume contractions (smaller interchain distances, smaller tan δ) [144, 317, 318], accompanied by defect formation [e.g. 144, 145, 153, 314, 315, 317]. Depending on stress and temperature fatigue lifetimes are often dominated by the time-to craze initiation [314, 320] and/or of craze propagation [314–316]. Since in most polymers crazes are preferentially initiated at surface irregularities there should be a strong influence of the state of the surface on fatigue behavior. In several of their research papers [321–323] and in their reviews [314] Sauer et al. have particularly explored this

Table 8.6. Effects of Surface Treatment on Fatigue Lifetime of PS* (Tabulated from data of Chen, Chheda, Sauer and Warty [321–323])

Surface treatment		Mean life [cycles]	Degree of enhancement
machined (reference of first group)		7'560	1
coated with 600 MW oligomer		55'300	7.3
chemically polished:			
exposed for 12–15 s	to benzene vapor		1.5
	to chloroform vapor		3.5
treated with swelling agents (acetone, diethyl ether)			2.2
mechanically polished (reference of second group)		18'800–24'000	1
wetted	with ethanol	7'530	0.3
	with n-propanol	1'230	0.05
	with Castor oil	47'000	2
aluminium or gold coated		48'000	2
natural rubber coat		80'600	4.3
vulcanized rubber coat		120–177'000	6.4–9.4
nitrile rubber coat		358–397'000	19–21

* M_w = 274'000, M_n = 100'000 fatigued at 1'250 cpm at 17.2 MPa.

interrelation. They indicate the following fatigue-lifetime reducing parameters:
- mechanical surface defects (scratches, inclusions)
- physico-chemical attack (crazing agents; solvants, which are the most "aggressive" if their solubility parameter is close to that of the polymer [321])
- chemical attack (including thermal and oxidative degradation [232] and crosslinking reactions)
- low sample molecular weight.

All of these parameters enhance craze- and crack-initiation, the latter three also craze-fibril breakdown which leads to more rapid fatigue crack propagation (see Chapter 9). Sauer et al. have also studied the parameters *increasing the fatigue resistance* of a specimen [314, 321–323].

They discuss specifically:
- mechanical and/or chemical polishing (for example by exposing PS to the vapors of benzene or chloroform, see Table 8.6)
- metallic surface coating (Al, Au, see Table 8.6)
- polar and hydrogen-bonding liquids (for PS it seems as if all liquids having a critical surface tension higher than PS had a beneficial effect [323])
- low molecular weight oligomers (a 600 MW oligomer PS gave a 5 to 7 fold enhancement of the mean fatigue life of PS)
- elastomer coatings (enhancement by a nitrile rubber up to 21 fold [322])

D. Plastic Deformation, Yielding, Necking, Drawing

In continuum mechanics plastic deformation designates the transition from one equilibrium state of the material to another. A perfectly plastic material should show no rate sensitivity and neither strain hardening nor softening. In accordance with the general practice in the polymer field [154–173] the term *plastic* will here be used in the larger sense of *non-elastic,* i.e. designating that part of the deformation which is not recovered at the temperature and within the time scale of the loading experiment. It is evident that plastic deformation in a polymer is brought about by some sliding motion of groups and/or chain segments past each other. The essential questions which arise concern the molecular nature of this motion, the correlation and interaction of the elementary steps, and their effect on sample structure, strain, strain rate, and fracture behavior. A comprehensive treatment of the plastic deformation of amorphous and semi-crystalline materials is given in ref. [325]. It may be stressed again at this point that the propensity of a specimen for plastic deformation favors its ductile failure. All parameter changes which reduce the extent of plastic deformation (lowering of T or M_w, increase of $\dot{\epsilon}$ or degree of crystallinity) will eventually provoke a ductile-brittle transition [292, 294, 296, 420].

The first question, the molecular nature of the sliding mechanisms, is of central interest in several areas of viscoelasticity: mechanical spectroscopy, stress relaxation, creep (see Section II B) and flow (see Chapter 4 IV B and C). Most theories are based on the *classical concept of Eyring* that a particle in a potential well passes over a barrier with the rate of flow steps depending on applied shear stress and on temperature. In his well-known theory of plasticity Robertson [162] considers the *orientation-dependent flexing of bonds.*

It is also recognized that (most) viscoelastic materials show stress and/or strain softening. Both, the increase in flow rate with shear stress and the structural weakening will give rise to a plastic instability which is called *yielding*.
Shear yielding, i.e. the onset of large-scale intersegmental displacements within a non-oriented thermoplastic polymer, has a distinct effect on the stress-strain curve. In tensile tests a drop in engineering stress is usually observed and the *yield point* defined as the point of maximum load (Fig. 2.11, curves b and c). In other tests, e.g. in a compression test, there may or may not be a load drop but a sudden decrease of $d\sigma/d\epsilon$ may be noticed. The important phenomenon of yielding has been intensively investigated. Review articles have appeared almost annually in recent years in the general literature [e.g. 114, 156–160, 163, 164, 415, 416].

The classical continuum mechanical criteria for shear yielding have been discussed in Chapter 3, Section III A in terms of the three-dimensional state of stress. At this point molecular interpretations are reviewed in an attempt to assess the role of molecular backbone chains.

Experiments at different temperatures, strain rates, and hydrostatic pressures have established that shear yielding is a thermally activated process [154–168, 170–173, 325–338]. According to Eyring's theory of flow (Chapter 3) strain rate can be expressed as

$$\dot{\epsilon} = A \exp - (U_0 - \gamma\sigma)/RT. \tag{8.33}$$

This gives the following dependence of yield stress σ_y on strain rate and temperature:

$$\frac{\sigma_y}{T} = \frac{U_0}{\gamma T} + \frac{R}{\gamma} \ln \frac{\dot{\epsilon}}{A}. \tag{8.34}$$

This equation was found to be valid for a number of polymers (PVC, PC, PMMA, PS, CA) in more or less extended regions of temperature and strain rate [154, 156, 158]. The (temperature-dependent) activation volumes γ had at room-temperature values between 1.4 (PMMA) and 17 nm^3 (CA). This means that according to this concept polymer deformation at the yield point is due to the thermally activated displacement of molecular domains over volumes which are between 10 (PMMA) and 120 times (PVC) as large as a monomer unit. It has been indicated by several authors [155–158, 160] that the above criterion (Eq. 8.34) corresponds to the pressure modified von Mises criterion $\tau_0 + \mu p$ = constant. The coefficient of friction μ is inversely proportional to γ. From an analysis of their experimental data (on PMMA, PVC, PC) according to Eq. (8.34) Bauwens-Crowet et al. [158] conclude that two flow processes exist. They relate these to an α-process (jumps of segments of the backbone chains) and to the β mechanical relaxation mechanism.

Argon and Bessonov [161] have recently presented a molecular model of shear yielding which involves the rotation of molecular segments against intra- and intermolecular forces and the systematic, pairwise reduction of kinks in a strained polymer. They calculate the free enthalpy of activation of a double kink in a small bundle of collectively acting molecules, ΔG^*, as

$$\Delta G^* = \frac{3\pi G\omega^2 a^3}{16(1-\nu)}\left[1 - 8.5\,(1-\nu)^{5/6}\left(\frac{\sigma}{G}\right)^{5/6}\right], \tag{8.35}$$

where G is the shear modulus, ν Poisson's ratio, ω the angle of bundle rotation during the activation, and a the mean bundle radius. Having examined a large number of glassy polymers Argon and Bessonov find that the bundles comprise between about one (PMMA, PS), two (PPO), three to four (PC, PETP) or seven (PI) molecular chains. The distances between double kinks along the molecule range from 0.8 to. 1.6 nm [162].

All of these proposed interpretations of yielding state that yielding is brought about by a displacement of neighboring chain segments under change of conformation. In the process of yielding of unoriented thermoplastics no large axial chain stresses are built up and no chain scission has ever been observed in strains *below* the yield strain ϵ_y. Yielding is equivalent to the onset of large scale orientational deformation. It is generally accompanied by a decrease of the material resistance against deformation, by a reduction of sample cross-section in a plane perpendicular to the direction of tensile plastic deformation and by a temperature increase due to the dissipated mechanical work. Material and thermal weakening at constant or increasing true stress give rise to a plastic instability. In uniaxial tensile straining this instability manifests itself as

- shear banding [154–157],
- neck formation [165–175, 334–338], see Fig. 8.32 b and d,
- homogeneous drawing [176–182], see Fig. 8.32 c.

The phenomenon of crazing can also be regarded as a (although localized) deformational instability. It will be discussed in the next chapter.

Neck formation has been investigated for a long time and under various aspects. For a non-linear elastic solid the states ahead and behind the neck can be derived from the conditions of continuity of mass, momentum and energy [338]. Bauwens [154] and G'Sell et al. [325, 335–337] have studied the influence of multi-dimensional stresses on onset and propagation of a neck.

Not persuing the continuum mechanical aspects any further some remarks will be made on the molecular rearrangements during yielding. The large scale "plastic" deformation associated with yielding quite naturally gives rise to shear and extensional deformation of molecular coils. The response of molecular chains in solution to shear or extensional flow has been described in Chapter 5IIE. It has been stated that chain breakages are obtained if strain rates are sufficiently large (generally above 600 s^{-1}). The strain rates in a yielding solid are considerably smaller. The local strain rates across the neck of a sample are typically 1 s^{-1} or less (Fig. 7.10). On the other hand molecular friction coefficients in a solid are so much larger than in a liquid that large forces will be transmitted once a chain molecule slips through the surrounding matrix. As discussed in 5IIE axial forces sufficient for chain breakage will be built up if the slipping chains have a highly extended conformation. It must be expected, therefore, that all those chain segments rupture as a consequence of necking or drawing which are properly oriented *and* which initially do have a highly extended conformation or assume such a conformation at the early stages of the yield process. With regard to chain breakage in necking or drawing the total orientational strain λ and the drawing temperature T_d are the most important variables.

There are very few reports on ESR investigations of chain rupture during necking and drawing [21, 169, 174–177]. These reports will be discussed in the following and be supplemented by observations with other methods: the IR spectroscopy of chain-end groups [178] and the determination of molecular weight distributions [179, 339].

Zaks et al. [169] investigated the neck formation in polycarbonate. As inferred from the reduction of sample cross-section in the neck, an orientational strain λ of about 2 had been imparted onto the material while "passing through the neck". At room temperature and at various sample stretching rates corresponding to strain rates in the neck of between 0.02 to 2 s^{-1}, the authors [169] observed a fairly stable but not very well resolved ESR spectrum. The intensity of this spectrum increased with the rate of passage of undrawn PC through the neck from $3 \cdot 10^{15}$ to $1.8 \cdot 10^{16}$ spins/g (Fig. 7.10). The authors also studied the behavior of stable nitroxide radicals and of radicals formed by photolysis during necking of LDPE and PC. The thus observed increased decay of the initially present radicals may be assigned to the reaction with newly formed radicals but, as well, to the strain-induced increased rate of recombination or decay of radicals already present. The existence of the latter phenomenon in highly oriented HDPE, PP, PA 12, and PETP was pointed out by Becht et al. [47].

The ESR technique had also been employed by Lishnevskii [174, 175] who found that during cold drawing of PP, PVC, PA 6, and aromatic polyamide fibers free radicals were formed in a concentration of between $8 \cdot 10^{14}$ to $3 \cdot 10^{16}$ g^{-1}. Matthies et al. [176] noted the intense coloring of different polyoxamide and hexamethylene isophthalamide fibers during cold drawing. They confirmed by ESR measurement that the formation of free radicals had given rise to the appearance of the color centers. In the case of the polyoxamide fibers they associated these color centers with the –NH–CO–CO–NH-group. The largest concentration of spins observed was $1 \cdot 10^{16}$ g^{-1}.

The above ESR experiments [174–176] were carried out on cold drawn samples immediately after completion of the drawing procedure. Thus the measured radical concentrations $[\dot{R}]$ are integral values. Nothing can be said about the form of $[\dot{R}(\lambda)]$, but, in all probability, most of the chain breakages have occurred during the final stages of the drawing procedure, i.e. within moderately to highly oriented and even microfibrillar structures. This assumption is supported by the ESR experiments of Levin et al. [21] and Chiang and Sibilia [177] with fibers partially oriented at elevated temperatures.

The hot drawing of PA 6 at temperatures of e.g. 150 to 210 °C permits one to apply orientational strains of up to $\lambda = 6$ without breaking an observable number of chains [21]. As discussed in detail in Chapters 7 and 8-I, free radicals are formed if such fibers are stretched at room temperature. For a PA-6 fiber hot stretched to $\lambda = 3.7$ radical formation began at an "additional cold" strain of 20%, i.e. at a total elongation of $\lambda = 4.44$ [21]. PETP fibers hot stretched to $\lambda = 3$ showed some radical formation at "cold" strains of between 5 and 20% but deformed by chain slippage thereafter – up to a strain of 50%, i.e. to a total elongation of $\lambda = 4.5$ [177]. In HDPE films no breakage of chains could be observed during cold drawing unless the draw ratio λ exceeded 5 [178].

An infra-red study of the latter material also revealed [178] that the number N of coiled rotary isomers (chains containing *gg* and *gtg* conformations) depends on the draw ratio λ. While $N(\lambda)$ is rather constant if $\lambda < 5$, N decreases linearly at higher draw ratios to reach about $0.2\,N(1)$ at $\lambda = 12$.

That chain scission occurs during drawing of PA-6 multifilament yarn becomes apparent from an investigation of the molecular weight. Using a fractional precipitation technique Sengupta et al. [179] determined the molecular weight distributions of undrawn yarn and of yarn drawn at 35 °C to $\lambda = 2.92$ and $\lambda = 4.20$ respectively. They observed that for the yarn drawn to $\lambda = 2.92$ the molecular weight distribution is unimodal as in the undrawn fiber, but the peak is smaller and slightly shifted toward smaller values of M. At $\lambda = 4.2$ this peak is further diminished and shifted and a second peak at low molecular weights (at less than half the viscosity number) appears [179].

It is a common observation [360] that repeatedly extruded polymers decrease their molecular weight. Generally it is assumed that imposed shear stresses and temperature simultaneously influence the chain scission reaction. In the case of PC extruded at temperatures between 275 and 320 °C Abbas [339] found that this reaction had an activation energy of 113 kJ mol^{-1}; it occurred to a much larger extent during extrusion than in simple static heating. However, he explained this difference by viscous heating and not as a direct consequence of stress-induced chain scission (see also IIIB2).

Indirect evidence of the likely occurrence of chain scission was proposed by some researchers. Davis et al. [170–173] studied the pressure sensitivity of the rate of plastic deformation of high M_w LDPE. From the pressure sensitivity of the rate of plastic deformation they derived a zero-pressure activation volume of 0.266 nm^3 which they suggested to correspond to the activation volume γ for the breakage of a PE chain. Oprea et al. [417] noted an increased solubility of PVC after stressing.

From the presented evidence it must be concluded that during the initial stages of necking and drawing ($\lambda < 3$) of semicrystalline polymers chain breakages do not occur in large numbers. This conclusion is completely in accord with the Peterlin model of the "plastic deformation" of semicrystalline polymers. As described in 2.II B the early stages of plastic deformation comprise the destruction of an original spherulitic texture and its transition to the new microfibrillar structure. During the transition interspherulitic tie molecules are most highly loaded and some are expected to break. The ties between the crystal blocks, however, are generally not severed at this stage and are instrumental in establishing the microfibrillar structure.

The influence of the supermolecular structure before drawing on drawing behavior and strengthening processes has also been discussed in Section I G and refs. [340–342].

If at room temperature a semicrystalline polymer is carried from moderate to higher extension ratios ($3 < \lambda < 10$) then chain breakages can occur [21, 169, 174–178]. All of the mentioned breakages must be considered to be of the static-loading type (cf. 5 II B and D). They become possible in large numbers because the *presence of laterally rigid crystalline regions permits:*

– the static transmission of large axial stresses onto chains,
– the balancing of ensuing irregularities of stress distribution, and thus
– the widespread attainment of critical chain stresses before macroscopic breakdown.

In recent years ultra-high modulus fibers with draw ratios up to and beyond 35 have received increased attention [180–182, 257–276]. Their highly extended chain structure and their extreme mechanical properties (elastic modulus E up to 67 GN m^{-2}, σ_b up to 7 GN m^{-2}) have been discussed in IG, and the techniques for their processing (drawing, hydrostatic extrusion, solution spinning) are outlined by Bigg [180] and more recently in the review paper by Ward [257b]. As yet there is no information on the extent of chain breakage during drawing of these fibers. Capaccio et al. [181] point out that in very highly drawn PE ($\lambda = 30$) much finer fibrils are present and these are characterized by an amazingly high concentration of longitudinal discontinuities in the form of microfibril ruptures. Molecular defects (chain ends, folds) are certainly present in extended chain crystals – but not in quantities indicating major chain breakages during processing [180–182].

With the exception of PC non-oriented amorphous polymers have not given rise to measurable quantities of broken chains during yielding in tension. This behavior is the result of their different morphology. In the absence of crystallites, large axial forces leading to chain scission can only be caused by frictional loading of slipping chain segments. The volume concentration of chain breakages (because of the larger size of the slipping segments) is predictably much smaller than in semicrystalline polymers. In addition (and because of the absence of the equalizing effect of a microfibrillar substructure) macroscopic tensile failure begins before stresses and strains have reached a level sufficient for widespread chain scission.

During necking or homogeneous drawing of many (transparent) polymers an intensive whitening is observed. The origin of this phenomenon are voids which are either formed in a correlated manner within crazes (cf. 9 II) or within shear bands or which appear in an uncorrelated manner within the deforming volume elements. Uncorrelated voids form in semicrystalline polymers (PE, PP) as well as in amorphous ones (PMMA, PVC) and in elastomers at cryogenic temperatures (BR, PI). Figure 8.45a shows an electron micrograph of a cross-section through the voided area which appeared during necking of a dry PA 6 monofilament [83]. The fact that the voiding starts in the center of the necking region and in dry monofilaments only, indicates that it is related to a triaxial state of stress. The coalescence of such voids may give rise to the fibrillar fracture shown in Fig. 9.49. The appearance of a thoroughly stress-whitened PMMA is shown in Fig. 8.45b [343]. Frank and Lehmann [344] have carefully analyzed the respective contributions of elastic and shear deformation and of void formation to the total tensile strain of PMMA. At small strain rates and at elevated temperatures there was practically no void formation. From the triaxiality of initiating stresses and from the absence of measurable intensities of free radicals in slightly drawn isotropic polymers ($\lambda < 3$), from the absence of conspicuous molecular weight changes [179, 220], and from the reversibility of the formation of holes it must be concluded that the hole formation depends more on conformational rearrangements of chains and on surface free energy than on the breakage of chains. If there are chain breakages, their number must be small. It seems to be quite con-

Fig. 8.45a. Electron micrograph of ultrathin cross-section through voided area of drawn 6 polyamide monofilament (Courtesy H. Hendus, [83]). Magnification 4800 x

ceivable that the kinetics and the significance of eventual chain scission events can be studied by the exoemission of charged particles during deformation. Preliminary investigations of Scherer and Fuhrmann [345] have already indicated the *extreme sensitivity* of this method (10^6 events per gram are traceable as opposed to 10^{14} by ESR). Measurements carried out with PE gave a strong indication that the emission of electrons during stretching was caused by the breakage of chains. A very strong maximum of exoemission was found to coincide with sample necking [345]. Somewhat similar qualitative conclusions were drawn by Betteridge et al. [346a] who studied acoustic emission (AE) and radical formation of some 10 different moulded thermoplastics and rubbers. They find that the more ductile samples are less acoustically active; in the others signals were emitted after stress-whitening had started and there was a certain "degree of comparability between acoustic and ESR results". In amorphous polymers (PMMA, PC) Matsushige et al. [346b] observed AE events over the whole deformation range; those events occurring before the yield point were attri-

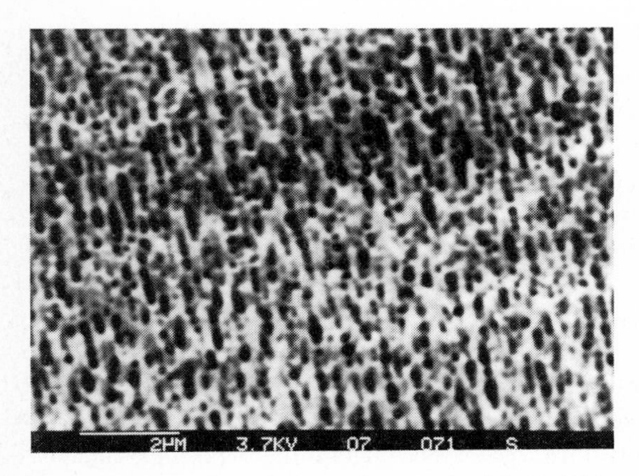

Fig. 8.45b. Scanning Electron Micrograph of Stress-Whitened PMMA [343]

buted to the development of submicrocracks and their combination, the others to the drastic reorientation, extension and frictional movement of molecular chains during necking [346, 411, 412].

The material behavior described above referred mainly to uniaxial tensile deformation. It will be different in techniques of working or forming of polymeric materials, such as rolling, sawing or grinding. These processes involve compressive, tensile, and shear deformation in a generally complex manner. Breakage of chains has been observed in various ways. In 6 IV B and 7 IC 3 a detailed account of the nature and concentration of free radicals formed in comminution has been given. Other mechano-chemical aspects will be discussed in 9 III D.

E. Rupture of Elastomers

In all preceding sections of this chapter polymer states have been treated where the forces acting between neighboring chain segments were of prime importance for the transfer of macroscopic stresses. In an (ideal) elastomer, however, intermolecular forces are small, the stress transfer occurs through the entropy-elastic deformation of the network chains. The basic principles and notations of rubber elasticity have been presented in Chapter 2 (Section II A), different fracture models in Chapter 3, and a discussion of the entropy-elastic deformation of a single chain in Chapter 5 (Section I A). If one speaks of a "chain" in relation to a filled and cross-linked polymer network one refers, of course, to the part of a molecule between adjacent points of attachment (filler particle or cross-link). At this point the role of such chains in network deformation, the limits of chain extensibility and the mechanisms of network rupture will have to be discussed.

The deformation of ideal or phantom networks has been intensively discussed in the literature beginning just 50 years ago with the classical works of Kuhn, Meyer and Ferry, Guth and Mark, James and Gut [347 a–d] and Flory [348 a]. The essential point certainly is that the macroscopic deformation corresponds to the uncoiling of the network chains; thus entropy-elastic forces are exercised.

From the general literature [348 a–c] it can be concluded that an average network chain influences the deformation in several ways:

- the originally random chain conformation changes to a more extended one at a rate determined by strain, temperature, and internal viscosity,
- in highly cross-linked or filled rubbers many (of the shortest) network chains have been completely extended and are broken at the fracture elongation (cf. 7 II),
- additional physical cross-links are introduced through crystallization at high elongations,
- voids or cavities can be formed due to non-homogeneous distribution of strains in filled or unfilled elastomers,
- the limits of extensibility are reached once the majority of the network chains have transformed into a highly extended conformation. In the case of a homogeneously cross-linked network with n random links per network chain the largest breaking elongation $\lambda_{b\,max}$ will approach $\sqrt{n}$. At that strain all average chains oriented in stretching direction are fully extended (cf. 5 I A).

Rupture of most technically important elastomers takes place at stretch ratios λ smaller or equal to about 8 [183–195]. As shown in Figure 3.6 and indicated in the literature the reduced ultimate stresses σ_b plotted versus elongation λ_b form a failure envelope. The elongation to break λ_b is shifted along the failure envelope (towards larger elongations if the temperature and/or the degree of cross-linking decrease or if the strain rate increases [183–195]. The macroscopic breaking stresses σ_b are of the order of 1 to 30 MN m^{-2} (depending on degree of filling or cross-linking), they decrease with increasing time-to-failure and degree of swelling.

It has been proposed by Mullins [183] some thirty years ago (and has since been known as Mullins effect) that the breakage of chains in a first loading cycle is responsible for the softening observable in subsequent load cycles. In that respect the phenomenology of the Mullins effect is comparable to that of the breakage of chains in the stretching of fibers.

For the breakdown of a stressed rubber two basic mechanisms are conceivable: the initiation and growth of a cavity in a moderately strained matrix or the accelerating, cooperative rupture of interconnected, highly loaded network chains. It has generally been concluded that the second mechanism is of dominating importance under those experimental conditions which permit the largest breaking elongation $\lambda_{b\ max}$ to be attained [187, 192, 194, 195]. In that case, the quantity $\lambda_{b\ max}$ was expected to be proportional to the inverse square root of the cross-link density ν_e; in fact an increase of $\lambda_{b\ max}$ with $(\nu_e)^{-0.5}$ to $(\nu_e)^{-0.78}$ was found experimentally for a variety of networks (Butyl, BR, EPR, SBR, SI, fluorinated rubber). Smith [187] concludes that the maximum elongation $\lambda_{b\ max}$ is the only feature of large deformation behavior which clearly depends only on network topology.

At ultimate strains which are smaller than $\lambda_{b\ max}$ some chains are already highly extended but their number is insufficient for cooperative accelerating breakdown. It is the general opinion that under such experimental conditions the final breakdown occurs through formation and growth of cavities [184, 185, 190]. It will not be further investigated at this point to what extent chain scission may accompany cavity formation and growth. It may be stated, however, that the chain scission mechanism does not determine the rate of cavity formation.

As a final point it should be emphasized that elastomeric materials are showing a higher strength if they exhibit a multiphase microstructure [183, 184, 186, 188, 195]. This may be achieved by incorporation of suitable filler materials (carbon black, silica); by crystallization under strain, or by blending or copolymerization with an incompatible polymer. The modulus and thus the load at a given deformation increase with the number of load carrying chains. The possible role of chain orientation, loading, and scission in these cases has been discussed in 7 II.

The fact that the tendency for *stress-induced* crystallization of an elastomer increases the stresses at comparable strains is clearly shown by Fig. 8.46a.

A self-reinforcement $S = (s - s_0)/s_0$ (through crystallization) can be defined through the deviation of the stress-strain curve from the Gaussian value (Fig. 8.46b). The higher λ the higher the degree of crystallinity and thus the possibility of load sharing between different chains. The resulting, strongly non-linear stress-strain curve has a very beneficial effect in the presence of cracks or other defects: the highest loads are supported in the zone of highest deformation, i.e. at the crack tip,

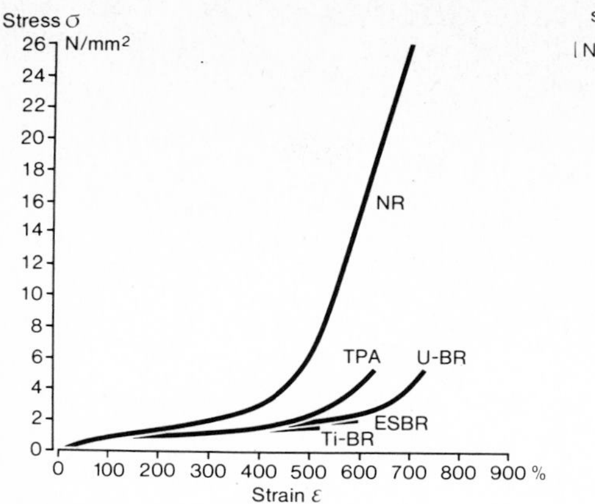

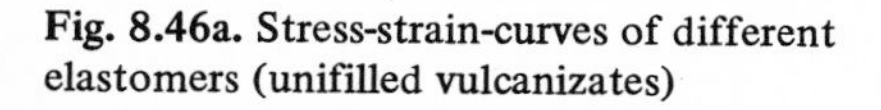

Fig. 8.46a. Stress-strain-curves of different elastomers (unifilled vulcanizates)

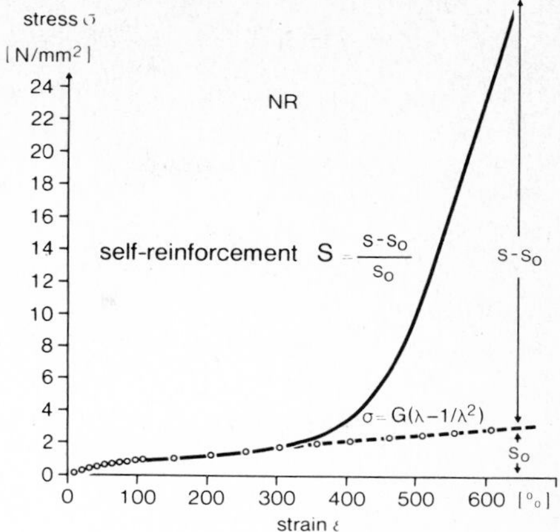

Fig. 8.46b. Definition of self-reinforcement S (both diagrammes courtesy Prof. R. Casper and Dr. U. Eisele [349a, b])

thus leading to the self-reinforcement of the crack tip (see Fig. 8.47). The self-reinforcement decreases with temperature (for NR from 9 at −40 °C to 1.5 at +100 °C [349a]. The temperature dependence of the tear strength of the elastomers shown in Fig. 8.46 closely parallels that of the self-reinforcement.

The former statement that the breaking strain depends on the limited extensibility of a network of given topology must be modified, therefore, by admitting the influence of strain-induced crystallization [348–350]. Both the concentration ν_e of permanent chemical cross-linking points and the intensity of temporary intermolecular interactions (such as caused by crystallites or entanglements) influence the stress at a given strain. In fact, ν_e is quite well correlated with the constant C_1 of the Mooney-Rivlin equation:

$$\sigma = 2\,C_1\left(\lambda - \frac{1}{\lambda^2}\right) + \frac{2\,C_2}{\lambda}\left(\lambda - \frac{1}{\lambda^2}\right). \qquad (8.36)$$

The constant C_2 can be taken as a measure of the transient intermolecular interaction, it is for instance decreased by swelling and by repeated extension [349, 350]. The ratio C_2/C_1 is the larger the more flexible (and spontaneously crystallizing) the chains are (Fig. 8.48). A comparison of Figs. 8.46 and 8.48 shows, however, that the self-reinforcement of the strain-crystallizing natural rubber is much more important than that of the more flexible, spontaneously crystallizing elastomers TPA and U-BR [349].

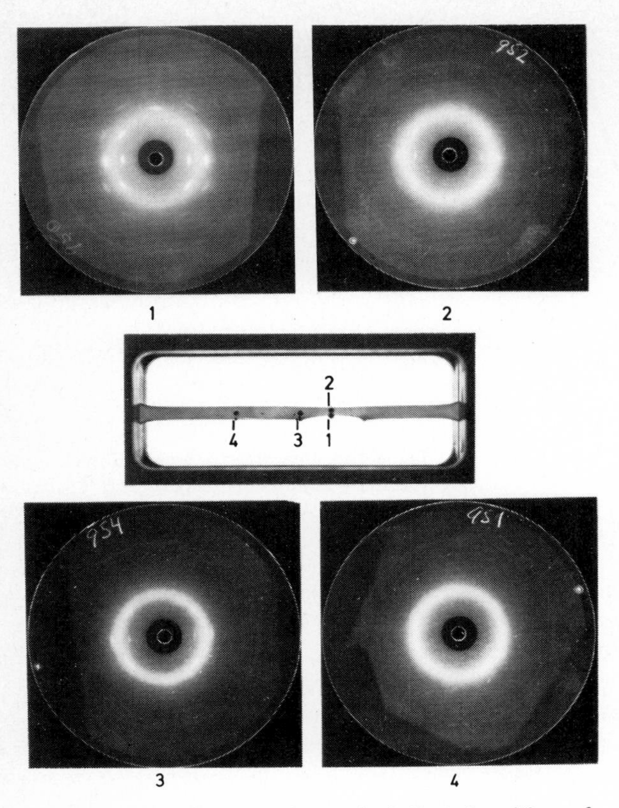

Fig. 8.47. X-ray diagrams taken at the indicated positions of a notched NR specimen showing a dramatic increase of crystallinity at the notch-tip (Courtesy Prof. R. Casper and Dr. U. Eisele [349a, b])

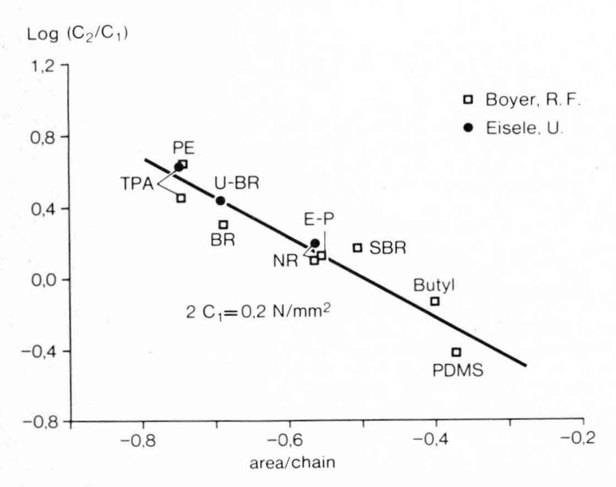

Fig. 8.48. C_2/C_1 at 2 C_1 = const. as a function of the logarithm of cross-sectional area per polymer chain (after Boyer [349c], courtesy Prof. R. Casper and Dr. U. Eisele [349a, b])

III. Environmental Degradation

A. Aspects to be Treated

Within the framework of this book the effect of *thermomechanical* chain scission on the mechanical properties of polymers are to be investigated. Up to this point it has been tried, therefore, to separate and exclude *environmental* effects whenever possible. In many cases it was tacitly assumed that the variables investigated (e.g. strain, stress, temperature, sample morphology, free radical concentration) had been the dominating variables in comparison to e.g. humidity, oxygen content, chemical attack, or effects of irradiation. It goes without saying that these environmental factors are of extreme importance to the service lives of polymer engineering components. A considerable number of comprehensive recent monographs and handbook articles treat the environmental degradation of polymers (e.g. [196–203, 351–354]). They deal in depth with those aspects of environmental degradation which cannot be discussed any further at this point: thermal degradation, fire and heat stability, chemical degradation, weathering, aging, moisture sensitivity, effects of electromagnetic and particle irradiation, cavitation and rain erosion, and biological degradation. For any detailed information on the above subjects and on methods and agents to protect and stabilize polymeric materials the reader is referred to the cited comprehensive references [196–203, 207–209, 353, 354].

It mainly remains to be discussed in this section the nature of the effect which the *simultaneous* action of mechanical and environmental parameters may have on mechanical properties. One of the most interesting phenomena in this field, the environmental stress crazing (ESC), will be treated in the next chapter.

B. Chemical Attack on Stressed Samples

1. Ozone Cracking

A characteristic stress-induced degradation process is that of unsaturated rubbers in the presence of ozone. The rates of crack initiation and growth, of formation of free radicals, and of stress relaxation and creep are amplified up to and more than a thousandfold by the action of ozone [196–197, 199, 201, 204–206]. The chemical reaction is not completely understood. It is normally assumed that the first steps of ozone degradation of unsaturated polymers follow the "Criegee mechanism".

Under the effect of molecular ozone the π-bonds of olefines are polarized and ozone joins the double bonds to form primary ozonides [196, 197a]. Primary ozonides, $\underset{\diagdown\;O_3\;\diagup}{-CH-CH-}$, are unstable, they either isomerize into iso-ozonide, $\underset{\diagdown\;O-O\;\diagup}{-HC-O-CH-}$,

or polymerize, or decompose into chains terminated by a zwitterion, $R_1C^{+}OO^{-}$, and a carbonyl group, R_2CO, respectively. If a bond attacked by ozone is under stress, decomposition can occur during the isomerization reation.

The reaction with ozone obviously is a surface reaction leading to the formation of a surface layer of ozonides and/or further reaction products. This layer grows in thickness in proportion to the square root of the time of exposure [199]. With increasing layer thickness ozone is gradually prevented from reaching undegraded rubber. The mechanical aspects of ozone cracking are extensively reviewed in the cited articles [196–197, 199, 201, 204–206]. It is unanimously reported that the degraded rubber material (natural rubber, SBR, acrylonitrile-butadiene, cis-polybutadiene) has a reduced strength and elasticity. Cracks open up and propagate at strains as low as 5 to 12%; it has been estimated [199] that even at the tip of a slowly propagating crack strains do not exceed 30% (as compared with 700% in an undegraded matrix). As a consequence of the chemical destruction of the material at the tip of a crack the mechanical energy expended in crack propagation (material resistance R) is exceedingly small. Employing the Griffith fracture criterion the various authors have obtained values of 0.05 to 0.12 J/m^2 from observation of macroscopic cracks [197a, 199a, 204, 205] or of 0.4 to 0.5 J/m^2 from micromorphological theories [206].

The formation of free radicals in unsaturated rubbers under simultaneous attack of ozone and stress is of particular interest. The primary steps of the above described reaction of ozone with unsaturated bonds do not lead to free radicals. In cis-polybutadiene (BR), natural rubber (NR), and acrylonitrile-butadiene (ABR), however, large quantities of peroxy radicals have been observed [206, 208]. As one possibility it was pointed out that these radicals may derive from the ozonides or zwitterions through unknown secondary steps possibly involving hydrogen abstraction or proton migration [197, 206, 208]. Another possibility certainly is that radicals are formed through scission of undegraded rubber molecules and that these primary radicals react with molecular oxygen. In BR and ABR the concentration of free radicals has shown the same dependency on strain and ozone concentration as the visual damage, i.e. the ozone cracking of the surface of the degrading rubber specimen. Notably the following observations may be mentioned [206, 208]:

- no (detectable quantities of) free radicals are formed below certain threshold strains (of between 5 and 12%)
- for strains beyond the threshold strain the rate of increase in free radical concentration increases initially linearly with strain and ozone concentration (in the range of 0.08 to 1.6 mg/l)
- increasing strain or ozone concentration lead to a preferential increase in number rather than size of microcracks.

The number of free radicals (up to $8 \cdot 10^{14}$ spins per cm^2 rubber surface) corresponds at a given ozone concentration of 2.8 mg/l to the number of ozone molecules in a zone of surrounding atmosphere of 0.02 cm depth. This number is very much smaller than the total number of ozone molecules available in the ambient atmosphere of the degraded specimens. The linear dependency of radical concentration on ozone concentration and strain, therefore, means that potential radical sites are created only after the application of stress and in the presence of ozone molecules. The appearance of free radicals simultaneously with crack opening certainly indicates the breakage of chains during the process of crack opening. In Section I E of this chapter an analysis of the energy content of an unsevered hydrocarbon chain strained elastically up to chain scission had been carried out. This analysis leads to the estimate that the $2 \cdot 10^{14}$ chains per cm^2 of new fracture surface can only be broken elastically if more than 9 J/m^2 were expended in forming that surface. Since the observed surface work parameters are much smaller one has to conclude that under ozone attack the breaking and radical forming chains break at small strains. This interpretation looks identical to that derived from analysis of the behavior of ozone cracks. Its importance lies in the statement, however, that during crack opening chains (of very low strength) are broken and not only broken chains are separated.

The action of ozone on SBR, studied by Mergler and Wendorff [355], is explained along the same lines.

In recent years the effects of an ozone atmosphere on chain breakage in strained PA-6 fibers [245] and PE fibers [246] have been investigated. In the former case there is a very distinct synergistic effect. The detailed mechanism, however, by which stressed chains react with ozone and break seems not yet well known. Excluding monomolecular chain scission Popov, Zaikov et al. [356] propose for HDPE, LDPE and isotactic PP an increased reactivity of a stretched macromolecule towards a hydrogen acceptor such as ozone. From their hydrogen abstraction experiments with i-membered cycloparaffins they estimate that the reactivity of a CH_2-group towards ozone should increase by one order of magnitude if the excess strain energy of this group attains 0.3 kcal/mol or $0.5 \cdot 10^{-21}$ kcal/bond [356b]. In fact they observe that the relative concentration of oxydation products in stressed polyolefins increases linearly with stress even at low stresses. The effect of stresses on elastomer oxidation has been reviewed by Popov and Zaikov [357].

2. Oxidation and Chemiluminescence

The oxidative degradation of polymers, especially in the presence of UV irradiation has been studied intensively; it has been treated comprehensively in a number of monographs [208, 209, 353, 354]. Most investigations in this field deal with the effect of degradation on the properties of polymers (e.g. 189, 202, 203, 207, 209e) or with an elucidation of the nature and kinetics of the intervening reactions [200, 201, 207, 208, 209a–d]. In the context of this book only the possibly synergistic effect of a simultaneous application of mechanical stress has to be investigated. Several methods have been employed to determine the rate and the effect of oxidation:

– detection of light quanta emitted during the oxidation reaction (chemiluminescence [358, 359]),
– detection of newly formed, oxygen containing groups by IR spectroscopy (see Sections Ib and c and [356, 357, 360, 418, 419]),
– formation of free radicals [245, 246, 360],
– decrease of residual strength [245, 246],
– change in other properties (M_n, MFI, elastic modulus).

The latter two methods are evidently indirect ones.

As the studies by George et al. [358] have shown chemiluminescence (in high strength PA 66 fibers at 40 °C) can already be traced in the unstressed state; at a stress level of 0.5 σ_b there is a five fold increase of oxidation rate with respect to the unstressed condition (see Fig. 8.49). Although the authors conclude that light emission arises from the bimolecular termination of alkyl peroxy macroradicals and that a

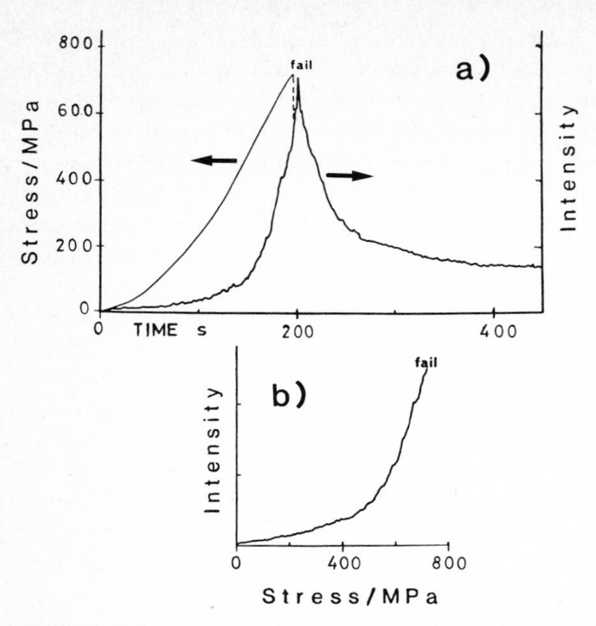

Fig. 8.49a. Change in chemiluminescence intensity from high tenacity PA 66 fibers at 40 °C in air with time of elongation. The measured fiber stress is also shown up to yarn failure. **b** Luminescence intensity as a function of stress (from George et al. [358])

moderately applied stress increases the oxidation rate due to internal and external frictional heating it may be useful to take also into account the oxidation mechanisms discussed in the preceding section. Stress-induced main-chain scission occurs at higher strains [358]. The sensitivity of the method permits shorttime testing of the effectiveness of additives (even of antioxidants added to foodstuffs [359]).

As one of the first Kim [210] searched for a possible stress effect on the rate of oxidative chain degradation [209, 210]. She employed a constant-load technique which permitted to derive the relative number of network chains, N(t)/N(O), by comparing the Mooney-Rivlin plot, $\sigma_0/(\lambda - 1/\lambda^2)$, of an aging sample with that of a fresh sample. Her interesting results may be summarized as follows:

- The creep rate of polybutadiene rubber films loaded in ambient atmosphere was about a thousandfold larger than the creep rate under dry nitrogen.
- A comparison of the relative potency of the chemical agents present in the ambient atmosphere showed that the few parts per hundred million of naturally occurring ozone had a much stronger effect than the oxygen.
- In ozone-containing air a behavior comparable to that described above for higher ozone concentrations had been found. A critical stress of 0.1 MN m^{-2} had to be exceeded in order to influence the initial creep rate. The (initial) rates of chain cleavage dN(t)/N(O)dt were found to be independent of stress within a stress range of $0.1 < \sigma_0 < 0.5$ MN m^{-2} and amounted to $2 \cdot 10^{-6} s^{-1}$. This phenomenon was explained in the same way as for higher ozone concentrations through formation and cracking of a degraded surface layer. Microscopical investigations of the sample surface confirmed such a statement.

- In the absence of ozone the rates of creep and chain scission were greatly reduced. The effect of oxygen *at room temperature* was small and hardly any differences between creep rates in dry nitrogen, dry air without ozone, and ozone-free air of 40% relative humidity had been observed. At *slightly higher temperatures* (49 °C), however, creep rates in dry air accelerated whereas those in inert atmosphere (Argon) remained constant.
- From her experiments Kim concludes that in the initial phase of degradation in ozone-free oxygen stress or strain have no effect on the rate of chain cleavage. After the chemical degradation had reached a stage, however, where cracks opened up under stress, the straining of a sample accelerated the failure process of the specimen through stress concentration and change in cross-linking behavior [210].

The general effect of cross-link density on the elastic modulus of an elastomer is indicated by Eq. 2.3. In their paper Landel and Fedors [189] consider the influence of a time-dependent cross-link density on the shape of the stress-strain curves of silicon, butyl, natural, and fluorinated rubbers. Introducing an additional shift-factor a_x related to the cross-link density, they were able to represent reduced breaking stresses as a function of reduced time in one common master curve.

Thermal and chemical degradation of cross-linked rubbers was also reviewed by Murakami [209c] who especially analyzed whether cleavage occurs along main chains or at cross-links. Depending on the nature of the chain and on that of the cross-links both mechanisms are observed. Natural rubber, radiation cross-linked to various degrees, showed at 20% strain and at temperatures of 80 to 130 °C random scission of network chains. An EPT rubber cross-linked by tetramethyl thiuram disulfide (TMTD) showed degradation at the junction points of network chains. There the S and S_2 linkages seemed to be preferred scission points. In TMTD-cured natural rubber both types of degradation occur simultaneously. The dependency of the stress relaxation curves on time, temperature, and initial cross-link density permitted discrimination between thermal and oxidative degradation. For differently cured natural rubber (peroxide-, irradiation-, TMTD-, and sulfur-cured) Murakami [209c] has tabulated the degradation mechanisms.

The accelerating effect of mechanical stresses on chemical stress relaxation clearly became apparent through the comparison of continuous and intermittent stress relaxation. The relative stress relaxation, $1\text{-}\sigma (t, \lambda)/\sigma (O, \lambda)$, under continuous loading was the more pronounced the higher λ and always consistently larger than under intermittent load application. Murakami [209c] also discussed possible increases of relative stresses due to cross-linking reactions and their partial prevention through radical acceptors.

In a recent review paper [360] Scott details the mechano-chemical degradation and stabilization of polymers. The strong influence of oxidative chain reactions on chain degradation in processing (of PP) becomes clear from the small changes in the melt-flow index (MFI) of melts protected by a stabilizer as opposed to the large changes of MFI of the unprotected material. Scott concludes that electron-accepting antioxidants which break the oxidative chain (CB-A) are relatively more effective in the case of mechanoradical formation than in thermal oxidation [360]. In her publications [418] and her extended review [410] on kinetics and mechanisms of PP autooxidation under stress Rapoport mentions a positive stress effect on the two oxidation initiation reactions (hydrogen abstraction and main-chain bond scission) and on the consumption of any inhibitor present. An intimate correlation between oxidative degradation and complete loss of load carrying capability of pipe materials at elevated temperatures [117b, 363] has already been observed and discussed in relation with Figure 8.39.

The influence of thermal oxidation on the mechanical behavior of polymers has been known for a long time. The principal effects are reduction of molecular weight and/or cross-linking both leading to an embrittlement. If, for instance, the extrusion conditions of polyethylene pipes are not well controlled, layers of oxidized material are formed at the inside pipe surfaces [362]. These layers, which are very brittle, have been studied in detail by Gedde, Jansson and Terselius [362a, b]; the intensity of the oxidation was traced by two types of bonds: –C=O and –C–O–C– by IR reflectance spectroscopy. The thoroughly oxidized layers extended to a depth of some 20 to 160 μm at the inside walls of some HDPE pipes extruded at 200 °C. These layers were extremely brittle and cracked upon drawing at strains of a few per cent. The times to fracture in static loading of internally pressurized pipes (average hoop stress $\overline{\sigma}_t = 4.2$ MPa, T = 80 °C) and of uniaxial creep samples, cut from the wall, were considerably shorter in the case of oxidized material. The final cracks seemed to originate at a distance of 0.2 to 0.3 mm from the oxidized layer [362a].

On the basis of these investigations four different layers could be distinguished in the oxidized HDPE pipes (distances s are with respect to the inner wall):

s = 0–100 μm thoroughly degraded, brittle material, $\sigma_t(s) = 0$
s = 150–250 μm somewhat degraded, readily creeping material, $\sigma_t(s)$ small
s = 300–500 μm slightly damaged zone, $\sigma_t(s) \sim \overline{\sigma}_t$
s > 500 μm normal material

The slightly damaged zone still contains numerous defects which favors creep craze initiation, and it also offers the required stress level. This combination results in a high probability for an early crack initiation in that zone. The cracks then propagate into the normal material. This mechanism explains well the observed shorter times to failure in oxidized pipes. In fact, Müller and Gaube report [362c] that by carefully removing the oxidized layer the times to failure could be considerably improved and were equal to those of undamaged pipes.

The above explanation correlates also very well with an observation discussed earlier, namely that the time to fracture of LDPE pipes turns out to be shorter than the average lifetime whenever the final crack had started close to the inside or outside surface of the pipe wall (see Section II B 4).

3. NO_x, SO_2

Igarashi and DeVries [245] had exposed strained oriented PA 6 fibers to NO_2 and SO_2 atmospheres. They performed subsequently tensile tests within an ESR cavity and also analyzed the sample molecular weight. They clearly demonstrated a distinct synergistic effect of sample straining during exposure to the different atmospheres; thus, residual strengths and toughness values were decreased (by up to 80% within 5 days) by exposure to SO_2, NO_x, O_3, (Fig. 8.50). They noted that the average sample molecular weight and the radical concentration produced in the final tensile tests of the NO_2 and O_3 degraded samples were clearly diminished. For the mildly strained samples exposed to SO_2 they observed, however, a gradual recovery of strength during subsequent annealing in air. From their studies they conclude that loaded tie-chains are attacked and broken within O_3 and NO_2 atmospheres; the SO_2 molecules, however, seem to attach themselves to the PA chains without breaking them if the stress levels did not exceede 70% of the yield stress [245]. This interpretation seems to be essentially correct. And some of the doubts forwarded by the authors can be resolved on the basis of the analysis given in the preceding chapter (7 IA). Thus, it must be expected that the relative decrease in free radicals produced in final fracture, $(N_0 - N_i)/N_0$, should be given by the slope of the radical concentration vs deformation curve (see Figs. 7.1, 7.4, 7.11); in view of the parabolic form of this slope

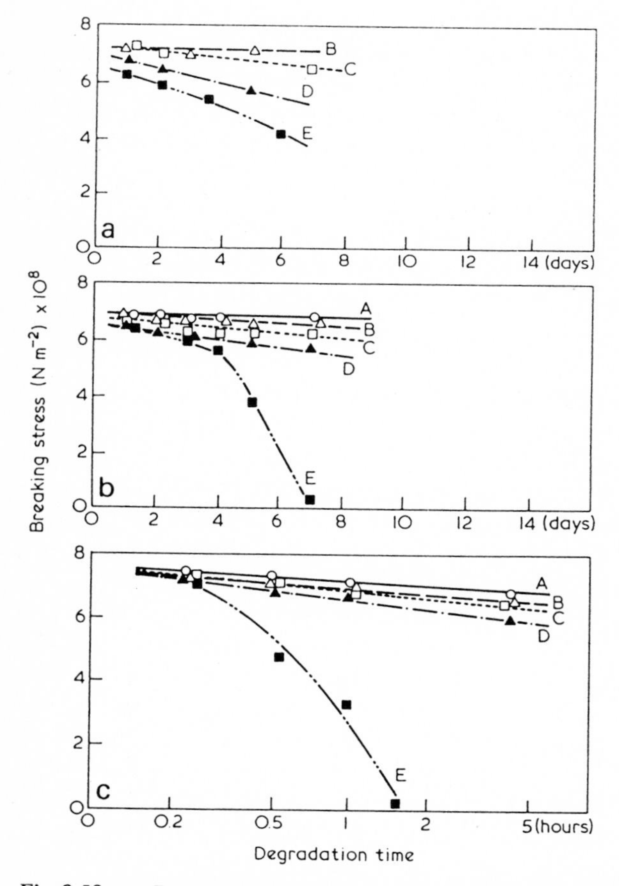

Fig. 8.50a–c. Degradation of strength of strained PA 6 fibers as a function of the time stored in an aggressive medium. **a)** 0.236 mol% O_3; **b)** 12 vol.% SO_2; **c)** 2.58 vol.% NO_2. During exposure the fibers were maintained at a constant strain; strain levels were chosen so that the stress observed immediately after loading corresponded to the following fractions of the fiber short time strength of 750 MPa: A (0%), B (60%), C (70%), D (80%), E (90%); from Igarashi and DeVries [245]

$(N_0 - N_i)/N_0$ will be much larger than the relative decrease of rupture strengths, $(\sigma_0 - \sigma_i)/\sigma_0$ where the subscript o refers to a reference sample and i to the sample exposed to the environment; as it should be, the authors measured a ratio of these two relative quantities of about 10 for the straining in air and values of between 3 and 6 for the aggressive media. One point needs to be strongly emphasized again. The fact that tensile strength σ_i and concentration of free radicals N_i are both adversely affected by an aggressive atmosphere must not mean that strength depends directly on the number of tie-chains present. As revealed by electron microscopy the chemical attack causes a surface degradation and thus an embrittlement of the sample (decrease of ϵ_b); this necessarily leads to a reduction of $\sigma_i = \sigma(\epsilon_b)$ and of $N_i = N(\epsilon_b)$. This point is also underlined by the fact that the reduction of molecular weight during chemical attack and the reduction of free radical concentration are not equivalent: for identical N_i ($3 \cdot 10^{16}$ spins/cm^3) different rupture concentrations have been determined from molecular weight changes, $1.4 \cdot 10^{18}$/cm^3 in SO_2 and $5.3 \cdot 10^{18}$/cm^3 in NO_2 [245]. These authors have also tested some non-oriented PA 66 rods for which they observe a notable decrease in "toughness" (defined as the integral over the stress-strain curve) at stresses which are about 5 times smaller than the stresses applied to the oriented fibers. This seems to explain why the observed chemical effect on the strength of the spherulitic polyamide was smaller [245].

Still smaller changes in strength were found by the same authors [246] for polyethylene. This seemed to be primarily due to the increased tendency of the PE molecules to slippage; this would prevent the build-up of molecular stresses large enough to further significantly the chemical attack.

4. H_2O, Other Liquids

With respect to the general effect of an active liquid environment on a polymer reference to the literature must be made [196–203]. This applies especially to quantitative data concerning the reduction in strength accompanying the swelling or plasticization of a material. The penetration of low molecular weight molecules into a polymer generally reduces the intermolecular attraction and facilitates chain slippage. In the case of hydrogen-bonded materials the possible interaction of the diffusing liquid with the hydrogen bonds has to be considered.

In a series of experiments Hearle et al. [211] studied the effect of air, water, hydrochloric acid, sodium hydroxide, and a laundering detergent on the strength and fatigue of PA-66 fibers. Their results show that the breaking strength of a PA-66 fiber immediately after immersion in water is 5% smaller than that in air. Pre-conditioning by immersion for 3 hours causes an additional strength reduction of 3%. Immersion in aqueous solutions of NaOH and HCl with ph-values larger than 2.5 has an effect similar to that of water. In the range of ph-values between 2 and 0, however, strength decreases linearly with decreasing ph up to a strength loss of 20%. Notable is the observation that the pre-conditioning of fibers before making a *fatigue test* (by rotation over a wire) does not reduce the fatigue life additionally. This may be explained by the fact that the slow growth rates of a fatigue crack starting from the surface permit the environmental agent to always penetrate the crack growth region [211].

A distinct synergistic effect of stress and humidity on the strength of PPTA fibers (Kevlar® 49) has been reported by Morgan et al. [421]. They suggested that hydrolytic cleavage of the amide linkage contributes to the – modest – degradation of strength at 65 °C: a loss of 0.9% per year of residual strength while stressed at 5% RH at 20% of ultimate stress; the loss increased to 18%/year at 100% RH and 20% ultimate stress; it amounted to 12%/year at 100% RH unstressed. As the authors point out, the principal mechanism of mechanical failure was not chain degradation but microvoid formation and coalescence [421].

The partial penetration of a liquid or vapor into a matrix establishes concentration gradients which do lead to a direct mechanical action through non-uniform swelling or to indirect actions through the nonhomogeneous relaxation or distribution of stresses. These actions are even enhanced in the presence of thermal gradients and may cause rapid crack or craze formation. In the case of a slowly penetrating environment in a well entangled, homogeneous matrix the induced stresses generally can be accommodated elastically or viscoelastically. For example in polycarbonate sheets exposed to artificial weathering no cracking becomes apparent even after severe cycling of temperature and humidity [212]. Within a relatively short period of 30 to 32 months of outdoor weathering, however, a network of surface microcracks developed on the side exposed to the solar radiation. By comparison with artificial UV irradiation the authors [212] were able to show that the photochemical degradation of the surface layer introduces defects and lowers the strength of the polymeric material to such an extent that physically induced non-homogeneous stresses cause microcracking rather than being accommodated. The physical effects of liquid environment on crack or craze formation will be further discussed in 9 II D. A special case of chemical stress cracking of acrylic fibers has been reported by Herms et al. [364a]. They observe the generation of periodic microscopic transverse cracks in oriented acrylic fibers immersed in hot alkaline hypochlorite solution. This phenomenon is greatly accelerated by external tensile stresses; it is ascribed to bond cleavage involving cyclization of nitrile groups followed immediately by N-chlorination and chain scission.

A whitening phenomenon induced by some non-solvents on highly oriented PMMA is described by Sung et al. [364b] and related to the entanglement density. Microvoids also form in PC exposed to different non-solvents [364b] or hot water [364c]. Shen et al. [422] have studied the effects of sorbed water on the mechanical properties of low and high molecular weight PMMA. They suggest that sorbed water up to a concentration of 1.1% acts as a mild plasticizer; at higher concentrations water clustering occurs and the samples fail in a brittle manner.

In the case of filled polymers evidently possible reactions of the environmental agent with the filler must be taken into account (see for example the stress corrosion data for glass-fiber reinforced PETP reported by Lhymn and Schultz [365]).

The references given in Section III of this chapter are deliberately incomplete. They serve only to underline the points under consideration. For details the reader is referred to the cited general literature.

C. Additional Physical Attack on Stressed Samples

1. Physical Aging

Although time and temperature are highly important parameters in environmental degradation their unique action – *physical aging* – will not be considered at this point. Nevertheless it should be indicated that aging influences environmentally important variables such as the rates of stress relaxation, creep and diffusion [366].

2. UV

Photo-chemical degradation is probably the most important factor of environment. The cited monographs [196–203, 207–209] treat in detail the primary processes of photon absorption, electron excitation, energy transfer through excitons, luminescence, phosphorescence, and radiationless transitions, chain scission and free radical formation, secondary reactions, stabilization and protection, and the effect of irradiation on the mechanical properties. Here only one point can be touched: the accelerating effect of UV radiation on microcrack formation in loaded polymers [74, 213–214].

As indicated by Tables 4.4 and 4.5 photon energy in the spectral range of $250 < \lambda < 360$ nm is larger than the dissociation energy of C–C bonds. The high energy tail of solar radiation extends to about $\lambda = 300$ nm. Chain scission in UV and solar light is, therefore, energetically possible. In the near UV in unprotected glassy polymers quantum yields of the order of $1 \cdot 10^{-3}$ to $60 \cdot 10^{-3}$ are observed [209b].

The principle mechanisms of the synergistic action of load and UV radiation may be discussed using the few data available. The simultaneous application of tensile load and UV irradiation to oriented polymers clearly accelerated the formation of free radicals and/or of micro and macro cracks in PA 6 [213–214, 367] and in natural silk, cotton, and "triacetate" fibers [213]. No effect was observed in PMMA [213]. The experiments on cotton and triacetate fibers revealed that at low tensile stresses ($\sigma_0 < 70$ MN m^{-2}) the UV irradiation reduced the lifetimes of the fibers by more than 4 orders of magnitude. Under these conditions the absence or presence of oxygen was of some, but minor, importance since the irradiation in vacuum led to only slightly longer lifetimes than the irradiation in air. In a range of stresses with $70 < \sigma_0 < 220$ MN m^{-2} no oxygen effect on the lifetime of triacetate fibers became apparent. In this stress range the irradiation effect decreased with increasing σ_0. For $\sigma_0 > 220$ MN m^{-2} the lifetime depended only on stress and not on the environmental factors UV irradiation and oxygen content. With cotton fibers a somewhat similar behavior was observed although the upper stress limit was smaller and depended on absence or presence of air [213]. The described behavior points to the existence of three failure mechanisms which are occurring simultaneously and at different rates: oxydation, UV degradation, and creep. An effect of oxydation was observed for the acetate fibers only at lifetimes $t_b > 5 \cdot 10^3$ s and in the presence of UV radiation. At shorter lifetimes, $100 < t_b < 5 \cdot 10^3$ s, failure was essentially due to irradiation. At very short lifetimes (and higher stresses) oxydation and UV degradation were irrelevant if compared to the stress effect.

The determination by X-ray scattering of the number of microcracks in a PA-6 fiber stressed in air at 128 MN m^{-2} gave the notable result [214] that the rate of microcrack accumulation increased almost instantaneously (from $5 \cdot 10^{16}$ m^{-3} s^{-1} to $110 \cdot 10^{16}$ m^{-3} s^{-1}) with the application of UV irradiation. The rate decreased to its initial value in the same abrupt and repeatable manner when UV was switched off after 10^4 s. Irradiation of an unstressed sample did not cause any microcrack formation and did not influence subsequent formation rates. In a previous study [213] it had been shown that UV irradiation of stressed PA-6 and natural silk fibers in a helium atmosphere increased the accumulation of free radicals. In that case the rate of radical accumulation at $200 < \sigma_0 < 600$ MN m^{-2} decreased with the length of the time of irradiation and reached a steady state concentration $N(\dot{R})$ after $5 \cdot 10^3$ s. In PA 6 at a stress of 600 MN m^{-2} the concentration $N(\dot{R})$ was of the order of 10^{24} m^{-3}; this is about the limiting concentration observed in purely stress-induced chain scission also.

From these results the following conclusions can be derived. The photodegradation of *unstressed* PA 6 in air most probably is a random oxidative chain scission process [215]. Based on a study of the formed ESR spectra Heuvel et al. [216] concluded that amide bonds in the amorphous regions are broken leading to R—$\dot{C}$O and $\dot{N}$H—R′ free radicals. Subramanian et al. [215] propose formation of hydroperoxides and keto-intermediates at the α—CH_2 group and a subsequent hydrolytic breakdown at the RCO—NHCOR′ bond. It must be assumed that in PA-6 in air under the combined action of UV and stress all chain segments in the amorphous regions are possible fracture sites and not only the most highly stressed ones. The rates of chain degradation and cross-linking may depend on chain stress through the mechanisms discussed below. In the absence of oxygen, UV irradiation accelerated the breakage of chains but did show a saturation at a concentration $N(\dot{R})$. This result can be explained if one assumes that in irradiated *and* stressed PA 6 preferentially those chains are broken and are prevented from recombination which are most highly stressed.

Several mechanisms are conceivable through which the UV degradation of highly extended and stressed chains would be facilitated. In the first place the decrease of bond dissociation energies through electronic excitation and tensile stress has to be mentioned (cf. 4IIB). In the case of ionized hydrocarbon chains, bond dissociation energies are as small as 100 kJ/mol (Table 4.6). As outlined in Chapter 7 bonds of such energy require rather small stresses for breakage (about 1/4 to 1/5 of the strength of a non-ionized chain). In addition to the reduction of the dissociation energy an increase in local thermal excitation – due to energy transfer or dissipation – may have an effect. Chain orientation and segment conformation also influence the effective stability of a chain. Very little is being said in the literature about the relation between molecular order and UV resistance. Reinisch et al. [217] investigated the influence of sample orientation on photo-chemical degradation. They found that highly drawn PA-6 fibers ($\lambda = 3.9$) had an even higher degradation resistance than undrawn material. Krüssmann et al. [218] indicate that the reactivity (in photooxidation) of N-vicinal methylen groups in trans conformation is 60 times that of those in gauche conformation. These data [213–218] were obtained for quite different specimens and experimental conditions. They do not yet permit a quantitative estimate as to the importance and interaction of the above mechanisms.

In recent years a few additional publications have appeared [245, 246, 367, 368–370]. For PA 6 a clear synergistic effect between orientation, state of stress, and UV irradiation is observed [245, 367]. The fact that non-oriented samples were much less sensitive to UV irradiation seems to be due to the smaller stresses exerted on the loaded chain segments. UV irradiation of PP films led to an accelerated decrease of remaining strength, formation of oxidation products and decrease of molecular weight; the latter parameter was in fact the most sensitive and could be used for damage control [368]. An interesting observation was made by Saidov et al. who studied the stress-induced shift of the 975 cm^{-1} IR band of UV irradiated PETP. They found that the "true tension of breaking bonds" during UV irradiation decreased from 11 to 10 GPa [369].

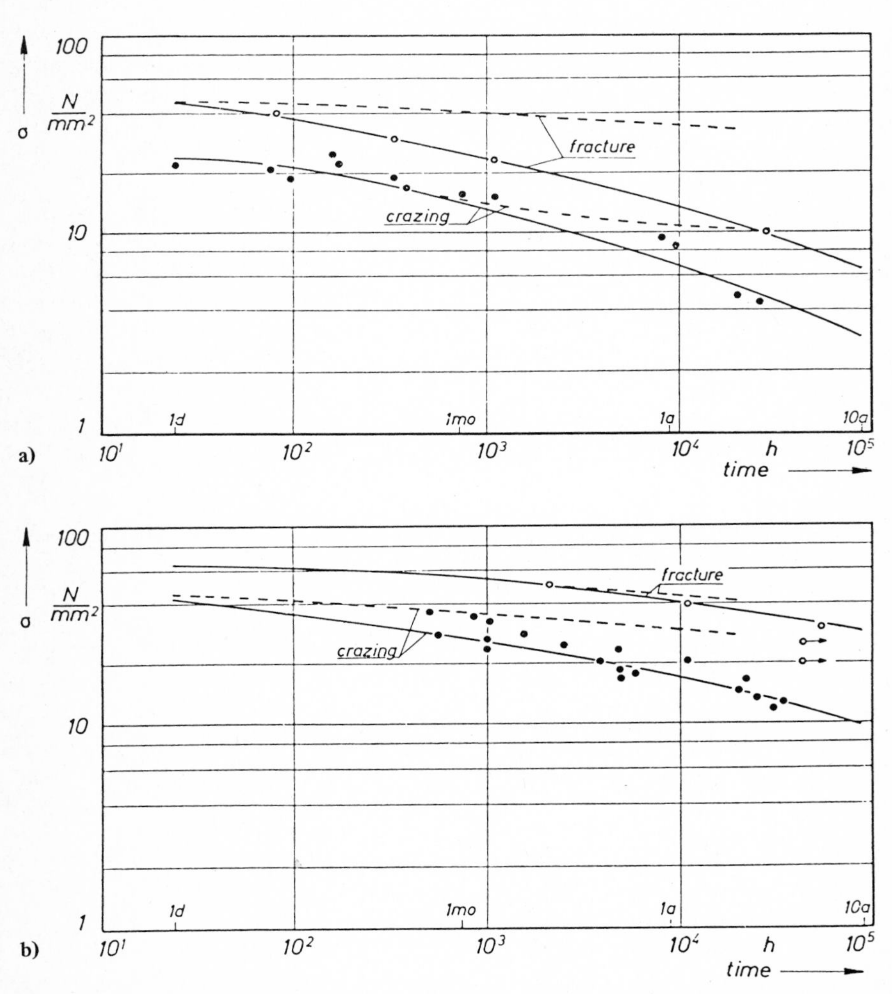

Fig. 8.51 a, b. Long term strength of Plexiglas 8H, 23 °C/50% r.h. (---), natural weathering (——); **a)** unoriented; **b)** biaxially stretched 70%. (from Hennig and Zaengler [370])

Hennig and Zaengler investigated the weathering behavior of stressed PMMA [370]. They report that craze initiation and fracture of loaded Plexiglas 8H were accelerated by natural weathering (see Fig. 8.51). Through biaxial orientation the fracture strength under natural weathering conditions could be made to practically coincide with the long term strength under laboratory conditions, i.e. 50% rh, 23 °C, no UV (Fig. 8.51 b). It seems as if the difference in sample deformation (which is smaller in the case of the oriented material) explained these results.

3. Particles and γ-Irradiation

Electronic excitation, ionization, radical formation, oxidation, and cross-linking are also the principal processes occurring in polymer solids subjected to nuclear radiation (α, β, γ, nucleons). In view of the fact that the molecular mobility influences the kinetics of degradation and cross-linking a synergistic stress effect is conceivable but not yet proven. The current investigations aim at an understanding of the interrelation between irradiation characteristics (dose and dose rate, atmosphere), network structure, degradation, and macroscopic properties after irradiation [198, 200, 219, 371–375]. A literature survey listing (the 124 most important) papers of accelerated-aging tests for predicting radiation degradation of organic materials has recently been prepared [373a]. A special word may be said with respect to electron-beam induced fracture of polymers [374, 375]. Dickinson et al. [374] irradiated the highly stressed material at the tip of a crack with a pulsed beam of 1.6 keV electrons. They noted an accelerated crack growth, increase of gas pressure and light emission in phase with the electron exposure. This observation points to the thermal and/or direct degradation through electron interaction of the stressed polymer (PI, IIR, PE) and to the break-up and removal of the degraded material in the highly stressed zone – in analogy to the ozone degradation of stressed rubbers (Section III, B 1). Selective rupture of only the most highly stressed chains has to be excluded in view of the rather low level of the applied *chain* stress. The radiation stability of unstressed aromatic group containing polymers has been studied by Sasuga et al. [375]. They established the following order of radiation stability:

polyimide > PEEK > polyamide > polyetherimide > polyarylate > PSU > PPO.

4. Electric Fields

In very high electric fields partial discharges and breakdown of polymer insulating materials is observed [e.g. 376–383]. This phenomenon, which in the presence of moisture occurs in the form of so-called "water-trees", is related to the accumulation of space charges and the formation of microcracks. It will be discussed in more detail in Chapter 9.

References for Chapter 8

1. S. N. Zhurkov, A. Ya. Savostin, E. E. Tomashevskii: Dokl. Akad. Nauk SSSR *159,* 303 (1964). Soviet Phys. "Doklady" (Engl. Transl.) *9,* 986 (1964).
2. H. H. Kausch, J. Becht: Rheol. Acta *9,* 137 (1970).
3. V. R. Regel, A. I. Slutsker, E. E. Tomashevskii: "The Kinetic Nature of the Strength of Solids" (in Russian), Moscow: Isdat. Nauka, 1974.
4. S. N. Zhurkov, V. I. Vettegren', I. I. Novak, K. N. Kashintseva: Doklady Akad. Nauk SSSR *176*/3, 623–626 (1967). Soviet Phys. "Doklady" (Engl. Transl.) *176*/3, 708–711 (1967).
5. S. N. Zhurkov, V. I. Vettegren', V. E. Korsukov, I. I. Novak: Fizika Tverdogo Tela *11*/2, 290–295 (1969). Soviet Phys. Solid State *11*/2, 233–237 (1969).
6. S. N. Zhurkow, V. I. Vettegren', V. E. Korsukov, I. I. Novak: Fracture 1969, paper IV/47.
7. A. I. Gubanov: Mekh. Polimerov *3*/4, 608–614 (1967).
8. A. I. Gubanov: Mekh. Polimerov *3*/5, 771–776 (1967).
9. A. I. Gubanov: Mekh. Polimerov *4*/4, 586–594 (1968).
10. V. E. Korsukov, V. I. Vettergren', I. I. Novak: Mekh. Polimerov *6*/1, 167–170 (1970). Polym. Mech. (USA) *6*/1, 156–159 (1972).
11. V. A. Kosobukin: Mekh. Polimerov *6*/6, 971–978 (1970) Polym. Mech. (USA) *6*/6, 846–852 (1970).
12. V. A. Kosobukin: Fizika Tverdogo Tela *14*/9, 2595–2602 (1972) Soviet Phys. Solid State *14*/9, 2246–2251 (1973).
13. V. A. Kosobukin: Mekh. Polimerov *8*/1, 3–11 (1972) Polym. Mech. (USA) *8*/1, 1–7 (1972).
14. V. A. Kosobukin: Opt. Spektrosk. *34*/2, 273–277 (1973). Opt. & Spectrosc. (USA) *34*/2, 154–156 (1973).
15. V. I. Vettegren', I. I. Novak: Fizika Tverdogo Tela *15,* 1417–1422 (1973). Sov. Phys. Solid State *15*/5, 957–960 (1973).
16. V. I. Vettegren', I. I. Novak, K. J. Friedland: Int. J. Fracture *11,* 789–801 (1975).
17. S. N. Zhurkov, V. A. Zakrevskii, V. E. Korsukov, V. S. Kuksenko: Fizika Tverdogo Tela *13*/7, 2004–2013 (1971). Soviet Phys. Solid. State *13*/7, 1680–1688 (1972). J. Polym. Sci.: A-2, *10,* 1509 (1972).
18. S. N. Zhurkov, V. E. Korsukov: Fizika Tverdogo Tela *15,* 2071–2080 (1973). Soviet Phys. Solid State *15*/7, 1379–1384 (1974)
    S. N. Zhurkov, V. E. Korsukov: J. Plym. Sic. (Polymer Phys. Ed.) *12*/385–398 (1974).
19. A. Ya. Savostin, E. E. Tomashevskii: Fizika Tverdogo Tela *12*/10, 2857–2864 (1970). Soviet Phys. Solid State *12*/10, 2307–2311 (1971).
20. V. A. Zakrevskii, V. Ye. Korsukov: Vysokomol. soyed. B*13,* 105 (1971). Vysokomol. soyed. A*14*/4, 955–961 (1972). Polym. Sci. USSR *14,* 1064–1071 (1972).
21. B. Ya. Levin, A. V. Savitskii, A. Ya. Savostin, E. Ye. Tomashevskii: Vysokomol. soyed. A*13*/4, 941–947 (1971). Polymer Sci. USSR *13,* 1061–1068 (1971).
22. S. I. Veliev, V. I. Vettegren', I. I. Novak: Mekh. Polimerov *6*/3, 433–436 (1970). Polym. Mech. (USA) *6*/3, 369–372 (1970).
23. U. G. Gafurov: Mekh. Polimerov 7/4, 649–653 (1971). Polym. Mech. (USA) 7/4, 578–581 (1971).
24. V. I. Vettegren', V. Ye. Korsukov, I. I. Novak: Plaste und Kautschuk *19*/2, 86–88 (1972).
25. V. E. Korsukov, V. I. Vettegren', I. I. Novak, A. Chmel: Mekh. Polimerov *4,* 621–625 (1972). Polym. Mech. (USA) *4,* 536–539 (1972).
26. V. Ye. Korsukov, V. I. Vettegren', I. I. Novak, L. P. Zaitseva: Vysokomol. soyed A*16*/7, 1538–1542 (1974). Polymer Sci. USSR *16,* 1781–1785 (1974). E. E. Tomashevskii, V. A. Zakrevskii, I. I. Novak, V. E. Korsukov, V. R. Regel, O. F. Pozdnyakov, A. I. Slutsker, V. S. Kuksenko: Int. J. of Fracture *11*/5, 803–815 (1975).
27. S. N. Zhurkov, V. S. Kuksenko, A. I. Slutsker: Fizika Tverdogo Tela *11*/2, 296–307 (1969). Soviet Phys. Solid State *11*/2, 238–246 (1969).

28. V. S. Kuksenko, A. I. Slutsker: Mekh. Polimerov *6*/1, 43–47 (1970). Polym. Mech. (USA) *6*/1, 36–40 (1970). V. M. Knopov, V. S. Kuksenko, A. I. Slutsker: Mekh. Polimerov *6*/3, 387–392 (1970). Polym. Mech. (USA) *6*/3, 329–333 (1970).
29. V. S. Kuksenko, V. S. Ryskin, V. I. Betekhtin, A. I. Slutsker: Int. J. of Fracture *11*/5, 829–840 (1975).
30. A. V. Amelin, O. F. Pozdnyakov, V. R. Regel: Mekh. Polimerov *4*/3, 467–473 (1968). Polym. Mech. (USA) *4*/3, 376–379 (1968). A. V. Amelin, O. F. Pozdnyakov, V. R. Regel, T. P. Sanfirova: Fizika Tverdogo Tela *12*/9, 2528–2534 (1970). Soviet Phys. Solid State *12*/9, 2034–2038 (1971). A. V. Amelin, Yu. A. Glagoleva, A. O. Podol'skii, O. F. Pozdnyakov, V. R. Regel, T. P. Sanfirova: Fizika Tverdogo Tela *13*/9, 2726–2731 (1971). Soviet Phys. Solid State *13*/9, 2279–2283 (1972).
31. E. E. Tomashevskii: Fizika Tverdogo Tela *12*/11, 3202–3207 (1971). Soviet Phys. Solid State *12*/11, 2588–2592 (1971). Yu. K. Godovskii, V. S. Papkov, A. I. Slutsker, E. E. Tomashevskii, and G. L. Slonimskii: Fizika Tverdogo Tela *13*/8, 2289–2295 (1972). Soviet Phys. Solid State *13*/8, 1918–1923 (1972).
32. E. E. Tomashevskii, E. A. Egorov, A. Ya. Savostin: Int. J. Fracture *11*/5, 817–827 (1975).
33. V. R. Regel, O. F. Pozdnyakov: Plaste u. Kautschuk *19*/2, 99–100 (1972).
34. D. K. Roylance, K. L. DeVries: Polymer Letters *9*, 443–447 (1971).
35. R. P. Wool: PhD Thesis, Univers. of Utah, Dept. of Materials Science and Engng., Salt Lake City, Utah 1974. R. P. Wool: J. Polymer Sci. *12*, 1575–1586 (1974).
36. R. P. Wool: J. Polymer Sci. *13*, 1795–1808 (1975).
37. R. P. Wool: J. Polymer Sci. *14*, 1921–1929 (1976).
38. R. P. Wool, H. H. Kausch: Unpublished results, Salt Lake City 1974.
39. R. P. Wool, W. O. Statton: In press.
40. K. K. R. Mocherla, W. O. Statton: Symposium for High Polymer Physics (1975), 1–13. K. K. R. Mocherla: PhD Thesis, Univers. of Utah, Dept. of Materials Science and Engng., Salt Lake City, Utah 1976.
41. H. Tadokoro et al.: J. chem. Phys. *42*, 4 (1965).
42. T. Miyazawa: J. Polymer Sci. C*2*, 59 (1974).
43. J. P. Luongo: J. Appl. Polymer Sci. *III*/9, 302–309 (1960).
44. J. Becht: Dissertation, Technische Hochschule Darmstadt, 1970.
45. J. Becht, H. Fischer: Kolloid-Z. u. Z. Polymere *240*, 766–774 (1970).
46. F. Szöcs, J. Becht, H. Fischer: Europ. Polymer J. *7*, 173–179 (1971).
47. J. Becht, H. Fischer: Angew. Makromol. Chem. *18*, 81–91 (1971).
48. H. H. Kausch, J. Becht: Deformation and Fracture of High Polymers, 317–333. J. A. Hassell, R. I. Jaffee, Plenum Press 1974.
49. K. L. DeVries, B. A. Lloyd, M. L. Williams: J. Appl. Physics *42*/12, 4644–4653 (1971).
50. B. A. Lloyd, K. L. DeVries, M. L. Williams: J. Appl. Polymer Sci. *10*/A-2, 1415–1445 (1972).
51. D. Klinkenberg: Dissertation, Technische Hochschule, Darmstadt, 1978.
52. T. Nagamura, K. Fukitani, M. Takayanagi: J. Polymer Sci. (Polymer Phys. Ed.) *13*, 1515–1532 (1975).
53. D. C. Prevorsek: J. Polymer Sci. A-2, *4*, 63–88 (1966).
54. H. H. Kausch, K. L. DeVries: Intern. J. of Fracture *11*/5, 727–759 (1975).
55. H. H. Kausch, J. Becht: Kolloid-Z. u. Z. Polymere *250*, 1048–1065 (1972).
56. W. O. Statton: J. Polymer Sci. C*32*, 219 (1971).
57. K. L. DeVries: J. Polymer Sci. C*32*, 325 (1971).
58. A. Peterlin: Internat. J. of Fracture *11*/5, 761–780 (1975).
59. F. H. Müller, A. Engelter: Kolloid-Z. u. Z. Polymere *149*/2-3, 126–127 (1956).
60. F. H. Müller, A. Engelter: Rheolog. Acta *1*, 39–53 (1958).
61. A. Engelter, F. H. Müller: Kolloid-Z. u. Z. Polymere *157*/2, 89–111 (1958).
62. F. H. Müller: Kunststoffe *49*/2, 67–71 (1959).
63. F. H. Müller, N. Weimann: J. Polymer Sci. C*6*, 117–124 (1964).
64. J. Stölting, F. H. Müller: Kolloid-Z. u. Z. Polymere *238*/1–2, 459–470 (1970). J. Stölting, F. H. Müller: Kolloid-Z. u. Z. Polymere *240*, 792–806 (1970).
65. W. Dick, F. H. Müller: Kolloid-Z. u. Z. Polymere *172*/1, 1–18 (1960).

66. W. Dick, A. Engelter, F. H. Müller: Rheol. Acta *1*/4, 6, 506–510 (1961).
67. F. H. Müller: J. Polymer Sci. C*20*, 61–76 (1967).
68. F. H. Müller, A. Engelter: Kolloid-Z. u. Z. Polymere *171*/2, 152–153 (1960).
69. F. H. Müller, J. Stölting: Kolloid-Z. u. Z. Polymere *240*, 790–791 (1970).
70. V. I. Vettegren', A. E. Chmel: Europ. Polymer J. *12*, 853–858 (1976).
71. A. C. Lunn, I. V. Yannas: J. Polymer Sci. (Polymer Phys. Ed.) *10*, 2189–2208 (1972).
72. W. F. Busse, E. T. Lessig, D. L. Loughborough, L. Larrick: J. Appl. Physics *13*, 715–724 (1942).
73. W. J. Lyons: Text. Res. J. *28*, 127 (1958).
74. V. R. Regel, A. M. Leksovskii: Int. J. Fracture Mechanics *3*/2, 99–109 (1967). V. R. Regel, A. M. Leksovskii: Polymer Mechanics *5*, 58–78 (1969) Mekhanika Polimerov *5*, 70–96 (1969). V. R. Regel: Mekhanika Polimerov, *7*, 98–112 (1971).
75. V. P. Tamush: Polymer Mechanics *5*/1, 79–87 (1969). Mekhanika Polimerov *5*/1, 97–107 (1969).
76. J. W. S. Hearle, E. A. Vaughn: Rheologica Acta *9*/1, 76–91 (1970).
77. A. R. Bunsell, J. W. S. Hearle: J. Materials Sci. *6*, 1303–1311 (1971).
78. D. C. Prevorsek, W. J. Lyons: Rubber Chemistry and Technol. *44*/1, 271–293 (1971).
79. A. R. Bunsell, J. W. S. Hearle: J. Appl. Polymer Sci. *18*, 267–291 (1974).
80. V. Regel, V. P. Tamush: Mekhanika Polimerov *3*, 458–478 (1977).
81. J. W. S. Hearle, B. S. Wong: J. Text. Inst. *68*/3, 89–94 (1977).
82. S. F. Calil, J. W. S. Hearle: Fracture 1977, Vol. 2, ICF4, Waterloo, Canada, June 1977.
83. J. W. S. Hearle: "An Atlas of Fibre Fracture". Text. Mfr. (1972), *99*, Jan., Feb., 14; March, 12; May, 20; Aug., 30; Sept., 16; Oct., 40; Nov., 12; Dec., 36; (1973), *100*, Jan., 24; March, 24; April, 34; May, 54; June, 44.
84. C. Oudet, A. R. Bunsell: Journal of Applied Polymer Science, Vol. *29*, 4363–4376 (1984).
85. C. Oudet: Thèse, Centre des Matériaux de l'Ecole des Mines de Paris, 1986.
86. M. C. Kenney, J. F. Mandell, F. J. McGarry: Research Report R84-2, MIT, april 1984.
87. J. R. Moraes d'Almeida, D. Hearn, A. R. Bunsell: Polymer Engineering and Science *24*, 42 (1984).
88. P. I. Vincent: Impact tests and service performance of thermoplastics, Plastics Inst., London, 1971.
89. C. B. Bucknall, K. V. Gotham, P. I. Vincent in: Polymer science, A. D. Jenkins (ed.), North-Holland Publ., Chapt. 10, 1972.
90. H. Oberst: Kunststoffe *52*/1, 4–11 (1962).
91. W. Retting: Kolloid-Z. u. Z. Polymere *210*/1, 54–63 (1966).
92. M. H. Litt, A. V. Tobolsky: J. Macromol. Sci.-Phys., B1(3), 433–443 (1976).
93. J. Heijboer: J. Polym. Sci. (C) *16*, 3755 (1968).
94. R. F. Boyer: Polym. Engng. Sci. *8*, 161 (1968).
95. J. A. Sauer: J. Polymer Sci. C*32*, 110–116 (1971).
96. P. I. Vincent: Polymer *15*, 111–116 (1974).
97. W. Retting: Materialprüfung *8*/2, 55–60 (1966).
98. H. Grimminger, G. Koch, J. Penzkofer, E. P. Petermann, W. Retting, J. Steinig: Kunststoffe *59*/6, 375–381 (1969).
99. K. Fujioka: J. Appl. Polymer Sci. *13*, 1421–1434 (1969).
100. H. H. Racké, T. Fett: Kunststoffe *64*/9, 481–487 (1974).
101. G. Hoff, G. Langbein: Kunststoffe *56*, 2–6 (1966).
102. H. Oberst: Kunststoffe *59*/4, 232–240 (1969).
103. W. B. Hillig: Impact Response Characteristics of Polymeric Materials, Techn. Inform. Series, General Electric Co., Report No. 76CRD271, New York 1976.
104. H. Saechtling: Kunststoff-Taschenbuch, 20. Ausg., 910–914, 516, 518, München: Carl Hanser 1977.
105. F. Ramsteiner: Kunststoffe *67*/9, 517–522 (1977).
106. a) E. Gaube, H. H. Kausch: Fracture Theories in Industrial Use of Thermoplastics and Glassfiber Reinforced Plastics, Intern. Conf. Fracture, Munich 1973, Vol. I, Pl VI-311.
106. b) E. Gaube, H. H. Kausch: Kunststoffe *63*/6, 391–397 (1973).

107. E. Gaube, W. Müller, C. Diedrich: Kunststoffe *56,* 673 (1966).
108. W. Retting: Angew. Makromol. Chem. *58/59,* 133–174 (1977).
109. T. T. Jones: Effect of Molecular Orientation on the Mechanical Properties of Polystyrene, Rep. IUPAC Working Party, IUPAC Internat. Symposium on Macromolecules, 10–14 Sept. 1973, 41–57.
110. W. Retting: The Effect of Molecular Orientation on the Mechanical Properties of Rubber-Modified-Polystyrene, to be published in "Pure and Applied Chemistry".
111. H. Niklas, H. H. Kausch: Kunststoffe *53*/12, 886–891 (1963).
112. I. V. Yannas: J. Polymer Sci. *9,* 163–190 (1974).
113. D. W. Hadley, I. M. Ward: Reports on Progress in Physics *38*/10, 1143–1215 (1975).
114. S. S. Sternstein in: Treatise on Materials Science and Technology, J. M. Schultz, ed., Vol. 10, New York: Academic Press 1977, 567–569.
115. E. N. C. da Andrade: Proc. Royal. Soc. (London) A 84, 1 (1910).
116. Taprogge, R.: Kunststoff-Rundschau *15*/5–12, 3–50 (1968).
117. a) E. Gaube, G. Diedrich, W. Müller: Kunststoffe *66*/1, 2–8 (1976).
b) E. Gaube, H. Gebler, W. Müller, C. Gondro: Kunststoffe *75*, 412 (1985).
c) E. Gaube, W. Müller: Kunststoffe *72*, 297 (1982).
118. a) P. Stockmayer: Stuttgarter Kunststoff-Kolloquium 197; Kunststoffe *67,* 470 (1977).
b) P. Stockmayer, S. Wintergerst: 3R international *20,* 274 (1981).
119. A. V. Savitskii, V. A. Mal'chevski, T. P. Sanfirova, L. P. Zosin: Polymer Sci. USSR *16,* 2470–2477 (1974). Vysokomol. soyed. A*16*/9, 2130–2135 (1974).
120. S. N. Zhurkov, B. N. Narzulayev: J. Techn. Phys. *23,* 677 (1953).
121. D. F. Kagan, A. M. Knebel'man, L. A. Kantor: Vysokomol. soyed A*14*/5, 1207–1214 (1972).
122. M. J. Mindel, N. Brown: J. Materials Sci. *9*, 1661–1669 (1974).
123. L. C. E. Struik: Polym. Eng. and Sci. *17*/3, (1977). L. C. E. Struik: Physical Aging in Amorphous Polymers and Other Materials Amsterdam/New York: Elsevier Scientific Publishing Co. 1978.
124. S. Matsuoka, H. E. Bair: J. Appl. Physics *48*/10, 4058–4062 (1977). S. Matsuoka, H. E. Bair, S. S. Bearder, H. E. Kern, J. T. Ryan: Analysis of Nonlinear Stress Relaxation in Polymeric Glasses, Bell Laboratories, Murray Hill, New Jersey 07974.
125. J. F. Jansson, B. Terselius: IUPAC Symposium on Long-Term Properties of Polymers, Stockholm, Sweden 30. 8. – 1. 9. 1976.
126. E. H. Andrews: Fatigue in Polymers, Testing of Polymers, Vol. IV, W. Brown, ed., New York: Interscience 1969, 237.
127. J. A. Manson, R. W. Hertzberg: CRC Critical Reviews in Macromol Sci. 433–500, 1973.
128. R. W. Hertzberg: Deformation and Fracture Mechanics of Engineering Materials, New York: Wiley 1976, Sec. Ed. 1983.
129. K. Oberbach: Kunststoffe *63,* 35–41 (1973).
130. K. Oberbach: Kunststoff-Kennwerte für Konstrukteure, München: Carl Hanser Verlag 1975, 87–96.
131. J. M. Schultz: Treatise on Materials Science and Technology, Vol. 10, Part B. New York: Academic Press 1973, 80–107.
132. G. Jacoby in: Neuzeitliche Verfahren der Werkstoffprüfung, Düsseldorf: Verlag Stahleisen 1973, 80–107.
133. C. E. Feltner, M. R. Mitchell: ASTM STP 465, Amer. Soc. Test. Mat., 27 (1969).
134. K. Oberbach, G. Heese: Materialprüfung *14*/6, 173–178 (1972).
135. Crawford, R. J., Benham, P. P.: J. Mech. Engng. Sci. (GB) *16*/3, 178–179 (1974).
136. M. E. Graf, M. Ya. Filatov: Probl. Prochn. (USSR) 7/9, 102–106 (1975).
137. R. J. Crawford, P. P. Benham: J. Materials Sci. *9,* 1297–1304 (1974).
138. V. Bouda, A. J. Staverman: J. Polymer Sci. Polymer Sci. (Polymer Phs. Ed.) *14,* 2313–2323 (1976).
139. V. Zilvar: J. Macromol. Sci.-Phys. B5/2, 273–284 (1971).
140. V. Zilvar: Plastics and Polymers, 328–332, October 1971.

141. E. Alf: Untersuchungen zum Verhalten ausgewählter Kunststoffe unter schwingender Beanspruchung (Dissertation), Technische Hochschule Aachen 1972.
142. S. Rabinowitz, A. R. Krause, P. Beardmore: Materials Sci. *8,* 11–22 (1973).
143. M. Schrager: J. of Polymer Sci. Part A-2 *8,* 1999–2014 (1970).
144. S. Sikka: PhD Thesis, Univers. of Utah, Salt Lake City, Utah 1976.
145. J. H. Wendorff: Personal communication, D. Kunststoff-Inst., Darmstadt (1977).
146. T. Nagamura, N. Kusumoto, M. Takayanagi: J. Polymer Sci. (Polymer Phys. Ed.) *11,* 2357–2369 (1973).
147. O. W. Brüller, N. Brand: Kunststoffe *67*/9, 527 (1977).
148. L. Konopasek, J. W. S. Hearle: J. Appl. Polymer Sci. *21,* 2791–2815 (1977).
149. H. H. Kausch: Materialprüfung *20,* 22–26 (1978)
150. K. V. Gotham: Plastics and Polymers *40,* 59–64 (1972).
151. N. T. Smotrin, V. M. Chebanov: Mekhanika Polimerov *6,* 453–467 (1970).
152. A. N. Machyulis, M. I. Pugina, A. A. Zhechyus, V. K. Kuchinskas, A. P. Stasyunas: Mekhanika Polimerov *2*/1, 60–66 (1966). V. K. Kuchinskas, A. N. Machyulis: Polymer Mechanics *4,* 538–543 (1968).
153. J. A. Sauer, A. D. McMaster, D. R. Morrow: J. Macromol. Sci.-Phys. B12(4), 535–562 (1976). J. A. Sauer, E. Foden, D. R. Morrow: Polymer Eng. and Sci. *17*/4, 246–250 (1977).
154. a) J. C. Bauwens: J. Polym. Sci., Part A-2 *5,* 1145 (1967).
b) J. C. Bauwens, Mem. Sci. Rev. Metall. LXV *4,* 355 (1968).
c) J. C. Bauwens: Rheol. Acta *13,* 93 (1974).
155. a) J. C. Bauwens, C. Bauwens-Crowet, G. Homes: J. Polym. Sci., Part A-2 *7,* 1745 (1969).
b) J. C. Bauwens: J. Polym. Sci., Part A2 *8,* 893 (1970).
c) Bauwens-Crowet, J. C. Bauwens, G. Homes: J. Mater. Sci. *7,* 176 (1972).
d) J. C. Bauwens: J. Mater. Sci. *7,* 577 (1972).
156. I. M. Ward: J. Materials Sci. *6,* 1397 (1971).
157. P. B. Bowden, J. A. Jukes: J. Materials Sci. *7,* 52–63 (1972).
158. C. Bauwens-Crowet, J. C. Bauwens, G. Homes: J. Materials Sci. *7,* 176–183 (1972).
159. R. Raghava, R. M. Caddell, G. S. Y. Yeh: J. Materials Sci. *8,* 225–232 (1973).
160. J. A. Sauer, K. D. Pae: Coll. and Polymer Sci. *252,* 680–695 (1974).
161. a) A. S. Argon, M. I. Bessonov: Philosophical Mag. *35*/4, 917–933 (1977).
b) A. S. Argon, M. I. Bessonov: Polymer Engng. Sci. *17*/3 (1977).
162. a) R. E. Robertson: J. Chem. Phys. *44,* 3950 (1966).
b) R. E. Robertson: Appl. Polym. Symp. *7,* 201 (1968).
163. R. P. Kambour, R. E. Robertson: The Mechanical Properties of Plastics, in Polymer Science, Chapter 11, Jenkins Ed., Amsterdam, London: North-Holland, 1972, 687–822.
164. T. E. Brady, G. S. Y. Yeh: J. Appl. Physics *42*/12, 4622–4630 (1971).
165. K. P. Grosskurth: Gummi/Asbest/Kunststoffe *25*/12, 1159–1164 (1972).
166. G. P. Andrianova, V. A. Kargin: Polymer Sci. USSR *12*/1, 1–8 (1970). Vysokomol. Soyed. A*12*/1, 3–9 (1970).
167. G. P. Andrianova, A. S. Kechekyan, V. A. Kargin: J. Polymer Sci. A-2/*9,* 1919–1933 (1971).
168. G. I. Barenblatt in: Deformation and Fracture of High Polymers, Kausch, Hassell, Jaffee Eds., New York: Plenum Press, 1972, 91–111.
169. B. Zaks, M. L. Lebedinskaya, V. N. Chaldize: Vysokomol. Soyed. A12/*12,* 2669–2679 (1970).
170. L. A. Davis, C. A. Pampillo: J. Appl. Physics *42*/12, 4659–4666 (1971).
171. L. A. Davis, C. A. Pampillo: J. Appl. Physics *43*/11, 4285–4293 (1972).
172. J. C. M. Li, C. A. Pampillo, L. A. Davis in: Deformation and Fracture of High Polymers, Kausch, Hassell, Jaffee Eds., New York: Plenum Press, 1972, 239–258.
173. L. A. Davis, R. H. Baughman, C. A. Pampillo: J. Polymer Sci.: Polymer Phys., Ed. *11,* 2441–2451 (1973).
174. a) V. A. Lishnevskii: Dokl. Akad. Nauk SSSR, *182*/3, 596–599 (1968).
b) V. A. Lishnevskii: Vysokomol. Soedin., Ser. B *11*/1, 44–49 (1969).
175. R. N. Haward, G. Thackray: Proc. Roy. Soc. A *302,* 453 (1968).
176. P. Matthies, J. Schlag, E. Schwartz: Angew. Chem. *77*/7, 323–327 (1965).

177. T. C. Chiang, J. P. Sibilia: J. Polymer Sci.: Polymer Phys. Ed. *10,* 2249–2257 (1972).
178. U. G. Gafurov: Vysokomol. Soyed. A14/*4,* 873–880 (1972).
179. A. K. Sengupta, R. K. Singh, A. Majumdar: Textile Res. J., 155–163, March (1973).
180. N. J. Capiati, R. S. Porter: J. Polymer Sci. (Polymer-Phys. Ed.) *13,* 1177–1186 (1975).
D. M. Bigg: Polymer Engineering Sci. *16*/11, 725–734 (1976).
181. G. Capaccio, I. M. Ward: Polymer *18*/9, 967–968 (1977).
182. Seminar on ultra-high modulus polymers; La Chimica e l'Industria *59,* Ottobre 1977, 728–735.
183. L. Mullins: Effects of Fillers in Rubber in: The Chemistry and Physics of Rubber-Like Substances, Chapter 11, Bateman Ed., New York: Wiley, 1963, 301–328.
184. T. L. Smith: Pure and Appl. Chem. *23,* 235–253 (1970).
185. N. Sekhar, B. M. E. van der Hoff: J. Appl. Polymer Sci. *15,* 169–182 (1971).
186. F. R. Eirich in Mechanical Behavior of Materials Vol. III, Kyoto: Society of Materials Sci., 1972, 405–418.
187. T. L. Smith, W. H. Chu: J. Polymer Sci. A-2/*10*/1, 133–150 (1972).
188. F. P. Baldwin, G. Ver Strate: Rubber Chemistry and Techn. *45*/3, 709–881 (1972).
189. R. F. Landel, R. F. Fedors in: Deformation and Fracture of High Polymers, Kausch, Hassell, Jaffee Eds., New York: Plenum Press, 1972, 131–148.
190. R. J. Morgan: J. Polymer Sci.: Polymer Phys. Ed. *11,* 1271–1284 (1973).
191. T. L. Smith in: Treatise on Materials Science and Technology, J. M. Schultz, Ed. Vol. 10, Part A New York: Academic Press, 1973, 369.
192. R. F. Fedors, R. F. Landel: J. Polymer Sci. (Polymer-Phys. Ed.) *13,* 419–429 (1975).
193. R. F. Fedors: J. Appl. Polymer Sci. *19,* 787–790 (1975).
194. R. F. Fedors, R. F. Landel: J. Appl. Polymer Sci. *19,* 2709–2715 (1975).
195. R. F. Fedors: The Stereo Rubbers, W. M. Saltman Ed., New York: John Wiley, 1977. 679–804.
196. Encyclopedia of Polymer Science and Technology, Mark Ed., New York: Interscience, 1965.
a. W. L. Cox, Vol. 2, 197. b. N. Grassie, Vol. 4, 702.
197. G. M. Bartenev, Y. S. Zuyev: Strength and Failure of Visco-Elastic Materials, Oxford: Pergamon Press 1968.
a. pg. 274.
198. D. V. Rosato, R. T. Schwartz: Environmental Effects on Polymeric Materials, Vol. I + II, New York: Interscience 1968.
199. E. H. Andrews: Fracture in Polymers, Edinburgh: Oliver and Boyd 1968.
a. pg. 168.
200. A. Charlesby, Radiation Effects in Polymers; in: Polymer Science Vol. 1, Chapter 23, Jenkins Ed., Amsterdam London: North-Holland, 1972, 1543–1559.
201. Polymer Stabilization, W. L. Hawkins Ed. New York: Wiley-Interscience, 1971.
202. Natürl. u. künstl. Alterung, Kunststoffe, Fortschrittsberichte, Vol. 1–3, München 1976: Carl Hanser.
203. F. H. Winslow: Environmental Degradation, in Treatise on Materials Science and Technology, J. M. Schultz, ed., Vol. 10, Part B, New York: Academic Press, 1977, 741–776.
Developments in Polymer Degradation, N. Grassie, ed., Vol. 1, London: Appl. Science Publ. 1977.
204. E. H. Andrews: J. Appl. Polymer Sci. *10,* 47 (1966).
205. G. Salomon, F. van Bloois: J. Appl. Polymer Sci. *8,* 1991 (1964).
206. K. L. DeVries, E. R. Simonson, M. L. Williams: J. Macromol. Sci.-Phys. B*4*/3, 671 (1970).
K. L. DeVries, E. R. Simonson, M. L. Williams: J. Appl. Polymer Sci. *14,* 3049 (1970).
207. Degradation and Stabilization of Polyolefins: B. Sedlacek, C. G. Overberger, H. F. Mark, T. G. Fox Eds: Polymer Symposia *57* (1976).
208. a) B. Ranby, J. F. Rabek: ESR Spectroscopy in Polymer Research, Berlin–Heidelberg–New York: Springer 1977.
b) B. Ranby, J. F. Rabek: Photodegradation, Photo-oxidation and Photostabilization of Polymers, London: Wiley 1975.

209. H. H. G. Jellinek Ed.: Aspects of Degradation and Stabilization of Polymers: Amsterdam-Oxford: Elsevier 1978.
a) Y. Kamiya, E. Niki: Oxidative Degradation, p. 79–148.
b) W. Schnabel, J. Kiwi: Photodegradation, p. 149–246.
c) I. Mita, Effect of Structure on Degradation and Stability of Polymers, p. 247–291.
d) K. Murakami: Mechanical Degradation, p. 296–392.
e) H. Kambe: The Effect of Degradation on Mechanical Properties of Polymers, p. 393–430.
210. C. S. Kim: Rubber Chemistry and Technology *42*/4, 1095–1121 (1969)
211. J. W. S. Hearle, B. S. Wong: J. of the Textile Institute *63*/4, 127–132 (1977).
212. A. Blaga, R. S. Yamasaki: J. Materials Sci. *11*, 1513–1520 (1976).
213. G. G. Samoilov, E. E. Tomashevskii: Fizika Tverdogo Tela *10*/4, 1094–1097 (1968). Soviet Physics-Solid State *10*/4, 866–869 (1968). T. B. Boboev, V. R. Regel': Mekh. Polimerov *5*, 929–931 (1969). Polym. Mech. (USA) *5*, 824–826 (1972).
214. Kh. Akimbekov, V. S. Kuksenko, S. Nizamidinov, A. I. Slutsker, A. A. Yastrebinskii: Fizika Tverdogo Tela *14*/9, 2708–2713 (1972). Soviet Physics-Solid State *14*/9, 2339–2343 (1973).
215. R. V. R. Subramanian, T. V. Talele: Textile Research J. *42*/4, 207–214 (1972).
216. H. M. Heuvel, K. C. J. B. Lind: J. Polymer Sci., A-2, *8*, 401–410 (1970).
217. G. Reinisch, W. Jaeger: Faserforschung und Textiltechnik *19*/8, 363–365 (1968).
218. H. Krüssmann, G. Valk, G. Heidemann, S. Dugal: Angew. Chemie *81*/6, 226–227 (1969).
219. H. Wilski: Physik-Grundlage der Technik (Physik 1974-Plenarvorträge), Weinheim 1974: Physik-Verlag, p. 183–198.
220. T. M. Stoeckel, J. Blasius, B. Crist: J. Polymer Sci., Polymer Physics Ed. *16*, 485–500 (1978).
221. R. P. Wool, R. H. Boyd: J. Appl. Phys. *51*, 5116 (1980).
222. a) R. S. Bretzlaff, R. P. Wool: J. Appl. Phys. *52*, 5964 (1981).
b) R. S. Bretzlaff, R. P. Wool: Macromol. *16*, 1907 (1983).
c) Y.-L. Lee, R. S. Bretzlaff, R. P. Wool: J. Polym. Sci., Phys. Ed. *22*, 681 (1984).
d) R. P. Wool, R. S. Bretzlaff, B. Y. Li, C. H. Wang, R. H. Boyd: Bull. Amer. Phys. Soc. *29*/3, 529 (1984). J. Polym. Sci.: Part B: Polymer Phys. *24*, 1039 (1986).
e) R. P. Wool: "Principles at Applications of FTIR of Stressed Polymers" presented at Amer. Chem. Soc., Polymer Preprints *25*, 2 (1984).
223. V. M. Voroboyev, I. V. Razumovskaya, V. I. Vettegren: Polymer *19*, 1267 (1978).
224. I. J. Hutchinson, I. M. Ward: Polymer *21*, 55 (1980).
225. G. Bayer, W. Hoffmann, H. W. Siesler: Polymer *21*, 235 (1980).
226. J. L. Koenig: Appl. Spectrosc. *29*, 293 (1975).
227. J. L. Koenig, M. K. Antoon: Appl. Opt. *17*, 1374 (1978).
228. H. W. Siesler, K. Holland-Moritz: Infrared and Raman Spectroscopy of Polymers, New York–Basel: Marcel Dekker 1980.
229. R. P. Wool: J. Polym. Sci. – Polym. Phys. Ed. *19*, 449 (1981).
230. D. O. Hummel, C. Votteler, M. Winter: Kunststoffe *73*, 193 (1983).
231. R. K. Popli, D. K. Roylance: MIT Report DAAG.29.76.C.0044 (1980), Polym. Eng. Sci. *22*, 1046 (1982).
232. R. P. Wool: Polym. Eng. Sci. *20*, 805 (1980).
233. a) K. L. DeVries, R. H. Smith, B. M. Fanconi: Polymer *21*, 949 (1980).
b) B. M. Fanconi, K. L. DeVries, R. H. Smith: Polymer *23*, 1027 (1982).
234. a) M. Dole, Polymer *22*, 1458 (1981).
b) K. L. DeVries, R. H. Smith, B. M. Fanconi: Polymer *22*, 1460 (1981).
235. B. M. Fanconi: J. Appl. Phys. *54*, 5577 (1983).
236. P. Fordyce, K. L. DeVries: Polym. Eng. Sci. *24*, 421 (1984).
237. F. Chao, S. Murthy, K. L. DeVries: Bulletin Am. Phys. Soc., March 1984, p. 531.
238. O. Frank: PhD Thesis, Technische Hochschule Darmstadt, 1984.
239. H. Weitkamp, R. Barth: Einführung in die quantitative Infrarot-Spektrophotometrie, Stuttgart: G. Thieme 1976.
240. K. Jud: PhD Thesis 413, Ecole Polytechnique Federale de Lausanne, 1981.

241. H. A. Gaur: Coll. & Polym. Sci. *256*, 64 (1978).
242. a) H. H. Kausch: Polym. Eng. Sci. *19*/2, 140 (1979).
b) H. H. Kausch: Coll. & Polym. Sci. *236*, 1 (1985).
243. O. Frank, J. H. Wendorff: Coll. & Polym. Sci. *259*, 70 (1981).
244. H. H. Kausch: in IUPAC Macromolecules, H. Benoit, P. Rempp Eds, Oxford–New York: Pergamon Press 1982, p. 211.
245. M. Igarashi, K. L. DeVries: Polymer *24*, 769 (1983).
246. M. Igarashi, K. L. DeVries: ibid. 1035.
247. K. J. Friedland, V. A. Marikhin, L. P. Myasnikova, V. I. Vettegren: J. Polym. Sci., Polym. Symp. *58*, 185 (1977).
248. V. V. Zhizhenkov, E. A. Egorov: J. Polym. Sci. Polym. Phys. Ed. *22*, 117 (1984).
249. S. P. Mishra, B. L. Deopura: J. Appl. Polym. Sci. *27*, 3211 (1982).
250. R. Bonart, F. Schultze-Gebhardt: Angew. Makromol. Chem. *22*, 41 (1972).
251. V. A. Marichin: Acta Polym. *30*, 507 (1979).
252. D. Roylance: Int. J. Fract. *21*, 107 (1983).
253. T. L. Nemzek, J. E. Guillet: Macromol. *10*, 94 (1977).
254. V. S. Kuksenko, A. I. Slutsker: J. Macromol. Sci. – Phys *B12*, 487 (1976).
255. J. H. Wendorff: Progr. Coll. Polym. Sci. *66*, 135 (1979).
256. a) I. M. Ward: Structure and Properties of Oriented Polymers, London: Applied Science Publishers 1975.
256. b) I. M. Ward: Developments in Oriented Polymers-1, London–New Jersey: Applied Science Publishers 1982.
257. a) A. Ciferri, I. M. Ward: Ultra-high Modulus Polymers, London: Applied Science Publishers 1979.
b) I. M. Ward: Adv. in Polymer Sci. *70*, 1 (1985).
258. 11th Europhysics Conference on Macromolecular Physics on "Thermal, Mechanical and Electrical Properties of Oriented Polymers", Leeds, Europhys. Conf. Abstr. 5B (1981).
259. A. G. Gibson, G. R. Davies, I. M. Ward: Polymer *19*, 683 (1978).
260. J. G. Rider, K. M. Watkinson: Polymer *19*, 645 (1978).
261. R. G. C. Arridge, P. J. Barham: ibid., 654 (1978).
262. a) G. Capaccio: Coll. & Polym. Sci. *259*, 23 (1981).
b) G. Capaccio, I. M. Ward: Coll. & Polym. Sci. *260*, 46 (1982).
c) G. Capaccio: Pure & Appl. Chem. *55*/5, 869 (1983).
263. a) E. S. Sherman, R. S. Porter, E. L. Thomas: Polymer *23*, 1069 (1982).
b) W. Wade Adams, R. M. Briber, E. S. Sherman, R. S. Porter, E. L. Thomas: Polymer *26*, 17 (1985).
264. a) T. Kanamoto, A. Tsuruta, K. Tanaka, M. Takeda, R. S. Porter: Polym. J. 15, 327 (1983).
b) T. Kanamoto, A. Tsuruta, K. Tanaka, M. Takeda: Polym. J. 16, 75 (1984).
265. J. A. Odell, A. Keller, M. J. Miles: Coll. & Polym. Sci. *262*, 683 (1984).
266. J. Smook, J. Pennings: Coll. & Polym. Sci. ibid., 712.
267. a) M. A. Wilding, I. M. Ward: Polymer *22*, 870 (1981).
b) I. M. Ward, M. A. Wilding: J. Polym. Sci., Polym. Phys. Ed. *22*, 561 (1984).
c) D. W. Woods, W. K. Busfield, I. M. Ward: Polym. Comm. *25*, 298 (1984).
268. P. F. van Hutten, C. E. Koning, A. J. Pennings: Coll. & Polym. Sci. *262*, 521 (1984).
269. T. Ohta: Polym. Eng. & Sci. *23*, 697 (1983).
270. D. L. M. Cansfield, I. M. Ward, D. W. Woods, A. Buckley, J. M. Pierce, J. L. Wesley: Polym. Comm. *24*, 130 (1983).
271. a) T. Kanamoto, R. S. Porter: J. Polym. Sci., Polym. Lett. Ed. *21*, 1005 (1983).
b) H. H. Chuah, R. S. Porter: J. Polym. Sci., Polym. Phys. Ed. *22*, 1353 (1984).
c) A. E. Zachariades, R. S. Porter: Polym. News *9*, 177 and 329 (1984).
d) A. E. Zachariades, R. S. Porter: Polym. News *10*, 13 (1984).
272. A. V. Savitsky, I. A. Gorshkova, I. L. Frolova, G. N. Shmikk, A. F. Ioffe: Polym. Bulletin *12*, 195 (1984).
273. V. A. Marichin, L. P. Mjasnikova, D. Zenke, R. Hirte, P. Weigel: Polym. Bulletin ibid. 287.
274. P. Smith, P. J. Lemstra, H. C. Booij: J. Polym. Sci., Polym. Phys. Ed. *19*, 877 (1981).

275. P. J. Lemstra, J. P. L. Pjipers: Europhys. Conf. Abstr. *6 G,* 88 (1982).
276. a) L. Fischer, R. Haschberger, A. Ziegeldorf, W. Ruland: Coll. & Polym. Sci. 260, 174 (1982).
b) L. Fischer, W. Ruland: Coll. & Polym. Sci. 261, 717 (1983).
c) W. Ruland: Makromol. Chem. Suppl. 6, 235 (1984).
277. R. J. Morgan, C. O. Pruneda, W. J. Steele: UCRL-86802 Preprint, Lawrence Livermore Laboratory, October 20, 1981.
278. J. R. White, T. J. Lardner: J. Mater. Sci. *19,* 2387 (1984).
279. S. J. Deteresa, S. R. Allen, R. J. Farris, R. S. Porter: J. Mater. Sci. *19,* 57 (1984).
280. P. Avakian, R. C. Blume, T. D. Gierke, H. H. Yang, M. Panar: Polym. Preprints (ACS) *21*/1, 8 (1980).
281. J. R. Brown, D. K. C. Hodgeman: Polymer *23,* 365 (1982).
282. I. M. Brown, T. C. Sandreczki, R. J. Morgan: Polymer *25,* 759 (1984).
283. T. Nagamura, M. Takayanagi: J. Polym. Sci., Polym. Phys. Ed. *12,* 2019 (1974).
284. H. C. Bach, F. Dobinson, K. R. Lea, J. H. Saunders: J. Appl. Polym. Sci. *23,* 2125 (1979).
285. F. Dobinson, C. A. Pelezo, W. B. Black, K. R. Lea, J. H. Saunders: J. Appl. Polym. Sci. *23,* 2189 (1979).
286. T. Nagamura, K. L. DeVries: Polymer *22,* 1267 (1981).
287. W. J. Welsh, D. Bhaumik, H. H. Jaffe, J. E. Mark: Polym. Eng. Sci. *24,* 218 (1984).
288. C. Galiotis, I. M. Robinson, R. J. Young, B. J. E. Smith, D. N. Batchelder: Polymer communications *26*, 354 (1985).
289. C. Galiotis, R. T. Read, P. H. J. Yeung, R. J. Young: J. Polymer Sci., Polymer Phys. Ed. *22*, 1589 (1984).
290. a) B. Stalder: These 586, EPF-Lausanne (1985).
b) B. Stalder, H. H. Kausch: J. Mater, Sci. *20*, 2873 (1985).
291. M. Rink, T. Ricco, W. Lubert, A. Pavan: J. Appl. Polym. Sci. *22,* 429 (1978).
292. F. Ramsteiner: Kunststoffe *73,* 148 (1983).
293. B. K. Daniels: J. Appl. Polym. Sci. *15,* 3109 (1971).
294. N. Brown, I. M. Ward: J. mater. Sci. *18,* 1405 (1983).
295. N. Brown: ibid, 2241.
296. S. Hashemi: PhD Thesis, Imperial College of Science and Technology, London 1984.
297. E. H. Andrews, P. Reed: Adv. in Polym. Sci. Vol. 27, Berlin–Heidelberg–New York, Springer-Verlag 1978.
298. T. Kusano, K. Kobayashi, K. Murakami: Rubb. Chem. & Techn. 5, 773 (1979).
300. O. D. Sherby, P. M. Burke: Prosg. Mat. Sci. *13*, 324 (1968).
301. F. K. G. Odqvist in: Inelastic Behavior of Solids, New York: McGraw-Hill 1970.
302. B. Ilschner: Hochtemperatur-Plastizität, Reine und angew. Metallk. in Einzeldarstell. Band 23, W. Köster Ed., Berlin–Heidelberg–New York: Springer Verlag 1973.
303. M. Yokouchi, Y. Hiromoto, Y. Kobayashi: J. Appl. Polym. Sci. *24*, 1965 (1979).
304. F. Schwarzl: Werkstoffkunde der Kunststoffe, Universität Erlangen–Nürnberg 1982.
305. F. R. Schwarzl, F. Zahradnik: Rheol. Acta *19,* 137 (1980).
306. C. Bauwens-Crowet, J. C. Bauwens: J. Macromol. Sci.-Phys. *B14*(2) 265 (1977).
307. R. H. Boyd, M. E. Robertsson, J. F. Jansson: J. Polym. Sci., Polym. Phys. Ed. *20,* 73 (1982).
308. D. G. Hunt, M. W. Darlington: Polymer *20,* 241 (1979).
309. M. Schlimmer: Kunstst. *70,* 778 (1980).
310. A. Takaku: J. Appl. Polym. Sci. *26,* 3565 (1981).
311. G. Sandilands, P. Kalman, J. Bowman, M. Bevis: Polym. Comm. *24,* 273 (1983).
312. L. Johansson: PhD Thesis, Lund Institute of Technology, Lund (1984).
313. A. Lustiger, R. L. Markham: Polymer *24,* 1647 (1983).
314. a) J. A. Sauer, G. C. Richardson: Int. J. Fract. *16,* 499 (1980).
b) J. A. Sauer, C. C. Chen: in Advances in Polymer Science 52/53, Berlin: Springer-Verlag 1983.
315. M. T. Takemori: Ann. Review of Mater. Sci. *14,* 171 (1984).
316. R. W. Hertzberg, J. A. Manson: Fatigue of Engineering Plastics, New York: Academic Press 1980.
317. S. Sikka: Mat. Sci. Eng. *41,* 265 (1979).
318. A. Takahara, K. Yamada, T. Kajiyama, M. Takayanagi: J. Appl. Polym. Sci. *25,* 597 (1980).
319. G. G. Trantina: Polym. Eng. Sci. *26*, 776 (1986).

320. J. F. Mandell, J.-P. F. Chevaillier, K. L. Smith, D. D. Huang: MIT-Report R84-1, February 1984.
321. S. Warty, D. R. Morrow, J. A. Sauer: Polymer *19*, 1465 (1978).
322. N. Chheda, C. Chen, J. A. Sauer: Adv. in Mater. Techn. in the Americas *1*, 43 (1980).
323. C. C. Chen, N. Chheda, J. A. Sauer: J. Macromol. Sci.-Phys. *B19*, 565 (1981).
324. B. Kleinemeier, G. Menges: Swiss Plast. *3*, 29 (1981).
325. B. Escaig, C. G'Sell: Plastic deformation of amorphous and semi-crystalline materials, les Ulis: Editions de Physique, 1982.
326. J. C. Bauwens: J. Mater. Sci. *13*, 1443 (1978).
327. J. C. Bauwens: Polymer *21*, 699 (1980).
328. a) C. Bauwens-Crowet, J. C. Bauwens: Polymer *23*, 1599 (1982).
b. C. Bauwens-Crowet, J. C. Bauwens: Polymer *24*, 921 (1983).
329. J. C. Bauwens: Polymer *25*, 1523 (1984).
330. C. Bauwens-Crowet, J. C. Bauwens: Proceedings of Int. Conf. on Deformation, Yield and Fracture of Polymers, Cambridge, 1985, p. 43.1.
331. E. Roeder, H. -G. Hilpert: Kunstst. *74*, 1 (1984).
332. J. M. Lefebvre, B. Escaig: Polymer *23*, 1751 (1982).
333. a) J. M. Lefebvre, C. Bultel, B. Escaig: to be published in J. Mater. Sci.
b) J. M. Lefebvre, B. Escaig: J. Mater. Sci. *20*, 438 (1985).
334. C. G'Sell, J. J. Jonas: J. Mater. Sci. *16*, 1956 (1981).
335. J. J. Jonas, N. Christodoulou, C. G'Sell: Scripta Metallurg. *12*, 565 (1978).
336. N. A. A. Helal: These, Institut National Polytechnique de Lorraine, Nancy, 1982.
337. C. G'Sell, P. Gilormini, J. J. Jonas: to be published.
338. J. W. Hutchinson, K. W. Neale: J. Mech. Phys. Solids *31*, 405 (1983).
339. K. B. Abbas: Polymer *22*, 836 (1981).
340. a) V. E. Korsukov, V. A. Marichin, L. P. Miasnikova, I. I. Novak: J. Polym. Sci., Symp. *42*, 847 (1973).
b) V. A. Marikhin, L. P. Myasnikova: J. Polym. Sci., Polym. Symp. *58*, 97 (1977).
341. R. G. C. Arridge, P. J. Barham: J. Polym. Phys. Ed. *16*, 1297 (1978).
342. G. Capaccio, I. M. Ward: J. Polym. Sci., Polym. Phys. Ed. *22*, 475 (1984).
343. A. de Brossin, M. Dettenmaier, H. H. Kausch: Helv. Phys. Acta *55*, 213 (1982).
344. O. Frank, J. Lehmann: Spring Meeting on Polymer Physics, Lausanne, 18.–20.3.1985.
345. H. G. Scherer, J. Fuhrmann: ibid.
346. a) D. Betteridge, J. V. Cridland, T. Lilley, N. R. Shoko, M. E. A. Cudby, D. G. M. Wood: Polymer *23*, 178 (1982).
b) Matsushige, S. Shirouzu, S. Shichijyo, K. Takahashi, T. Takemura: Reprints 1st SPSJ International Polymer Conference, Kyoto, 20.–24.8.1984, p. 254.
347. a) W. Kuhn: Kolloid-Z. *68*, 2 (1934).
b) ibid *76*, 258 (1936).
c) K. H. Meyer, C. Ferri: Helv. Chim. Acta *18*, 570 (1935).
d) H. M. James, E. Guth: J. Chem. Phys. *11*, 455 (1943).
348. a) P. J. Flory: Principles of Polymer Chemistry, Ithaca–London: Cornell University Press, 1953.
b) L. R. G. Treloar: The Physics of Rubber Elasticity, 2nd Ed. Oxford: Clarendon Press, 1958.
c) F. R. Eirich: Science and Technology of Rubber, New York– San Francisco–London: Academic Press, 1978.
349. a) U. Eisele: Kautschuk + Gummi. Kunststoffe *33*, 165 (1980).
b) R. Casper: VIII. Int. Makrom. Kollequium Interlaken, 6.–7. Sept. 1984.
c) R. F. Boyer: Polymer Letters *17*, 925 (1976).
350. J. E. Mark: Polym. Eng. Sci. *19*, 258 (1979).
351. A. Casale, R. S. Porter: Polymer Stress Reactions, Vol. 1, Introduction, 1978; Vol. 2, Experiments, 1979; New York: Academic Press.
352. R. S. Porter, A. Casale: Polymer Eng. Sci. *25*, 129 (1985).
353. W. Schnabel: Polymer Degradation, New York–Toronto: Hanser International 1981.
354. W. L. Hawkins: Polymer Degradation and Stabilization, Berlin: Springer-Verlag 1984.

355. R. Mergler, J. H. Wendorff: Coll. & Polym. Sci. *259*, 894 (1981).
356. a) A. A. Popov, B. E. Krisyuk, N. N. Blinov, G. E. Zaikov: Europ. Polym. J. *17*, 169 (1981).
b) A. A. Popov, N. N. Blinov, B. E. Krisyuk, S. G. Karpova, L. G. Privalova, G. E. Zaikov: J. Polym. Sci. – Polym. Phys. Ed. *21*, 1017 (1983).
c) J. D. Razumovsii, G. E. Zaikov: Ozone and its reaction with organic compounds, Amsterdam: Elsevier 1984, p. 405.
d) N. M. Emmanuel, G. E. Zaikov, Z. K. Marizus: Oxidation of organic compounds. Effect of medium, Oxford: Pergamon Press 1984, p. 650.
357. A A. Popov, G. E. Zaikov: Rev. Macrom. Chem. Phys. *C23*, 1 (1983).
358. a) G. A. George, G. T. Egglestone, S. Z. Riddell: J. Appl. Polym. Sci. *27*, 3999 (1982).
b) G. A. George, G. T. Egglestone, S. Z. Riddell: Polym. Eng. Sci. *23*, 412 (1983).
359. S. B. Monaco, J. H. Richardson: Polym. News *9*, 230 (1984).
360. G. Scott: Polym. Eng. Sci. 24, 1007 (1984).
361. D. Putz, G. Menges: Br. Polym. J. (GB) 10, 69 (1978).
362. a) U. W. Gedde: PhD Thesis, Royal Institute of Technology, Stockholm (1980).
b) U. W. Gedde, B. Terselius, H.-F. Jansson: Polym. Test *2*, 209 (1981).
c) W. Müller, E. Gaube: Kunststoffe *72*, 297 (1982).
363. a) E. Kramer, J. Koppelmann: Kunststoffe *73*, 11 (1983).
b) J. Koppelmann, E. Kramer, J. Dobrowsky: Spring Meeting on Polymer Physics, Lausanne, 18–20.3.1985.
c) J. Koppelmann, G. Steiner: Spring Meeting on Polymer Physics, Lausanne, 18–20.3.1985.
364. a) J. Herms, L. H. Peebles, Jr., D. R. Uhlmann: J. Mater. Sci. 18, 2517 (1983).
b) N. H. Sung, R. E. Gahan, R. E. Haven: Polym. Eng. Sci. 23, 328 (1983).
c) M. Narkis, L. Nicolais, A. Apicella: Polym. Eng. Sci. 24, 211 (1984).
365. C. Lhymn, J. M. Schultz: Polym. Eng. Sci. 24, 1064 (1984).
366. L. C. E. Struik: Physical Aging in Amorphous Polymers and Other Materials, Amsterdam: Elsevier Scientific Publ. 1978.
367. Y. Fujiwara: J. Appl. Polym. Sci. *27*, 2773 (1982).
368. J. Binder, M. Stangl: Ku. Fortschr. *3* (C), München: Hanser 1976.
369. D. Saidov, R. M. Marupov, H. Habibulloev, V. M. Malishev: Spectroscop. Letters. *11*, 747 (1978).
370. J. Hennig: Chemische Kunststoffe Aktuell, 10. Donauländergespräch – Die Natürliche und Künstliche Alterung von Kunststoffen *32*, 123 (1987).
371. H. Wilski, E. Gaube, S. Rösinger: Coll. & Polym. Sci. *260*, 559 (1982).
372. D. T. Grubb: Ultramicroscopy *12*/4, 279 (1983).
373. a) R. L. Clough, K. T. Gillen, J.-L. Campan, G. Gaussens, H. Schönbacher, T. Seguchi, H. Wilski, S. Machi: Nuclear Safety *25*, 238 (1984).
b) H. Schönbacher: Swiss Plastics *7*, 29 (1985).
374. J. T. Dickinson, M. L. Klakken, M. H. Miles, L. C. Jensen: J. Polym. Sci. Polymer Phys. Ed. *23*, 2273 (1985) and J. of Vac. Sci. and Techn. A4, 1501 (1986).
375. T. Sasuga, N. Hayakawa, K. Yoshida, M. Hagiwara: Polymer *26*, 1039 (1985).
376. J. Schirr: PhD Thesis, Technische Universität Carolo-Wilhelmina, Braunschweig, 1974.
377. R. Patsch: Coll. & Polym. Sci. *259*, 885 (1981).
378. V. Ya. Ushakov, A. L. Robezhko, G. V. Efremova: Sov. Phys. – Solid State *26*, 25 (1984).
379. L. Niemeyer, L. Pietronero, H. J. Wiesmann: Phys. Rev. Lett. s52, 1033 (1984).
380. P. Pfluger, H. R. Zeller: J. Bernasconi, Phys. Rev. Lett. *53*, 94 (1984).
381. H. R. Zeller, W. R. Schneider: J. Appl. Phys. *56*, 455 (1984).
382. L. Pietronero, H. J. Wiesmann: J. Stat. Phys. *36*, 909 (1984).
383. A. E. Tschmel, V. I. Vettegren, V. M. Zolotarev: Macromol. Sci.-Phys. *B21* (2), 243 (1982).
384. V. I. Vettegren: Sov. Phys. Solid State *26* (6), 1030 (1984).
385. V. I. Vettegren, R. R. Abdul'manov: Sov. Phys. Solid State *26*(11), 1964 (1984).
386. V. I. Vettegren, S. V. Bronnikov, S. Ya. Frenkel: Vysokomol. soyed. A26, *5*, 1046 (1984).
387. S. N. Zhurkov, V. S. Kuksenko, V. A. Petrov: Theoretical and Applied Fracture Mechanics *1*, 271 (1984).
388. A. I. Slutsker, Kh. Aidarov: Polymer Science USSR *26*, 2034 (1984).

389. a) E. E. Tomaskevskii: personnal communication, Leningrad, 1985.
b) B. B. Narzullaev, N. G. Kvachadze, E. E. Tomashevskii, A. I. Slutsker: Fiz. Tverd. Tela *23*, 429 (1981).
390. E. A. Egorov, V. V. Zhizhenkov: Journal of Polymer Science *20*, 1089 (1982).
391. G. Dadobayev, K. Ismonkulov, A. I. Slutsker: Vysokomol. soyed. A25, *1*, 37 (1983).
392. N. G. Kvachadze, A. V. Savitskii: Solid State Soviet Physics *26*, 282 (1984).
393. V. A. Marichin, L. P. Mjasnikova, Z. Pelzbauer: J. Macromol. Sci.-Phys. *B22*(1), 111 (1983).
394. A. M. Leksovskil, B. L. Baskin, A. Ya. Gorenberg, G. Kh. Usmanov, V. R. Regel: Sov. Phys. Solid State *25*(4), 630 (1983).
395. V. A. Marikhin: Makromol. Chem. Suppl. *7* 147 (1984).
396. V. A. Bershtein, A. V. Savitsky, V. M. Egorov, I. A. Gorshkova, V. P. Demicheva: Polymer Bulletin *12*, 165 (1984).
397. Popli and Roylance (Polymer Eng. and Sci. *25*, 828 (1985)) have recently predicted from a thermodynamic model that in PA 6 fibers about 5% of the tie-chains are sufficiently taut to be broken in a tensile experiment.
398. R. H. Ericksen: Polymer *26*, 733 (1985).
399. Y. Termonia, P. Meakin, P. Smith: Macromolecules *19*, 154 (1986).
400. C. Pruneda, R. J. Morgan, R. Lim, L. J. Gregory, J. W. Fischer: SAMPE, Meeting Anaheim, CA, 1477 (1985).
401. Q. Ying, B. Chu, R. Qian, J. Bao, J. Zhang, C. Xu: Polymer *26*, 1401 (1985).
402. B. Chu, Q. Ying, C. Wu, J. R. Ford, H. S. Dhadal: Polymer *26*, 1408 (1985).
403. G. Hinrichsen, P. Wolbring, H. Springer: Interrelations between Processing Structure and Properties of Polymeric Materials, 195 (1984).
404. C. Sawatari, M. Matsuo: Colloid & Polymer Science *263*, 783 (1985).
405. P J. Barham, A. Keller: J. Mater. Sci. *20*(7), 2281 (1985).
406. S. Gogolewski, A. J. Pennings: Polymer *26*, 1394 (1985).
407. P. Smith, H. D. Chanzy, B. P. Rotzinger: Polymer Communications *26*, 258 (1985).
408. A. Misra, B. Dutta, V. Kali Prasad: Journal of Applied Polymer Science *31*, 441 (1986).
409. T. Kunugi, C. Ichinose, A. Suzuki: Journal of Applied Polymer Science *31*, 429 (1986).
410. The first stages of defect formation during creep have more recently been studied by analysis of acoustic emissions [346, 387, 411, 412], SAXS [17, 255, 412], Brillouin scattering [413] and photoacoustic spectroscopy [414]. The results have tentatively been interpreted in terms of submicrocrack formation.
411. D. Betteridge, P. A. Connors, T. Lilley, N. R. Shoko, M. E. A. Cudby, D. G. M. Wood: Polymer *24*, 1206 (1983).
412. S. Shichijyo, S Shirouzu, S. Taki, K. Matsushige: Japanese Journal of Applied Physics *22*, 1315 (1983).
413. B.-C. Yap, S. Shichijyo, K. Matsushige, T. Takemura: Japanese Journal of Applied Physics *21*, L523 (1982).
414. M. E. Abu-Zeid, E. E. Nofal, L. A. Tahseen, F. A. Abdul-Rasoul: Journal of Applied Polymer Science *30*, 3791 (1985).
415. I. M. Ward: Polymer Engineering and Science *24*, 724 (1984).
416. G. W. Halldin, Y. C. Lo: Polymer Engineering and Science *25*, 323 (1985).
417. C. V. Oprea, M. Popa, A. Ioanid, P. Bfrsanescu: Colloid & Polymer Science *263*, 738 (1985).
418. N. Ya. Rapport, G. E. Zaikov: Eur. Polym. *20*, 409 (1984).
419. N. Ya. Rapoport, in "Developments in Polymer Degradation", N. Grassie, ed., Appl. Sci. Publ., London 207 (1985).
420. a) H R. Brown: Journal of Material Science *17*, 469 (1982).
b) B. Z. Jang, D. R. Uhlmann, J. B. Vander Sande: J. Appl. Polymer Sci. *29*, 3409 (1984).
421. R. J. Morgan, C. O. Pruneda, N. Butler, F.-M. Kong, L. Caley, R. L. Moore: 29th National Society for the Advancement of Material and Process Engineering (SAMPE), April 3–5, 1984, in Reno, Nevada, 891.
422. J. Shen, C. C. Chen, J. A. Sauer: Polymer *26*, 511 (1985).

423. F. Szöcs (personal communication, Bratislava 1986) concludes that radical recombination plays an important role even at liquid nitrogen temperatures. If added radical traps prevented recombination the measured ESR intensities in ground PMMA increased up to 10 times.

Chapter 9

Molecular Chains in Heterogeneous Fracture

In Chapter 8 principally the contribution of spatially homogeneously distributed molecular rearrangements to polymer fracture was studied. The term *spatially homogeneous* referred to the *absence* of flaws, inclusions, cracks, or notches of a size sufficient to act as a stress concentrator. Under those conditions during an initial phase of external loading damage development or growth is homogeneously distributed on a macroscopic scale. *Heterogeneous fracture* now is defined as the converse of homogeneous fracture or briefly as *fracture through crack propagation*. In this case cracks, notches, inclusions, or accumulated crack nuclei act as macroscopic stress concentrations and essentially confine further damage development to the close proximity of the then existing defect(s). The phenomenon of *crazing* has been included in this chapter because of the well observable structural irregularities and despite the fact that with increasing stress new crazes can be formed at arbitrary nucleation sites.

The development of fracture through cracks will be treated in this chapter following the *fracture mechanics approach*. In view of the extensive, competent, and recent literature on this subject prime attention will be given to the effect of chain properties (length, structure, strength) on crack propagation and on fracture mechanics parameters.

I. Linear Elastic Fracture Mechanics (LEFM)

A. Stress Concentration

It is intuitively clear that the presence of a crack or a defect leads to a local increase of stresses, strains and stored elastic energy. The linear-elastic fracture mechanics approach describes this influence quantitatively by providing an analytical procedure. To a first approximation at some distance from a crack the solid under consideration is treated as an elastic continuum. Griffith [1] was the first to formulate and apply the fracture mechanics concept. From the balance of *elastically* stored energy (U) and energy to produce new surface area ($\gamma_c A$) he derived his well-known fracture criterion for isotropic materials containing an elliptical crack of length 2 a (Eq. 3.13). In the last 50 years this *fracture mechanics approach* has been systematically developed to account for partially anelastic and/or plastic behavior of solids for various crack and sample geometries, and even for material heterogeneities: In all cases it has been – and still is – the aim of the fracture mechanics analysis to derive *generally valid quantitative criteria* for crack stability and crack propagation behavior. Criteria are being sought which are as much as possible independent of the state of external and internal stresses and of crack and sample geometry and which depend principally on material functions such as material resistance R (Rißausbreitungswiderstand), critical stress intensity factor K_c (kritischer Spannungsintensitätsfaktor), and elastic modulus E.

A detailed account of the development of the theory of fracture mechanics is given in the series Mechanics of Fracture [2]. From the extensive general literature on this subject only a few works may be cited which deal with the deformation and fracture of engineering materials [3] or specifically polymers [4–6, 229] and standards [8]. A clear and rigorous presentation of the fracture mechanics of polymers is given by Williams [229].

For the determination of the material functions R and K_c three experimental methods of controlled crack propagation are predominantly used (Fig. 9.1):

- Mode I, crack opening or tensile mode (einfache Rißöffnung)
- Mode II, sliding or in-plane shear mode
- Mode III, tearing or antiplane shear mode.

For details of these methods, their evaluation, and possible influences of sample geometry on the resulting "material" functions the reader is referred to the general

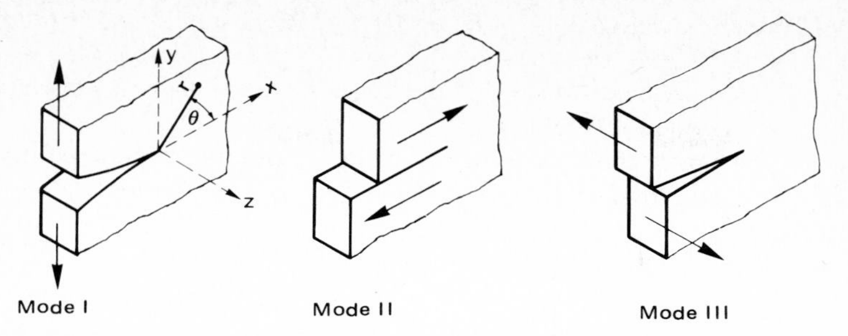

Fig. 9.1. Experimentally used modes of controlled crack propagation: I crack opening or tensile mode, II sliding or in-plane shear mode, III tearing or anti-plane shear mode

literature cited [2–8]. In this Section the most frequently used crack opening mode, namely mode (I), will be considered (for nomenclature see [8c)–f)]).

For a uniaxially stressed isotropic, elastic, *thin* plate containing an infinitely sharp edge crack a state of *plane stress* arises which has a singularity at the crack tip [3–8], for $r \ll a$ the components of stress are expressed as

$$\sigma_x = \frac{K_I}{\sqrt{2\pi r}} \cos\frac{\theta}{2}\left(1 - \sin\frac{\theta}{2}\sin\frac{3}{2}\theta\right) \tag{9.1}$$

$$\sigma_y = \frac{K_I}{\sqrt{2\pi r}} \cos\frac{\theta}{2}\left(1 + \sin\frac{\theta}{2}\sin\frac{3}{2}\theta\right) \tag{9.2}$$

$$\tau_{xy} = \frac{K_I}{\sqrt{2\pi r}} \sin\frac{\theta}{2}\cos\frac{\theta}{2}\cos\frac{3}{2}\theta \tag{9.3}$$

The coordinates r and θ are indicated in Figure 9.1. The quantity K_I determines the intensity of the stress field without affecting its shape. It is accordingly termed *stress intensity factor.* The stress intensity factor takes into account both the external (uniaxial) stress σ_0 and the crack length a:

$$K_I = \sigma_0\sqrt{\pi a} \cdot f(a/D) \tag{9.4}$$

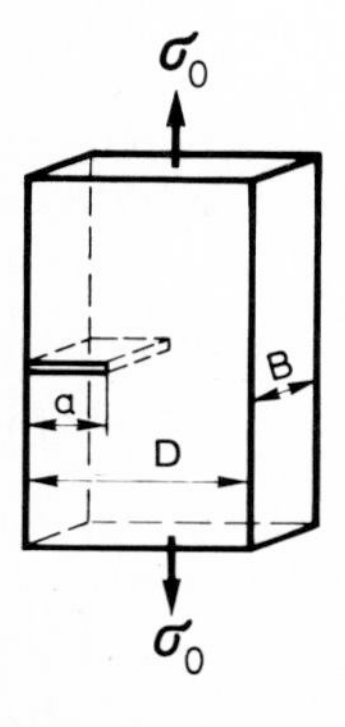

Fig. 9.2. Tensile specimen with single edge crack

where $f(a/D)$ is a correction factor of the order of unity which derives from the finite specimen dimensions and crack configuration [3–8]. The stress intensity factor can be considered a much more general measure of the criticality of the state of a stressed, cracked plate than σ_0 or a individually.

In the case of a *thick* plate the lateral contraction of the material at the crack tip is hindered; a state of *plane strain* prevails and in addition to σ_x and σ_y a normal stress σ_z exists:

$$\sigma_z = \nu(\sigma_x + \sigma_y). \tag{9.5}$$

In plane strain and for $r \ll a$ the components of displacement in the x-direction (u) and the y-direction (v) are obtained [6] as:

$$u = \frac{2\,K_I(1+\nu)}{E}\sqrt{\frac{r}{2\pi}}\cos\frac{\theta}{2}\left(1 - 2\,\nu + \sin^2\frac{\theta}{2}\right) \tag{9.6}$$

$$v = \frac{2\,K_I(1+\nu)}{E}\sqrt{\frac{r}{2\pi}}\sin\frac{\theta}{2}\left(2 - 2\,\nu - \cos^2\frac{\theta}{2}\right). \tag{9.7}$$

For $\theta = \pi$ Eq. (9.7) yields the crack opening displacement

$$2\,v = \frac{8\,K_I}{E}\sqrt{\frac{r}{2\pi}}\,(1-\nu^2). \tag{9.8}$$

Whereas the stresses show a singularity for $r \to 0$ the displacements go to zero. Through the introduction of a crack into a (thick) plate under constant uniaxial stress σ_0 the elastically stored energy in the vicinity of the crack tip is *increased* by the finite amount W:

$$\frac{W}{B} = \frac{\pi a^2\,\sigma_0^2}{2\,E}(1-\nu^2). \tag{9.9}$$

Concurrently work 2 W is expended by the external system to maintain the uniaxial stress σ_0. The total elastic energy U_e decreases, therefore, by W. This apparent paradox was pointed out by Eshelby [9]. If the crack is extended by Δa then the stress field is shifted by Δa in the x-direction and its energy content *increased* by $(\delta W/\delta a)\Delta a$; an energy increment of the same magnitude is set free by the stress release accompanying the failure of the highly stressed crack tip material. It is the latter energy increment which is termed *energy release rate* G_I:

$$G_I = -\frac{\partial U_e}{B\partial a} = \frac{\pi a\,\sigma_0^2}{E(T,t)}(1-\nu^2)\,f^2(a/D) = \frac{K_I^2}{E(T,t)} \tag{9.10}$$

For a thin plate (state of plane stress) the terms $(1+\nu)$ in Eqs. 9.6 and 9.7 and $(1-\nu^2)$ in Eqs. 9.8 to 9.10 have to be replaced by unity. In view of the discussion later it is important to determine how the energy is distributed around the crack tip. Following Irwin [10] the elastic strain energy within a cylinder of radius r_1 around the crack tip is (per unit of plate thickness):

$$\frac{W(r_1)}{B} = \frac{\pi a^2\, \sigma_0^2}{2\,E}\left(\frac{5}{4} - \frac{3}{4}\,\nu - 2\,\nu^2\right)\frac{r_1}{a}. \tag{9.11}$$

Whereas the energy content $W(r_1)$ increases proportionally to r_1 the local energy density $w(r)$ decreases as $1/r$; for large distances from the crack tip it attains $w_0 = \sigma_0^2/2\,E$, the strain energy density of an isotropic, elastic solid. From Figure 9.3 it is readily seen that increasing a by $\Delta a = r_1$ sets free an amount of stored energy $G_I\,\Delta a$ which is a factor of $(8 - 8\,\nu)/(5 - 8\,\nu)$ larger than the elastic energy stored within a cylinder of radius r_1 around the crack tip. This ratio is independent of the absolute value of r_1. In other words any crack propagation step in an ideal elastic material – even the smallest – releases elastic energy in a material volume several times larger than that immediately surrounding the crack tip (volume $\pi \Delta a^2 B$). Experimentally it is seen that a triangular zone bounded by a plane connecting the crack tip with the lateral face is unloaded.

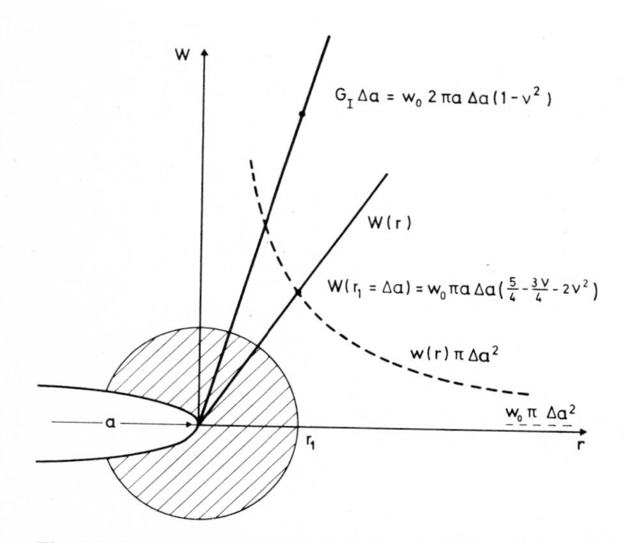

Fig. 9.3. Distribution of elastically stored energy in front of a plane crack (mode I). $G_I\ \Delta a$ energy release (per unit specimen thickness) though propagation of crack by Δa, W(r) energy stored within cylinder with radius r about crack tip, w(r) energy density

B. Crack Tip Plastic Deformation

Plastic deformation is especially pronounced in polymeric materials. The electron micrographs presented in this Chapter provide ample evidence of this fact. It will be necessary, therefore, to study whether and how the fracture mechanics approach derived for an elastic material can be applied to elastic-plastic solids as well. The effect of plastic deformation on the stress distribution at a crack tip is well documented [3–7]. An elastic-plastic material under quasi-elastic conditions, for instance, begins to deform plastically wherever the state of stress meets the yield or flow criterion. Plastic deformation begins in the region of largest stresses, i.e. in the proximity of the crack tip; it limits the components of stress to the yield stress σ_F. In order to maintain the mechanical equilibrium stresses must be raised (up to the yield stress) in more distant regions. Thus the plastic deformation has the effect of increasing the effective crack length [3–7]. Two common approaches to calculate the effective crack enlargement through plastic deformation have been proposed: the von Mises shear yield criterion [4, 6, 229] and the normal stress criterion (Dugdale model [7, 229]).

From the von Mises yield criterion together with Eqs. (9.1) to (9.3) one derives the shape of the plastic zone ($r < r_F$) for a state of plane stress as [6]:

$$r_{F,Sp} = \frac{K_I^2}{2\pi\sigma_F^2} \cos^2\frac{\phi}{2}\left(1 + 3\sin^2\frac{\phi}{2}\right) \tag{9.12}$$

and for plane strain as

$$r_{F,V} = \frac{K_I^2}{2\pi\sigma_F^2} \cos^2\frac{\phi}{2}\left[\left(1 + 3\sin^2\frac{\phi}{2}\right) - 4\nu(1-\nu)\right]. \tag{9.13}$$

The plastic zone extends in the plane of the crack by

$$(r_{F,Sp})_{\phi=0} = \frac{K_I^2}{2\pi\sigma_F^2} \quad \text{for plane stress;} \tag{9.14}$$

and by

$$(r_{F,V})_{\phi=0} = \frac{K_I^2}{2\pi\sigma_F^2}(1-2\nu)^2 \quad \text{for plane strain.} \tag{9.15}$$

In a moderately thick plate one finds a state of plane strain in the center and of plane stress at the lateral faces. This leads to the well-known dog-bone shape of the plastic zone.

The Dugdate model [7] recognizes the effect of the plastic deformation in front of the crack tip in a different way. It considers a virtual crack which includes the plastic zone – of size r_p – and it treats the yield stress as a superimposed stress acting normally on the boundary between the elastic and the "plastic" zone. The superimposed stress tends to bend the surface of the virtual crack. A consideration of mechanical equilibrium gives [6, 7, 229]:

$$r_p = \frac{\pi a \sigma_0^2}{8\,\sigma_F^2}, \tag{9.16}$$

$$K_I = \sigma_0 \sqrt{\pi (a + r_p)} \tag{9.17}$$

and a crack opening displacement

$$2\,v_a = \frac{K_I^2}{E \sigma_F} \tag{9.18}$$

The occurrence of *small scale yielding* in a material has two consequences:
- it limits the elastic energy density to $\sigma_F^2/2E$
- it may render uncritical a given K_I (a given combination of σ_0 and a).

The previous considerations do *not* answer three important questions:
- are there any limits to the extent of plastic deformation if K_I is increased?
- what happens to the size r_p of a plastic zone or to the length of a crack if K_I is maintained constant for some time?
- what are the limits of validity of the K_I-concept at large plastic deformations?

From experiment one knows that there are indeed limits to the extent of plastic deformation, that a crack can grow in a stable manner at constant K_I, and that *crack blunting* limits the validity of the K_I-concept [229, 232].

It must be one of the main objectives of fracture mechanics to give quantitative answers also to these three questions.

C. Material Resistance and Crack Propagation

At small values of K_I the elastic-plastic deformations described above lead to mechanical equilibrium, that is crack length a, crack opening displacement $2\,v_a$ and the size of a possibly present plastic zone remain practically constant. However, if K_I exceeds a certain threshold value K_i then small variations caused by thermal fluctuations, stress relaxation, creep, and/or material weakening can disturb this equilibrium and cause the plastic zone to widen and/or to partially disintegrate: the crack grows.

Propagation of a crack occurs if the energy release rate G_I is sufficient to account for the material resistance R, i.e. for all modes of energy consumption related to the propagating crack. The condition for crack growth is written, therefore, as

$$G_I(a, t) \geqslant R(a, t). \tag{9.19}$$

At low crack propagation rates kinetic energy terms in R can be neglected. Then the material resistance will include the specific surface energy $2\,\gamma$ (to overcome the cohesion of atoms or molecules across the newly formed surface area), and the energy terms V_{re} of elastic retraction of stressed molecules, V_{pl} of plastic deformation, and V_{ch} of chemical reactions including bond scission. The release of internal stresses (U_i) or chemical reactions with the environment (U_{ch}) may subtract from R:

$$R = 2\gamma + \frac{\partial V_{re}}{B\partial a} + \frac{\partial V_{pl}}{B\partial a} + \frac{\partial V_{ch}}{B\partial a} + \ldots - \left[\frac{\partial U_i}{B\partial a} + \frac{\partial U_{ch}}{B\partial a}\right]. \tag{9.20}$$

In general – and especially with polymeric materials – R will be a function of crack length, specimen geometry, and extent and *rate* of plastic deformation at the crack tip. A schematic representation of the change of $R(a)$ in *static loading* of a ductile, strain hardening material is given in Figure 9.4. A crack a_0 grows if $G_I(a_0) > R(a_0)$; if $G_I(a_0) = R(a_0)$ it continues to grow if the slope of the line $G_I(a)$, i.e. $\partial G/\partial a$, is larger than the change of material resistance with crack length, $\partial R/\partial a$. For an arbitrary energy release rate G_I with $R(a_0) < G_I(a_0) < G_c(a_0)$ crack stability will again be achieved at a new value $a_1 > a_0$ where $G(a)$ intersects $R(a)$ (cf. Fig. 9.4). If, however, $G_I(a)$ is equal to or larger than that $G_c(a)$ which forms the tangent to the static $R(a)$-curve, then the rate of energy release will always be sufficient to account for the material resistance: the crack will grow continually. In this case the provided energy release rate $G_I(a, t)$ exceeds the *static* material resistance $R(\dot{a} \sim 0)$ and the rate of plastic deformation at the crack tip and consequently R will be increased until a new equilibrium is found:

$$R(\dot{a} \neq 0) = G_c(\dot{a}) = G_I(t). \tag{9.21}$$

The term G_c is the *(critical) energy release rate for slow crack growth*. The larger G, the rate of release of provided elastic energy, the larger $\dot{a}$, the rate of crack propagation. This behavior is especially well demonstrated by PMMA. In Figure 9.5 the data of Döll [30] are represented. It is also noted that the crack speed shows an instability at a critical value G_{Ic}, termed *critical energy release rate for unstable fracture* (in this book the term G_{cc} will also be used [8g]).

Thus, in most polymeric materials three G zones of crack propagation behavior can be distinguished:

- no crack growth at $G_I < G_i$ (stable crack)
- slow crack growth at $G_i < G_I < G_{cc}$ (subcritical crack growth)
- rapid crack propagation at $G_I \geqslant G_{cc}$.

The latter two regimes will be discussed in the following sections.

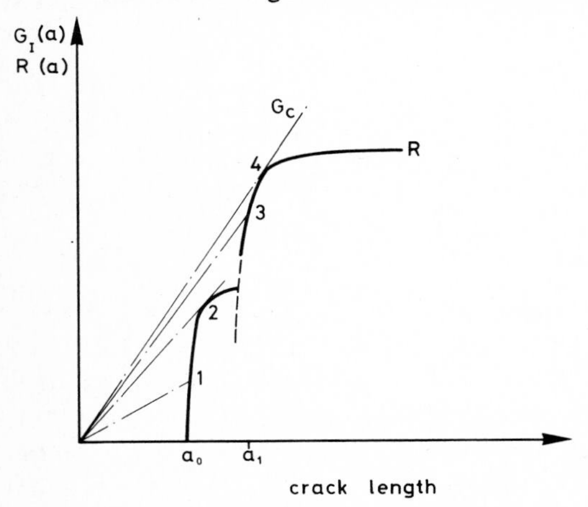

Fig. 9.4. Schematic representation of crack stability in ductile, strain-hardening material: 1. $G < R(a_0)$, 2. $G(a_0) = R(a_0)$, 3. $G(a_1) < G_c$, 4. $G = G_c$

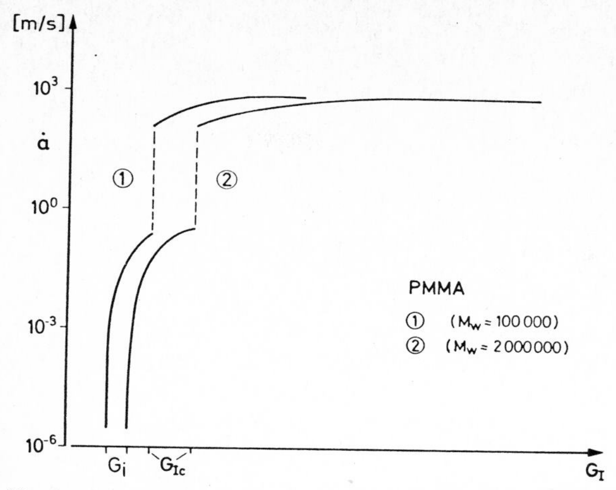

Fig. 9.5. Crack speed as a function of energy release rate G_I (after [30])

D. Subcritical, Stable Crack Growth

The instantaneous loading of a sample to any value of K_I above the K_i threshold provokes the following steps:

- elastic loading of the crack tip region (in impact in the form of stress waves),
- viscoelastic, time-dependent response of the material,
- structural reorganization (including plastic deformation) in the vicinity of the crack tip within the so-called "fracture process zone" which in an inorganic crystal may ideally be as small as an interbond distance (0.3–0.6 nm); in brittle polymers (PMMA, epoxy resins) it is of the order of some μm, in plasticized or ductile polymers it can be of the order of the specimen dimensions [229, 240, 245].

Subcritical crack growth occurs through material weakening within the fracture process zone and by displacement of the latter. The fact that constantly new material will be affected by the process zone, material, which will resist for a certain time, explains the relative stability of the phase of subcritical crack growth. Evidently, the rate of crack growth is governed by the laws of material deformation and breakdown. As an important point it must be considered that the macroscopically determined laws may not be valid at molecular levels. On an atomic scale these laws have been studied preferentially for non-polymers such as silica glass [246] or metals [247]. At such a level the *continuum description* breaks down (parabolic crack-tip contours of atomic dimensions are physically meaningless). Such studies must be done, therefore, by *computer simulation* [247]. Using a lattice model with a discrete "cut-off distance" for the cohesive forces Esterling [248] discusses equilibrium and kinetic aspects of brittle fracture. He concludes that at this level the condition $\delta G = 2\gamma\delta a$ is no longer valid. As to be expected crack growth (or closure) will be governed by the "kinetic barrier", i.e. by the energy it takes to bring the next atom pair to the cut-off distance.

In view of the much larger stress transfer lengths of chain molecules fracture-process zones in polymer materials will be more extended. In many cases they are amenable to direct observation by transmission electron microscopy [249] or to indirect observation through traces left on the fracture surface, see Section 9 II and III). In principle the failure mechanisms at a crack tip are the same as in the bulk under comparable conditions of state of stress, loading rate etc.: flow and failure of a ligament, formation of voids or secondary cracks and their coalescence, craze formation and rupture, chain orientation, scission and slippage. All these mechanisms have been observed and all of them are sttess-activated.

The subcritical growth data obtained by several investigators [12–15] for PMMA have been compiled in Figure 9.6. The double-logarithmic plot reveals an extended linear section (growth rates $\dot{a}$ from 10^{-8} to 10^{-2} m/s for K-values[1] from 0.8 to 1.4 MN $m^{-3/2}$). Several relations between K_I and $\dot{a}$ have been established. One of these has been found to describe the subcritical crack growth of many polymers [3, 12, 13]:

$$\dot{a} = AK_I^b \tag{9.22}$$

with A and b being empirical constants, depending on material, environment, temperature, etc. The constant b has been related to $1/\tan\delta$ of the β relaxation [12].

For the linear portion the exponent b in Eq. (9.22) can be determined to be 25 ± 2 [13].

From Eqs. (9.22) and (9.4) one can determine the time Δt necessary for the growth of a crack a_1 to length a_2. Neglecting here and in the following $f(a/D)$ one obtains for σ_0 = constant:

$$\Delta t = \int_{a_1}^{a_2} \frac{da}{\dot{a}} = \int_{K_{I1}}^{K_{I2}} \frac{2\,K_I dK_I}{\sigma_0^2 \pi A K_I^b} = \frac{2(K_{I1}^{2-b} - K_{I2}^{2-b})}{(b-2)A\pi\sigma_0^2}. \tag{9.23}$$

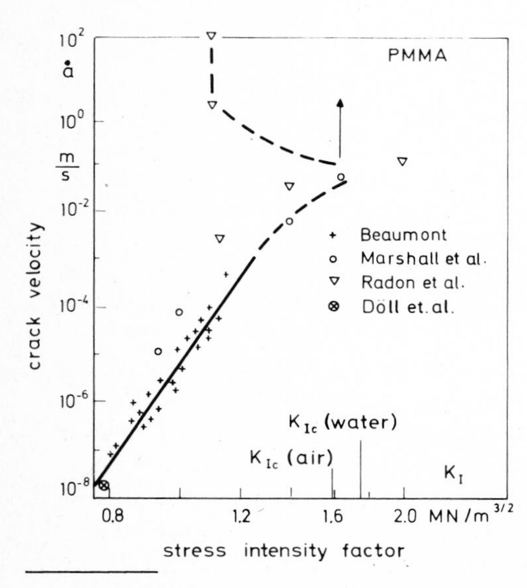

Fig. 9.6. Slow crack propagation in PMMA at room temperature (after [11], including data from ref. [12–15])

1 In slow growth experiments any imposed $K_I > K_i$ evidently constitutes a K_c.

If one chooses as upper limit K_{cc} [8g] and as lower limit K_{Im}, i.e. that value of K which corresponds to the largest flaw or defect present in the sample, then one obtains as Δt the lifetime t_b of the sample under constant stress. The extremely strong exponential dependence of Δt on K_{Im} makes t_b very sensitive to flaw size. Increasing the size of the largest flaw, a_m, by only 10% decreases the lifetime t_b by a factor of 3.

The above considerations on the dependence of t_b on the size of inherent flaws have been extended by various authors [13, 16–17] to a method of predicting the creep life of glassy polymers. In a first approach it was assumed that the size a_m of the largest intrinsic flaw can be determined from the breaking stress σ_b in short time loading according to Eq. (9.4):

$$a_m = K_{cc}^2/\sigma_b^2\pi. \qquad (9.24)$$

Broutman and McGarry [17] thus obtained intrinsic flaw sizes of 50 μm for PMMA and of 620 μm for PS. Berry [40] reports values of between 60 and 225 μm for PMMA. These values seem to be very high since no direct optical observations of such flaws have been reported. They are also two to three orders of magnitude larger than the diameters of microcracks found by Zhurkov et al. in PMMA and listed in Table 8.3.

It must be concluded, therefore, that the flaw sizes inferred from a tensile experiment according to Eq. (9.24) are by no means inherent but the result of stable crack growth during the experiment.

Starting from this consideration Döll and Könczöl [250] carefully studied the fracture surfaces of a large number of long-time tested PMMA dumbbell specimens. The fracture surfaces (Fig. 9.7) offered the usual pattern: a semi-circular mirror zone (slow crack growth) and a subsequent region covered with parabolae (rapid crack propagation). In the center of the mirror regions they frequently found semi-circular markings having radii of

79 ± 31 μm (23 °C) and
108 ± 55 μm (50 °C)

which could very well correspond to former craze zones (Fig. 9.8). The development of an existing defect into an unstable crack in PMMA would then occur in the following manner:

> a preexisting defect (a few μm or even less develops within Δt_1 into a craze zone (fairly rapid development up to the temperature dependent size of 70 to 160 μm) followed by a period Δt_2, the growth of the mirror zone (slow growth up to several 100 μm, depending on stress); the final breakdown occurs by rapid crack propagation (Δt_3).

If the time Δt_2 is taken to account essentially for the total time-to-failure t_b of a loaded specimen then the length a_1 of the craze zone can be *calculated* from Eqs. 9.23 and 9.4 for each pair of values σ_0 and t_b (Fig. 9.9). In doing so Döll and Könczöl determined a_1 to be 70 to 100 μm (23 °C), 100 to 140 μm (50 °C) and 114 to 160 μm (60 °C). These values are obviously in excellent agreement with those found for the markings on the fracture surfaces; this coincidence strongly supports the proposed fracture mechanics concept of the rupture of PMMA and the feasibility of predicting long time strength from comparable short time experiments.

The higher the crack propagation rate in a sample gets ($\dot{a} > 10^{-3}$ m/s) the more the relation between $\dot{a}$ and K_I – and even the measured value of K_{Ic} – depend on the parameters of the experiment. As shown in Figure 9.6 K_c increases monotonically up to a K_{Ic} of 1.6 MN/m$^{-3/2}$ if σ_0 is kept constant [12, 13]. In other experiments (Charpy impact tests, crack initiation and arrest in DCB specimens) K-values at higher crack speeds (2 to 300 m/s) are obtained which lie below the K_{cc} [14, 18].

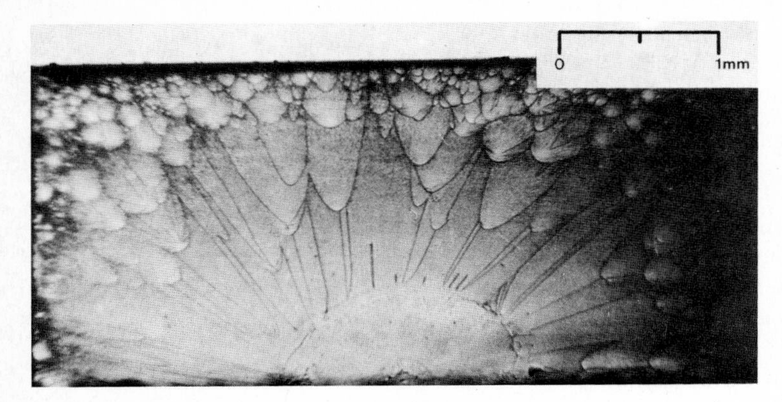

Fig. 9.7. PMMA fracture surface showing a semi-circular mirror (slow growth) and a rapid crack propagation zone (parabolic pattern); from Döll and Könczöl [250]

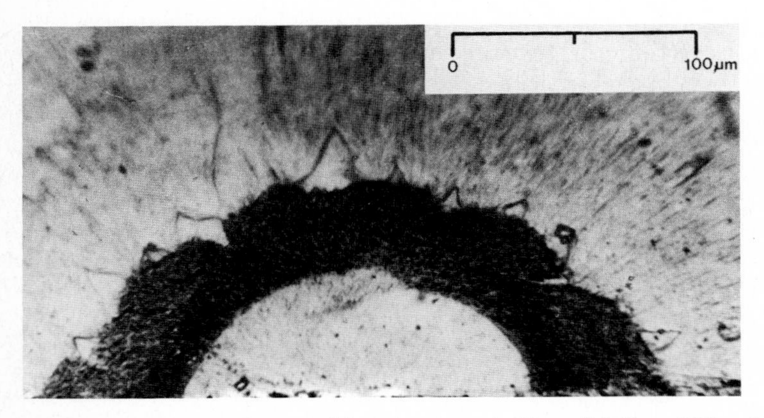

Fig. 9.8. Enlargement of a part of the mirror zone in Figure 9.9 (from Döll and Könczöl [250])

The different points on the $\dot{a} - K_I$ curve in Figure 9.6 represent quite different energy release rates G_I. According to Eqs. (9.4) and (9.10) G_I increases with the square of K_I. For the material employed one may use an average elastic modulus E of 4.5 GN/m^2 and a ν of 0.36 thus obtaining G_I values of 70 J m^{-2} at the lower range of K_c (0.6 MN m$^{-3/2}$) and 500 J m^{-2} for a K_{Ic} of 1.6 MN m$^{-3/2}$. These G values are equal to the corresponding material resistance R under the given experimental conditions. Their significance and interpretation will be discussed in the following Section.

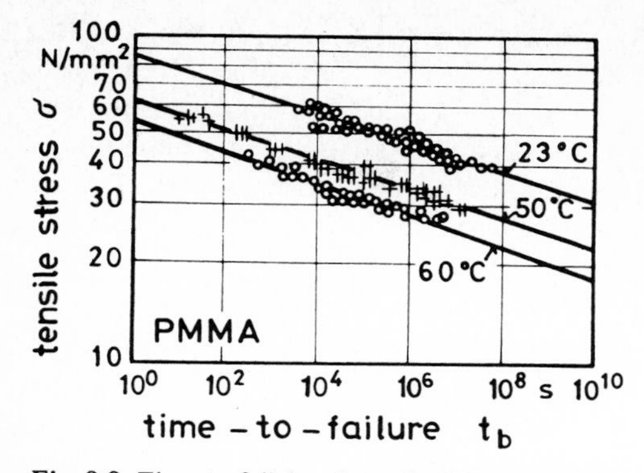

Fig. 9.9. Time to failure of tensile PMMA specimens as a function of applied stress σ. The data points were obtained experimentally with unnotched tensile specimens by Döll and Könczöl [250] who had predicted (solid curves) the time to failure from Eq. 9.23 after having determined A and b from $v(K_I)$ studies with CT specimens

The slow crack growth behavior of polycarbonate, also a glassy polymer, is even more complex than that of PMMA. At low temperatures ($T < -40$ °C) K_c values between 2.6 and 3.4 MN $m^{-3/2}$ have been obtained [19] from DT specimen which were independent of crack speed at small crack speeds ($\dot{a} < 10^{-3}$m/s) but dependent on sample thickness and temperature. At higher crack speeds ($\dot{a} \simeq 10^{-1}$ m/s) K_c values slowly increased. Double-side grooved DCB specimen (DG-DCB) however showed a decrease of K_c with increasing $\dot{a}$ in the region of $\dot{a} < 10^{-1}$ m/s [20]. This behavior was confirmed by Kambour et al. [21], who observed a G_c of 8.2 kJ/m² at $\dot{a} = 2.5 \cdot 10^{-5}$ m/s and of 12 J/m² at $\dot{a} = 300$ m/s. The microphotographical investigation of the slowly fractured sheets (B = 12.7 mm) revealed the – energy-consuming – formation of shear lips 0.4 mm wide (mixed mode crack propagation). At high speeds no shear lips were found. By copolymerisation of PC with silicone blocks the authors [21] were able to increase the fracture toughness in the whole temperature range $T > -110$ °C. With this material mixed mode crack propagation occurred fairly independently of crack velocity.

The increase of crosshead speed (over 7 decades) in static three point bending of PC resulted [22] only in a slight decrease of the calculated K_c from 3.8 to 3.0 MN $m^{-3/2}$. From the three point bending test a K_{cc} is derived [14] from:

$$K_I^2 = \frac{EP^2}{2\,B_N} \cdot \frac{\partial C}{\partial(a/D)} \tag{9.25}$$

if P is taken as the largest load measured, $\partial C/\partial a$ as the proper change of specimen compliance with growth of a (perpendicular) crack, and B_N as the thickness of the load carrying section. The quantity $\partial C/\partial a$ increases with increasing crack length. At the beginning of crack growth the values of P and $\partial C/\partial a$ (and the corresponding K_i) will be smaller than the K_{cc} at catastrophic failure. Arad et al. [22] point out that K_i and K_{cc} are the closer together the less material ductility and/or slow crack growth are favored by the experimental conditions.

In relation to studies of the fracture surface energy as a function of loading rate the early and wide application of the tear test (modes III and I) must be mentioned (e.g. [5, 24–28]). In this fracture mode the material in the crack tip zone is subjected to a complex and highly plastic deformation. Without going into details it may be mentioned that rate effects on the extent of plastic deformation (and thus on fracture surface or tearing energies) have been found [24–29] and correlated with β- and γ-relaxation maxima [5, 24–26]. The tearing energies of thermoplastics and rubbers are generally quite large: e.g. up to 1 kJ/m^2 for PS, 20–200 kJ/m^2 for PE, and 0.1–500 kJ/m^2 for various butadiene copolymers [24–26]. For highly oriented PE (draw ratio between 10 and 20) the G_c values are between 0.7 and about 3 kJ/m^2 [252]. For elastomers Thomas [27] and Ahagon and Gent [28] report that after adjustment for the changing effective fracture area a common *threshold fracture energy* T_0 of 40 to 80 J/m^2 could be determined under different experimental conditions. This energy was shown to be independent of temperature and degree of swelling for various types of swelling liquid. The threshold energy decreased slightly with increasing degree of crosslinking (of the polybutadiene samples). In aggressive environment (oxygen, ozone) T_0 is substantially reduced.

An experimental and theoretical analysis of the steady propagation of a crack in a viscoelastic continuum is due to Knauss [29]. Employing a polyurethane elastomer (solithane 113) as material Knauss studied crack propagation under pure shear. From his extensive analysis, Knauss derives a solution to the viscoelastic-boundary-value problem represented by a crack moving in an isotropic, homogeneous, incompressible solid. He finds that the stress singularity at the crack tip vanishes. Under these conditions the stress-intensity factor only describes the far-field loading conditions. Knauss observes that the rate-dependent fracture energy is essentially the *product* of an "intrinsic fracture energy", presumably of molecular origin, and of a non-dimensional function which relates to the rheology of the material surrounding the crack tip. For the polyurethane elastomer this intrinsic fracture energy amounted to 2.45 J/m^2. As criteria for crack stability are concerned, he found that the criterion of ultimate, constant strain (crack-opening displacement) is as applicable as a generalized energy criterion.

As stated in the introduction to this section the subcritical crack growth in a polymer is due to the thermomechanical activation of different molecular deformation processes such as chain slip and orientation or void opening. The energy dissipated depends on the frequency, nature, kinetics, and interaction of these processes. But there are many and notable attempts to treat the subcritical crack propagation as one thermally activated multi-step process characterized by one enthalpy or energy of activation and one activation volume. Several of these kinetic theories of fracture have been treated in Chapters 3 and 8.

The concept of thermal activation has been extended to the fracture mechanical analysis of crack propagation. Thus Pollet and Burns [179] have discussed earlier approaches and the significance of various variables. Commonly the energy release rate G_I is considered as the mechanical *intensive* parameter. In their treatment, the crack front is thus represented by a line on which a force per unit length, G_I, tends to move it in the forward direction; its motion is, however, restrained by the presence of "thermal" obstacles, i.e., obstacles, or barriers, which can be overcome by thermal activation. Whether the physical origins of those energy barriers are individual obstacles placed on a unique surface along the fracture path or whether they are located somewhere in the surrounding region, does not affect the generality of the treatment if G_I can be considered as an intensive variable.

With the assumption of thermally activated barriers, the average crack velocity can be written according to a general Arrhenius-type equation as:

$$\dot{a} = v_0 \exp\left[-\Delta G(G_I, T)/kT\right] \tag{9.26}$$

where the quantity v_0 can be thought of as the maximum attainable crack velocity and ΔG as the free enthalpy of activation to overcome one "obstacle".

As observed by various investigators [12–15, 180] an approximately linear relation holds between the logarithms of crack velocity and crack extension force:

$$\ell n \frac{\dot{a}}{v_0} = \frac{C(T)}{kT} \ell n \frac{G_I}{G_{Ic}} . \tag{9.27}$$

Pollet and Burns find the data for PMMA taken by Atkins et al. [180] at various temperatures to comply with Eq. (9.27). They determine from C(T) an area of activation $A^* = C(T)/G_I$ of the order of $0.7 \cdot 10^{-4}$ to $2 \cdot 10^{-4}$ nm^2. They interpret this small value as indicating that the thermal obstacles are not lined up on a single surface along the fracture path but distributed in space in a region of material near the apparent fracture path.

The question of the uniqueness of the $K_1 - \dot{a}$ curves has been raised repeatedly and for different reasons. Leevers et al. and Stalder et al. [253] point out that no unique speed can be assigned to the propagation of all the elements of a *curved* crack front (in double torsion specimens). They show that the crack shape does not vary with crack speed whereas the specimen geometry has a strong influence on it. Pavan [254] has recently presented a unifying approach, namely the two-dimensional displacement of the crack contour. Dally et al. [255] invoke the three-dimensional state of stress and individual errors.

E. Critical Energy Release Rates

It has been discussed in the previous Section that an increase in stress intensity factor or in G_I, the driving force for crack extension, promotes subcritical crack growth (Figs. 9.5 and 9.6). If G_I increases then R will have to increase as well in order to establish a new equilibrium. Since R depends most strongly on the plastic energy term $\partial V_{pl}/B\, \partial a$ an increase in R can be accomplished in two ways:

– through increase of the volume element which deforms plastically during crack propagation or
– through increase of the rate of plastic deformation.

As pointed out in the last Section it is generally the latter phenomenon, the rate-sensitivity of plastic deformation and of craze formation which assures that an increase in G_I will initially be accommodated. However, with increasing rate of crack propagation the time Δt available for the plastic deformation decreases. Taking a typical plastic zone size of 0.1 mm and a modest crack speed of 1 cm/s one arrives at a time interval of $\Delta t = s/a = 10^{-2}$ s. This means that thermal equilibrium will most probably not be attained. At higher crack speeds the crack tip region will even behave *adiabatically*. In that case adiabatic cooling due to the elastic expansion of the material in front of the crack tip (of the order of 1 to 10 K [256]) is being followed by adiabatic heating due to the plastic deformation. The ensuing temperature rise is of the order of several tens or even hundreds of K. Such a temperature rise rapidly counterbalances the rate-dependent increase in R; with decreasing material resistance, however, the crack accelerates further: the crack becomes unstable. The value of G_I where this transition from stable to unstable crack growth occurs is termed *critical energy release rate* G_{cc} [8g]. This transition is especially pronounced in glassy thermoplastics; in PMMA the crack speed changes from ~ 0.1 to more than 100 m/s (cf. Fig. 9.5).

Thermal effects associated with crack propagation have been studied especially by Williams et al. [12, 64, 229, 257] and by Döll et al. [30, 50, 61, 66, 67, 197, 256]. Weichert and Schönert have calculated the temperature distribution around a moving circular heat source [258]. Using their model Döll [197] estimated temperature rises in PMMA of 30 to 80 K at crack velocities between 0.06 and 1 m/s and of 230 K at the instability [30].

One important consequence of the adiabatic temperature rise is the local material softening (or even degradation) and the strong decrease of material resistance leading to crack instability. As another consequence, the modification of the stress field through the development of thermal stresses has been mentioned in the literature [259–261].

As indicated in Figure 9.5 an unstable crack accelerates rapidly to a level of velocities where inertia forces and the finite speed v_e of elastic waves have to be considered [67, 181–182]. So far the contribution of the kinetic energy of the receding fracture surfaces to R has been neglected. At the point of incipient crack instability in PMMA with crack velocities of about 0.1 m/s the kinetic energy term amounts to 6 J/m^3. At these velocities this term is, therefore, an insignificant fraction of the average strain energy density of 100 to 500 kJ/m^3. At velocities of between 100 and 1000 m s^{-1}, however, an increasingly larger fraction of the stored elastic energy $G_{I(static)}$ is necessary to account for the kinetic energy of the crack front – reaching 100% at a crack speed equal to the Rayleigh surface wave velocity v_R. The dynamic strain energy release rate G_d is, therefore, only a fraction F of $G_{I(static)}$:

$$G_d = \frac{\pi \sigma_0^2 a (1 - \nu^2)}{E} F(\dot{a}, v_e) \qquad (9.28)$$

Broberg [181] derived the function $F(\dot{a}, v_e)$ for an infinite, isotropic, elastic plate. His expression may be approximated by $(1 - \dot{a}/v_R)$.

Recently a special issue of the International Journal of Fracture has been devoted to dynamic fracture [262, see also 263–269].

Attention should also be drawn to the important phenomenon of crack bifurcation as a means of energy dissipation (see especially the recent work of Theocaris [269]).

Table 9.1. Molecular and Mechanical Data of Selected Polymers (after [11, 31] if not indicated otherwise), Average values for non-modified, unoriented polymers at room temperature

Quantity	Symbol	Unit	HDPE	PP	PA66	PS	PMMA	PVC	PC
Density	ρ	10^3 kg/m^3	0.96	0.91	1.14	1.06	1.19	1.39	1.20
Young's modulus	E	MN/m^2	1200	1300	1300	3100	3300	2600	2350
"Chemical surface energy"	U/qN_L	J/m^2	2.9	1.3	1.6	0.5	0.4	0.5	0.6
Surface tension	γ_{exp}	J/m^2	0.031	0.029	0.046	0.040	0.039	0.039	0.042
Monomer molecular weight	M_m	g/mol	28	42	226	104	100	62.5	254
Activation energy for bond scission	U_0	kJ/mol	335	272	188	230	175	(84)	117
Monomer length	l_m	nm	0.253	0.217	1.73	0.221	0.211	0.255	1.075
Monomer cross-section	$q = M_m/l_m \rho N_L$	nm$^2 \cdot 10^{-2}$	19.0	35.1	18.9	73.2	65.7	29.1	32.8
Solubility parameter	δ^2	MJ/m^3	268	354	(774)	332	502	427	402[c]
Undisturbed statistical chain end-to-end distance	$h_0\ 10^3 M^{-1/2}$	nm(mol/g)$^{1/2}$	108[a]	83[a]	89	67[a]	76[a]	100[a]	110[b]
Chain rigidity	$s = h_0/h_{0f}$		1.63	1.61	1.63	2.3	2.14	1.83	1.1

[a] E. W. Fischer, M. Dettenmaier: J. Non-Cryst. Solids *31*, 181 (1978)
[b] W. Gawrisch, M. G. Brereton, E. W. Fischer: Polymer Bulletin *4*, 687 (1981)
[c] R. P. Kambour, C. L. Ginnes, E. E. Romagosa: Macromolecules 7, 248 (1974)

Similar considerations apply to the slowdown and eventual arrest of a rapidly propagating crack [18, 182]. An understanding of this phenomenon is of considerable importance in the use of high pressure pipes where the propagation of a crack over long distances must be prevented under any circumstances.

The transition from subcritical to rapid crack propagation is especially pronounced in PMMA but more or less clearly observed in all polymers capable of brittle fracture. The corresponding critical quantities K_{Ic} and G_{Ic} characterize the fracture behavior of a material. In Table 9.2 an (incomplete) listing of measured critical values (G_{Ic} and K_{Ic}) and of energy release rates referring to particular experimental conditions (G_c, G_d, G_{III}) are given. The main experimental parameters and their ranges of variation have been indicated.

Of the various terms of Eq. (9.13) contributing to the material resistance the plastic deformation, $\partial V_{pl}/B\partial a$, is generally by far the largest. It has been attempted in Table 9.1 to identify also surface tension γ and the (hypothetical) energy density of chain breakage U/qN_L. The smallest term is that of the specific surface energy γ (free surface energy, surface tension). The γ values of polymers lie in the range of 0.020 to 0.046 J/m^2 [31]. The energy of elastic retraction is the product of strain energy density and width of the retracting layers. If the layers consist of fully oriented chains stressed to break and thus having the highest possible energy density (of the order of 1000 MJ/m^3) the energy of retraction of chain ends of length $L = 5$ nm amounts to 8 J/m^2.

The chemical surface energy, $\partial V_{ch}/B\partial a = U_0/qN_L$ also contributes comparatively modestly to R, even if one chain breakage event per molecular cross-section q is considered: 0.4 J/m^2 in PMMA, 2.9 J/m^2 in HDPE. The terms U_i and U_{ch} cannot be evaluated in all generality. The physico-chemical data used for the above and subsequent calculations are collected in Table 9.1 for seven semicrystalline or glassy polymers.

F. Fracture Criteria

As fracture criteria derived from LEFM principally have been proposed [229]:

- the critical stress intensity factor K_{cc} for unstable crack growth and the corresponding energy release rate G_{cc} [8g]
- the crack opening displacement (COD)

$$\delta_t = K_I^2/E\sigma_F \quad \text{(plane stress)} \quad (= K_I^2(1-\nu^2)/E\sigma_F \text{ in plane strain})$$

- the two-parameter criterion critical stress at critical distance [229, 232, 395]
- the J contour integral;

for materials with a pronounced tendency to anelastic deformation and large scale yielding the load-deflection curves will be non-linear. In such cases the J contour integral proposed by Rice [230] may advantageously replace the energy release rate G_c. J has the same physical significance, namely to describe the flow of energy into the crack tip region [229–234], although in tough materials at high stresses only a fraction of that energy may be used for the actual crack extension.

G. Fracture Mechanics Specimen and Selected Values of K_c, G_c, K_{Ic}, G_{Ic} (K_{cc}, G_{cc})

At present several different experimental methods of working in the crack opening mode are employed [7–151, 229]. Flat samples with a single edge notch (SEN) are preferentially used under constant load, under slowly increasing strain or under cyclic tensile loading (cf. Fig. 9.10). Under these conditions a subcritical crack will grow at the root of the notch leading eventually to instability. To determine the energy release rate Eq. (9.10) can be used if the appropriate long-term relaxation modulus E(T, t) is employed. The same is true for plates with a central notch (CN), compact tensile (CT) specimens, Charpy notched samples in slow bending, and wedge open loaded (WOL) plates at small loading rates.

The specimen types shown (Fig. 9.10) lend themselves very well to "instrumentation" by the Stalder-graphite gauge. The method has been described in Section 8 II A 1, an example of an instrumented specimen is given in Fig. 8.26. This way slow (*and* rapid) crack growth can be monitored in real time [235].

In *shock loading* the short-time moduli are relevant to describe the energy release rate of pre-cracked samples; crack instability is reached primarily because of the increasing stresses and at unchanged crack length. The contribution of thermally activated crack growth to an increase of K_I is, therefore, negligible (the situation is different, though, with uncracked samples and flaws of molecular dimensions as has been shown in Chapter 8).

The WOL and the double-cantilever beam (DCB) specimens are suitable to determine the arrest of an initially unstable crack; the double-torsion (DT) specimen permits the investigation of crack propagation under constant K_I [7–15, 23].

Values of various types of energy release rates G and of fracture toughness values K_c of polymeric materials have been gathered from the literature [12–75, 236–244] and are compiled in Table 9.2. The experimental method and the principal parameters and observations have also been indicated. However, the materials and parameters studied have become so numerous that a systematic collection of all published data had to be abandoned. Only selected values have been incorporated in Table 9.2 which are thought to be representative for a material and for the effect of a particular parameter (molecular weight, temperature, rate of loading etc.).

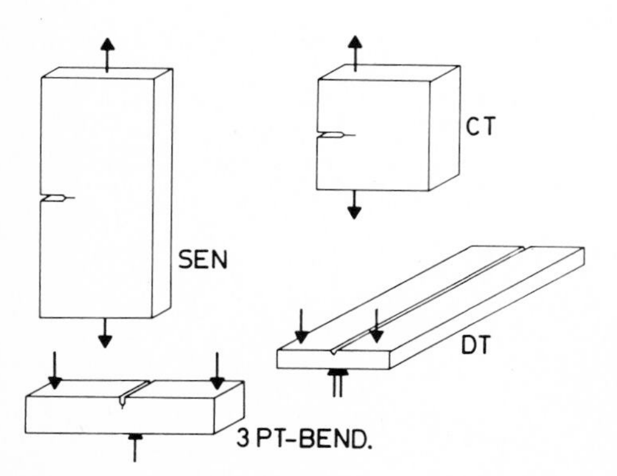

Fig. 9.10. Some of the most common samples currently used in fracture-mechanics experiments: single edge notch (SEN), compact tension (CT), three-point bending and double torsion (DT)

Table 9.2. Energy Release Rates G and Fracture Toughness Values K of Selected Polymeric Materials

* As indicated in the fifth column, not always the critical values G_{cc} or G_{Ic} and K_{cc} or K_{Ic} are listed.

Polymer (standard abbrev.)	Method and Specimen	G_{Ic}* J/m^2	K_{Ic}* $MN\ m^{-3/2}$	Parameters of Test, Observations	Ref.
Thermoplastic Materials					
PE					
(LDPE)	impact	5000		G_{c_1} plane strain value	[64]
		(35000)		G_{c_2} plane stress value	
(medium density)	impact	1300		G_{c_1}	[64]
		11900		G_{c_2}	
(medium density)	impact	14200		plastic zone correction applied	[69]
($\rho = 0.95\ g\ cm^{-3}$, pigmented)	SEN		1.7–4	linearly increasing with "time to crack initiation" (inverse of load rate)	[57]
($\rho = 0.96\ g\ cm^{-3}$)	mode III tearing	$2 \cdot 10^4$–$20 \cdot 10^4$		G_{III} shows maxima and minima which are related to the α- and β-relaxation maxima	[24]
(PE hard, black)	SEN, CT	3600–20000	3.0–6.5	K_c	[70]
	CT		0.24–0.54	K_Q at 23 °C, B = 50 mm crack opening rates 0.1–100 mm/min	[236]
PP					
(ICI, compression moulded)	SEN, SN		6.5–3.6	K_{c_2} decreases with T, ($-180 < T < 20$ °C), change in slope at −60 °C related to β-relaxation peak	[68]
			2.8	K_{c_1} essentially constant	
(BASF Novolen 1120 Lx)	CT		5.2–4.5	$v = 0.08\ mm\ s^{-1}$ -40 °C $< T <$ 20 °C	[183]
			3.3–3.75	$v = 8.6\ mm\ s^{-1}$ -40 °C $< T <$ 20 °C	
PB 1					
(non-modified)	SEN		2.9–1.2	K_c gradually decreasing with "time to crack initiation"	[57]
(pigmented)	SEN		1.2–3.2	distinct maximum of K_c at "times to crack initiation" around 0.1 to 10 s	[57]

Table 9.2. (Continued)

Polymer (standard abbrev.)	Method and Specimen	G_{Ic}* J/m^2	K_{Ic}* $MN\ m^{-3/2}$	Parameters of Test, Observations	Ref.
PS					
–	SEN	1700 ± 600			[34]
–	DCB	300–2500		increasing with decreasing $\dot{a}$	[17, 35–37]
(Shell Chemical)	DCB	200–7.5		rapidly decreasing with increasing degree of orientation	[45]
–	SEN		0.6–2.3	$10^{-7} < \dot{a} < 10^{-2}$ m/s	[51]
	tapered cleavage		3.7–7	$10^{-9} < \dot{a} < 10^{-2}$ m/s	[51]
(HIPS)	impact	350		G_{c_1}	[64]
		(900)		G_{c_2}	
	impact	1000		G_{c_1},	
		15000		G_{c_2}	
PVC					
(BP, England)	Charpy notched		2.3–6	$-200 < T < 50$ °C	[63]
	three point bending		2.2–5.6	T, rate of bending	
(Cobex)	DCB	3600		undrawn sheet	[46]
	DCB	850–50		decrease of G_c with increasing orientation ($1.2 \cdot 10^{-3} < \Delta n < 3.5 \cdot 10^{-3}$)	
–	impact	1230		G_{c_1}	[64]
		1440		G_{c_2}	[64]
ICI Darvic	DEN		1.94–2.57	K_c	[71]
(modified)	impact	10000		G_c	[64]
hard PVC	DEN, TPB		2.6–2.7	K_{Ic} at -43 °C	[238]
(type 1120)	CT, SEN		3.3–3.6		[238]
PMMA					
(ICI)	SEN, DCB, impact, three point bending		1–3.5	$-197 < T < 21$ °C; $2 \cdot 10^{-8} < \dot{a} < 0.08$ m/s; distinct peak of K_{Ic} related to side chain (γ) relaxation	[14]

Table 9.2. (Continued)

Polymer (standard abbrev.)	Method and Specimen	G_{Ic}* J/m^2	K_{Ic}* $MN\ m^{-3/2}$	Parameters of Test, Observations	Ref.
–	SEN	200			[33]
–	SEN		1.7		[52]
(Plexiglas, Röhm)	SEN		1.2–1.8	$0.1 \cdot 10^6 < \bar{M}_v < 8 \cdot 10^6$ K_{Ic} increasing with molecular weight	[15, 66]
(Plexiglas, Röhm)	SEN	1000–15000		G_d increasing with $\dot{a}$ and M $200 < \dot{a} < 700$ m/s $0.1 \cdot 10^6 < M_w < 8 \cdot 10^6$	[50, 67]
–	cleavage	120–600		various	[17, 21, 35–37, 45, 48]
		237		G_A	[48]
(Perspex)	(DCB)	100–35		decreasing with increasing orientation $4 \cdot 10^{-4} < \Delta n < 14 \cdot 10^{-4}$	[45]
(Plexiglas G)		0.75–100		log G_{Ic} increases linearly with log $\bar{M}_v$ $(2 \cdot 10^4 < \bar{M}_v < 10 \cdot 10^4$	[65]
		120 ± 20		G_{Ic} essentially independent of molecular weight $(\bar{M}_v > 10 \cdot 10^4)$	
(Perspex)	CT		1.3–0.75	decreasing with increasing T	[74]
(Perspex)	DT		1.6–1.7		[13]
(Simplex)			1.8–2.1		[13]
–			1.9–2.9	various $\dot{a}$, T	[12]
–	impact	1000 1300		G_{c_1} G_{c_2}	[64]
(Perspex)			1.7–2	notch tip radius	[58]
ABS					
	impact	12000–16250		G_c increasing with decreasing specimen width	[53]
–	impact	1000–50000		steep increase of G_c between –20 and +10 °C	[64]
–	impact	23000			[69]
(Lustran 244)	impact	47000–49000		G_c	[64]

Table 9.2. (Continued)

Polymer (standard abbrev.)	Method and Specimen	G_{Ic}* J/m^2	K_{Ic}* MN m$^{-3/2}$	Parameters of Test, Observations	Ref.
PETP					
(medium M_v)	Charpy notched	4300		G_c is calculated from impact resistance of differently notched specimens	[53]
(high M_v)		5200			[53]
(fiber grade)		7200		G_c values for amorphous and crystallized fiber grade PETP are the same	[53]
	SEN, DEN		6.1–5.3–3.8	K_{c_2}	
			5.1–4.6–3.7	K_{c_1} crystalline, 20 °C, –50 °C, –100 °C	[49]
PC					
(Makrolon)	three point bending	5000	3.8–2.5	decreasing with increasing rate of deflection ($10^{-6} < v < 10$ m/s)	[22]
(Makrolon)	three point bending		10–2.5	decreasing with increasing temperature ($-200 < T < -100$ °C), slight maximum at 70 °C	[54]
(Makrolon)	impact		4–2	decreasing with increasing T (T < 50 °C)	[54]
			2–15	increasing with increasing T (T > 50 °C)	
(Makrolon)	impact	3500		G_{c_1}	[64]
		5000		G_{c_2}	[64]
		4000		G_c calculated from impact resistance	[53]
(Lexan 101)		1500–2700	3.8–2.3	distinct maximum of G_c at –40 °C; decrease of K_{Ic} between –60 and +20 °C; correlation of G_c with tan δ and fracture surface morphology	[62]
–	cleavage	8000–170		G_{Ic} decreases rapidly with $\dot{a}$ ($0.25 \cdot 10^{-4} < \dot{a} < 10^{-2}$ m/s)	[21, 20]
	(DCB)	170–12		G_{Ic} decreases gradually with $\dot{a}$ ($10^{-2} < \dot{a} < 300$ m/s)	[21]

Table 9.2. (Continued)

Polymer (standard abbrev.)	Method and Specimen	G_{Ic}* J/m^2	K_{Ic}* MN m$^{-3/2}$		Parameters of Test, Observations	Ref.
(Makrolon)	DT		2.6–3.8		various temperatures and crack speeds	[19]
(Makrolon)	SEN		2.2	K_{c_1}	(corresponds to K_c of crack initiation)	[19]
			8–5	K_{c_2}	decreases with increasing T ($-140 < T < -30$ °C)	
Polyamide 66						
(Maranyl)	three point bending	8800	4.4–2.3		gradually decreasing with higher rates of deflection	[22]
(Maranyl, 0.2% H_2O)	SEN		4–7	K_c	increasing with T ($-160 < T < +40$ °C)	[68]
	SN		4	K_c	constant in same temperature region,	
			3–4	K_{c_1}	little variation with T	
			8–10	K_{c_2}	maxima corresponding to γ- and β-relaxation peaks are apparent at -140 and -80 °C	
(Nylon 66, dry)	impact	250		G_{c_1}		[64]
		4150		G_{c_2}		
(Nylon)	impact	5000–5300		G_c		[64]
Thermosetting Resins						
PUR						
(unsaturated vinyl-urethane)	impact	1300		G_c		[53]
UP						
(non-modified)	SEN	440	1.2	G_c		[56]
	cleavage	42 ± 6				[59]
EP						
(non-modified)	DT, CT		0.55–1.7		$-48 < T < +77$ °C, various rates of deformation	[75]
(non-modified)	SEN	700	1.54	G_c		[56]
	cleavage	80 ± 5				[59]
–	DT		0.6–2.2	K_c	at 20 °C, 0 to 50 Vol.-% alumina	[239]

Table 9.2. (Continued)

Polymer (standard abbrev.)	Method and Specimen	G_{Ic}* J/m^2	K_{Ic}* MN m$^{-3/2}$	Parameters of Test, Observations	Ref.
(non-modified)	DCB-tapered	50–200		adhesive layer between aluminium substrates	[73]
	DT	210 ± 48			[41]
(rubber-modified)	cleavage	310 ± 10			[59]
	CT		1.5–4	K_{Ic} for T increasing from −80 to +60 °C	[240]
Phenolics					
(phenol-formaldeyhde)	SEN, CN	0.42–50		G_{Ic} calculated from rupture stresses G_{Ic} decreases with T ($20 < T < 205$ °C)	[55]
Composite Materials					
Bakelite SR + 40 Vol. % aligned high modulus carbon fiber (Morganite I)	Charpy	20000–36000		G_c values depend on fiber surface treatment (HNO_3, silane, brominated, silicone oil treatments resulting in interlaminar shear strengths τ between 12 and 30 MN/m^2	[47]
		8800		morganite treatment with τ = 55 MN/m^2	
Epoxy + aligned RAE carbon fibers	SEN	5000–50000	13.5–19	$v_f = 0.4$ } G_c and K_c for various crack geometries and rates of deformation	[56]
		10000–15000	20–28	$v_f = 0.6$ }	
UP + randomly oriented E glass fibers	SEN	8000–19000		$v_f = 0.15$ }	
Epoxy + boron fibers	SEN	35000	47	$v_f = 0.66$	
Plate Glass	SEN	8.3	0.8		[6, 49, 60]

Table 9.2. (Continued)

Polymer (standard abbrev.)	Method and Specimen	G_{Ic}* J/m^2	K_{Ic}* MN m$^{-3/2}$		Parameters of Test, Observations	Ref.
Elastomers						
BR						
(mixed cis-1,4,trans-1,4, and 1,2)	mode III tearing	18–1000		G_{III}	increasing with effective rate of tear $a_T\dot{a}$ ($10^{-22} < a_T\dot{a} < 10^{-14}$ m/s)	[28]
		40–78			threshold fracture energy T_0 increasing with M_c ($2400 < M_c < 50000$); $\lambda_s{}^2 T_0$ independent of test temperature, degree of swelling, type of swelling liquid	
(cis-1,4)	mode III tearing	30–1000		G_{III}	for various load rates, degrees of crosslinking, and swelling	[28]
		58 ± 8			threshold fracture energy for M_c = 3200	
		81 ± 8			threshold fracture energy for M_c = 14000	
(Polysar)	mode III tearing	1000–3000		G_{III}	(for trousers and angled test pieces alike) increases with rate of crack growth ($10^{-7} < \dot{a} < 10^{-4}$ m/s)	[26]
(92% cis)	DT	8000 ± 2000		G_{Ic}	at −180 °C for M_c between 3000 and 50000	[241]
		~ 10000–2000		G_{Ic}	at −180 °C for crosshead-speeds between 0.05 and 100 mm/min	[241]
(butadiene-styrene and butadiene-acrylonitrile copolymers)	mode III tearing	100–100000		G_{III}	increasing with rate of tear ($10^{-7} < \dot{a} < 0.1$ m/s), decreasing with temperature after passing through a maximum	[25]
(polyurethane-solithane 113)	pure shear	2.45			intrinsic fracture energy derived from visco-elastic analysis of crack propagation	[29]

Table 9.2. (Continued)

Polymer (standard abbrev.)	Method and Specimen	G_{Ic}* J/m^2	K_{Ic}* $MN\ m^{-3/2}$	Parameters of Test, Observations	Ref.
Other "polymeric materials"					
Apples (Granny Smith)	compression	5.4		G_{Ic} calculated values from energy of bruising	[242]
		6.2		in quasi-static and impact loading	
	tension	200			[242]
Potatoes (Sebago)	compression	208		cleavage fracture	[242]
		770		slip	[242]
Pine Wood (Pinus radiata)	SEN		0.6	K_{Ic} crack plane perpendicular to radius	[243]
			0.3	circumference	
			2.5	axis	
For comparison:					
Ice	plane strain	0.7–0.9		G_c at $T \leqslant -5\ °C$	[244]
		0.7–0.2		energy of adhesive fracture which prevails at $-1 > T > -5\ °C$	

* As indicated in the fifth column, not always critical values G_{Ic} or K_{Ic} are listed.

II. Crazing

A. Phenomenology

The phenomenon of crazing has been observed in many glassy polymers, and also in some crystalline polymers, when subjected to tensile stress. Conventional polymer crazes (Figs. 9.11, 9.12a, b) in their appearance are similar to very fine cracks known for a long time to occur on the surfaces of inorganic materials such as ceramics. However, there is a difference between the crazes and the cracks in that crazes have a continuity of material across the craze plane (Figs. 9.13–9.17) whereas cracks do not possess any continuity. Consequently crazed zones are capable of bearing loads as opposed to cracked ones. In the last thirty years a growing number of publications has been devoted to the phenomenon of crazing. Two comprehensive reviews [76–77] on this subject have appeared in 1973.

The bibliography on crazing compiled in the first edition of this monograph [78–178] was mostly to be understood as an extension of these earlier reviews. In 1983 an entire volume of the "Advances in Polymer Science" was devoted to "Crazing in Polymers" [270]. That volume must be considered as the most up-to-date and explicit reference on crazing. Thus, the reader will frequently be referred to it for more detailed information. Nevertheless, the section on crazing in this monograph will be maintained as an introduction and in view of the importance of crazing phenomena in crack propagation, stress whitening and fracture of (glassy) polymers.

The subject of crazing has been extensively researched for various reasons:

– in brittle fracture of a crazeable material crazes generally precede cracks so that subcritical crack growth is essentially controlled by the formation, propagation and breakdown of crazes (see previous sections and refs. [270–338]);

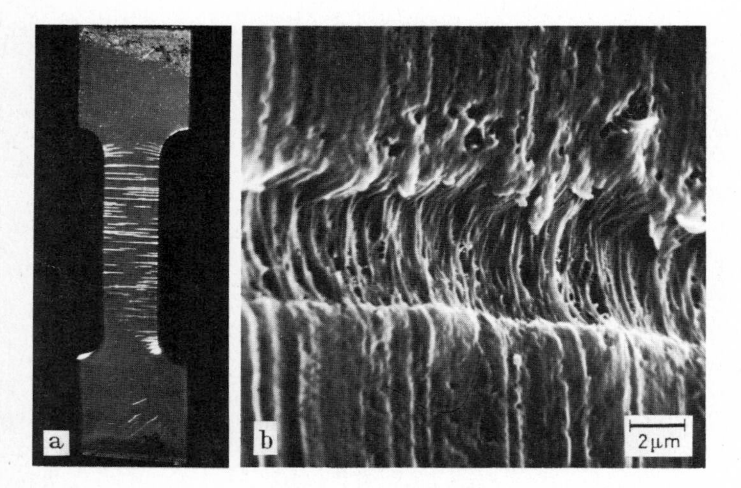

Fig. 9.11. Crazes in semicrystalline polymers.
a Conventional crazes in poly(trifluoro chlorethylene) under constant load (from [130]), **b** electron micrograph of one of these crazes showing the fibrillar craze material which connects rather straight boundaries (from [130])

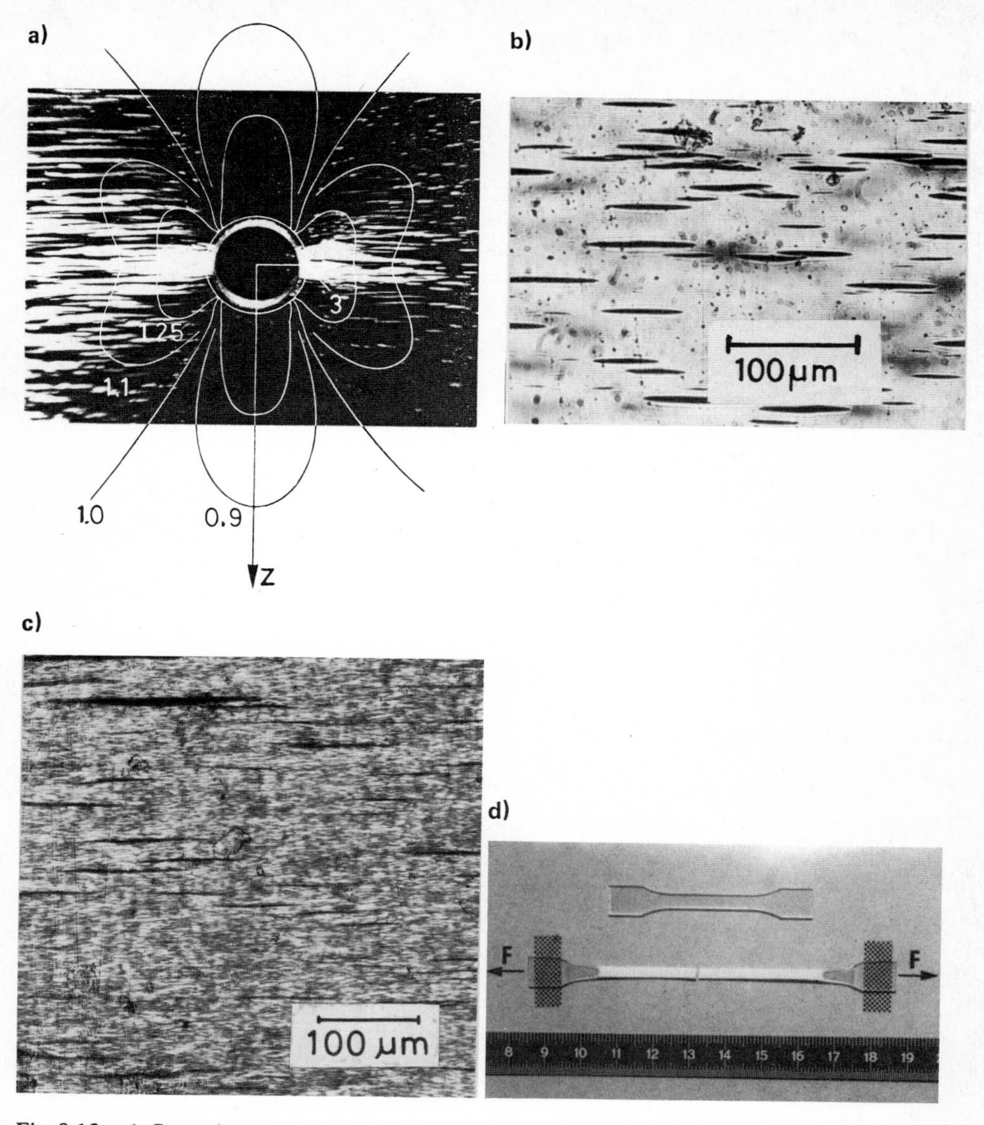

Fig. 9.12a–d. Crazes in amorphous polymers.
a conventional crazes in amorphous PMMA created through stressing in z-direction at 60 °C; overlaid on the micrograph are the major principal stress contours taken from Sternstein et al. [86]; it is well seen that in this case crazes are only formed at a principal stress $\sigma_3 > 1.0\ \sigma_0$, **b** conventional crazes in (oriented) PC (from [271]), **c** conventional, extrinsic crazes and crazes II (intrinsic crazes) in a preoriented PC specimen strained at 110 °C to an additional strain of 57% (from [271]), **d** PC, stress whitened

– the large deformations occurring *within* a craze are a powerful source of energy dissipation and thus a potential toughening mechanism (already well exploited in rubber-modified, "otherwise brittle" polymers [339–349]);

- the formation of crazes under stress, and/or in presence of an active environment, effects (and generally impairs) mechanical and optical surface properties and permeability [350–384];
- crazes trace and make visible the statistical and geometrical distribution of the variables of their initiation (states of strain or stress, molecular properties, structural defects).

Since the publication on *intrinsic crazing* by Dettenmaier and Kausch [271] in 1980 it is known that crazes exist in two different forms:

- conventional or extrinsic crazes (crazes I) which are preferentially formed at sample surfaces and under the influence of defects or stress concentrators (scratches, dust particles, finger prints, frozen-in tensions); therefore, in non-modified polymers, crazes I appear individually; their characteristic morphology (Figs. 9.12 to 9.15) is well described in recent publications [270–295];
- intrinsic crazes (crazes II which are formed after notable stretching in the whole sample volume and as a consequence of the breakdown of the entanglement network; intrinsically crazed samples (of PC, PMMA, PS) appear thoroughly whitened; they have *stress whitened* (Figs. 9.12d); as shown by Dettenmaier [272] the morphology of intrinsic crazes is characterized by the large thickness (100 to 200 nm) and regularity of their fibrils and by the high volume content of craze matter (~ 75%).

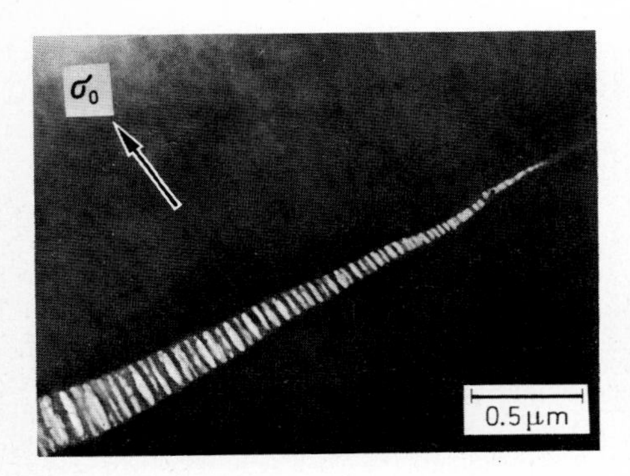

Fig. 9.13a. Newly formed craze in thin slice cut from uncrazed bulk polystyrene; craze growing from left to right in a direction perpendicular to that of the uniaxial tensile stress (Courtesy D. Hull [106])

Fig. 9.13b. Electron micrograph of the central section of a craze grown as that in a (Courtesy D. Hull [106])

It remains to be discussed whether the crazes formed at modifier particles (and thus in the whole volume of) rubber-toughened glassy polymers are of the first or of the second kind.

Conventional or extrinsic crazes (crazes I) occur most commonly in amorphous, glassy polymers such as PS, SAN, PMMA, PVC, PSU, PPO, and PC but also in semi-crystalline ones (PE, PP, PETP, and POM). Crazes have also been claimed for epoxy [131] and phenoxy resins [109]. With reference to the cited literature (e.g. [76–78, 85, 270]) it can be said that conventional crazes are sharply bounded regions filled with "craze matter". The craze matter consists of oriented strands interspersed with voids and has an average density of 40 to 60% of the matrix density (Figs. 9.13–14). Kambour [76, 85] has described the structure of crazes to be similar to an opencelled foam in which the average diameter of the holes and polymer elements is about

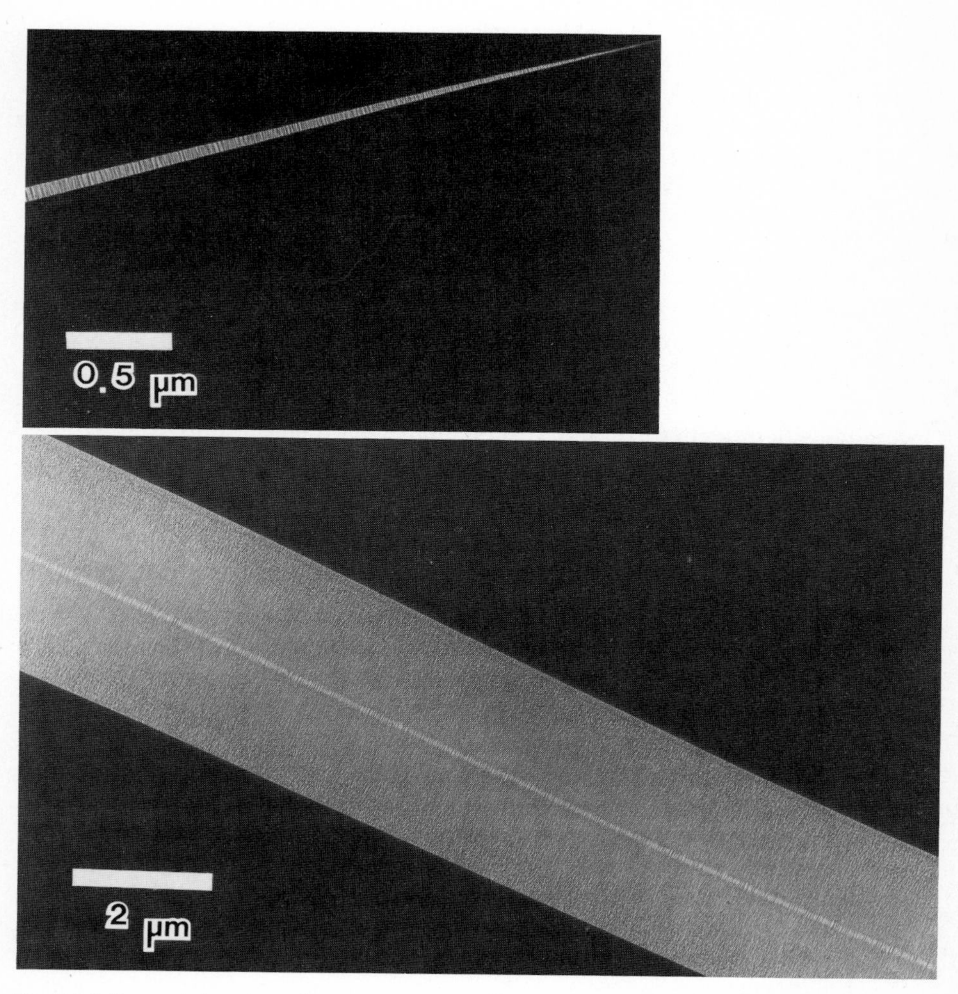

Fig. 9.14. Transmission electron micrographs of a craze in thin solvent cast PS film (Courtesy L. Berger and E. J. Kramer, Ithaca)

20 nm. There is much evidence that the polymer elements are oriented primarily along the direction of the principal tensile stress. Generally a fairly uniform fibril structure exists over the whole length of a tapered craze (Fig. 9.13–14). From the uniformity of this structure, one concludes that *the increase in the width of the craze with sample extension is mostly due to fibrillation of further matrix material rather than to an increase in fibril extension*.

The mechanical response of fibrils in the direction of the applied stress has been determined for polycarbonate [83] and polystyrene [120]. Figure 9.15 shows a stress-strain curve of craze material obtained by Kambour and Kopp [83] for PC. It may be noted that the craze material can sustain tensile stresses only slightly less than the yield stress σ_F of the bulk material. The strains, however, are enormously greater in the craze – between 40 and 140% as compared to the bulk yield strain of about 2%. It follows from this behavior that the energy of deformation of a thoroughly crazed sample can attain very large values.

In recent years an enormous amount of information concerning the morphology of crazes and fibrils has been obtained [270–293]. An especially powerful source has been the systematic mass thickness contrast analysis of thin films by transmission electron microscopy (TEM) used by Kramer, Lauterwasser, Donald, Brown et al. at Cornell University in Ithaca [275–284]. A comprehensive account of their work is given by Kramer in the cited volume on crazing [249]. Michler in Halle (Institute of Solid State Physics and Electron Microscopy of the Academy of Sciences) has intensively studied thicker films (up to and exceeding 5 μm) by high voltage electron microscopy (HVEM) with an accelerating voltage of 1000 kV [285–287]. In view of this considerable progress the original compilation of observations on craze morphology, given in Table 9.3, would have to be substantially enlarged. However, in order to limit duplication with Kramer's and Michler's articles, only their principle conclusions with respect to midrib formation, fibril elongation and stress distribution

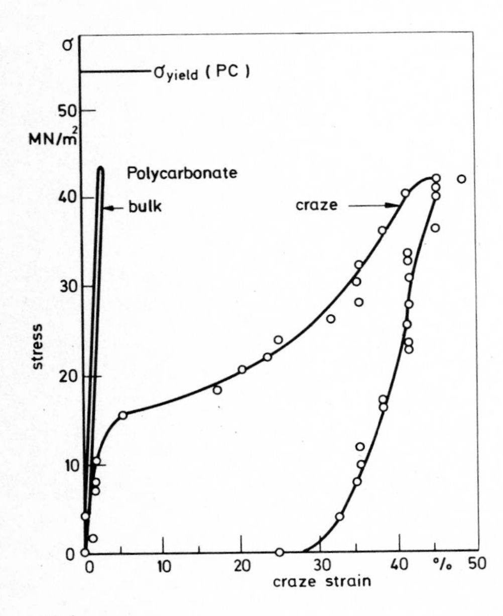

Fig. 9.15. Stress-strain curve of craze material in polycarbonate (after [83])

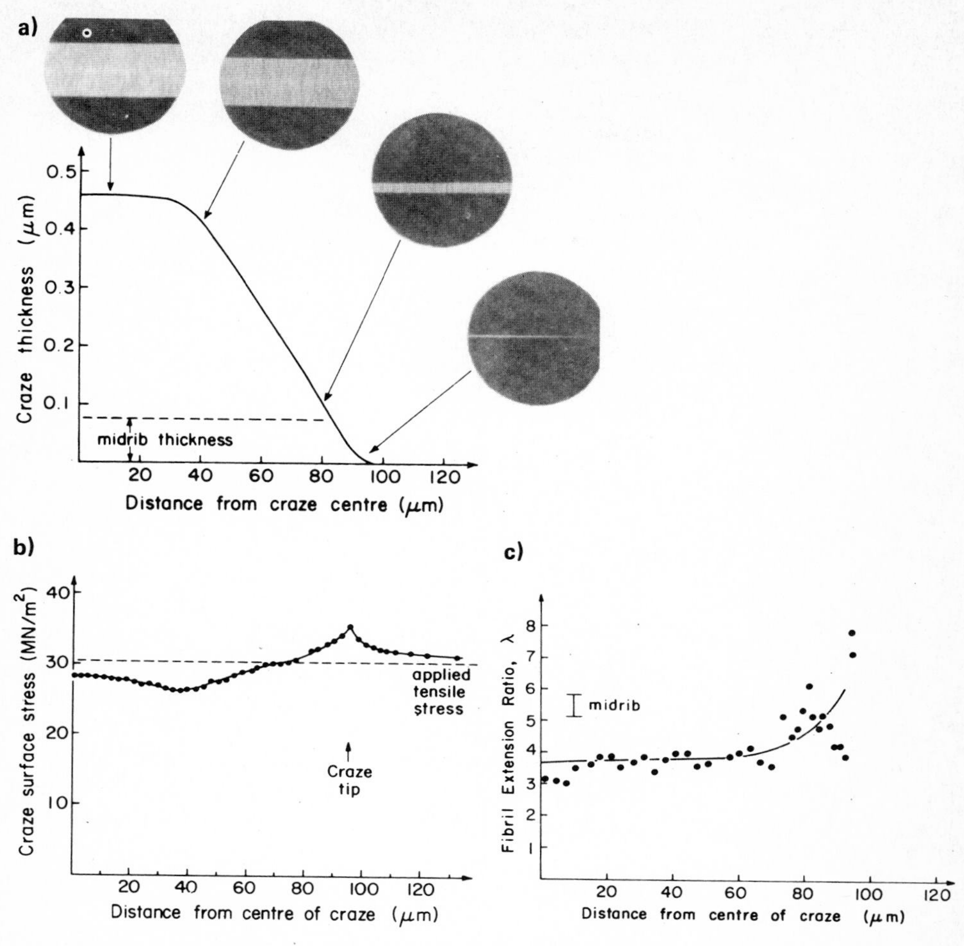

Fig. 9.16a–c. Analysis of craze morphology by the electron beam imaging technique (from Lauterwasser and Kramer [276]).
a Craze thickness as a function of distance from the center of a craze in a thin PS film, **b** Stress profile deduced from **a**, **c** Craze fibril extension ratio profile deduced from **a**

(see Figs. 9.16 and 9.17) have been incorporated into Table 9.3. Another extremely fruitful source of information has been the light optical interference method which permits the study of craze geometry and deformation in bulk specimens (Fig. 9.18). This method has been greatly furthered by Kambour at General Electric and Döll, Könczöl et al. at the Fraunhofer Institut in Freiburg/Br. A competent review has appeared in the above mentioned volume [273]. Again, in order to avoid duplication, only some principal observations have been included in Table 9.3. A similarly restricted reference is given to crazing and shear banding in semi-crystalline polymers in view of the extensive review article by Friedrich [274]. Valuable information on craze microstructure has also been obtained by small angle X-ray scattering [272, 288–291] and small angle electron scattering [276, 292, 293].

Table 9.3. Morphology of Crazes in Glassy Polymers

Observations	Conclusions	Material and References
Optical and mechanical study of the effect of various variables, such as magnitude and duration of stress on initiation and development of crazing.	Crazing is a mechanical separation of polymer chains or groups of chains under tensile stress. Crazing does not occur in materials oriented along the draw direction.	Hsiao and Sauer [78] PS
X-ray scattering and electron microscopy suggests that void content of the craze is dispersed in the form of inter-connected spheroidal holes having a common dimension of 10–20 nm. Stress-strain curves of crazes have been determined showing a yield stress followed by a recoverable flow to roughly 40 to 50% strain at 41 to 55 MN/m^2. Returned to zero-stress, the craze exhibits creep recovery at a decelerating rate.	Crazes resemble on open-celled foam, the holes and polymer elements of these average about 20 nm in diameter. Craze-tensile behavior is an extension of the craze growth process. Decrease in elastic modulus and yield stress with increasing strain are rationalized in terms of strain-induced decrease in density and resultant increase in stress concentration factor on the microscopic polymer elements of the craze. Polymer surface tension and large internal specific surface areas of the craze are suggested to be important factors in large creep recovery rates of craze.	Kambour [85] PC
Microstructure of crazes in thin films by transmission electron microscope. In thin PS films ϵ (craze initiation) = 1%	Fibrils of 25–50 nm are a common feature of the microstructure of crazes in cast and bulky films. Major fibrils diameter = 20–30 nm. Minor fibrils connecting major fibrils have a diameter >10 nm and they tend to orient normally to major fibrils. At large total deformation at low $\dot{\epsilon}$, there is a gradual transition from coarse to fine microstructure of crazes.	Beahan, Bevis & Hull [115] PS
Optical and micromechanical study of PS craze behavior: – average craze thickness = 0.19 μm – spacing between the crazes = 38 μm – crazes start forming at 0.5 σ_F and ϵ = 1.0%	The large number of parallel crazes formed permitted the determination of the stress-strain curve of a craze; it is similar to that found by Kambour and Kopp [83] for solvent-induced crazes in PC (Fig. 9.12)	Hoare and Hull [120] PS

Table 9.3. (continued)

Observations	Conclusions	Material and References
From electron imaging studies in a transmission electron microscope quantitatively fibril density and craze width in (thin) film crazes are obtained. Generally a *midrib*, a layer of higher drawn fibrils (and consequently of lower craze matter density), is observed in the center of the craze; this midrib typically has a width of some 50 to 80 nm (see Fig. 9.16).	The density profiles in the x (craze plane) and y (perpendicular to the craze plane) directions permit the calculation of fibril draw ratio $\lambda(x, y)$ and of surface stress profiles $S(x)$. Fibril draw ratios and surface stresses show pronounced maxima at the craze tip (see Fig. 9.16).	Kramer [249] ~ 20 diff. polymers Michler [286] PS
Higher energy electron microscopy (of thicker films) reveals narrow, long pre-craze zones containing "weak domains" some 10 to 15 nm in diameter.	Molecular heterogeneities are amplified by non-homogeneous deformation; the pre-craze zones mostly situated at the sample surfaces transform into the bulk fibrillar crazes.	Michler [286, 287] PS
Small angle X-ray scattering patterns of crazes (in PS, PMMA, PC) are strongly anisotropic with the equatorial scattering curves exhibiting a well pronounced interfibrillar interference maximum.	The average fibril diameter D can be evaluated from the position of the scattering maximum D (PS) ~ 4 to 10 nm D (PMMA) ~ 20 to 30 nm D (PC) ~ 20 to 60 nm D depends strongly on the stress under which fibrillation occurs.	Brown/Kramer [289] PS Paredes/Fischer [288] PMMA, PC Dettenmaier [272] PC
Interference optical microscopy permits the determination of the profile (width, length) of crazes in bulk, transparent (amorphous) polymers (see Fig. 9.18).	Craze profiles observed in bulk PMMA and PC conform well to the Dugdale model; craze stresses and their relaxation has been calculated; the influences of temperature, molecular weight and load history (fatigue) on craze deformation have been evaluated.	Döll [273] PMMA, PC
Scanning electron microscopy of crazed and/or fractured semi-crystalline materials.	Crazes (in PP, PETP) are showing the fibrillar microstructure; at ambient temperature crazes develop along certain morphological components (interspherulitic boundaries, radially oriented lamellae); thus parameters which influence morphology (tacticity, cooling rates) influence crazing and/or the occurrence of shear banding.	Friedrich [274] PP, PETP

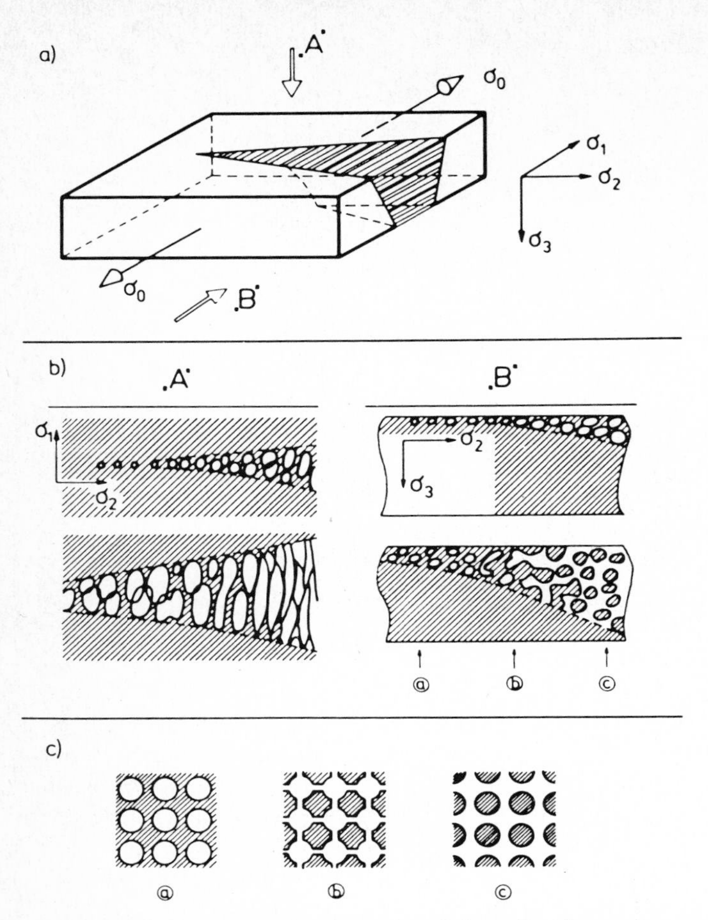

Fig. 9.17a–c. Model of the transformation of the surface furrow (pre-craze zone) into a craze in a thick PS film developped by G. H. Michler on the basis of HVEM-investigations (Courtesy G. H. Michler, Halle [287])

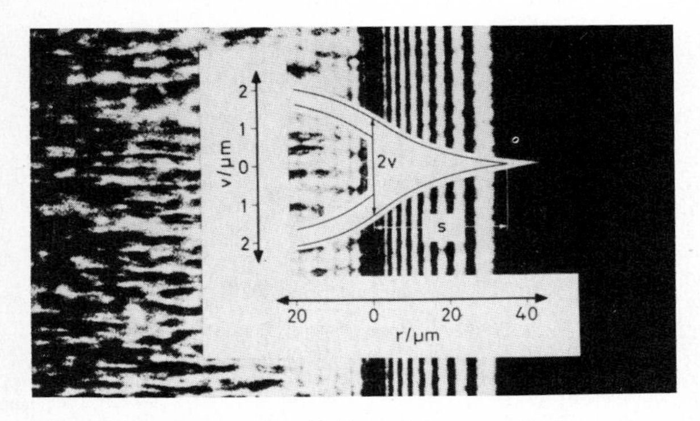

Fig. 9.18. Principle of optical interference method: appearance and spacing of interference fringes permit to identify craze width 2v(r) and length s (from Döll et al. [250], see [273] for details of the method)

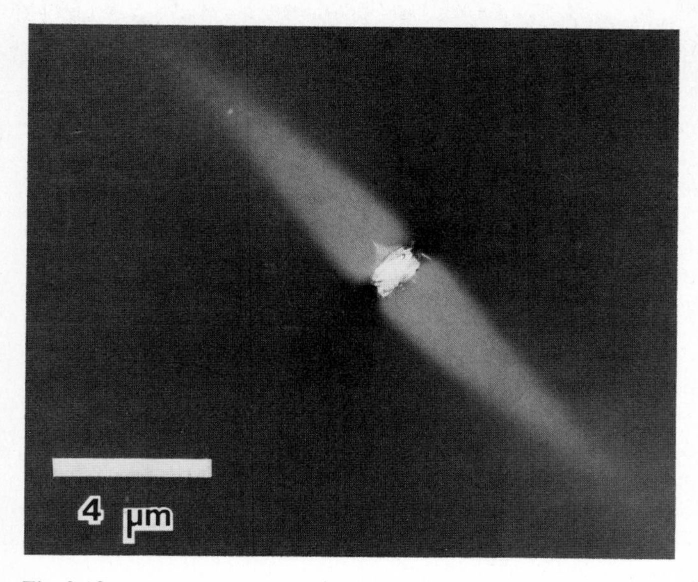

Fig. 9.19. Deformation zones in solvent cast thin films of PC (Courtesy L. Berger and E. J. Kramer, Ithaca)

The determination of the craze surface stress profile S(x) from craze displacement profiles (through Fourier transform analysis or dislocation theory), i.e. the *craze micromechanics* has especially been addressed by Kramer and collaborators (reviewed in [249]). Finite-element and boundary-element calculations of the displacement profile for a given surface stress distribution [295] are in adequate agreement with the above approach.

It should be pointed out that in regions of high stress concentration as for instance at the tip of a crack besides the predominant microfibrillar craze structures also other forms of morphology can be observed. These range from homogeneous deformation zones (DZ) in solvent cast thin films of some ten different homo- and copolymers, such as PS, SAN, PC (Fig. 9.19) and PPO [249] or PS, SAN and PC

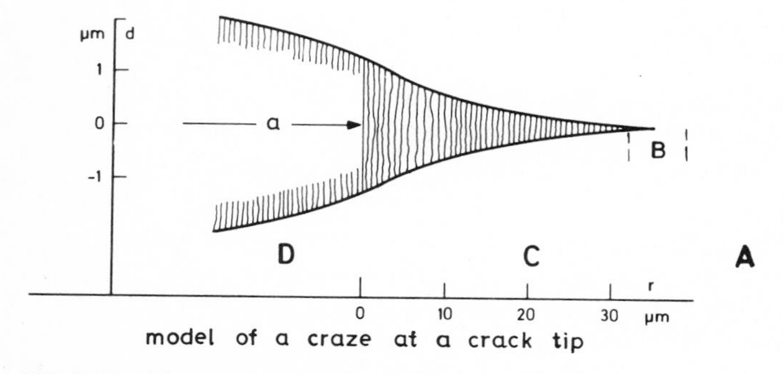

Fig. 9.20. Model of a craze at a crack tip.
A elastically strained, glassy polymer, *B* "process zone" of craze initiation, *C* craze growth or propagation zone, *D* breakdown of fibrils leads to transformation of the craze into a crack

microtome sections [285–287] to networks with or without more heavily deformed center or boundary regions, and to crazeless fracture surfaces in oriented polystyrene [155].

It has already become clear from the above discussion of craze morphology that there are three distinct phases in the life of a craze: *its initiation or nucleation, its propagation or growth, and its breakdown.* The model of a craze ahead of a crack tip representing all three phases is shown in Figure 9.20. The molecular mechanisms controlling each phase and the role of chains therein are different. In the following two sections an account of observations and criteria will be given.

B. Initiation of Extrinsic Crazes

Until about 1980 a considerable amount of research had been conducted on various aspects of crazing. However, there had been no exact agreement on the mechanism of craze initiation. No theoretical model was available which could predict in all generality whether a certain polymer will craze or not under a given set of conditions. And, if it did craze, what effect the temperature and rate of deformation would have on craze formation and propagation. This was certainly due to the fact that the initiation of a craze depends simultaneously on three groups of variables characterizing respectively the *macroscopic state of strain and stress*, the nature of defect or *heterogeneities* in the matrix, and the *molecular behavior* of the polymer in the given thermal and chemical environment. In focussing on particular variables five conceptually different approaches to describe the initiation of conventional crazes emerge. These are based on stress bias, critical strain, fracture mechanics, molecular orientation, and molecular mobility respectively. The principal observations and the craze initiation criteria proposed on the basis of the above concepts are listed in Table 9.4.

These criteria describe different states of local excitation and deformation of chain segments. Following 1968, Sternstein et al. [86, 86a, 139] had developed their famous stress bias criterion which refers to two mechanisms: *cavitation* in a dilatational stress field and growth of cavities through a deviatoric stress component. These mechanisms have been more explicitely considered in the mathematical model of cavity expansion in a rigid plastic by Haward et al. [137] and in the molecular model by Argon [152].

Before discussing these criteria in more detail, it seems to be useful to present a general *molecular* model of the initiation of extrinsic (conventional) crazes which permits to reconcile (most of) the theóries and hypotheses proposed so far [167, 249, 270, 296, 300].

An amorphous thermoplastic polymer can be represented by a statistical arrangement of interpenetrating random coils (see Fig. 9.21). The most important geometrical data of such an arrangement are known: the radius of gyration for instance from neutron scattering [298] and the entanglement molecular weight M_e from viscoelastic analysis of the plateau modulus [299]. In Figure 9.21 the data on PMMA have been taken as an example. Its characteristic features are an M_e of 9150 g mol^{-1}, a radius of gyration of the segments between entanglement points of 2.5 nm and an

Table 9.4. Criteria of Craze Initiation

Observation	Conclusion	Materials References
Stress-based criteria		
Analysis of craze initiation stresses in tensile samples containing circular holes, in thin-walled biaxial tube samples, and for rods in combined tension-torsion tests	*Stress bias criterion:* $\lvert \sigma_1 - \sigma_2 \rvert \geqslant A(T) + B(T)/I_1$ A, B material constants depending on thermal history and environment, $I_1 = \sigma_1 + \sigma_2 + \sigma_3 > 0$; hydrostatic pressure suppresses crazing	Sternstein et al. [86. 139] PMMA
Analysis of location and direction of crazes formed – in sheets compressed by a die or wedge – during three-point bending of a round-notched bar	Crazes are initiated at the intersection of lines where shear bands are formed. Since crazes can be initiated before or after shear band formation it is assumed that large hydrostatic tensions rather than strains cause the craze initiation. Critical hydrostatic tensions: 87 MN m^{-2} for slowly cooled PC 89 MN m^{-2} for quenched PC	Camwell et al. [141] PS Ishikawa et al. [172] PC
At excellently adhering glass beads in a PS matrix crazes form at the poles.	Stress analysis shows that crazes form in regions of maximum dilatation and of maximum principal stress.	Dekkers [311] PS
Consideration of the three-dimensional state of stress at rubber particle – matrix interfaces.	Extension of the Sternstein criterion into three dimensions (τ^* = octahedral shearing stress): $6\,\tau^{*2} = (B/3I_1 - A)^2 + 3I_1^2$	Breuer/Stabenow [309] Retting [310]
Internal crazing in thick notched specimens of 14 different glassy polymers.	$\sigma_{yy} \sim$ cohesive energy density x $(T_g - T_{test})$	Kambour/Farraye [297]
Strain-based criteria		
Analysis of location and direction of crazes formed in tension at the interface between steel balls in PS and rubber balls in PS	Criteria of *principal strain* and of *strain energy* agree best with test data, maximum principle stress, maximum principle shear, and stress bias criterion correspond less well, maximum dilation the least	Wang, Matsuo, Kwei [105] PS

Table 9.4. (Continued)

Observation	Conclusion	Materials References
Craze formation in long-term tensile testing	Time-dependent critical principal strain; limiting values ($t \to \infty$), in %: PS 0.3 [110]; SAN 0.7 [110]; PMMA 0.7–0.9 [104, 110]; PVC 0.8–0.9 [104]; POM 1.5–2 [104, 121]; PP 2 [121]; PC 2 [110]	Menges, Pohrt, Schmidt et al. [104, 110, 113, 121]
Craze formation in biaxial stress tests "second quadrant"	$\epsilon_c = A'(T) + B'(T)/I_1$	Oxborough et al. [142] PS
Determination of critical surface strain ϵ_c for initiation of "dry" crazes within 20 h for about 30 glassy polymers.	$\epsilon_c = 0.075 \left[\frac{\%}{K}\right] \frac{\text{cohesive energy density x } (T_g - T_{test})}{E}$ (see Fig. 9.22)	Kambour [296]
Fracture mechanics criteria		
Study of craze behavior in sharply notched PMMA immersed in methanol Craze initiation and growth characteristics are controlled by the initial stress intensity factor K_0 not by applied stress: $K_0 < K_m$ no crazing $K_0 < K_n$ craze initiation and arrest $K_m = 0.06$ MN m$^{-3/2}$ $K_n = 0.2$–0.3 MN m$^{-3/2}$ (depending on sample thickness)	Craze growth follows one of two distinct patterns. One leads to eventual craze arrest, the other to a constant speed of craze propagation and final failure. Kinetics of craze growth in methanol explained by liquid flow through the porous crazed material (to which a void spacing of 0.25 μm, a void size of 72 nm and a craze yield stress of 9 MN m^{-2} are assigned).	Marshall et al. [102] PMMA
Study of crack arrest in tensile loading of cast PMMA sheets in various organic liquids	In a plot of σ_t^2 (square of the threshold stresses for crack propagation) versus reciprocal crack length *a* a distinct lower bound was found which defined a *critical energy release rate* G_I; whereas G_I initially decreases with temperature it becomes independent of temperature at $T > T_c$ where T_c roughly corresponds to the T_g of PMMA in the particular environment.	Andrews et al. [124] PMMA

Table 9.4. (Continued)

Observation	Conclusion	Materials References
Photoelastic study of craze initiation and propagation in kerosene and near a Griffith crack. Initial value of K_0 is the governing factor for craze initiation and propagation $x = At^n$ where x = craze length A = constant depends on K_0 n = 0.53 for PMMA = 0.24 for PC obtained from craze growth histories	Stress concentration at *crack tip* is sensibly reduced with craze initiation and *not* transferred to *craze tip*. There is a critical stress intensity factor K_m below which no crazing occurs: $K_m = 0.13$ MN $m^{-3/2}$ PMMA $K_m = 0.29$ MN $m^{-3/2}$ PC The difference in sorption kinetics is held responsible for the higher critical stress intensity factor and the lower exponent n in PC, relative to PMMA.	Narisawa et al. [127] PMMA PC
Structural and morphological criteria		
Study of craze initiation stress σ_i, of location of crazes within bulk samples and of change of birefringence R across sample thickness x $\sigma_i = 46$ MN m^{-2} for $dR/dx = 0$ $\sigma_i = 38$ MN m^{-2} for $dR/dx = 2 \cdot 10^{-3}$	Craze initiation occurs in regions where chains were oriented transversely to direction of stress (birefringence positive)	Haward et al. [89] PS
Study of birefringence and of tensile and compressive behavior of hot-drawn sheets at various angles to draw direction	The crazing stress increases with amount of orientation when the tensile axis is parallel to the draw direction, and decreases with amount of orientation when the axis is normal to the draw direction. At high draw ratios the crazing stress parallel to the draw direction is higher than the shear yielding stress and the material undergoes a yielding process similar to compression yielding. Molecular orientation has only a small effect on the orientation of the craze plane. The effect of molecular orientation on the crazing behavior of polystyrene is primarily associated with the nucleation stage of the crazing process.	Hull et al. [153] PS

Table 9.4. (Continued)

Observations	Conclusions	Materials References
Statistical nature of initiation of stress crazing is explored; a plot of histograms of the time to craze initiation follows approximately an exponential distribution.	The time required for the formation of crazes is an inherent, but statistical, characteristic of the material itself involving a stress depending rate process reflecting to some extent the initial microscopic flaw distribution.	Narisawa et al. [118] PMMA
Molecular criteria		
Relation between yield stress σ_F and craze initiation stress σ_i is investigated: $\sigma_F(T)\ \dot{\epsilon}$ = constant $\sigma_i\ (T)\ \dot{\epsilon}$ = constant where σ_i decreases less than σ_F at high temperature. Also $\left[\frac{\partial \sigma_F}{\partial(\log \dot{\epsilon})}\right]$ and $\left[\frac{\partial \sigma_i}{\partial(\log \dot{\epsilon})}\right]$ are represented by the same function f(T)	– Both yield stress and crazing stress vary with temperature, but crazing stress varies less than the yield stress. This indicates that crazing has to provide the nucleation surface energy for voids. – Similar response of crazing and yielding data to temperature indicates that in both cases similar molecular conformational changes and backbone motions are involved. – Materials having highest crazing tendency (PMMA, PS, SAN) also have high yield stress and thus a large energy associated with local plastic instability (which is necessary to account for the nucleation and growth of the tiny voids).	Haward et al. [108] PS
Analysis of craze initiation in tubular specimens under combined tension and torsion considering the mechanism of craze growth as one in which craze tufts are produced by repeated break-up of concave air-polymer interfaces at the craze tip (meniscus instability). For a surface energy $\gamma = 0.05\ J/m^2$ and a tensile plastic resistance of $\sigma_y = 10^2\ MN\ m^{-2}$, an extension ratio $\lambda_n = 2$ of craze matter gives a craze tuft diameter = 0.05 μm.	Crazes initiate at surface or interface stress concentrations where localized plastic flow produces microcavities by intense inhomogeneous plastic shear at a molecular scale. The rate of cavity formation depends primarily on the activation free energy ΔG^* (cf. Eq. 8.35) and on the local concentrated deviatoric stress while the subsequent plastic expansion into craze nuclei is in response to global negative pressure.	Argon et al. [152, 166] PS

end-to-end distance of 6.1 nm. The average distance between entanglement points is 3 nm. It is important to point out that the end-to-end distance of a typical segment M_e is larger than the average distance between entanglement points. This means that on the average the sequences of entanglement points are not connected with each other over ten to 15 nm (see Fig. 9.21). If such an arrangement of sequences of stiff entangled segments is subjected to tensile forces then there will initially be an instantaneous (elastic) and an anelastic displacement of the segments which evidently leads to the storage of elastic energy. This energy can be dispersed by two principally different modes of deformation: by the local shear displacement of segments or by their orientational and conformational reorganization. It is proposed that the *dominant* mechanism in the nucleation of conventional crazes (crazes I) is an (unstable) reorientation of segments in stress direction, a reorganization which seems to preserve the entanglements previously existing in that direction (Fig. 9.21). The important question, what happens to the entanglements in lateral direction, will temporarily be postponed. Figure 9.21 permits an evaluation of the contributions of rotation and elongation to the orientation of stretched matter.

The effect of segment rotation in craze nucleation becomes clear from the heavy line which has been traced from one entanglement point to the next always following that statistical segment which points the closest in stress direction. (In a sequence of loaded and interconnected entanglement points as in Figure 9.21 there must always be two segments pointing "upwards" and two "downwards".) Each of these segments had an average end-to-end distance of $\ell_s = \langle r_e{}^2 \rangle^{1/2}$. The length of the traced

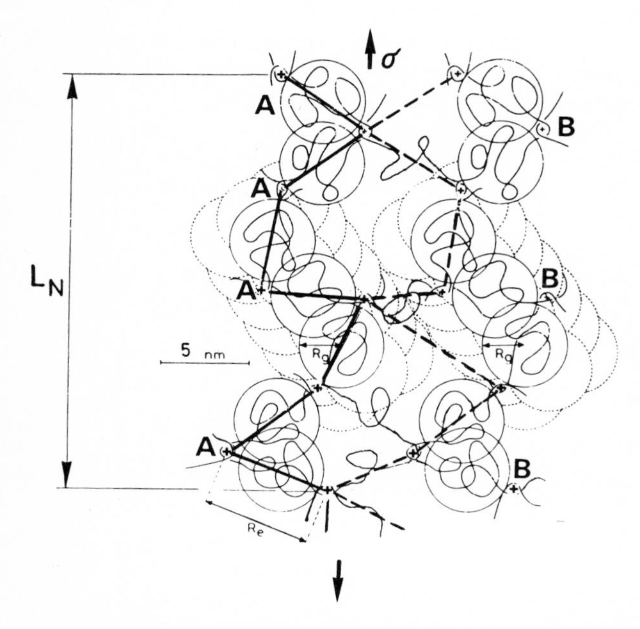

Fig. 9.21. The heavy solid lines represent (plane) sequences of entanglement points. It is quite conceivable that other sequences (for instance those attached in A or B) are not lying in the plane of reference. In that case there would be very little lateral constraint to a tensile deformation of the shown element

line which consists of N such segments is, evidently, $N\ell_s$; the end-to-end distance of this line is $N\ell_s \overline{\cos\theta}$, where θ is the angle between segment and stress direction. The probability to find in an isotropic polymer a segment with direction θ_1 is assumed to be equal to $\overline{\sin\theta_1 d\theta_1}$. According to the model one has always to follow the segment which is closest to the stress axis, that is element no. 1 if $\theta_2 > \theta_1$ and element no. 2 if $\theta_2 \leqslant \theta_1$. Since

$$p(\theta_2 \geqslant \theta_1) = \cos\theta \tag{9.27}$$

one obtains

$$\overline{\cos\theta_1} \,|_{\theta_2 \geqslant \theta_1} = \int_0^{\pi/2} \cos^2\theta \sin\theta d\theta = \frac{1}{3} \tag{9.28}$$

It can be shown that the inverse situation ($\theta_1 \geqslant \theta_2$) gives a similar contribution, the total average, $\overline{\cos\theta}$, therefore is equal to 0.67. Thus, the simple reorientation of segments in a non-oriented polymer would already permit the elongation of a single, non-attached sequence of entanglement points by a factor of 1.5. It must be assumed, however, that in addition the molecular segments of length ℓ_s will uncoil and will partly assume their extended length ℓ_e. If all N segments would *completely* uncoil, then the maximum extension of $N\ell_e$ could be attained. This, evidently, corresponds to a draw ratio of $N\ell_e/0.667\, N\ell_s$ which is equal to $1.5\,\lambda^e$ where λ^e is the natural draw ratio of a segment of entanglement molecular weight M_e. This value of $1.5\,\lambda^e$ must be considered as the maximum elongation in the absence of lateral constraints. In a real and especially in a bulk polymer, however, these constraints are notable and the maximum extension must be expected to be smaller than $1.5\,\lambda^e$.

Another approach is to consider the zone of width L_N (Fig. 9.21) as a three-dimensional network [249]. The uncoiling of the entangled segments would extend their projected lengths in each of the three coordinate directions by a factor of λ^e:

$$x \rightarrow \lambda^e x.$$

According to the Gaussian theory of rubber elasticity (see Chapter 5A) one can deform such a network by a maximum uniaxial network strain λ^{net}:

$$(\lambda^{net})^2 + \frac{2}{\lambda^{net}} = 3(\lambda^e)^2. \tag{9.30}$$

If λ^e is > 2 one has (to better than 1%) the approximation

$$\lambda^{net} = \sqrt{3}\,\lambda^e. \tag{9.31}$$

It has to be expected, therefore, that the combined effect of segment rotation and uncoiling *while maintaining all entanglements* will lead to a local extension of between $1.5\,\lambda^e$ and $1.73\,\lambda^e$.

In their experiments on craze microstructure Kramer et al. [249] determined the fibril extension ratio λ_{fib} for a large number of homo- and copolymers. They found for instance for air crazes in PS immediately at the craze tip λ_{fib} values of 7 to 8 which correspond to 1.71–1.95 λ^e (see Fig. 9.16c). In the midrib region – under slightly decreased stresses – the λ_{fib} values "dropped" to 5–6 (1.22–1.46 λ^e) which is still significantly higher than the natural draw ratio $\lambda^e = 4.1$ of PS. Outside of the midrib region for solvent cast films Kramer found λ values between 3 and 4 (Fig. 9.16c). For most of the nearly 20 polymers he investigated the fibril extension ratios corresponded to or were slightly less than λ^e [249]. Values smaller than λ^e are often reported in the literature for thicker films e.g. by Kambour [76], Grosskurth [176] and Michler [286, 287]. The effect of film thickness (and of the absence of plastic contraints in film thickness direction) has especially been discussed by Kramer et al. [249, 278, 279].

If the volume of the deforming matter is conserved in the craze initiation process then the stretching by a factor of λ^{net} in stress direction must necessarily be accompanied by a lateral contraction of $1/\sqrt{\lambda^{net}}$ in the perpendicular directions. Even in thin films such a contraction cannot entirely be accommodated. The few examples presented previously (Figs. 9.13–9.14) already show that this will give rise to sample discontinuity in the form of voids and fibrils.

In recent years several papers have been devoted specifically to the question of craze initiation [270, 282, 283, 287, 296, 297, 300–312]. The principal observations and criteria concerning the formation of conventional crazes have been compiled in Table 9.4, they will be discussed subsequently.

The strain criteria [104, 110, 121, 296] specify that no crazing will occur if the *largest principal strains* – even at extended times – remain below a limiting value, which was found to be some 2% or less. At short times craze initiation strains ϵ_c may, in some polymers, be of the order of 3 to 5% [104, 113]. In the absence of intensive strain concentration strains of this order are insufficient to cause chain scission but they tend to decrease the intermolecular attraction (cf. 8 II B). This is recognized by Oxborough and Bowden [142] who also use the tensile hydrostatic stress (I_1) in their general critical-strain criterion. In biaxial stress fields the stress bias criterion and their criterion differ only little; as Oxborough and Bowden point out Sternstein's data and their own can be explained by either one of the two criteria.

In one of the most comprehensive studies on this subject, Kambour [296] has determined in a short-time bending experiment, the initiation strains ϵ_c of surface crazes of some 30 different polymers. He has established a linear correlation between ϵ_c and the quantity (CED) $\Delta T/E$, where CED is the cohesive energy density [see e.g. 297] and $\Delta T = T_g - T_{test}$ (Fig. 9.22).

In view of the close relationship between the tensile yield stress σ_F and the modulus E, Kambour also obtains an excellent correlation between the short-time crazing strain ϵ_c and (CED) $\Delta T/\sigma_F$.

As it was to be expected, the strain at which crazes are initiated is load- and time-dependent; in a (uniaxial) creep experiment at $\sigma < \sigma_F$, crazes are initiated after an incubation period and at a strain which is generally smaller than the short-time value of ϵ_c [103, 104, 324a, 337]. This observation can be explained through a certain strain-softening of the matrix which permits that critical displacements leading to matrix break-up (in much fewer sites) be achieved at smaller stresses. In fact, Fischer [11, 337] found that crazes (in PC) were initiated whenever the relative compliance $d\epsilon/d\sigma$ assumed a critical value.

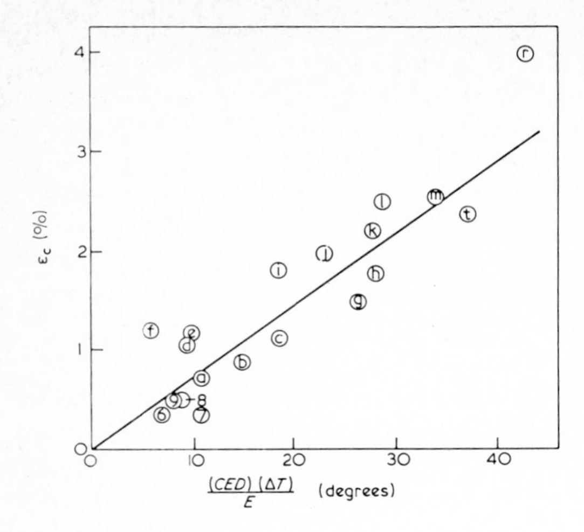

Fig. 9.22. Critical strain for craze initiation ϵ_c vs. CED x $\Delta T/E$ where CED is cohesive energy density and $\Delta T = T_g - T_{test}$ (from Kambour [296])

If one would apply one of the above criteria to the craze initiation in a sharply notched plate loaded in tension, one would have to expect instantaneous crazing since both $|\sigma_1 - \sigma_2|$ and ϵ show a singularity at an infinitely sharp crack tip (cf. Eqs. 9.1 to 9.3). Such an expectation would be contrary to the experimental findings. Marshall et al. [102] and Narisawa et al. [127] have established that it is the *initial stress intensity factor* K_0 which controls craze initiation at the crack tip. In the case of PMMA and PC immersed in methanol or kerosene critical values K_m exist below which no craze initiation and growth occurs. This behavior can be understood in view of the discrete sizes of the chain segments and of any voids to be formed, in view of the fact that the density of stored elastic energy is limited (Fig. 9.3), and in view of plastic deformations eliminating the stress singularity. Marshall et al. [102] conclude from their data that crazing occurs when the material at the crack tip reaches a critical strain or crack opening displacement.

The effect of sample structure and morphology on craze initiation has been recognized since the earliest investigations [78]. The structure of the sample surface is for various reasons especially important:

- *defects and/or contaminations* are preferentially found at the surface enhancing craze initiation there [173, 176];
- injection molded or machined samples mostly contain a *surface structure* which differs from that of the bulk material [89, 175, 176];
- molecular chain segments in surfaces have a higher degree of *mobility* [112];
- *environmental attack* (diffusion, plastification, degradation) begins at the surface and penetrates from there [350–384].

The investigations of the effect of chain orientation on craze initiation show that a transverse orientation of the chains with respect to the direction of principal stress enhances craze initiation [89, 153]. Because there are fewer chain segments pointing in the direction of principal stress critical local strains are obtained at smaller stresses (cf. 3 IV E). On the other hand the craze initiation stress increases with the degree of alignment of the chains in the stress direction (increasing degree of orientation, small angles θ between draw direction and principal stress). In the case of good alignment craze initiation stresses will be higher than the stress for shear yielding so that no crazing is observed. For PS at 20 °C yielding in tension occurs in samples drawn to $\lambda = 2.6$ or more and at a $\theta(\lambda)$ smaller than 20 to 30° [153]. Particularly noteworthy is the observation of Hull and Hoarse [153] that the molecular orientation has only a small effect on the orientation of the craze plane.

This indicates that the transfer of stresses along the axes of the oriented chain segments does not determine the direction of craze extension.

As an effect of sample structure and morphology the enhancement of craze initiation through surface flaws [118, 152, 173, 176], impurities [155, 161], and heterophase inclusions (reviewed in [114, 168, 190, 191] has also to be mentioned.

The molecular criteria generally take into consideration molecular weight, chain entanglement and the local mobility of chain segments at the given temperature and chemical environment [11, 15, 50, 79, 146, 165–167, 173]. The effect of the chemical environment will be discussed in Section D. The following investigations generally have been conducted in standard atmosphere. An interpretation of craze initiation in terms of local strain softening has been given by Rusch and Beck [95]. They propose that there exists a critical strain for crazing which is dependent on the level of frozen free volume initially distributed in the bulk of the material. Through the imposed dilatant strain sufficient free volume is added to bring the polymer to a state similar to the one reached by it at its T_g. Gent [96], too, suggests that craze initiation results from lowering of the T_g in local high stress fields at surface flaws. However, as shown by various authors [89, 129, 158] surface crazes occur in PS, PMMA, and PC at temperatures of –75 °C and below. This raises a doubt whether the polymer reaches a state characteristic of its T_g at such low temperatures or not.

Lipatov and Fabulyak [112] point to the importance of low temperature relaxation processes due to the motions of side chains. These relaxations shift towards lower temperatures in samples with high surface to volume ratios. This behavior has been interpreted in terms of lower segmental packing on the surface and, therefore, easier molecular motions. It is claimed that this facilitates crazing. Shifts of molecular relaxations towards lower temperature (in PC) have also been observed by Sikka [163] who suggested that this shift may have its origin in the formation of microvoids.

Studies of molecular weight dependence of crazing and of breaking stresses have been carried out by Fellers and Kee [146]. These authors suggest that craze *initiation* stresses are independent of molecular weight when $\overline{M_n} > 2\ M_e$ whereas craze *development* and *breakdown* are clearly molecular weight dependent. This is evident from the plot of stresses versus molecular weight, shown in Figure 9.23. Since the measured elastic moduli (1.5 GN m^{-2}) did not depend on molecular weight the craze initiation strains are also constant (2%). From these results it may be concluded that craze initiation is an event which depends primarily on the interaction between chain segments. Contrary to the apparent independence of stress and strain at craze initiation to molecular weight the T_g of the different PS fractions rises monotonically from 88 °C at $M_n = 70000$ to 105 °C at $M_n = 150000$. Another observation made by these and other authors is that crazes in high molecular weight samples are very

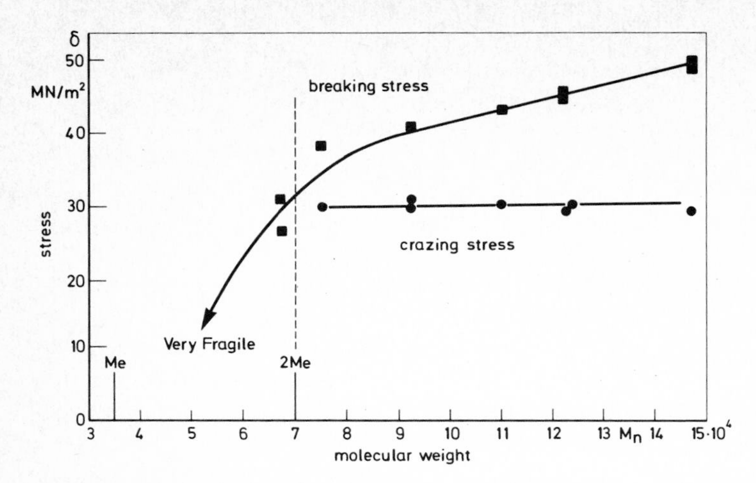

Fig. 9.23. Breaking stress and crazing stress as a function of molecular weight M_n of polystyrene at 25 °C (after [146]). M_e entanglement molecular weight

fine, *numerous*, long, and straight as compared to the ones in lower-molecular weight polymers, which are coarser in texture and somewhat shorter in length. This phenomenon seems to be related to the homogeneity of the stress field which is the better the more numerous and the stronger the fibrils [146, 313].

Based on these studies [11, 76, 77, 146, 249, 271–294, 296, 300, 305] and anticipating the essential results of the following Sections C and D, a *model of craze initiation* can be presented:

- a glassy amorphous polymer is considered as an entangled solid network (Fig. 9.21); well below T_g, the coiled subchains between entanglement points (called *segments*) have a very small internal mobility (they are "stiff") and also small external mobility (the intermolecular barriers opposing segmental motion are especially pronounced in PS and PMMA [305]);
- when strained, local stresses will develop between segments; in view of the segmental stiffness, the inception of incremental flow processes or of "bond-flexing" which eventually could lead to shear yielding or flowing is highly restricted; however, bending stresses can be transmitted by a segment (this seems to be an essential point in craze initiation); consequently, the local stress distribution will be more uneven in a glass than in a more ductile polymer; during the elastic straining of a glassy matrix (region A in Fig. 9.20), the exact length of the chains (provided $M_w > 2$ Me) and the presence or absence of *some* cross-linking points are of minor importance;
- the elastic straining activates two mechanisms: an *anelastic dilatational deformation* which is especially furthered by the concentration and triaxial state of the stress field (at surface defects or a crack tip) and a *(plastic) shear deformation* which is favored by hydrostatic pressure; the first mechanism is predominant at high strain rates and at low temperatures, the latter at longer times and/or at higher chain mobilities [324].

The dilatational deformation in the most highly stressed matrix sites is brought about by a limited collective reorientation of the (stiff) segments within the sequences shown in Figure 9.21; the initially modest dilatational reorganization preserves existing entanglements but it "breaks up" secondary bonds and thus activates the matrix material for one of the following steps. In case of a low entanglement concentration (low M_w [313b], blending with low M_w-material [313c, d], prior orientation perpendicular to the loading direction [40, 155, 294]), catastrophic fracture (even without crazing) will rapidly occur. In a well entangled network, however, the stretching in a preferential direction necessitates a contraction of the matrix in lateral directions, which could more easily occur at the specimen surface, at an interface with soft inclusions, or (in the case of very thin or more ductile specimens), by contraction of the whole sample; at this stage where no voids have formed as yet, it still remains open how the material will respond to a continued straining (e.g. by crazing, formation of homogeneous deformation zones, or shear yielding);

– with continued straining of a well entangled network, the *homogeneous* local deformation through segments turning into stress direction and/or the accomodation of the lateral contraction can become increasingly difficult; in such a case there are at least three principally different types of material response to be envisioned:
  a) *break-up of the continuous matrix* in the form of pre-craze surface furrows or by advance of a craze tip in the form of a "meniscus" (protruding fingers) [166, 287, 321] or by cavitation (void formation); it remains to be discussed whether and how disentanglement and/or chain scission influence this phenomenon [287, 307, 308, 330];
  b) *localized plastic deformation* in the form of homogeneous deformation zones (DZ, see Fig. 9.19, cf. [280, 283, 287, 306]);
  c) *slow-down (and stoppage) of the collective segment reorientation* and general increase of stress level (which could lead to stress relief through initiation of a) or b) in other sites or to other forms of deformation (e.g. shear yielding [283, 305, 318b]).

Evidently, if homogeneous (dilatational) deformation is followed by matrix break-up, the craze is born.

Craze initiation, therefore, can be defined as the localized collective dilatational reorganization of stiff chain segments accompanied by matrix break-up. The influence of the most important molecular and experimental parameters on craze initiation can be summarized as follows:

Molecular weight

If $M_n > 2$ Me, it has very little influence on the elastic straining behavior and on the stress or strain at which crazes are initiated; however, after craze initiation, fibrils are formed which are the stronger and which can develop, under increasing stress and time, to greater lengths the higher M_n; consequently, the higher M_n the more region C (Fig. 9.20) is extended; thus, the force transmitted by region C grows with M_n and so does the stress at break.

Concentration of chain ends
A large concentration of chain ends facilitates craze initiation but reduces craze stability.

Concentration of entanglements or cross-linking points
A large density of entanglements renders matrix break-up difficult or even impossible, however, it increases the stability of any formed craze; except for kinetic aspects, the influence of cross-linking points is analogous.

Molecular mobility, temperature
To a certain extent, molecular mobility facilitates craze initiation (smaller ϵ_c). However, inception of internal rotation increases the tendency towards yielding. No crazes are initiated at $T \sim T_g$.

Pressure and/or Preorientation reduce the propensity for crazing of a glassy polymer.

Craze initiation in *Creep* is retarded but occurs at smaller strains.

Active environments greatly enhance craze formation (see II E).

C. Intrinsic Crazing and Stress-Whitening

It has previously been mentioned that the initiation of the ordinary type of crazing is generally controlled by foreign particles and surface grooves which both act as stress concentrators. Under these conditions, in amorphous single phase polymers a small number of extrinsic crazes grow and usually cause the premature rupture of the specimen.

A large number of crazes distributed throughout the sample volume are observed in multiphase systems such as rubber-modified polymers where numerous rubber particles act as stress concentrators. The intensive light-scattering from the multitude of crazes gives rise to the phenomenon of stress-whitening. Stress-whitening is also observed for semicrystalline polymers where the transformation of the spherulitic into the fibrillar microstructure is often accompanied by intensive void formation.

Stress-whitening due to intrinsic crazing has been reported in a few cases for amorphous single-phase polymers. For instance, Goldbach and Rehage [82] observed intrinsic crazes in poly(methyl methacrylate) (PMMA), plasticized at the surface to avoid premature fracture by surface crazes. Hull et al. [98, 120], Lainchbury and Bevis [162] and Argon and Hannoosh [166] reported on craze yielding of polystyrene (PS) by formation of numerous intrinsic crazes. Like extrinsic crazing, this phenomenon occurred at small strains of the orderof a few percent.

Recently, Dettenmaier and Kausch [271, 272, 300, 314, 315] have observed an intrinsic craze phenomenon in bisphenol-A polycarbonate (PC), drwan to high stresses and strains in a temperature region close to the glass transition temperature, T_g. This type of crazing is not only initiated under extremely well defined conditions which reflect specific intrinsic properties of the polymer but also produces numerous crazes of a very regular fibrillar structure. These crazes were called crazes II in order to distinguish them from the extrinsic type of craze, called craze I.

Figure 9.24 shows a nominal stress-strain curve of PC measured at T = 129 °C. Extrinsic crazes (crazes I) are initiated at an early stage of deformation, well below the yield point. They start growing from surface defects, in particular, if the sample has come into contact with some crazing agent. Crazes I are shown in the optical (Fig. 9.12b) and scanning-electron micrographs (Fig. 9.25). These crazes are largely separated from each other and it is clearly visible that they are situated at the surface of the specimen (Fig. 9.25). At high stresses and strains, well above the yield point, numerous crazes (crazes II) are initiated throughout the sample volume. Their initiation is preceded by some strain hardening of the material and is followed by a strain softening mode leading to the second peak in the stress-strain curve of Figure 9.24. The specific volume of PC increases notably when crazes II are initiated, resulting in a loss in density of approximately 8% at the rupture of the specimen (Fig. 9.24). The large number of crazes II giving rise to the phenomenon of stress-whitening and to the observed loss in density is shown by the optical and scanning electron micrographs in Figures 9.12c and d and 9.26, respectively. In Figure 9.12c, crazes I and II can clearly be seen together, the large and isolated crazes I are easily distinguished from the dense pattern of very fine craues II.

Dettenmaier and Kausch [272, 315] have derived strong evidence that the initiation of intrinsic crazes is governed by a temperature and stress activated instability of the entanglement network frozen-in during the glass transition. The effect of pre-orientation on intrinsic craze initiation gave considerable support to this model. In fact, intrinsic craze initiation intimately reflects the modification of the entanglement network induced by pre-stretching above T_g. In principle, the instability of the entanglement network observed in PC may arise from the activation of several mechanisms such as chain slippage, disintegration of entanglement points or chain rupture. However, GPC-measurements of crazed samples did not reveal any appreciable amount of chain rupture [316].

Dettenmaier [272] has analyzed in detail the microstructure of intrinsic crazes by small-angle X-ray scattering (SAXS). The fibrillar microstructure detailed analysis

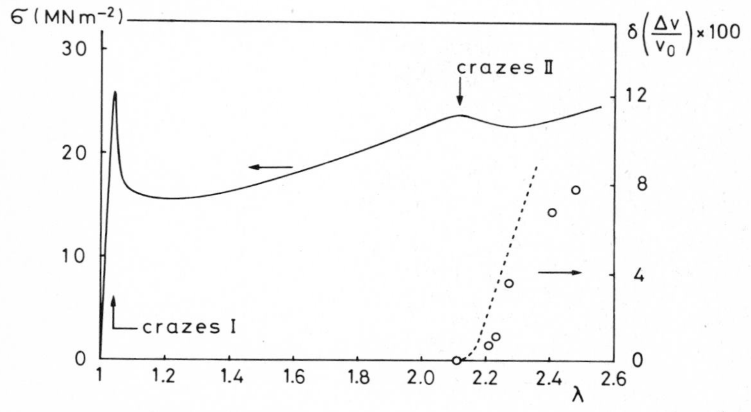

Fig. 9.24. Stress-strain curve of PC at 130 °C; the relative volume change, $\delta\left(\frac{\Delta v}{v_0}\right)$, was determined from strain measurements (dotted line) and from density (circles) (from [272])

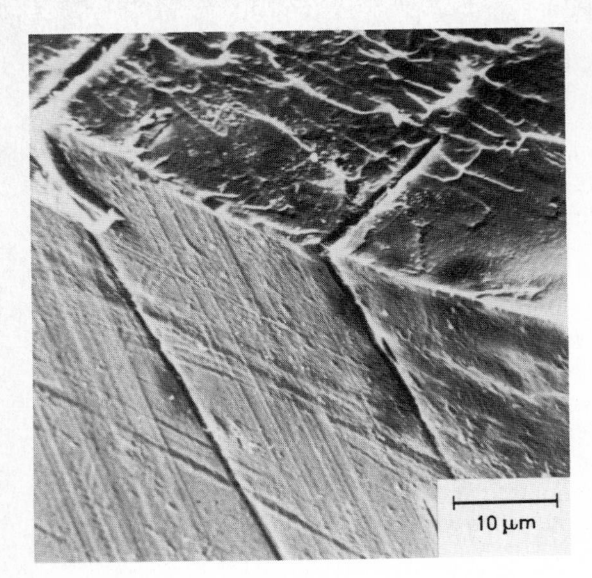

Fig. 9.25. Scanning electron micrograph of crazes I in PC; the original specimen surface is seen on the bottom left hand side (from 272)

of the craze structure in terms of both the volume fraction of craze fibrils and of the fibril diameter. This analysis showed that the microstructure of intrinsic and extrinsic crazes is considerably different. Depending on strain rate, drawing temperature and pre-orientation the average fibril diameters cover a range of 100–200 nm (see Fig. 9.27). As for extrinsic crazes [288] the product of the average fibril diameter and the stress at craze initiation is found to be independent of the strain rate and the drawing temperature. The volume fraction of craze fibrils is approximately constant and amounts to $V_f^{II} \approx 0.75–0.80$ as compared to $V_f^{I} \approx 0.5$ for crazes I. The fact that

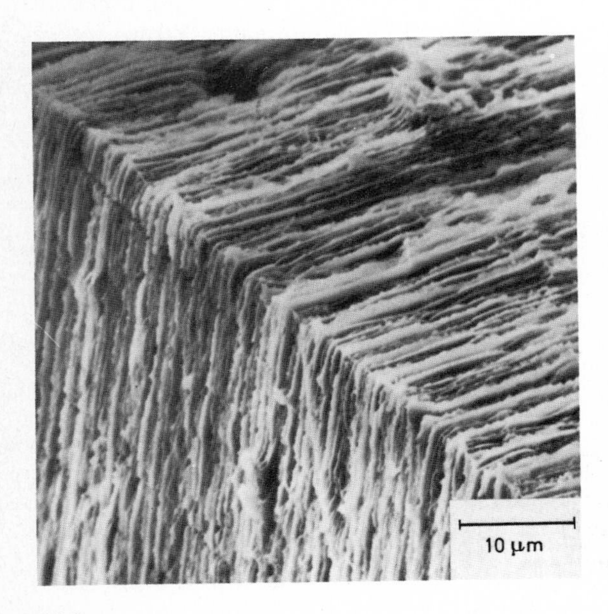

Fig. 9.26. Scanning electron micrograph of crazes II; the original specimen surface is at left hand bottom side (from 272)

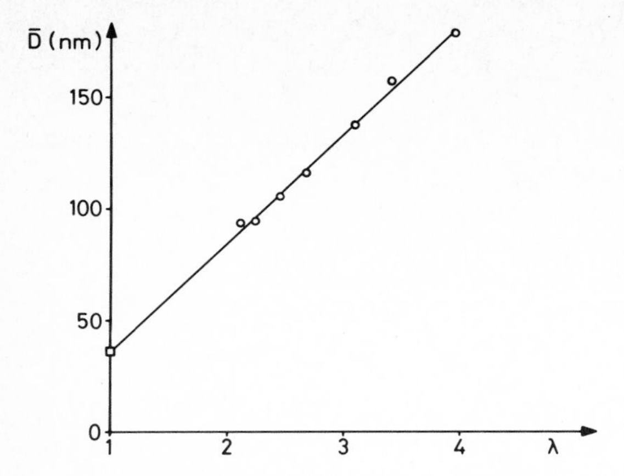

Fig. 9.27. Average fibril diameter, $\bar{D}$, as a function of the total extension ratio λ at craze initiation in unoriented and pore-oriented PC: crazes I (□), crazes II (○) (from 272)

$V_f^{II} > V_f^I$ implies that the additional extension of the material within the craze II fibrils during fibrillation is smaller than in crazes I; however, the total draw ratio λ_{fib}^{II} is larger than λ_{fib}^{I}.

There is some evidence that the intrinsic crazing of PC is related to the existence of a general mode of cavitational plasticity in highly extended polymers. PMMA and PS are further examples of polymers in which cavitation and stress-whitening occur at high stresses and strains [272]. Intrinsic crazing or related phenomena have also been observed in crystalline polymers. For example, Friedrich [274] reported on intrinsic craze formation in highly deformed polypropylene. There is also some evidence that instability phenomena which occur under certain drawing conditions in several polymers such as PETP, PE, PP and PA may have the same origin. In tensile tests these instabilities result in stress oscillations associated with the formation of stress-whitened zones of a fibrillar and voided microstructure [317].

D. Molecular Interpretation of Craze Propagation und Breakdown

The foregoing discussion on the phenomenology of craze initiation has also given a molecular interpretation of the propagation of an individual craze. Taking into consideration the formation of many crazes either simultaneously, or in sequence, a particular craze grows

- in length through the displacement of the craze tip into uncrazed, "activated" material ahead of the craze tip by breaking up the so far contiguous matrix [166, 249, 287, 301, 321, 326];
- laterally by an increase in fibril length through the transformation of matrix material into (highly) drawn matter in what may be called a micro-yielding process.

Seen from the craze tip, the craze preserves its shape during growth.

There are three aspects of craze propagation which will be discussed at this point: the kinetics of craze growth, the stress at the craze-matrix interface (region C in Fig. 9.20), and the craze breakdown.

The rate of longitudinal craze growth has been studied by a number of authors. The many problems still existing with regard to the transition of matrix material into craze matter and with the rheological properties of the latter are found to be of double weight in any quantitative description of craze propagation. For this reason no detailed description of the various approaches will be given here – but the basic concepts should be mentioned. The fracture mechanical investigations on PMMA [15, 50, 102, 127, 133] and PC [127, 144] have led to empirical expressions for $d(a + r_p)/dt$, the rate of craze growth, in terms of *stress intensity factors*. Kambour [76] and Marshall et al. [102, 133] stress the importance of *environmental flow* through the porous craze material. Verheulpen-Heymans [156] formulates a model for craze growth based on stress and strain analysis around a craze and on the *rheological properties of the craze matter*. In those cases where the craze length was found proportional to the crack length [15, 144, 177] the *empirical law for crack growth* (e.g. Eq. 9.22) also describes craze growth.

Using holographic interferometry Peterson et al. [148] determined the distribution of the incremental stresses caused by a strain increment applied to a craze in PC. At low prestrains (1.3–1.7%) they observed that the craze material near the tip of the craze actually supported a stress increment $\Delta\sigma_y$ well in excess of the average stress increment $\overline{\Delta\sigma_y}$ ($\Delta\sigma_y/\overline{\Delta\sigma_y}$ equal to 1.2 to 1.45). The peaks in the stress increments occurred at a distance of 0.2 to 0.4 of the craze length *behind* the craze tip. In their photoelastic studies of PMMA and PC crack tips Narisawa et al. [127] noted high stress concentrations just before the initiation of a craze and very modest stresses 60 min later. No noticeable new stress concentration appeared at the craze tip.

In recent years, the craze micro-mechanics has been tremendously developed, giving detailed information on craze shape and craze stress profile (see Section IIA and refs. [249, 278–295, 323, 324]). Notwithstanding some differences in the modellisation of the craze-matrix interface [329, 331] the shape of these interfaces conformed well to the predicted shape according to the Dugdale-Muskelishvili plastic zone model [7]. In that model it is assumed that a normal stress equal to the matrix yield stress σ_F is acting on the interfaces. This stress σ_F is derived from Eqs. (9.18) and (9.20) as $\pi E v_a/4r_p$. Using the approximately molecular-weight independent ratio of $r_p/2v_a \simeq 27$ as given by Weidmann and Döll [15] for PMMA and using an E of 4.5 GN m^{-2} one arrives at an σ_F of 65 MN m^{-2}. This value corresponds to the strength of the isotropic PMMA (Table 1.1). Fraser and Ward [177] obtained in the same way an interfacial stress of 58 MN m^{-2} (at 22 °C) and 130 MN m^{-2} (at – 130 °C) for PC.

Another method, Knight's analysis of stress distribution along a craze, has been used in a number of investigations (cf. [76, 77]). Verheulpen-Heymans [157] quite recently pointed out, however, that the – mostly unknown – rheological behavior of craze material and of the craze tip zone has such a strong effect on the calculated stress field that at the present time the results of this method cannot be interpreted unambigously.

In a growing craze molecular strands are formed (at B in Fig. 9.20) and extended (zone C in Fig. 9.20). The growth of a craze is slowed down or even halted
- if the craze hits upon an obstacle which prevents further fibrillation of matrix material (e.g. a heterophase inclusion; a zone of higher orientation, larger flexibility or lower stresses; the end of the specimen) [349];
- if the craze has grown to such an extent that the initiating crack or flaw is blunted and rendered uncritical [318b, 321, 329];
- if the craze interacts with other crazes or with the environment in such a way as to change the conditions which have led to its initiation and/or growth (stress relaxation of the sample because of craze-related creep deformation, penetration into a lower-stress zone adjacent to another craze) [319, 324, 327a, 338].

On the other hand craze growth will continue or be resumed if the molecular strands fail (transition from craze to crack, zone D in Fig. 9.20). It becomes immediately clear from Figure 9.20 and Eq. (9.18) that the energy dissipated in the opening up of a craze is the higher the longer and wider the craze (zone C) that is the more resistant the molecular strands and the higher σ_F. As already indicated earlier, the breakdown of the molecular strands will occur through:
- chain scission (rapid loading of well entangled networks at "low" temperature, cross-linked networks)
- chain slip (shorter chains and/or higher temperatures)
- chain disentanglement (long-time and/or fatigue loading).

These mechanisms have been clearly confirmed [283, 284, 308, 313, 318a, 330, 335].

The presence of an active environment evidently favors the latter two mechanisms (see later).

Apart from electron microscopical observations, the phenomenon of the loading and breakdown of molecular strands had also been studied by thermal measurements [31, 50, 184–186], from analysis of the influence of molecular weight on crazing [11, 15, 65, 79, 146, 178], through acoustic emission [174, 188], and by the ESR technique [189–190] respectively.

Döll [31, 50] determined the amount of heat Q dissipated in PMMA at the tip of a rapidly propagating crack. To derive Q they measured the temperature rise $\Delta T_1(t)$ at a thermocouple placed on the specimen surface close to the prospective fracture plane. Considering geometry, thermal conductivity, and crack speed $\dot{a}$ they calculated the heat Q dissipated per unit are of crack surface. As expected Q increased with $\dot{a}$ and M_w (from 0.15 to about 5 kJ/m^2) and it was found proportional to the dynamic energy release rate G_d. Extrapolating $Q(G_d)$ down to G_{Ic} (at $\dot{a} = 140\ m\ s^{-1}$) they observed that the dissipated heat Q ($\dot{a} = 140\ m\ s^{-1}$) = Q_0 accounted for only 57% of the released mechanical energy G_{Ic}. They ascribe this difference to the fact that the highly stressed molecular strands conserve an amount of energy of the order of 0.43 G_{Ic}, that this energy is released at the moment of fibril rupture and that it is dissipated over wide distances in the form of small shock waves which escape measurement by the employed difference technique [30]. The acoustic emission analysis (AEA) of a crazing PMMA sample by Roeder [188] permits similar conclusions. The formation of crazes (transition A → C in Fig. 9.14) did not give rise to any discernible acoustic signal. Whereas the transition from a craze to a crack (C → D) could be traced by AEA even at an earlier stage than by optical microscopy. In PS, however, craze formation is accompanied by sharp cracking sounds at strains of 0.6% or less [78].

The thermal measurements are not only interesting in view of the energy balance of the crazing process but also because they permit the calculation of the local temperature rise ΔT_0 caused by the opening up and breakdown of the craze in PMMA. Döll [30] assumed that Q_0 was initially confined to the zone of crazed material. Using a density of 0.6 g cm^{-3}, a specific heat of 1.46 J g^{-1} K^{-1}, a craze layer thickness 2 v of 1.65 μm, and a Q_0 of 335 J m^{-2} he arrives at a ΔT_0 = 230 K. This value is in accord with theoretical estimates of Weichert and Schönert [185] and with infrared measurements of Fuller et al. [184] on PMMA. The latter determined within a range of $\dot{a}$ of 200 to 640 m s^{-1} a constant ΔT of 500 K. The simultaneously observed increase in $Q(\dot{a})$ implied that the plastic deformation at the crack tip became more extensive at higher crack speeds. Preliminary experiments on PS showed a ΔT of 400 K and smaller heat values [184]. These temperatures are certainly large although not completely unreasonable. They indicate that not only melting but thermal degradation must occur under these conditions. And they are in proportion with the much higher temperature rises – of several thousand K – which were deduced by Weichert [186] from the spectroscopical analysis of light emitted during fracture of glass.

It was noted at an early stage that sample molecular weight is an important variable in craze propagation and breakdown. Rudd [79] studied the stress relaxation of PS films in contact with the crazing environment butanol. A plot of $\sigma(t)/\sigma(0)$ versus logarithm of time revealed that the time to reach the 40% stress level was three orders of magnitude larger for the higher molecular weight samples. The final decay from 0.4 $\sigma(0)$ to zero stress occurred rather rapidly for all molecular weights [79].

The observation of Fellers and Kee [146] that the breaking stress of PS only gradually increases once $M_n > 2\ M_e$ has already been mentioned in the previous section (Fig. 9.23). Their results conform quite well to those of Döll and Weidmann [15, 50]. These authors determined the shape of a craze, the released heat Q and the material resistance R for a series of PMMA samples having well defined molecular weights M_w between $1.1 \cdot 10^5$ and $8 \cdot 10^6$. Measuring the crack opening 2 v, the craze width 2 v_c, and the craze length r_p at a crack speed of 10^{-8} m s^{-1} they noted that these parameters of the craze shape increased with M_w up to an M_w of about $2 \cdot 10^5$. At higher M_w hardly any changes of 2 v and r_p and very small increases of craze width were observed [15, 329]. This means that initially ($M_w < 1.6 \cdot 10^5$) craze widths increase with chain lengths. At molecular weights comparable to M_e the interpenetration and entanglement of the molecular coils hardly allows the formation of fibrils [11, 146, 187]. At higher molecular weights (up to $M_w = 2 \cdot 10^5$) the coils are larger in size and the absolute number of entanglements per coil increases; the fibrils become stronger and more resistant to failure by slippage or disentanglement [282, 283, 308, 313, 330]. At even higher molecular weights ($M_w > 2 \cdot 10^5$) and under constant load the fibril strength seems to be more determined by the strength than by the flow behavior of the molecules. Under oscillating loads disentanglement may even occur at such molecular weight levels. Recent experiments of Skibo, Hertzberg, and Manson [191] strongly support the contention that the breakdown of molecular strands as a consequence of their disentanglement triggers the stepwise growth of fatigue cracks. This phenomenon will be discussed in more detail in Section 9 III D.

The critical energy release rates G_{Ic} show a behavior similar to that of the craze shape. For small M_w the G_{Ic} are strongly dependent on M_w. Thus they increase from $1.4\ J\ m^{-2}$ at $M_w = 2 \cdot 10^4$ to $110\ J\ m^{-2}$ at $M_w = 12$ to $15 \cdot 10^4$ [65]. At higher molecular weights – up to $8 \cdot 10^6$ – only a gradual increase of G_{Ic} to values of between 160 and $600\ J\ m^{-2}$ is observed [30, 65]. This corresponds to what has been said above on the effect of molecular weight on fibril length and strength.

The occurrence of chain rupture in craze breakdown has been inferred from the above thermal, mechanical and acoustical evidence. Its direct observation by ESR technique is difficult because of the limited number of crazes which can simultaneously be carried to breakdown. The number of chain breakages generally corresponds to that of a single fracture plane. No quantitative data on free radicals formed in crazing have been reported. DeVries et al. [189] have observed a very weak ESR signal in a crazed and cut PMMA fiber. Nielsen et al. [190] did not find any evidence of free radicals in thoroughly crazed – but unbroken – samples of PS and ABS. Popli et al. [335] also do not find an ESR signal in crazed PS, they detect, however, a 20% molecular weight decrease using GPC.

The above considerations mostly dealt with an individual craze, with the conditions of its initiation, propagation and breakdown in a brittle polymer. Generally a larger number of crazes is formed in a stressed specimen (cf. Fig. 9.12). If those crazes are sufficiently distant from each other then they will grow freely. Opfermann [175, 327a] found that in PMMA a lateral distance between crazes of 80 μm assured their unimpeded longitudinal growth. The macroscopic mechanical properties (strain at break, strength at short or long times, energy to break) depend to some extent on the number of crazes per (surface) area, but they are still comparable to those of a brittle solid with a strain at break of some 4 to 5% and a low fracture energy. In order to sensibly increase the macroscopic creep compliance and the energy to break, crazes must be initiated in large numbers throughout the sample volume and they must be prevented from premature breakdown. Both objectives have been achieved with heterophase copolymers or blends.

The obvious technological importance of the toughening of brittle polymers through controlled crazing is reflected by the considerable amount of work done in this area. In recent monographs [92, 339, 340] a comprehensive elaboration of the physical, chemical, and technological aspects of polymer mixtures, rubber-toughened plastics, block copolymers, grafted copolymers, interpenetrating networks, and polymer blends has been given. The reader is referred to these monographs for any detailed information on these systems, on their characteristic behavior and on the specific materials science problems they pose. The fact that several international meetings are annually devoted to the synthesis, compatibility, and mechanical properties of copolymers and blends [193, 341] testifies to the importance of this class of polymer materials and to the number of open problems which cannot even be touched upon in the context of this book. The few references given focus on individual morphological aspects of the crazing of heterogeneous glassy polymers [94, 97, 143, 152, 168, 342–349]. The observation that rubber-toughened polymers are capable to withstand substantial (shock) loads without traceable damage (that is *without craze-initiation*) points to the importance of the phase of homogeneous tensile dilatation which preceeds matrix break-up. If crazes are initiated they impart plastic behavior to the (stress-whitening) samples [342–349].

Valuable information on the role of crazing in breakdown of a polymer is obtained, of course, from fractography. Relevant investigations [30, 49, 61, 66, 132, 150, 155, 169, 194–204] on this subject will be discussed in Section III A.

E. Response to Environment

In the previous sections on crazing primarily mechanical and molecular parameters have been discussed while the chemical environment was not considered as a variable. At this point an overview over the physico-chemical response of a crazable material to an active chemical environment will be given. (The *chemical* response to an active *physical* environment such as photodegradation or ozonolysis has been treated or referenced in 8 III).

The physico-chemical actions of a (gaseous or liquid) chemical agent on a polymer involve adsorption and absorption of the agent, the swelling and/or plasticization of the matrix, the reduction of surface energies, and/or chemical reactions such as the hydrolytic depolymerization. These aspects of environmental stress cracking (ESC) have been treated in numerous original articles [119–162, 352–381] and some reviews (e.g. [76–77, 80, 123, 171, 275, 350, 351, 382–384]). From the extremely large body of experimental observations only a few can be discussed in the following.

The physico-chemical actions of a liquid environment may influence the initiation, propagation, or breakdown of a craze in a thermoplastic polymer. There seems to be agreement that a liquid must be able to diffuse into a polymer in order to influence craze initiation. Narisawa [119] determined the critical stresses σ_i for craze initiation in thin films of PS and PC in contact with various alcohols and hydrocrabons. He observed that crazes appear without significant delay and that σ_i decreases with decreasing chain length of the solvent (from 45 to 20 MN m^{-2} for PS, from 70 to 50 MN m^{-2} for PC). From these results he concludes that a slight swelling of a microscopic surface area is necessary and sufficient to effect craze initiation. He derived a criterion for σ_i in the form of Eq. (8.29) with activation volumes of the order of 1.0 to .13 nm^3, activation energies of 109 to 130 kJ/mol, and rate constants of 1 to $10 \cdot 10^{-38}$ s^{-1} (PS) and 2 to $50 \cdot 10^{-45}$ s^{-1} (PC).

The solvent crazing in terms of polymer and liquid solubility parameters has been studied in detail by Andrews et al. [124, 126] and Kambour et al. [125, 128]. Kambour employed a wide range of swelling liquids with solubility parameters δ_S between 5.34 and 19.2 $cal^{1/2}$ $cm^{-3/2}$. He determined the equilibrium solubility S_v, expressed as volume of liquid absorbed per unit volume of polymer, for PS, PPO, and PSU. The swelling of PPO across the whole spectrum of organic liquids was found to be inversely correlated by $|\delta_S - \delta_{PPO}|$ and, consequently, the craze resistance was correlated by $|\delta_S - \delta_{PPO}|$. In PS and PSU craze resistance was less well correlated by the solubility parameters. With all three polymers, however, the equilibrium solubility S_v gave a good measure of the polymer-solvent interaction. Using S_v as independent variable, unique plots of T_g and of craze initiation strain ϵ_i were obtained for two sets of polystyrene data. One set was obtained from samples preplasticized to various degrees by o-dichlorobenzene; the other from "dry" samples in contact with the swelling agent (ϵ_i) or swollen films (T_g). Kambour concludes from these results that the presence or absence of a liquid/polymer interface is immaterial to the crazing effectiveness of a given crazing agent.

The crazing agent thus acts through its presence within the polymer matrix. In increasing the chain mobility (lowering T_g) it facilitates the primary and secondary steps of craze initiation: nucleation and stabilization of a craze. This leads to the lowering of σ_i and ϵ_i in brittle polymers such as PS. Easier nucleation and stabilization even cause the appearance of crazes in otherwise ductile materials such as PPO, PSU, PVC, or PC.

Andrews et al. [124, 126] also studied the equilibrium swelling, of PMMA, in various alcohols and related it to the observed changes in yield stress, σ_F, glass transition temperature, T_g, and material resistance R. They report the interesting phenomenon that R above a certain critical temperature T_c became independent of temperature. Within experimental error T_c corresponded to the T_g of the PMMA under the particular conditions of swelling. They propose a relation for R involving the surface energy of the nucleated cavities and σ_F. They then interpret the constancy of $R(T > T_c)$ with the vanishing of $\sigma_F(T > T_c)$. The absolute values of $R(T > T_c)$ are rather small: about 0.1 J m^{-2} for PMMA in isobutanol, carbon tetrachloride, n-propyl and isopropyl alcohols, 0.3 J m^{-2} for ethanol, 0.45 J m^{-2} for methanol [124].

The swelling, crazing, and cracking behavior of PMMA, PVC, and PSU in contact with some 70 different liquids was analyzed by Vincent and Raha [123]. They considered not only the solubility parameter δ_S but also the value of the hydrogen bonding parameter H_{OD} based on the displacement of the OD infra-red absorption band in CH_3OD in presence of the particular liquid and benzene. Better (but still no unique), correlation between the mode of failure and δ_S and H_{OD} was obtained.

An extensive investigation of the role of metal salts in stress cracking of various polyamides was carried out by Dunn and Sansom [90–93]. Metal halides permitted to distinguish two types of action: the formation of complexes between the metal and the carbonyl oxygen and interference with the hydrogen bonding (found for the chlorides of Zn, Co, Cu, Mn) or with solvent cracking, as with Li Cl, Ca Cl_2, Mg Cl_2, or Li Br [90–91]. The action of metal thiocyanates on polyamide 6 was similar to that of the corresponding metal halides [92]. Of various nitrates $Cu(NO_3)_2$ had the strongest effect on the stress cracking of PA6 films [93].

In the investigations of Marshall and Williams et al. [52, 102, 133, 151, 164] Narisawa et al. [127], Kitagawa et al. [144], and Krenz et al. [159–160] craze growth and creep were related to fracture mechanical quantities. Graham et al. [164] model crazes in PMMA as a line plastic zone subjected to a craze stress σ_{cr}. In the presence of active liquids σ_{cr} is reduced from its air value of 100 MN m^{-2} to e.g. 7 (methanol), 5 (ethanol and propanol) or 10 MN m^{-2} (butanol). A clear linear correlation between K_m and σ_{cr} was found to exist. For the craze size at initiation (i.e. just before the beginning of crack growth) the authors found a unique value of 11.5 μm [164].

Similar studies of solvent crazes in polystyrene were carried out by holographic interferometry by Krenz et al. [159–160].

These and later investigations [e.g. 367–370, 381], have confirmed the fracture mechanics approach of environmental crack propagation proposed especially by Williams et al. [102, 133, 151]. In terms of the applied stress intensity factor K_c and of the rate $\dot{a}$ of craze and crack propagation, three zones of behavior are identified:

- if $\dot{a}$ is small, the craze tip is always in perfect contact with the environmental agent and $K_c(\dot{a})$ is influenced by the described mechanisms of matrix and fibril plasticization (low-R regime);
- if $\dot{a}$ becomes comparable to the speed of capillary flow of the agent through the craze material, then $\dot{a}$ may largely be controlled by the latter (leading to a strong viscosity effect of ESC in this "transition regime");
- if $\dot{a}$ is large, the crack-tip "escapes" contact with the stress cracking agent and $K_c(\dot{a})$ may not at all depend on its presence (high-R regime).

A similar three-stage behavior is found in static environmental loading (of LDPE) [367–370]:

- if K_c is slightly smaller than K_{cc}, the times-to-failure t_b in air and in an ESC agent are identical (incubation period);
- if K_c is clearly smaller than K_{cc} a strong dependence of t_b on K_c and on the ESC agent is observed; the failure mode changes from brittle to semi-ductile;
- at small K_c, t_b approaches infinity, failure is brittle again and seems to be caused by ESC-assisted chain disentanglement [369, 384].

Studies of the micromechanics of deformation in semicrystalline polyethylene in environmental stress cracking agents [107, 154, 170, 171, 204] have elucidated the role of spherulites and of the mosaic block structure and its disintegration into independent nonuniform fibrils. The morphology of crazes in amorphous PVC in liquid and vapor environment was examined by Driesen [122] and Martin et al. [198].

It has not always been recognized that liquid gases such as N_2 and Ar are also acting as environmentally active agents. Parrish and Brown et al. [116, 129] have been the first to report that the fracture stresses of PE and PTFE immersed in helium at 78 K are respectively 21 and 33% larger than those obtained in liquid N_2. Later studies of Brown et al. [134, 157–158], Peterlin et al. [135, 137–138, 145] and Kastelic and Baer [140] concern N_2, Ar, O_2, CO_2, and He and a variety of materials (PP, PTFE, PETP, PC, PMMA). It is the common consensus that

- N_2 and Ar exert an environmental effect on the tensile stress-strain curves of all polymers at low temperatures
- the environmental effect is the larger the closer a gas is to its condensation point and the greater its thermodynamic activity [145]
- N_2, Ar, O_2, and CO_2 usually cause craze yielding whereas in He or *in vacuo* the polymers undergo brittle fracture without crazing [352–355]
- the gases have to act at or close to a free surface
- as crazing mechanisms gas absorption and matrix plasticization are proposed as well as gas adsorption and surface energy reduction which facilitates cavity formation.

The above mentioned aspects of interaction between a stressed polymer and an active environment have, in recent years, continued to be the object of strong interest [350–384]. Thus, the effects on craze behavior of changing surface tension [356], of craze fibril plasticization [357], and of different molecular, structural, and environmental parameters have been investigated for e.g. PS [356–358], PVC [324c], PMMA [359–363], LDPE [364–370], HDPE [371–377], and PC [378–380].

The important influence of viscosity and capillary force on the $K_c(\dot{a})$ relationship in ABS has been analyzed by Kambour and Yee [381]. They found that the average linear flow rate of an ESC agent in an ABS craze was proportional to the ratio of liquid surface tension to viscosity. The literature up to 1982 on the mechanical behavior in gaseous environments [382] and on the thermodynamics [383] and phenomenology [384] of environmental stress cracking (ESC) has been covered and discussed recently.

From the discussion in this section a number of craze *suppression* mechanisms have become apparent which may be summarized here:

- imposition of hydrostatic pressure in addition to tensile stresses [86, 139, 142]
- introduction of compressive stresses into the surface [77, 174, 176]
- orientation of molecules in the surface layer [89, 153]
- plasticization, preferentially of the surface [82, 162, 178]
- coating of the surface by an oligomer of the sample material, e.g. coating of PS by a 600 molecular weight polystyrene oligomer [178] polishing of the surface [173, 178]
- increase of sample molecular weight.

III. Molecular and Morphological Aspects in Crack Propagation

A. Fracture Surfaces and Molecular Mechanisms

1. Thermosetting Resins

The term thermosetting resins covers a wide range of crosslinked polymers. Amongst these the fracture properties of epoxide resins [385–389] and unsaturated polyesters [386, 390, 391] have been most extensively studied although some work has been devoted to polyurethanes [392], phenol-formaldehyde [393] and polyimides [235]. Direct comparison of the work carried out in different laboratories is often complicated by the wide choice of resin and hardener chemistry, curing procedures and even batch-to-batch variability. As a result characteristic behavior is only found within quite broad limits.

Thermosetting resins have been traditionally considered to be extremely brittle materials it being assumed that the crosslinked state precludes extensive viscous flow. Indeed in uniaxial tensile tests well below the glass transition temperature the stress-strain diagrams are essentially linear to rupture with deformations at break of some 2 to 5%. However, it is now realised that even these highly crosslinked materials are capable of considerable plastic deformation which is highly localised at the crack tip. In fracture mechanics testing using stable specimen geometries where the compliance is linear with the crack length (for example double torsion, tapered double cantilever beam) crack propagation in thermosetting resins may be characterised by either of the load-displacement diagrams shown in Figure 9.28. For resins with high yield stresses (> 100 MPa) crack tip plasticity is limited and stable propagation is observed. However, for resins with lower yield stresses, material at the crack tip may plastically flow blunting the crack sufficiently to cause arrest. The sample must then be reloaded to reinitiate a sharp crack. The load-displacement diagrams have the distinctive saw tooth appearance shown in Fig. 9.28 and values of K_c appropriate to crack initiation (K_{ci}) and crack arrest (K_{ca}) may be determined. Since this flow is time dependent, values of K_{ci} vary considerably with testing rate and at higher cross head speeds propagation may become continuous and stable [394]. The magnitude of K_{ci} has been shown to be dependent on the radius of the blunted crack [395].

These modes of crack propagation have obvious repercussions on the fracture surface morphology. When crack propagation is stable and continuous, fracture surfaces are smooth and relatively featureless as shown in the extreme left of Figure 9.29. In this micrograph a typical arrest line may be observed followed by a region of closely spaced striations parallel to the direction of crack growth. These striations are characteristic of slow propagation after crack arrest and have been observed in epoxide resins [385], polyimides [235] and phenol-formaldehyde resins [396].

Lee [397] has suggested that these structures result from the superposition of reflected elastic stress waves which create local compressive stresses causing crack branching. Another explanation has been given by Atsuta and Turner [396] who have used an approach developed to explain similar effects in silica glasses [398]. In a perfectly elastic system a crack will propagate in a plane perpendicular to the local tensile stress. If the crack moves into a region where the principal stress is oriented at a slightly different angle the crack cannot adapt instantaneously and instead may divide into several segments. Indeed Melin [399] has shown theoretically that two cracks in a fragile material cannot meet due to the stress state in their proximity. They start by avoiding each other and then joint forming a lip which is partially attached to each fracture surface. The ultimate separation of the two surfaces de-

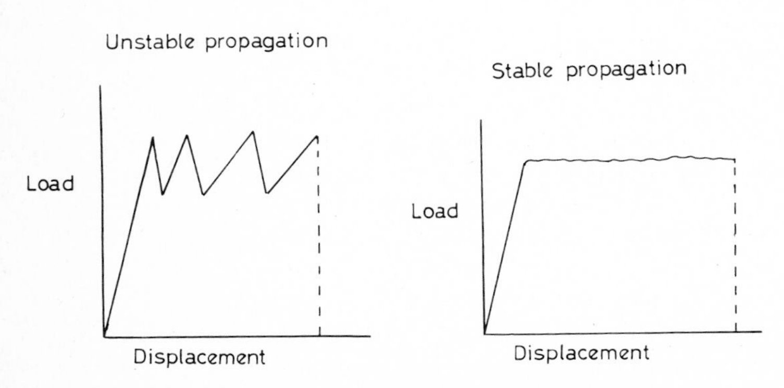

Fig. 9.28. Typical load displacement diagrams for an epoxy resin

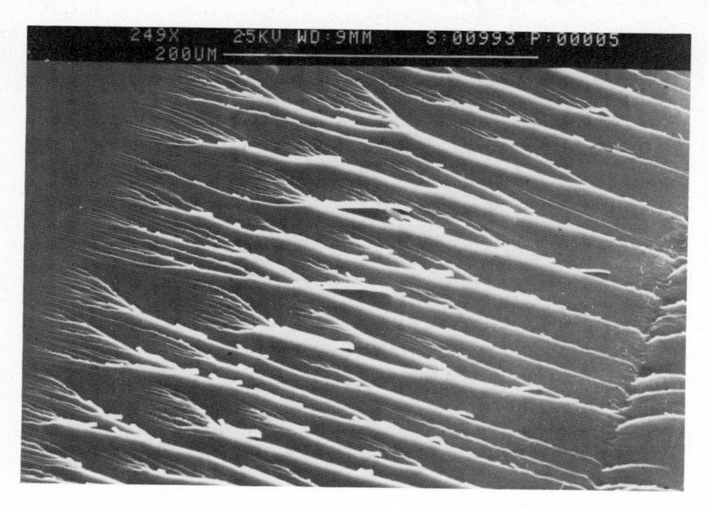

Fig. 9.29. Stable crack propagation (extreme left) crack arrest line and subsequent striated region

forms and finally ruptures the fibres. This latter phenomenon has been observed directly during in situ fracture tests in the scanning electron microscope [400]. A further explanation has been postulated more recently for the existence of these striations that is based on the theory of meniscus instability [401]. This theory is founded on the observation of a meniscus between two fluids of differing density. Under certain conditions the originally smooth meniscus may break up into a series of fingers [402, 403]. Craze formation in thermoplastics has been explained on the basis of this theory (see Section II). The phase of lower density being a wedge-shaped zone of plastically deformed and strain softened material; whereas the zone of higher density is the undeformed material neighbouring the craze. Furthermore the theory of meniscus instability has recently been invoked to explain ductile crack growth in epoxide polymers [404] tested at elevated temperatures and slow crosshead speeds. The finger-like furrows observed on the fracture surface are said to be formed by the break-up of the crack front as the crack grows through the plastic zone which is under constraint from the elastically deformed material outside. Fracture surfaces characteristic of ductile fracture may also be observed for specimens tested as very thin sheets [232] or in under-cured resins [405].

There has been considerable debate in the literature as to whether crazing may occur in thermosetting polymers. The evidence for crazing is not very substantial and comes from rather specific cases such as in straining very thin films [38], in under-cured resins [406] or for systems with very low crosslink density which thus approach thermoplastic behavior. Indeed Donald and Kramer [318b] have shown in thermoplastics that there is a transition from crazing to shear yielding when the distance between physical entanglements decreases to below ~ 20 nm. There is an increasing amount of evidence that the primary mode of deformation certainly in epoxide resins, and probably for other thermosetting polymers as well is by shear yielding both in unmodified resins [407], rubber modified systems [240, 408] and in composite materials [409] with epoxide matrices.

In many publications on fracture surface morphologies specific features have been attributed to certain crack velocities. However, such a correlaton may be questioned since the smooth featureless surfaces may be observed for crack speeds varying as widely as 10^{-7} ms^{-1} (in a double torsion test) to above 300 ms^{-1} (in a single edge notch test with a sharp pre-crack). This surface morphology is then related to stable continuous cracking at speeds which are defined by the specimen geometry and the inherent material characteristics. In contrast, Fig. 9.30 shows the fracture surface of an unfilled epoxide resin tested in the single edge notch geometry with a *blunt* pre-crack. The surfaces are rough with conical markings characteristic of the formation of multiple secondary cracks ahead of the main crack front. In many specimens tested under these conditions macroscopic crack bifurcation is also observed. The crack speed in this test, as measured by the sensitive graphite gauge technique [235], was also 300 ms^{-1}. Thus, the crack speed under these conditions is independent of the pre-crack geometry and does, in fact, approach the limit for the propagation of stress waves in these materials [410]. Identical surface features are seen for unnotched tensile tests in both epoxide resins and polyesters [390].

Thus, for specimens without pre-cracks or with blunt notches excess elastic energy is stored in the sample at the moment of rupture which is expended in the creation of increased fracture surface area giving rise to rough surfaces. Narisawa et al. [245] calculate that in a blunt notched epoxy specimen an "ideal fracture stress" of up to 150 MPa is reached which is comparatively larger than that of glassy thermoplastics. Leksovskij et al. [411] report on the "explosive nucleation of microcracks" in ED-20 epoxy resin. It may be concluded then that the fracture surface morphology of these crosslinked materials is essentially determined by the specimen geometry, the crack tip radius and the applied stress intensity factor rather than being uniquely defined by the crack speed.

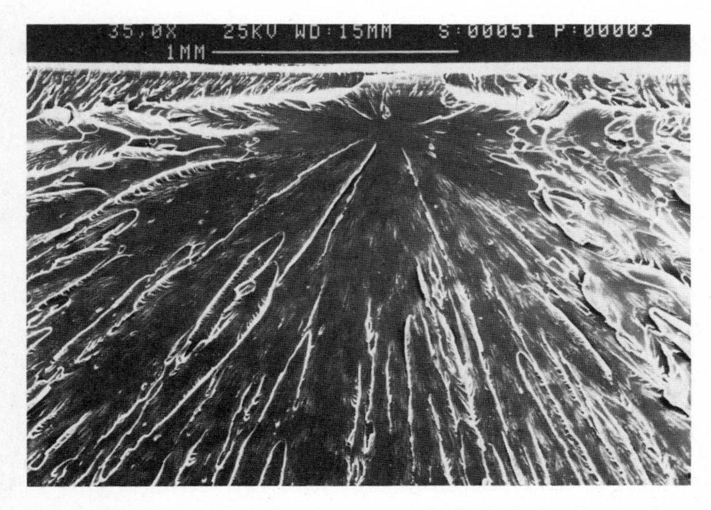

Fig. 9.30. Fracture of an unfilled epoxy resin from single edge notch specimen with a blunt pre-crack

2. Thermoplastics

Fractography, the study of the morphology of fracture surfaces, is an obvious tool to elucidate the origin and the mode of propagation of a crack. At this point some remarks are necessary concerning particularly the effect of chain length and intermolecular attraction on the fracture surface morphology of thermoplastics.

The discussion in 9 I C of the various terms contributing to the critical energy release rate G_{Ic} and the data assembled in Tables 9.1 and 9.2 have made it clear that a fracture surface is obviously not simply formed by the breakage of primary and/or secondary bonds across a fracture plane of molecular dimensions. There is always a plastic deformation of the crack tip zone and, consequently, of the ensuing fracture surface. It is to be expected that the extent of plastic deformation is the smaller the smaller the segmental mobility, i.e. the lower the temperature. At liquid nitrogen temperature most high polymers resemble glasses and fracture in a brittle manner. Viewed without magnification the fracture surfaces (e.g. Fig. 9.31) shows a macroscopic roughness but they appear to be locally smooth although not shiny. This indicates that the surfaces contain structural irregularities larger than the wavelength of light. This is the case with e.g. PE, PP, PVC, PS, and also with PMMA which, however, has a very smooth surface.

The macroscopic features of the shown fracture surface (Fig. 9.31) can be ascribed to the propagation of a cleavage crack at high speed normal to the direction of *local* tensile stress. The local stress field is strongly affected by elastic waves generated at earlier stages of crack development and by the initiation of secondary cracks. The fracture surface was obtained by bending a notched polyethylene sample at liquid nitrogen temperature [130]. The surface is locally smooth but otherwise full of steps and ridges. The intersection of wavefronts and crack planes under different (e.g. right) angles leads to the curious patterns as shown in the bottom right of Figure 9.32. The characteristic properties of thermoplastics, molecular anisotropy and strong rate dependency of deformation become apparent to a limited extent only.

The seemingly smooth parts of the cleavage surface deserve two further comments. Figure 9.33 shows that the material disintegration of the "smooth" surface is not restricted to molecular dimensions but has a honeycomb-structure with cells of 100 to 600 nm width. This is in accordance with the earlier statement on the size of surface irregularities. It is also of interest to note that these dimensions are only a

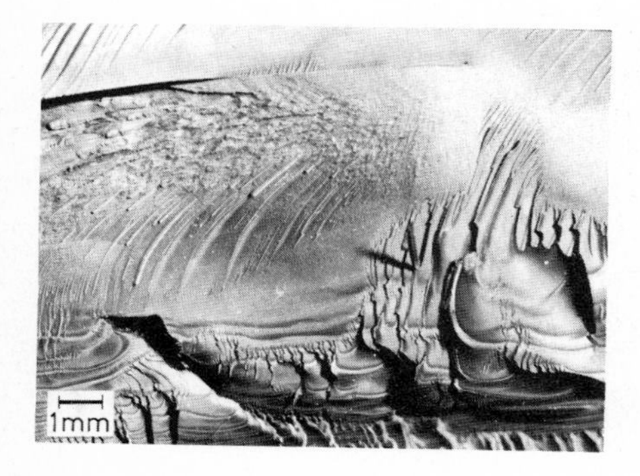

Fig. 9.31. Central section of a fracture surface from a notched HDPE specimen broken in bending at liquid nitrogen temperature (from [130])

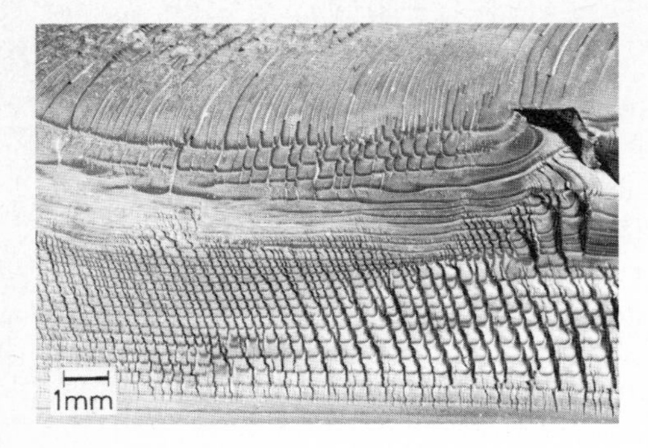

Fig. 9.32. Section of the fracture surface in the originally compressive zone from the same specimen shown in Fig. 9.31 (from [130])

few times larger than the radius of gyration of the chain molecules. The ridges between cells are highly deformed. As their width is comparable to that of the cells a plastic deformation of about 100% is indicated. This is in view of the test temperature quite remarkable. It is definitely caused by the fact that the comparatively long and strong molecules are not highly oriented but have a coiled conformation corresponding in all probability to the "solidification model" of semicrystalline polymers (cf. 2 I C). Under these circumstances axial forces are only built up if a segment is subjected to a sufficiently large shear displacement. Material separation thus necessitates the prior and substantial deformation of several entangled molecular *coils*.

At room temperature molecular mobilities are increased and intermolecular attraction is decreased. The extent of plastic deformation leading to material separation is, therefore, increased rendering some of the formerly brittle plastics tough (cf. 8 II A). In HDPE at room temperature no cleavage fracture (unstable crack propagation) can be obtained by bending, it occurs only in notched specimens of low aspect ratio or if the rate of loading is high (impact loading). In samples of HDPE

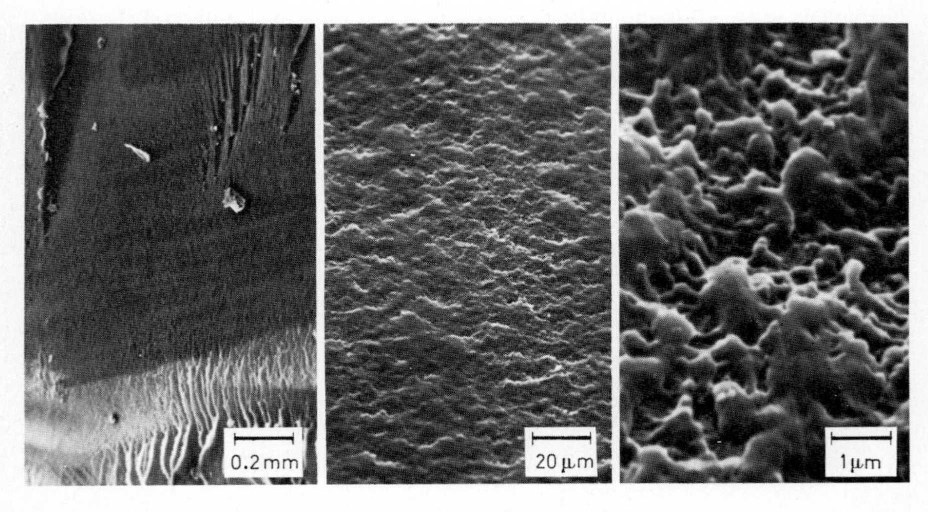

Fig. 9.33. Successive enlargements of the apparently smooth surface in the center of Fig. 9.31 (from [130])

with extremely high molecular weight ($M_w > 10^6$) cleavage fracture does not occur at all. Material separation must be enforced through large strains. Visual inspection gives already an impression of the very large plastic deformation of such a specimen. The fracture surface contains numerous rosettes and ridges which are drawn by several hundred per cent. The energy consumed in forming new surface area is accordingly high. Specific surface features are observed on broken crazeable thermoplastics (see later). One can state that in short time loading of isotropic bulk polymers chain molecules do not significantly contribute to the fracture surface energy through the energies of bond breakage and elastic retraction. Large values of chain length and chain strength reduce, however, the effect of natural of artificial flaws as stress concentrators due to the larger amount of plastic deformation necessary before material separation occurs. Vice versa, at low molecular weights samples are increasingly more brittle, fracture surfaces are smoother, secondary events are rare or absent, and the fracture surface energy accordingly small.

The above considerations apply to isotropic homogeneous materials, i.e. to amorphous [61, 198, 200] and to semicrystalline polymers having no well defined microstructure [130]. There the direction of the fracture path is more or less determined by the local stress field. In any event no traces of a microstructure became apparent through the fracture surface morphology. Distinct fracture surface features are obtained, however, from polymers with a pronounced microstructure as introduced by extended-chain crystallites or spherulites. There the material resistance depends strongly on the relative orientation of the fracture plane with respect to the structural element.

Extended-chain crystallites have been studied by several authors (cf. 2 I C and 8 I G).

In Figures 9.34–9.36 electron micrographs of replicas taken by Gogolewski and Pennings et al. [201–203] from the fracture surface of pressure crystallized poly-

Fig. 9.34. Electron micrograph of a replica of a fracture surface of 6 polyamide crystallized at 295 °C for 48 h under a pressure of 6.5 kbar. The arrow shows the growth direction of the spherulite (Courtesy S. Gogolewski [202]). (Reproduced by permission of the publishers, IPC Business Press Ltd. ©)

Fig. 9.35. Transmission electron micrograph of a replica of another part of the fracture surface of the sample shown in Fig. 9.19 (Courtesy S. Gogolewski [202]). (Reproduced by permission of the publishers, IPC Business Press Ltd. ©)

amide 6 are reproduced [202]. The micrographs reveal stacks of lamellae which have a thickness of up to 700 nm. The authors conclude from their extensive IR, WAXS, and electron microscopical studies that the lamellae represent extended chains. As they have proposed in Figure 9.37 a crack may preferentially proceed either along the (010) planes (which contain the chain ends as well as impurities rejected from the crystal growth front) or along the (002) hydrogen bonded sheets of the lamellae. Both processes do not involve the breakage of main chain bonds or of hydrogen bonds.

With respect to extended chain crystallization under high pressure ($p > 4$ kbar) the authors discuss two mechanisms [201–202]. Based on the observed change in molecular weight distribution and decrease in molecular weight concurrent with chain extension they suggest that the thermal treatment under high pressure makes possible a transamidation reaction between –NH and –CO groups of broken chain folds belonging to adjacent lamellae [201]. On the other hand the appearance of

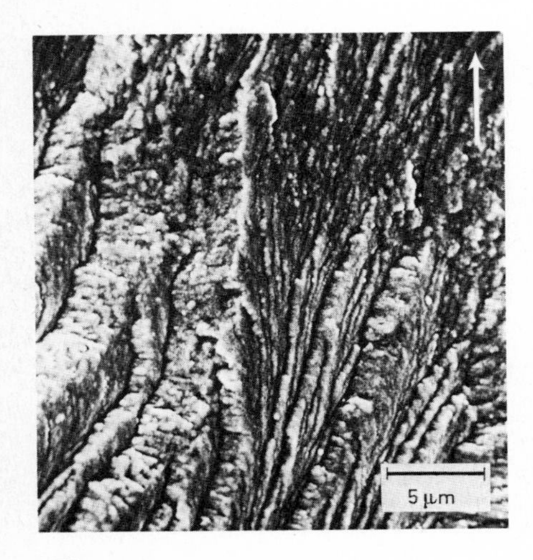

Fig 9.36. Scanning electron micrograph of a fracture surface of 6 polyamide crystallized at 295 °C for 48 h under a pressure of 6.5 kbar. (Courtesy S. Gogolewski [202]). (Reproduced by permission of the publishers, IPC Business Press Ltd. ©)

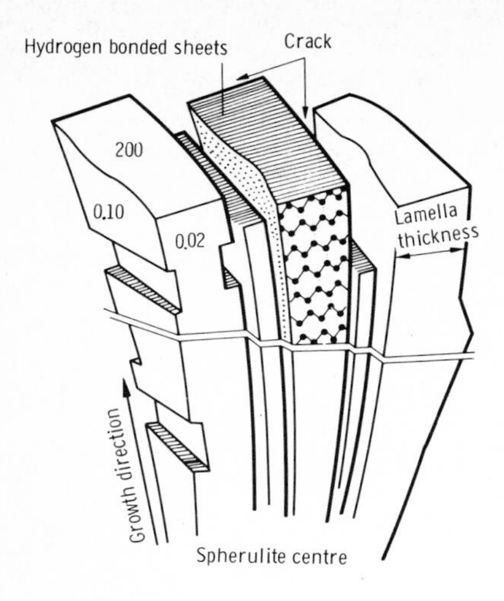

Fig. 9.37. Model proposed for fracture behavior and surface morphology of chain-extended crystals of polyamide (Courtesy S. Gogolewski [202]). (Reproduced by permission of the publishers, IPC Business Press Ltd. ©)

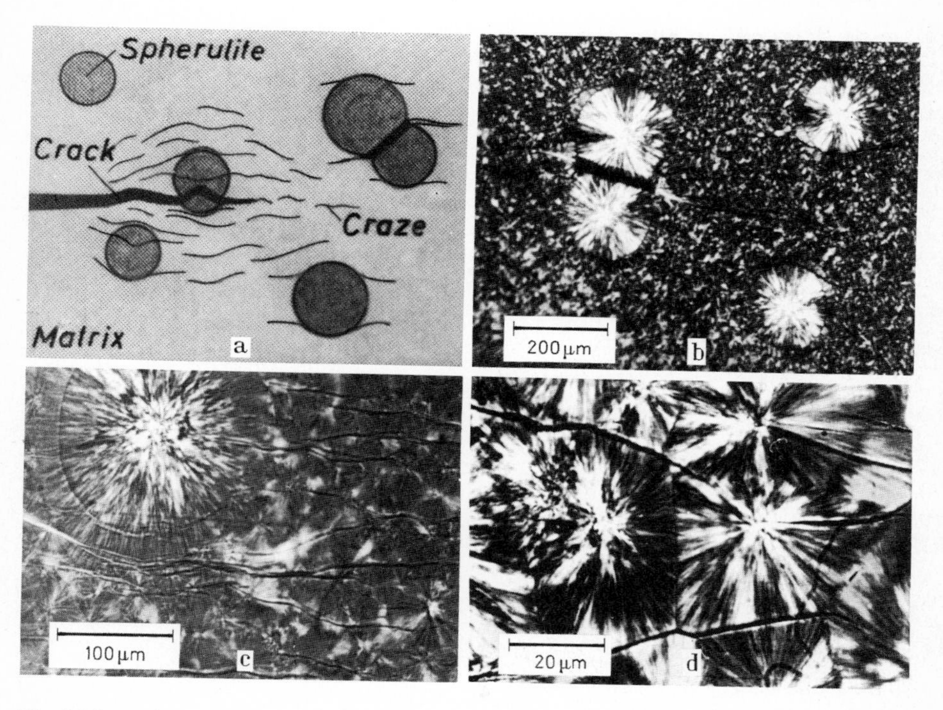

Fig. 9.38a–d. Slow crack growth following craze formation in bulk polypropylene containing coarse and fine spherulites (Courtesy K. Friedrich, Bochum, now Hamburg-Harburg).
a schematic representation, **b** crazes at the interface between coarse spherulites, **c** trans- and interspherulitic crazing, **d** crazing of finely spherulitic matrix

lamellae with a thickness much smaller than the average chain length has led them to the conclusion that fractionation is accompanying the pressure crystallization [202]. Gedde et al. observe fractionation in linear PE [412].

The influence of a spherulitic structure of a semicrystalline polymer on its mechanical properties has been implicitely and explicitely recognized for a long time. Only a few references can be given here [183, 204–207]. Clark and Garber [205] review the effects of industrial processing on the morphology of crystalline polymers. Patel et al. [206] and Andrews [207] are especially concerned with a stress-strain analysis of spherulitic HDPE. They find a maximum of sample elastic modulus of 1.2 GN m^{-2} at a spherulite radius of 13 μm. Studies of the failure morphology of HDPE in an active environment [204] revealed

- that failure occurred almost entirely in a brittle mode
- that failure within an individual spherulite is interlamellar, and
- that the mode of crack propagation depended on the position of the spherulite center with respect to the crack front (leading to interspherulitic fracture in the case of poorly matching interfaces [204].

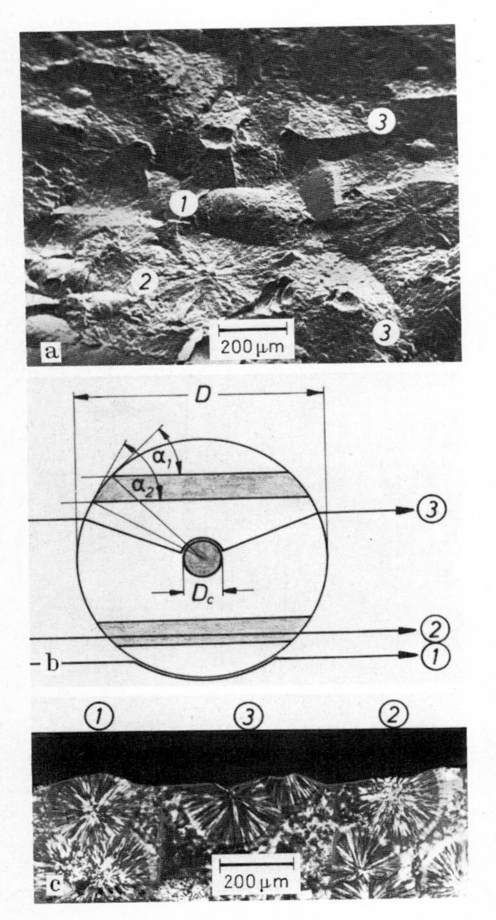

Fig. 9.39a–c. Rapid crack propagation in polypropylene. (Courtesy K. Friedrich, Bochum, now Hamburg-Harburg). **a** scanning electron micrograph of fracture surface showing inter-(1) and transspherulitic fracture (2 and 3), **b** model representation, **c** optical micrographs of sections transverse to the fracture surface showing the same phenomenon

Extensive morphological studies of polypropylene fracture surfaces have been carried out by Menges et al. [121, 136] and quite recently by Friedrich [183]. Using polypropylenes of different molecular weight and tacticity subjected to different thermal treatment Friedrich [183] obtained samples of different microstructure:
I fine spherulites ($\bar{D} = 20\ \mu m$)
II coarse spherulites embedded in a matrix of finer spherulites
III fully grown coarse spherulites ($\bar{D}$ up to $500\ \mu m$).

The slow crack growth in CT specimens of morphology II is schematically represented in Figure 9.38a. Crazes were seen to develop ahead of the crack preferentially at the interfaces between coarse spherulites (Fig. 9.38b), within larger spherulites (c) and at the boundaries of small ones (d). The behavior of a crack in the zone of rapid crack propagation is illustrated in Figure 9.39. A scanning electron micrograph of the fracture surface reveals the structure-induced deviations of the fracture path from a plane (Fig. 9.39a). The points 1 to 3 identify corresponding fracture paths in Figure 9.39a, b, and c. These paths follow a spherulite boundary (1) or they cut off respectively, a segment (2), or a cone (3). From his extensive morphological studies Friedrich [183] derived an order of the partial material resistance values which decrease from that of the

- separation of finely spherulitic matrix, which is the largest, to
- straight paths through coarse spherulites and their center,
- radial and tangential planes in coarse spherulites, and
- interfaces between coarse spherulites and fine matrix, to that of
- polygonal interfaces between coarse spherulites.

The K_c values were found to decrease linearly with $\bar{D}$, increase with decreasing strain rate, and increase with molecular weight, the latter effect being especially pronounced with the coarsely spherulitic morphology. Tentative absolute values of R are determined for the separation of finely spherulitic matrix as $22\ kJ\ m^{-2}$, and for interspherulitic rupture as $4\ kJ\ m^{-2}$ [183]. The first value is of the same order of magnitude as that calculated by Kausch [11] for an oriented PP film (cf. further below).

Special features of fracture surface have been studied by various methods. Rose et al. [413] identify by TEM intra- and interlamellar fractures as well as shear zones of several μm depths in HDPE broken at liquid nitrogen temperature. Marichǐn, Mjasnikova and Pelzbauer [414, 415] noted the bending of microfibrils at kink band boundaries in stress whitened PE and the formation of submicrocracks at these boundaries, a phenomenon thus far known to occur in fibers subjected to compression (see 9 III A 3).

Besides by REM, TEM [413, 416] and optical microscopy [417], structural irregularities have also been analyzed by microhardness tests [418] and surface analysis techniques [419] such as photoelectron spectroscopy (XPS) or secondary ion mass spectroscopy (SIMS).

Obviously, most of these studies are carried out with one principal objective in mind: to learn more about the structural causes of a failure and about the influence of molecular and environmental parameters, material microstructure, and processing variables on crack initiation and propagation. For the failure of semi-crystalline polymers in traction, Schultz [420] has drawn a useful failure map relating strain rate and temperature with failure mode (ductile, spherulite boundary or intercrystallite cracking) and material and processing data (M_w, number of defective chains, crystallisation temperature and annealing treatment).

One of the most important applications of thermoplastic construction materials is certainly their use as pipe material for water, gas and chemically agressive liquids. The technological problems which arise in their fabrication and use and the employed test methods are summarized in Table 9.5. Characteristic stress-strain diagrams and failure morphologies of PVC (Figs. 1.1–1.4) and HDPE (Figs. 1.5–1.8 and 8.36 to 8.39) have been presented earlier in this text.

Table 9.5. Technological Problems and Test Methods in Pipe Fabrication and Use

Material and processing-related problems
- choice of material in view of operation conditions
- prediction of lifetime at given (fluctuating) stresses
- control of morphology, crystallinity
- influence of cracks and internal defects, crack arrest
- degradation (also during processing) and aging

Construction-related problems
- influence of external defects, shock loads, etc.
- strength reduction through welding
- bending stresses

Test methods
- long-term hydrostatic pressure tests at 20 °C (possibly also: 40 °C, 60 °C);
- short-term hydrostatic pressure tests at 80 °C
- one-hour pressure test followed by burst strength evaluation
- tensile properties of samples taken from the pipe
- heat reversion test (control of excess stretching during extrusion)
- fracture mechanics measurements

A large number of publications deal with the (static) long-term behavior of pipe materials; the few references chosen as an example [421–430] strongly refer to the works of Gaube, Müller et al. These studies concern obtention and evaluation of stress-rupture data in general [421], analysis of the morphology, including that of (defect-containing) fracture surfaces [422], the influence of processing variables and nucleation agents [423], specifications for testing [424], failure criteria based on creep analysis [425] and (linear elastic) fracture mechanics [426], guidelines for the choice of material, the dimensioning, and designing of plastic pipes [427], the long-time behavior of weldings, fittings and other elements [428], and the influence of temperature and oxygen [429] or of the presence of chemically aggressive environments [430] on the performance of plastic pipes.

Some of the principal observations mentioned in the cited references have already been discussed in earlier sections while treating e.g. impact behavior (8 II A 1), creep failure mechanisms (8 II B 3) and fracture through slow crack growth (9 I D). For convenience, the essential conclusions with respect to pipe failure may be briefly summarized as follows:

- hoop-stress vs. time-to-failure diagrams of polyolefins at modest temperatures generally reveal two branches corresponding respectively to (ductile) creep and (brittle) crack growth failure;
- although for PVC in principle both mechanisms operate, the two branches are sometimes [421a, 422k] but not always [421d, 422a] observed (see Fig. 1.4 as an example of the latter case).
- at higher temperatures oxidative degradation constitutes an additional failure mechanism resulting in a third, almost vertical branch of the log σ_v-log t_b curve (Fig. 8.39);
- one way of predicting the 50 years-20 °C behavior of stressed pipes is the representation of creep and crack-growth mechanisms by two different Arrhenius equations and extrapolation from the short-time high temperature results [e.g. 421g, 424, 427];
- another way is offered by fracture mechanics which, in principle, permits calculating the time of growth of "inherent flaws" to a critical size; this concept has been treated in detail in Section 9 I D, using PMMA as an example. Gray et al. [426e] have found it valid to predict the times to brittle rupture of HDPE with (calculated) flaw sizes of 10 to 100 μm;
- the presence of welds and fittings or of aggressive environments necessitates the use of an appropriately larger safety factor [426–430]
- the presence of notches, crack, and/or flaws affects the principal pipe materials differently; proposed criteria have been compiled in Table 9.6;

Table 9.6. Notch Sensitivity and Fracture Criteria

Polymer (test method)	Notch sensitivity (critical param.)	Fracture criterion	
		at short times	at long times
PVC hard (DEN, SEN, TPB, CT, DCB)	large if $T < T_c(v)$ T_c (25 mm/s) = 15 °C	K_{Ic}	yield behavior
PE LD (pressured pipes)	small if $T > 0$ °C	defect induced limits of extensibility	oxidation?
PE MD (DEN, TPB)	notable if $T < T_c$ (crystallinity, MW)	Dugdale	yield behavior
PE HD (tensile, DEN, pipes)	only at low MW (and if $T < -100$ °C)	brittle fr.: internal inhomogeneities ductile fr.: yield	disentanglement of chains, oxidation
PP (fatigue, CT, pipes)	some at low MW (tacticity, morphol.)	K_c	creep strain, yield behavior
PB-1 (pipes)	notch strengthening	yield behavior	oxidation (elution of antioxidant?)

T_c: critical temperature below which LEFM is valid [238, 426h]

– despite all progress in fracture mechanics, the visual inspection of the fracture surfaces obtained in short-time experiments for control of the extent of ductile deformation seems to be indispensable for an adequate rating of the long-time behavior of pipe materials (see also Section 9 III B and the references cited there).

The latter statement is underlined by taking a closer look at the calculation of lifetimes from inherent flaw sizes. Although this concept is basically promising, it still gives rise to some ambiguity because of the strong dependence of t_b on the unknown size a_1 of the largest "inherent flaw" and thus on $K_{I_1}^{2-b}$ in Eq. (9.23). Taking Gray's data, one starts from the relation:

$$\dot{a} = AK_I^4 \tag{9.32}$$

which is in excellent agreement with measurements of Chan and Williams [426g]. Integration of Eq. (9.32) should then lead to:

$$\Delta t = t_b \sim K_{I_1}^{-2} = (Y^2 a_1 \sigma_v)^{-2} \tag{9.33}$$

It has been observed, however, that the slope of the log t_b vs. log σ_v diagram for HDPE varies around –4 (it is between –3 and –4 for Gray's data) and between –4 and –5 for Gaube's [421g, 427a]. A possible explanation of the extremely noteworthy σ_v dependence of t_b may come from the disentanglement model presented earlier (8 II B4). If it is assumed that a creep crack is preceded by a craze of length r_p and that the crack grows once the most highly strained fibril fails through disentanglement at a time τ after its formation, then one has:

$$\dot{a} = r_p/\tau. \tag{9.34}$$

Substitution from Eqs. (9.16) and (9.32) results in:

$$\tau \sim K_I^{-2} \tag{9.35}$$

This seems to be plausible since with decreasing fibril stress, a decrease in the rate of disentanglement should also be observed (as shown for PMMA [329, 330]).

If it is finally assumed that the essential portion of a crack in a pipe grows at a rate determined by Eq. (9.34), one arrives at:

$$\frac{h}{t_b} = \dot{a} \sim K^4 \sim \sigma_v^4. \tag{9.36}$$

Although the exact forms of Eqs. (9.35) and (9.36) have as yet to be derived, the latter expression gives the correct relation between t_b and σ_v.

Concluding this topic, the special characteristics of the principal pipe materials, together with typical applications, have been compiled in Table 9.7, enabling a first material selection.

In the first part of this section, semicrystalline polymers (HDPE, PP, PA) have been discussed. The fracture surface morphology of glassy polymers has received no

Table 9.7. Choice of Pipe Material

Polymer	Special characteristics	Typical applications
PVC		
– unplast.	Low material and installation cost, high strength, relatively poor toughness	Principle pipes in water distribution, effluent disposal, process piping, cable protection, land drainage
– chlorin.	higher strength and serv. temp.	Hot water and waste water piping
– high imp.	lower serv. temp. ~ – 30 °C	Low pressure gas distribution
PE		
– high dens.	Tough, available in large lengths, no solvent-welding	Water pipes in rough ground, secondary mains, canalization, linings
– med. dens.	better weldability and toughness	Gas pipes
– low dens.	good flexibility, weaker	Water distribution, also in agricult.
– crosslink.	better creep and temp. resist.	Floor heating, warm water
PP	Higher strength than PE, less tough	Hot waste water, chemical plant piping
PB	Tough, strong (even at elev. temp.)	Hot water distrib., outdoor piping
ABS	Tough (even at low temp.), strong	High press. air lines, food process., well casing

less attention in the literature. In fact many of the investigations of crazing [76–177] use fractography as a means to elucidate the propagation and breakdown of crazes. Earlier studies of fracture processes concerning PS are especially reported in refs. [106, 115, 132, 150, 155, 169, 191, 194, 199], concerning PMMA in [61, 66, 197, 200], PVC in [198, 208], and PC in [196], newer ones have been discussed in Section 9 II.

A feature common to practically all fracture surfaces of glassy polymers are the remnants of craze layers. At low speeds of crack propagation craze breakdown will generally occur in the center of the craze material leaving a more or less homogeneous layer on each fracture surface [15, 50, 150, 194, 199]. In PS at intermediate and higher crack speeds decohesion at the craze/matrix interface becomes possible at room temperature. Beahan et al. [150] have investigated this phenomenon more closely; Figure 9.40 shows their micrograph of (the rather isolated) event of a crack which after having propagated at intermediate speeds was stopped in the craze region. The

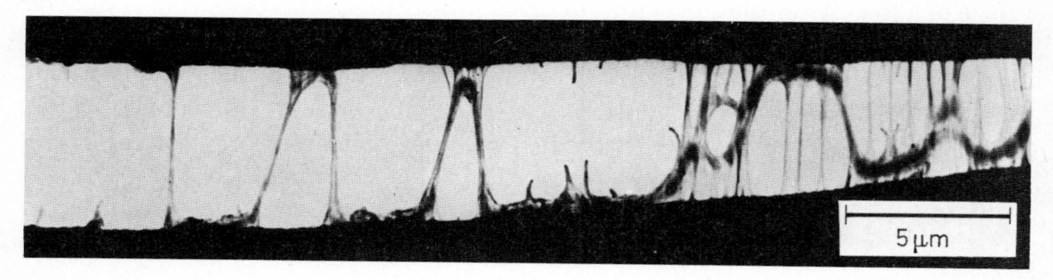

Fig. 9.40. Transmission electron micrograph of the craze region ahead of a crack advancing from left to right in polystyrene. (Courtesy D. Hull, Liverpool)

micrograph permits recognition of the decohesion of highly strained craze material at the interface and also the oscillation of this phenomenon between the opposite interfaces. The regular oscillation of the decohesion of craze material creates "mackerel patterns" with typical band repeat distances of between 10 and 70 μm. Hull et al. [150, 194, 199] and more recently Doyle [431] discuss the dynamic conditions of their formation and the effect of sample temperature. They observe that at temperatures below 235 K no mackerel patterns are formed whereas the initiation of secondary fractures is greatly enhanced.

A more pronounced transition from slow to fast crack propagation is reported for PMMA [61, 66, 197, 200] but the surface features are roughly comparable to those in PS. As observed by Döll and Weidmann [61, 66] at slow crack speeds ($\dot{a} < 0.1$ ms^{-1}), independent of molecular weight, a rather smooth fracture surface is formed which even carries on accidental deviations from the plane in the form of parallel markings (left hand side of Fig. 9.41). After the transition to rapid crack propagation the fracture surface morphology changes completely. A pattern of ribs or lines running approximately parallel to the crack front becomes visible. Although length, direction, and spacing of the ribs are not very regular (Fig. 9.41, right hand side) an average spacing of 90 μm has been measured [61]. At a somewhat higher molecular weight (M_w = 163000) a spacing of about 220 μm is observed (Fig. 9.42, right hand side). At very low molecular weight (M_w = 51000) rib spacing is 20 to 24 μm [200]. In view of the symmetry of these markings on the two corresponding fracture surfaces, rupture of the craze material in the center plane seems to be indicated. This phenomenon is somewhat different in origin, therefore, from the mackerel pattern but also related to the formation and breakdown of crazes. The mackerel pattern was caused by the oscillating occurrence of craze/matrix decohesion. The rib markings, however, are better interpreted as *hesitation lines,* also designated as *stick-slip,* i.e. as the alternating development and breakdown of craze material [61, 196, 200]. This is explained by referring to the craze model presented in Figure 9.20. It may be assumed that at $t = t_0$ the crack front just faces a fully developed craze whose width and length (region C) comply with the Dugdale model (Eqs. 9.16 and 9.18). The propagating stress field and the stresses which can be supported by the craze zone are momentarily in equilibrium. At $t = t_0 + \Delta t$, however, the craze is supposed to open further and to support loads which are beyond the load carrying capability of the molecular strands. Consequently the most advanced strands break thus initiating the rupture of neighboring strands as well. The catastrophic breakdown is slowed down once the crack has advanced within region C so far as to reach little-drawn craze material. After the slow-down of crack propagation redevelopment of the craze begins which may be completed at, say, $t = t_0 + t_1$. Considering the average crack speed ($v = 400$ ms^{-1}) and the rib spacing in Figure 9.42 (220 μm) one can calculate that the time interval t_1 for catastrophic breakdown and redevelopment is 0.55 μs. The cyclic crack propagation seems to be influenced, however, by additional mechanisms which are not yet fully understood.

Thus the PMMA samples having the slightly higher molecular weight reveal at crack speeds of 200 to 300 ms^{-1} a rather smooth fracture surface containing no ribs but a large number of parabolic markings (Fig. 9.42, left hand side). These markings are seen in higher magnification in Figure 9.43, they derive from the coalescence of

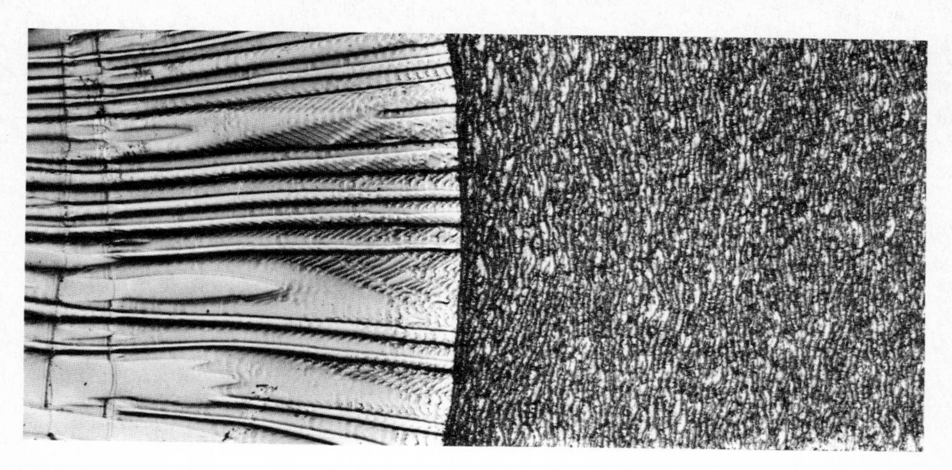

Fig. 9.41. Fracture surface of polymethylmethacrylate (M_W = 115000) showing the transition from slow ($\dot{a} \simeq 0.1$ ms^{-1}) to fast ($\dot{a} > 200$ ms^{-1}) crack propagation; crack propagates from left to right. (Courtesy W. Döll, IFKM Freiburg), Specimen thickness 4.75 mm

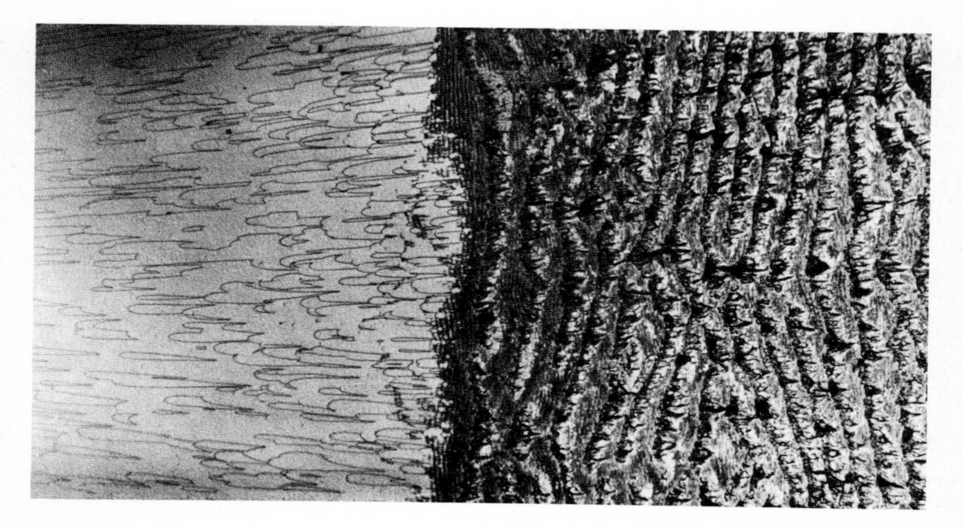

Fig. 9.42. Fracture surface of polymethylmethacrylate (M_W = 163000) showing the transition from a smooth surface (crack speed $\dot{a}$ = 300 ms^{-1}) to a coarse one ($\dot{a} > 400$ ms^{-1}); crack propagation is from left to right. (Courtesy W. Döll, IFKM Freiburg), Specimen thickness 3.75 mm

primary and secondary crack fronts travelling in one craze zone but in slightly different planes. The fracture surface of polycarbonate – broken in tension at room temperature – reveals a strikingly similar pattern [196]. In that case the distance between the planes of primary and secondary cracks has been found to be 0.43 μm at room temperature increasing to 0.75 μm at -196 °C.

Döll [61] reports that PMMA samples with a higher molecular weight ($M_w \geqslant 490000$) do not show ribs on the high-speed fracture surface (Fig. 9.44). Secondary fractures (this time in different craze zones) may be initiated, however, leading to parabolic markings in that surface region (Fig. 9.45). Kusy et al. [200] find ribs *and* parabolas in the molecular weight range $92000 \leqslant M_v < 270000$. A certain molecular coil size is (quite obviously) necessary to permit the transfer of stresses sufficient for the initiation and temporary propagation of a secondary fracture plane. Cottrell [432] has noted rather early that the density of parabolic markings is correlated with the energy consumed and thus with fracture toughness.

In the referenced literature further details are discussed which may be obtained from fractographic analysis, such as the effect of Wallner lines on the disposition of the ribs [61, 196, 200], "crazeless fracture" in low molecular weight PS [155], the retarded breakdown of crazes in fatigue (cf. III D), the ductile fracture of PS at lower loading rates and at temperatures close to T_g through the growth of one or more diamond shaped cavities [169], the globular appearance of ion-etched PS craze matter [132] and PVC surfaces [208], and craze-like features on the surface of broken phenol-formaldeyhde samples [195].

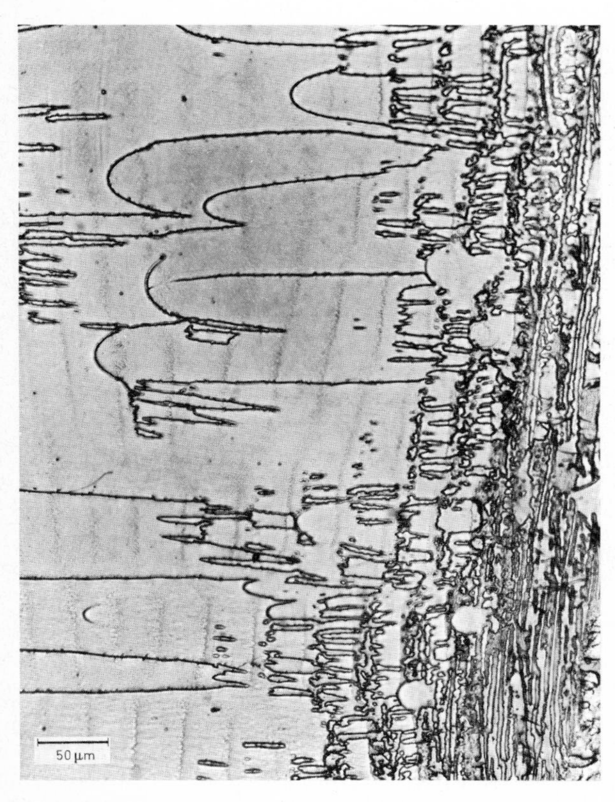

Fig. 9.43. Higher magnification of the middle section of Fig. 9.27 showing parabolic markings. (Courtesy W. Döll, IFKM Freiburg)

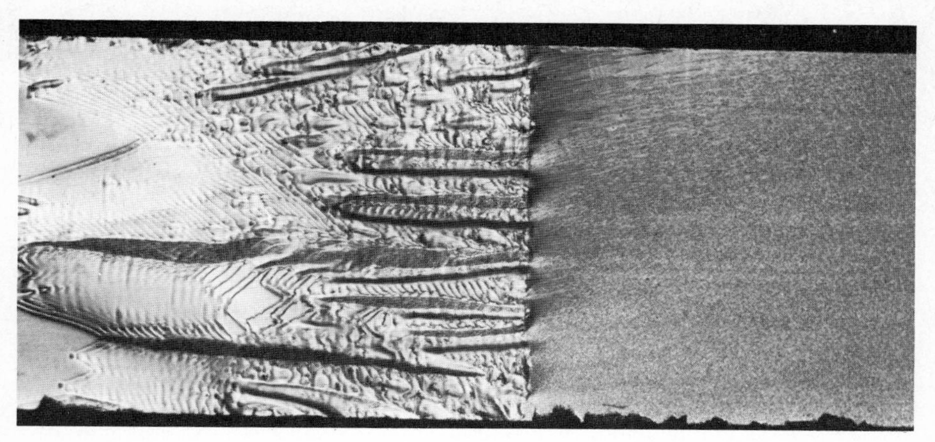

Fig. 9.44. Fracture surface of PMMA (M_w = 8 000 000) showing the transition from slow ($a \simeq 0.1$ ms^{-1}) to fast ($\dot{a} > 100$ ms^{-1}) crack propagation; crack propagates from left to right. (Courtesy W. Döll, IFKM Freiburg), Specimen thickness 4.166 mm

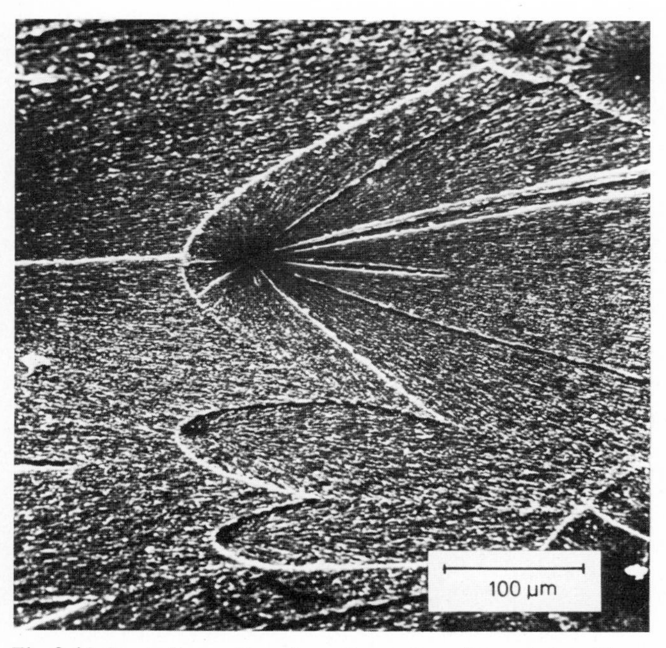

Fig. 9.45. Parabolic markings from the surface of a crack travelling at 240 ms^{-1} in PMMA ($M_w = 1.2 \cdot 10^6$); crack propagates from left to right. (Courtesy W. Döll, IFKM Freiburg)

In this section the morphology of fracture surfaces has been discussed, a morphology which reflects the local modes of material separation. The *microscopic* size of the structural elements to be broken or separated has been recognized: molecular strands, fibrils or coils, ribs, crystalline lamellae, spherulites. In talking about their ultimate response, however, macroscopic terms had been used: breakage, shear deformation, limits of plastic deformation, material resistance. For good reasons no molecular criteria for material separation had been given. For individual molecules such criteria exist: the temperature of thermal degradation and the stress or strain which causes the chain to scission. For the mentioned structural elements no such simple criteria exist. Perhaps one may mention the critical role of temperature in the transition to rapid crack propagation [30, 50, 184–186, 197] and the constant value of local strain ϵ_y in the extension direction (Fig. 9.46) which was found – independent of crack length – to reach about 60% at the cracktip in biaxially oriented PETP film [209]. One may also mention the critical concentration of chain end groups N_{IR} determined by microscopic infra-red spectroscopical investigation of the near-crack zones of oriented PP film (Fig. 9.47) and of the fracture surface of PE bulk material [210]. Both materials are tough and strong. The given stress distribution in front of the crack in the PP film permits calculation of K_c as $\sigma(r)\sqrt{2\pi r} = 8.3 \pm 2$ MN m$^{-3/2}$ and G_c as 30 ± 17 kJ m^{-2} [11]. These values, in connection with Table 9.1, indicate quite clearly that chain scission is accompanied by severe plastic deformation. The possible role of chain scission during or after the application of large orientational strains has been discussed in detail in Chapter 8.

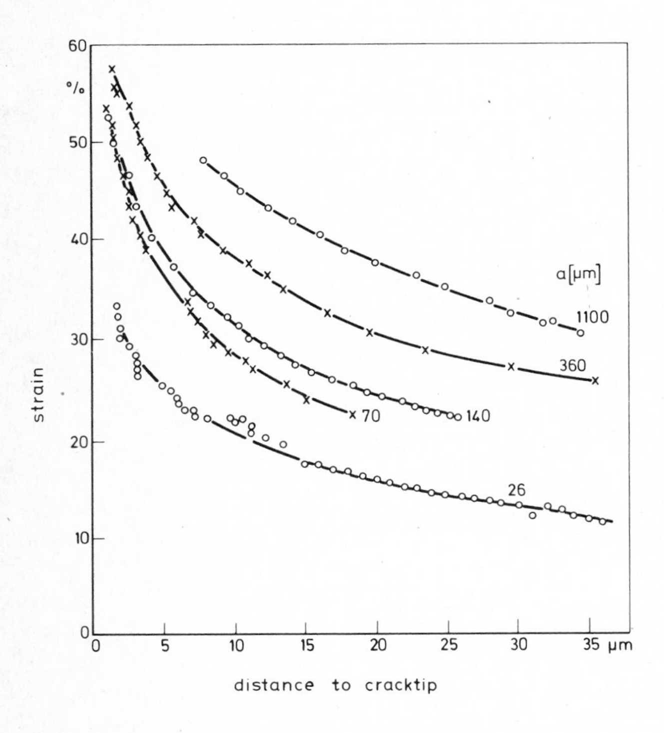

Fig. 9.46. Distribution of principal strain in front of a plane crack in biaxially drawn PETP film (after [202])

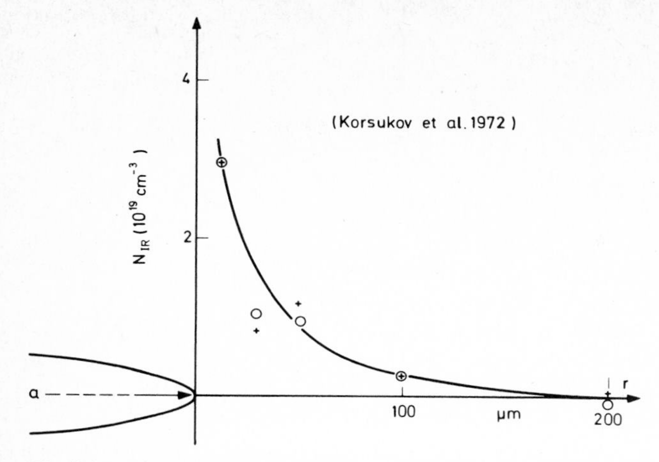

Fig. 9.47. Distribution of chain end concentration in front of a propagating plane crack at $t = 0.2\ \tau$ (+) and $0.75\ \tau$ (○) in polypropylene film (after [11, 210])

3. Fibers

Important aspects of fiber fracture have already been treated in Sections I, F and G of chapter 8 (fracture mechanisms in relation to sample morphology). At this point reference to fractographic studies will be made. The visual inspection and the study of broken samples by means of optical or scanning electron microscopy (SEM) are important tools of fracture analysis. They obviously aid
– in the detection of possible causes of crack initiation and
– in the interpretation of the fracture process.

The appearance of a fracture surface undoubtedly is the most convincing evidence that the fracture process has reached the phase of non-homogeneous deformation. Frequently the surface and morphology of a failed specimen reveal whether or not a phase of homogeneous material deformation has contributed to the fracture event. Figures 9.48 and 9.49 serve to illustrate this point. These fracture surfaces are more or less an arbitrary result of a large number of homogeneously distributed scission and slip processes of chains and microfibrils. The fracture surface was developed within a very small fraction of the total lifetime and at a location not predictable in advance.

Hendus and Penzel [433] investigated the fracture morphology of polyamide 6 monofilaments. The regularly spun and drawn monofilaments were subsequently subjected to tensile tests at various strain rates. Characteristic fracture surfaces are reproduced in Figures 9.50 and 9.51. At small strain rates ($\dot{\epsilon} = 0.033\ s^{-1}$) frequently v-shaped notches are observed (Fig. 9.50). Such a notch is formed through a crack which initiates at a flaw or material inhomogeneity contained in the filament surface or in a zone close to the surface. While the crack grows slowly the remaining fiber cross-section continues to deform plasticly. At a point determined by the sizes of the crack and of the remaining cross-section and by the material properties rapid transverse crack propagation occurs. The measured strength of the monofilament is

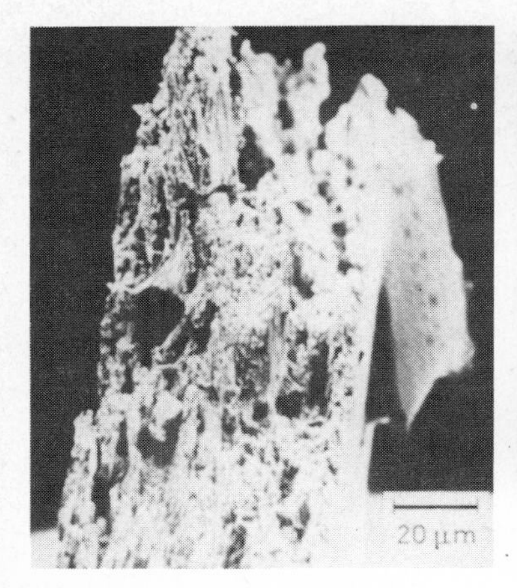

Fig. 9.48. Fracture surface of dry 66 polyamide banding material [434]

Fig. 9.49. Fracture surface of 66 polyamide banding material, before fracture stored to equilibrium at 65% relative humidity [434]

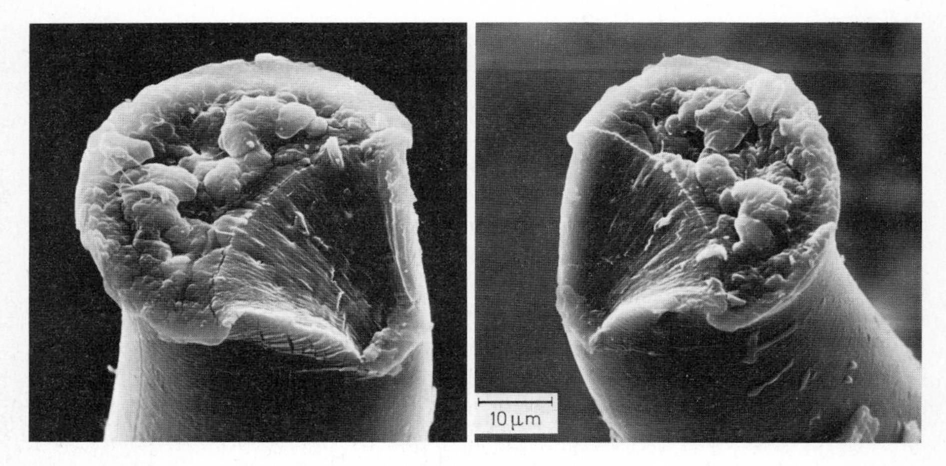

Fig. 9.50. Corresponding ends of PA 6 monofil broken at a strain rate of 0.033 s^{-1} (Courtesy H. Hendus, [435])

the higher the smaller the v-shaped notch [433]. Filaments of highest strength contained barely visible small voids.

Figure 9.50 also shows an enlargement of the head of the monofilament. This enlargement corresponds to a shrinkage of the oriented material caused by the warming of the filament during the plastic deformation. If the rate of deformation is increased to 50 s^{-1} then the heat generated through the plastic deformation of the remaining cross-section cannot be dissapted fast enough. Locally the melt temperature is exceeded, the filament heads expand almost to the diameter of the undrawn material (Fig. 9.51).

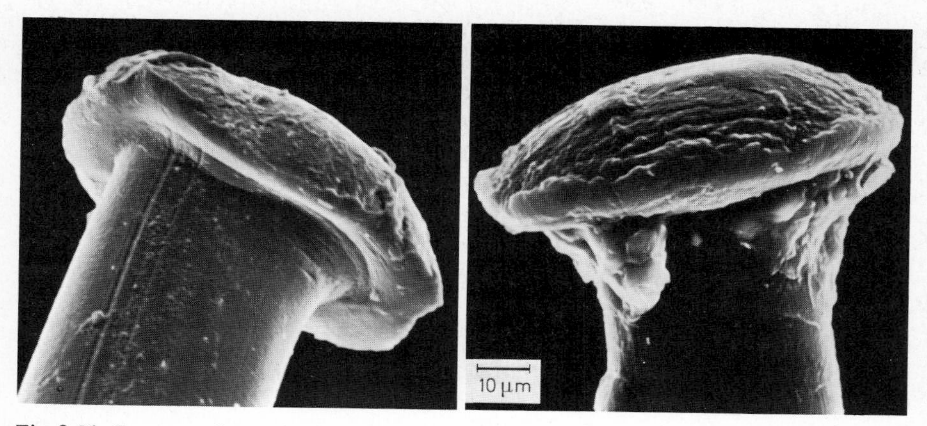

Fig. 9.51. Corresponding ends of a PA 6 monofil broken at a strain rate of 50 s^{-1} (Courtesy H. Hendus, [435])

The fibrillar nature of drawn monofilaments becomes apparent in Figure 9.52. The filament ends of two different fracture events show a strong axial splitting. As concluded from the formation of small heads at the ends of some microfibrils the fibrillation must have occurred before the catastrophic fracture took place. Strong axial splitting is also known from polyamide 66, Kevlar [438a], PETP, acrylic fibers, wool, human hair, and cotton fibers (see 8 I F and G and [436, 438]).

On the basis of fractographic studies Bunsell and Hearle [438b] indirectly identified a fatigue mechanism in polyamide 66. Oriented PA-66 fibers failed in a tensile type manner if they were fatigued uniaxially between an upper stress level equal to the long-time static strength and a lower level of clearly non-zero tensile stresses. These fatigue failures had a morphology similar to that shown in Figure 9.50. The breakage apparently did not involve any specific fatigue effect. If the loading conditions were changed, however, so as to include zero tensile or even compressive

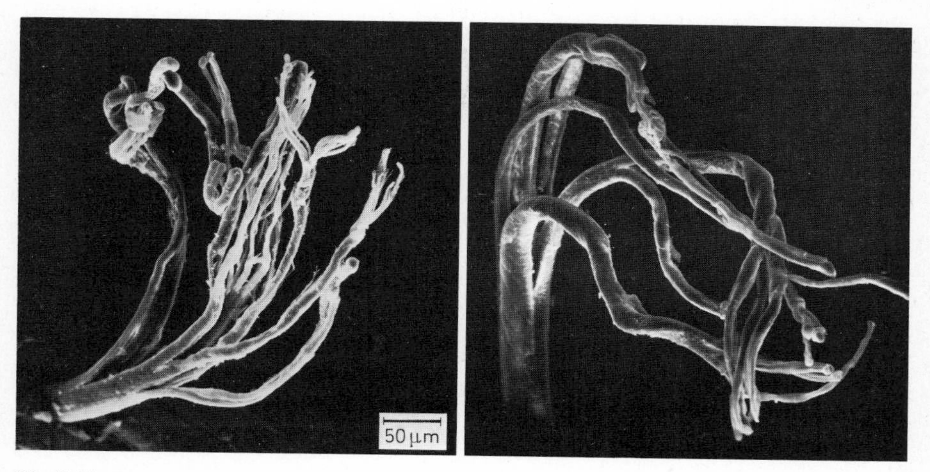

Fig. 9.52. Two different PA-6 fiber fractures under tension showing strong axial splitting (Courtesy H. Hendus, [435])

stresses a specific fatigue fracture morphology became apparent. Those breaks, occurring after 10^5 cycles, were characterized by the propagation of a crack, almost parallel to the direction of the fiber axis. As shown by Figure 9.53 [437] this leaves a long tail of material at one fiber end. As discurred in 8 I F Oudet, Bunsell et al. [438c] have recently studied a similar phenomenon in PETP fibers. It may be concluded from Bunsell's and Hearle's observations [436, 438b] that compressive yielding and/or buckling of microfibrils or fibrils takes place on a microscopic scale. The repeated lateral rearrangement leads to a gradual amorphisation and to loss of interfibrillar cohesion. A crack which has started at some flaw in a direction normal to the tensile stress σ will turn, therefore, into a direction u almost parallel to σ – the angle between u and σ being a function of the lateral stress transfer which still remains. The extent of fibrillation thus depends on the degree of orientation and crystallinity [438d].

On the basis of his extensive investigations Hearle [436] has established a classification of the main features of fiber fracture morphologies:

A. Transverse Elastic Crack Propagation
This is the classic form of failure in elastic materials and consists of a relatively smooth "mirror zone" of crack propagation leading to a rougher zone of final failure due to multiple crack-initiation (comparable in morphology to Fig. 1.7).

B. Ductile Transverse Crack Propagation
This is a form of fracture in which a crack propagates stably across a fiber under increasing load and/or strain and is opened out into a V-notch by the continued plastic yielding (the final stages of the drawing process) of the remaining material; the crack leads into a region of final catastrophic failure, occurring when the stress in the remaining reduced cross-section reaches a critical level (Fig. 9.50).

C. Transverse "Fibrous" Break
This type of break runs perpendicularly across the fiber, with a rough texture and no evidence of any propagating cracks; the whole structure appears to be ready to fail at the same time, and the breaks are very similar in appearance to

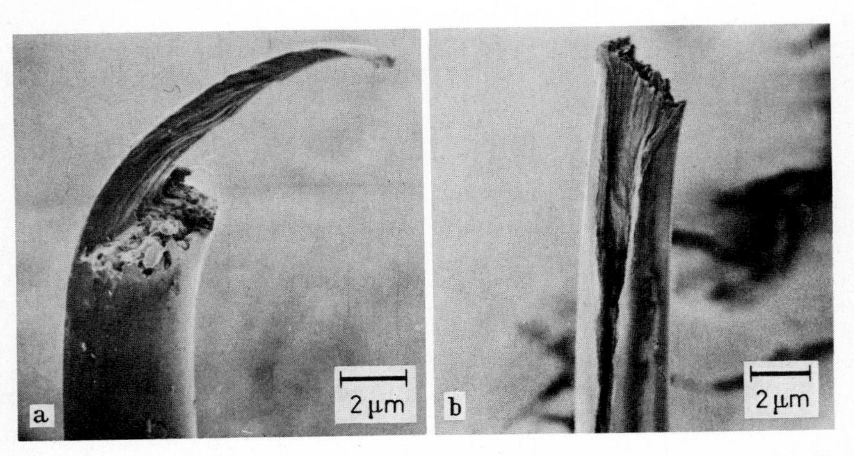

Fig. 9.53a, b. Corresponding ends of fatigue failure of polyamide 66 under oscillating load dropping to zero in each cycle (Courtesy J. W. S. Hearle, Manchester) [437]

lower-magnification views of the fracture of fiber-reinforced composites (Fig. 9.48 and 9.49).

D. Axial Splitting under Tension
In some fibers failure in axial tension occurs by cracking, or splitting, along planes close to the fiber axis. Due to a nonhomogeneous fibrillar substructure there will be shear stresses in these planes. With reduced interfibrillar cohesion a break is produced characterized by multiple axial splitting over a long length (equal to many fiber diameters) (Fig. 9.53).

E. Splitting Due to Torsion
Under shear stresses generated by torsion, splitting can occur along lines determined by the directions of stress and of material weakness.

F. Axial-fatigue Cracks
Particular localized cracks, deviating slightly from the fiber-axial direction, develop in many fibers as a result of tensile fatigue: they are associated with shear stresses at discontinuities. The resulting breaks show a long tail on one end, a strip off the other end, and a final catastrophic-failure region (Fig. 9.53).

G. Fatigue-cracking along Kink-bands
Kink-bands at angles of about 45° develop in many fibers in compression, for example, or on the inside of bends, and, in repeated flexing, these eventually turn into cracks and lead to fiber failure.

H. Multiple Splitting in Fatigue
In torsional- or flexural-fatigue situations, multiple splitting frequently occurs and can eventually lead to failure.

For a more detailed discussion of *fiber* fracture morphology the reader is referred to the series of publications by Hearle, especially to "An Atlas of Fibre Fracture" [436b].

B. Defects and Inherent Flaws

Almost any brittle fracture surface contains traces of a flaw or an inclusion which automatically will be blamed for having started the final crack and reduced the strength or time to failure of the specimen. Whereas the first point (origin) can generally be determined unambigously, the second is less clear, particularly if the flaws or inclusions originally are small.

This section can not attempt to review the abundant literature on stress concentration effects of cracks or particles and on the numerous electron microscopical investigations as to the origin of polymer failures. But a few remarks concerning the identification, nature and location of flaws and inclusions and their effect on a stressed sample will be made.

The value of an optical and electron-optical inspection of a fracture surface for defect recognition has already been discussed earlier (see Section 9 III A, especially refs. [251, 422d, e, 439–442]). Defect analysis using surface analytical methods [419], microtome slices [422a, 443] and positional X-ray scattering technique (PSAXS) [444] must also be mentioned. The role of flaws as possible stress concentrators and the notion of inherent flaws has been discussed in detail in Section

9 I D (see also [440, 445]). Without any doubt, the presence of particles (impurities or fillers) reduces the strength and the time-to-failure of a specimen; the scatter of these variables grows with particle size [422d, 440, 446–448].

In order to identify systematically the nature and position of defects in (LDPE) pipes Stockmayer and Wintergerst [422a] developed a cutting technique which permitted to transform the entire pipe wall into a continous thin peeling of 0.06 to 0.15 mm thickness. By inspecting this film they were able to correlated nature, frequency and position of defects with time to failure t_b of the pipes. The irregularities they found were clear zones not mixed with carbon black in the form of points, linear marks, hooks (Fig. 9.54) and parallel striations. They claim that with a certain preference the final creep crazes originated in those areas of the pipe wall where the irregularities were more frequent. It was particularly noted by them that the lifetimes t_b of those pipes where the creep crazes had started close to the inner or outer surface of the pipe wall were an order of magnitude smaller ($\bar{t}_b$ = 740 h) than those where the crack had originated in the center ($\bar{t}_b$ = 7400 h). Although the differences in stress distribution within the pipe wall, due to frozen-in tensions (see next section), may influence crack initiation it seems to be more probable that network defects are mostly responsible for this observation. As discussed in detail in Chapter 8 II B the inner walls of extruded pipes down to a depth of 0.5 mm are the first to experience an eventual network degradation. It was precisely in this region that Stockmayer and Wintergerst found the origins of the most rapid failures [422a, see also 439, 442].

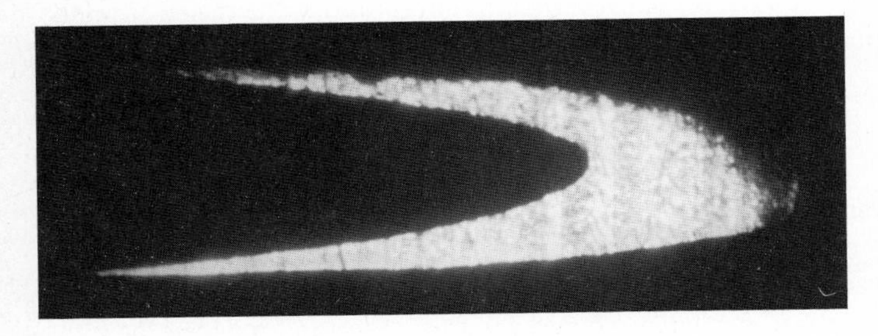

Fig. 9.54. Zone of about 3 mm length in a LDPE pipe which escaped mixing with carbon black (from Stockmayer [422a])

To test the influence of the observed irregularities further the authors performed tensile drawing and static loading experiments on samples cut from the peelings. In these experiments, there was practically *no correlation between yield stress and time to failure* on the one hand and *kind and concentration of flaws and inclusions* on the other. However, the effective elongation of films containing defects was much smaller since the necking samples generally broke whenever the neck had reached the defect. This behavior confirms the statements made before [439] and in the previous section, namely that very often creep craze nuclei must be considered as network defects which are not detectable in short time loading tests unless the defective zone is locally subjected to large deformations.

Evidently the question on the influence of inclusions added in the form of blending components or fillers is of vital importance to the fabrication and use of polymer composites. This problem cannot be discussed, however, in the context of this monograph.

Apart from the – material-related – flaws, voids, inclusions and imperfections of mixing, there are other weakening effects which are related to the structure as a whole such as irregularities of the geometry, internal tensions, weld lines, and orientation distributions; although the latter effects will be discussed subsequently, it may already be indicated that an analysis of the nature of decisive defects in POM and PA [443] revealed that material inhomogeneities and voids had been the most frequent.

Geometry

A simple variation in sample cross-section can turn out to be a decisive defect. Thus, the *ductile failure* of uniaxially or biaxially loaded samples generally takes place at the smallest cross-section (Fig. 1.2). It should be noted that the failing section may be perfectly safe at higher stresses (brittle fracture) or longer times (creep crazing), because the breakdown in those cases is initiated by network defects or flaws. The weakening influence of bends, joints, sleeves, and fittings on pipe constructions had already been indicated above [428].

Internal Tensions

Any viscoelastic body cooled from outside from above its solidification temperature (T_g or T_m) to a lower temperature T_c deforms non-homogeneously and either bends or contains frozen-in tensions. The distribution and absolute value of such tensions can be determined from a layer removal experiment [442]. In pipes they generally show a parabolic distribution, compressive at the outside, smaller and tensile at the wall inside. Williams [449] derived the thermal, circumferential stresses $\sigma_R(s)$ as:

$$\sigma_R(s) = \frac{\alpha E(t')}{1-\nu}(T_m - T_c)\left(\frac{Nu}{2 + Nu}\right)\left[\frac{1}{3} - \left(\frac{s}{W}\right)^2\right]$$

with:

s	distance from inside wall
α	coefficient of thermal expansion
$E(t')$	Young's modulus at time after cooling down
ν	Poisson's ratio
Nu	Nusselt number
W	wall thickness

For a HDPE pipe he had determined (compressive) stresses of –3 MPa at the outside and a tensile component of +1 MPa at the inside. Compared with the externally applied values of 15 to 20 MPa the frozen-in tensions are not entirely negligible. Mittal et al. [450] and Lee et al. [451] use the birefringence of transparent polymers to measure residual stresses.

Orientation distributions

Structure anisotropy is one of the characteristic properties of chain molecules and of any oriented polymer matrix. Many injection molded or extruded samples show signs of anisotropy (birefringence, extent of shrinkage). The HDPE pipe represented in Figure 1.6 gives an excellent example. The heavily deformed zone consists of molecules highly oriented in circumferential direction. In this direction fracture strength has considerably increased (perhaps by a factor of 10). Although the largest stress component still acts in this direction rupture occurs in axial direction, perpendicularly to the chain axis. Thus, a growing crack preferentially separates chains instead of breaking them (the fibrils bridging the crack only underline this statement).

C. Impact Fracture of Notched Specimens

In Chapter 8 tensile and impact behavior of *unnotched* specimens have been discussed. The principal changes in behavior to be expected from the presence of a notch would have to be due to the changed state and intensity of stress in the crack tip zone, to the essential confinement of fracture development to a limited region, and to the increase in rate of local deformation. An analytical description of these effects has been attempted in terms of linear elastic fracture mechanics (9 I). Neglecting the geometrical correction term the uniaxial brittle strength of a cracked thick plate is derived from Eq. (9.10) as:

$$\sigma_b(a) = K_{Ic}/(\pi a)^{1/2} = (E\,G_{Ic}/\pi a\,[1 - \nu^2])^{1/2}. \tag{9.37}$$

In this case the effect of a notch of length a is clearly indicated, therefore. In the general case of rupture involving plastic deformation, however, one would have to use K_c (or G_c) and a corrected crack length $a + f(r_p)$ in Eq. (9.37). All these quantities depend heavily on the degree of plastic deformation at the crack tip, which in turn is influenced by crack length, notch tip radius, and loading conditions. In the case of heavy plastic deformation the rupture stress of a specimen would depend little on the initial crack geometry but strongly on the "strength" of the plastically deforming matrix. Three questions are raised, therefore, which will be discussed in the following: which notch sensitivity is actually observed with different polymers, when are geometrical and when are material parameters of critical importance, and what is the influence of chain length and mobility?

The notch sensitivity of polymeric materials has been comprehensively investigated by Takano and Nielsen [211]. The authors define a notch sensitivity factor k_S of the yield or breaking strength of tensile bars as

$$k_S = \frac{\sigma_F(o)\,A(a)}{\sigma_F(a)\,A(o)} \tag{9.38}$$

with A being the (residual) sample cross-section. In the same way a notch sensitivity factor k_T of the energy to break (area under stress-strain curve) is defined.

The authors tested some 40 different materials at strain rates of 2.5 mm/min for rigid materials and of 25 mm/min for soft elastomers employing six different crack geometries. The "most severe" was a single notch of 3 mm depth (in a 12 mm wide sample) and of 0.25 mm notch tip radius (notch *s*). The "least severe" was a double notch of 3 mm depth and 1.5 mm notch tip radius (notch *d*). The authors measured – where possible – the Izod impact strengths, the yield (σ_F) and breaking strengths (σ_b), and Young's modulus of notched and unnotched specimens.

From their measurements a sequence of $k_S(s)$ values can be established:
PMMA (4.71), Kraton ®-1101 (3.68), SAN (2.71), PSU (2.36), EPDM rubber (2.12) and various particle filled thermoplastics (2.14–1.33). The $k_S(d)$ values exhibit a similar sequence:
PMMA (1.64), Kraton ®-1101 (<1.69), SAN (1.62), and PPO (1.48). There are, however, quite a number of (ductile) materials with $k_S(s)$ values *smaller* than one, notably HDPE (0.88), PTMT (0.96), PTFE (0.94), PA 6 (0.93), and Hytrel 4055. Values below unity of $k_S(d)$ are more numerous than those of $k_S(s)$.

Notch sensitivity factors for energy, k_T, show a somewhat different order. With some rare exceptions (PP + 20% glass, PE + 40% glass) k_T values are always larger than unity. The factor $k_T(s)$ reaches values of 625 (semi-transparent PETP), 241 (PSU), 61 (PA 66 + 0.6% H_2O), 48 (PP and PMMA), and 16 to 17 (mineral filled PA 66, PC). For $k_T(d)$ the authors determine 81 (PA 66 + 0.6% H_2O), 34 (PPO), 29 (semi-transparent PETP), 19 (PP), and 14 (annealed HDPE and transparent PETP). All other k_T values measured are smaller than 13.

The data of Pagano and Nielsen [211] confirm the observation that the presence of a crack *not necessarily reduces* the tensile strength of a sample or its energy to break. The authors also note that in some cases (notably in HDPE, PTMT, PTFE, PA 66 + 0.56% H_2O) the "more severe" notch *s* is the less dangerous and has a smaller k_T value than notch *d*. The reason for this unexpected reversal is not clear but it may be related to the volume of polymer under elevated stress in the notched region. This volume at the tip of a blunt notch is larger than that at the tip of a sharp notch [211]. Regarding the experimental evidence (e.g. [58, 64, 211, 214, 452]) one may say that in the case of a very sharp notch ($\rho_a < 50\ \mu m$) the onset of rapid crack propagation is determined predominantly by notch length *a* and material fracture toughness K_c according to Eq. (9.37). In a blunt notch specimen, however, the tangential stress σ_y at the root of the notch has to be considered the critical quantitiy [58, 64]. If the shape of a blunt notch is approximated by an ellipse the well known Neuber-Inglis formula [4] can be used to calculate σ_y:

$$\sigma_y = \sigma_0 (1 + 2\sqrt{a/\rho_a}). \qquad (9.39)$$

The stress concentration increases with decreasing ρ_a whereas the apparent critical energy release rate $(G_c)_{blunt}$ decreases. Considering also a plastic zone of size r_p Plati and Williams [64] obtain:

$$(G_c)_{blunt} = G_c \frac{(1 + \rho_a/2\,r_p)^3}{(1 + \rho_a/r_p)^2}. \qquad (9.40)$$

Whenever the measurement of the energy to break a specimen in tension or flexion is used to determine critical energy release rates, it must be recognized, that this energy is the sum of a number of quite different terms. As outlined in 8 II A, the energy loss A_n, which a pendulum incurs in striking and breaking a specimen, is a measure of the elastically stored energy W_e, of the fracture surface energy W_s, of the kinetic energy W_{kin} of the broken pieces, and of otherwise dissipated energy. The elastic energy to bend the specimen to a deflection δ under load P amounts to: $W_e = \frac{1}{2}\delta P = \frac{1}{2}P^2C$, with C being the bending compliance. The energy to propagate failure should principally correspond to the elastic energy stored at the point of onset of rapid crack propagation, which can be expressed now as:

$$A_n \simeq W_{el} = RBD\phi \tag{9.41}$$

A plot of A_n over $BD\phi$ should give, therefore, a straight line with slope $R = G_c$. Testing nine different brittle and ductile polymers (from PS to PE) by the Charpy and the Izod methods, Plati and Williams [64] were able to arrive at fairly unique values of G_c (cf. Table 9.2). They point out, however, that in order to obtain linear $BD\phi$ plots they had to consider an effective crack length consisting of the initial crack length and the size of suitably chosen plastic zone (see also [453–455]).

If the condition for unstable crack propagation is not fulfilled or maintained throughout the available range of crack lengths then a continued supply of mechanical energy is necessary to propagate the crack through the material. In that case the failure energy A_n should be mainly proportional to the ligament area:

$$A_n \simeq W_s = RB(D-a_0). \tag{9.42}$$

Corresponding plots of A_n over $B(D\text{-}a_0)$ for HiPS and ABS [53, 64, 69] revealed this linear relationship.

The fact that even the rapid deformation of a glassy polymer under concentrated stresses entails considerable local plastic deformation immediately suggests that the molecular properties, which influence yielding and flow, are also effecting G_c and thus the impact strength. The data compiled in Table 9-2 reveal this dependency of G_c on temperature, rate of deformation, and molecular properties. A possible relation between molecular relaxation processes and fracture energy of polymers has been pointed out in many of the cited references (e.g. [14, 19, 22, 24, 25, 54, 63, 64, 212–214, 235, 324d, 453–455]).

In 8 II A an account of the role of mechanical relaxation mechanisms in the impact loading of *unnotched* specimens has been given. The reasons for an expected positive correlation (and for the also observed deviations) between impact strength and magnitude of mechanical losses have been indicated there. At this point the molecular aspects of *notched* impact strength are to be investigated. Sauer [213] and Vincent [214] have reviewed the impact and stress-relaxation data from a large number of publications including their own extensive works and those referenced in 8 II A. They arrive at various general conclusions which may be listed in the following:

– The *impact resistance rating* of a polymer is generally determined by the storage component E′ of its dynamic modulus. For some twenty different polymers tested, 65% of the data complied with the following correlation [214]:

Impact resistance rating	Modulus
– brittle	$E' > 4.49\ GN\ m^{-2}$
– brittle if bluntly notched	$E' \simeq 3\ GN\ m^{-2}$
– brittle if sharply notched	$E' \simeq 2.2\ GN\ m^{-2}$
– tough but crack propagating	$E' < 1.5\ GN\ m^{-2}$

Deviations from this scheme were either related to molecular structure (bulky side groups of PMMA), to mechanical loss peaks (PTFE, HDPE, PC, PPO), to morphology (injection molded PP), or to heterogeneous reinforcements (short glass fibers, particulate fillers).

- Polymers with high impact strength at room temperature also have a significant low temperature loss peak ($\tan \delta \simeq 5 \cdot 10^{-2}$ or higher). Polymers in this class include, PE, PC, PB, PTFE, PCTFE, POM, and the polyamides.
- Polymers with low impact strength at room temperature usually have no significant low temperature loss peaks in the 100 K to 300 K range (PS, PMMA).
- Impact strengths have been measured as a function of temperature in several polymers and it has been shown that in many cases one gets a rise in brittle impact strength in the vicinity of the low temperature γ-relaxation (PTFE, PE, POM). For example, in PTFE at about 210 K, in PE at 150 K. Good correlation has also been observed between the magnitude of the low temperature γ-loss peak in polysulfone and impact resistance for a series of specimens containing various amounts of an antiplasticizer.
- An increase in molecular weight will generally affect the impact strength through the increase in G_c; this is most notably observed in HDPE (cf. 8 II A).
- In high impact polystyrene (HiPS) the normally brittle PS matrix has been blended with a suitable rubber-type component such as a styrene-butadiene copolymer which has a glass transition well below room temperature. The impact strength increases in proportion to the amount of the second component added. In this case of a two-phase polymer, however, impact resistance is increased through the initiation of a myriad of fine crazes and not through the generally increased extensibility of a homogeneously deforming matrix.

D. Fatigue Cracks

The characteristics of fatigue loading, i.e. of the repeated application of varying stress or strain amplitudes have been introduced in detail in Section 8 II C. In that discussion one point was left open: that of the mechanism of fatigue crack propagation. A comprehensive elaboration of this subject has quite recently been given by Hertzberg in his book on "Deformation and Fracture Mechanics of Engineering Materials" [3]. The reader is referred to this work or to the review articles by Plumbridge [217] and Manson and Hertzberg [218] for a detailed discussion of the different stages of fatigue crack growth, of the distinct characteristics of fatigue fracture surfaces, of the various theoretical approaches to derive crack growth-rate equations, and of the S–N curves of a wide variety of homogeneous and fiber reinforced polymers – and of metals, for that matter [3, 217, 218]. At this point only some recent observations will be discussed which seem to be directly related to the chain-like nature of macromolecules [173, 178, 191, 215–220].

As to be expected, the growth of a crack under constant or increasing load and fatigue crack propagation have quite a few features in common. Thus both the static and the fatigue fracture surfaces reveal a slow growth and a rapid crack propagation zone. The stress intensity factors reached in both cases at the transition from stable to unstable crack propagation correlate very well [218]. The rate of stable fatigue crack growth is empirically expressed by most authors [218, 456–460] in form of

$$\frac{da}{dN_F} = A(\Delta K)^n, \tag{9.43}$$

with A and n material parameters and ΔK the applied stress intensity factor range. This is in complete analogy to the expression for the rate of crack propagation under static loading (Eq. 9.22). Marshall et al. [133] and Radon et al. [219] substitute ΔK by the difference of K^2 and K^3 terms. Andrews et al. [215] employ in their analysis of the fatigue of PE the energy release rate G_c (instead of ΔK) as independent parameter. Such approaches are not significantly different because of the relation between K and G and of the form of Eq. (9.43). In their detailed reports [3, 218] Hertzberg and Manson analyse the various theoretical approaches, furnish fatigue crack growth data for some 20 materials at different temperatures and frequencies, and discuss the effect of environmental and material parameters.

Skibo, Hertzberg, and Manson [191] studied fatigue crack growth characteristics in polystyrene as a function of stress intensity factor range and cyclic frequency. Precracked single edge notched and compact-tension type specimens made from commercially available polystyrene sheet (mol. wt. = $2.7 \cdot 10^5$) were cycled under constant load at frequencies of 0.1, 1, 10 and 100 Hz, producing growth rates ranging from $4 \cdot 10^{-7}$ to $4 \cdot 10^{-3}$ cm/cycle. For a given stress intensity level, fatigue crack growth rates were found to decrease with increasing frequency, the effect being strongest at high stress intensity values. The variable frequency sensitivity of this polymer over the test range studied was explained in terms of a variable creep component. The macroscopic appearance of the fracture surface showed two distinct regions. At low stress intensity values, a highly reflective, mirror-like surface was observed which transformed to a rougher, cloudy surface structure with increasing stress intensity level. Raising the test frequency shifted the transition between these areas to higher values of stress intensity. The microscopic appearance of the mirror region revealed evidence of crack propagation through a single craze while the appearance of the rough region indicated crack growth through many crazes, all nominally normal to the applied stress axis. Electron fractographic examination of the mirror region revealed many parallel bands perpendicular to the direction of crack growth, each formed by a discontinuous crack growth process as a result of many fatigue cycles. The size of these bands was found to be consistent with the dimension of the crack tip plastic zone as computed by the Dugdale model. At high stress intensity levels a new set of parallel markings was found in the cloudy region which corresponded to the incremental crack extension for an individual loading cycle [191].

In these experiments two chain-related phenomena have been observed: the discontinuous growth of a fatigue crack and the frequency sensitivity of the growth rate. The discontinuous growth of a fatigue crack had been noted earlier for other polymers as well: PVC [216], PMMA, PC, and PSU [191]. Elinck et al. [216] report that, depending on frequency, between 130 and 370 load cycles were necessary to form one "arrest line" in PVC; Skibo et al. [191] found 600 to 1400 cycles for PS. The width of these growth bands corresponded well to the plastic zone size r_p calculated from the Dugdale model. Based on these findings Skibo et al. [191] propose as a possible crack growth mechanism the gradual development of a craze of width r_p and its subsequent and rapid breakdown. Such a mechanism would also be supported by the model of craze formation given in G II C: once the critical and frequency-dependent fibril extension is reached at the crack tip cooperative breakdown of the fibrils occurs and the crack propagates from D/C almost to B (Fig. 9.20).

This behavior has been well detailed in a series of intriguing experiments by Schinker, Könczöl and Döll [461–463]. These authors took a video-film of the interference fringe patterns of a discontinuously growing crack (in PVC); they succeeded in observing the very moment of crack jumping [462]. They also showed that the so-called continuous crack growth (i.e. one striation formed per cycle) occurs in reality within about 20% of the loading period (at peak load). Using synchrotron radiation as a powerful X-ray source, the breakdown of craze fibrils (in PS) during fatigue can also be traced in real time [464].

The frequency-dependent rate of disentanglement of the molecular coils in the fatigued fibrils seems to account for part of the frequency effect on growth rate. In addition hysteresis heating occurs in the strained craze material. Both effects combine to give rise to a distinct frequency sensitivity of A for a variety of materials such as PC and PMMA [219, 220], and PPO, PVC, PA 66, PC, PVDF, and PSU [220]. As has been noted by Skibo et al. [220] the frequency sensitivity varies with temperature. It reaches a maximum at that temperature where the external (fatigue) frequency corresponds to the internal segmental jump frequency (of the β-relaxation process).

The effects of strain amplitude, sample molecular weight, environment, and surface coating on the fatigue properties of PE and PS have been studied by Sauer et al. [173, 178]. Their results have been discussed in 8 II C. It is especially noteworthy that the initiation of a fatigue crack can be retarded by a decade or more through the application of a compatible, viscous coating. A 600 molecular weight PS oligomer served this purpose for both polished and unpolished surfaces of cylindrical PS specimens [178].

The relation between microstructure of LDPE and fatigue crack growth was investigated by Andrews et al. [215]. They found growth rates according to Eq. 9.43 in two regions: that of the initial brittle transpherulitic crack propagation and that of the ductile crack propagation at higher K values. Between these two regions a distinct transition region was found where da/dN_F depended little on ΔK [465, 466].

Concluding this section, some recent references on additional aspects of fatigue crack growth will be given: analysis of the propagation in viscoelastic media [467], lifetime predictions [468], the observation of an epsilon crack tip plastic zone [469] and fatigue damage analysis [470, 438c]. The latter investigations have clearly shown that the microstructure (of PVDF, PA 66, POM, and PETP) in advance of the crack front is different from the original morphology.

E. Fracture-related Phenomena

1. Electro-Fracture Mechanics

A fracture mechanics concept has been used by Zeller et al. [471] to describe the mechanism of dielectric aging in solids. It had been noted already earlier by Schirr [472] that the growth direction of partial discharges (PD) in an electrically *and* mechanically stressed solid (epoxy resin) depended on both fields: the discharge channels tried to avoid the plane perpendicular to the largest compressive stress component; on the other hand tensile stresses increased their growth rate in perpendicular direction [472, 473].

In order to better understand this phenomenon several approaches have been taken. A criterion analogous to the Griffith one can be set up for the growth of partial discharge channels in a dielectric. The formation of a PD channel requires a formation energy W_f. W_f contains a surface energy term and a plastic deformation energy term because it is supposed that channel formation requires plastic deformation. It can be shown that under normal circumstances the plastic deformation term is dominating and the surface energy may be neglected.

The initiation or growth of a PC channel leads to a release of electrostatic energy W_{es}. Growth is energetically possible only if

$$\frac{\partial W_f}{\partial x} \leqslant \frac{\partial W_{es}}{\partial x}, \tag{9.44}$$

where x is the growth coordinate and the derivative is to be taken at the growth front. Equation (9.44) is universally correct independent of the detailed growth mechanisms. In the literature [471] models have been introduced to calculate both the left- and right-hand side of Eq. (9.44).

An important feature of discharge channels certainly is their tendency to branching into complicated stochastic patterns [471–475]; however, similar fractal structures will be obtained in non-equilibrium particle aggregation which, therefore, has been used for the modellisation of (planar) discharge patterns [474–476].

The above problem is not yet fully understood although many contributing phenomena, such as electron injection and transport [477–479], the kinetics of damage development in electric fields [473, 480–483], the influence of polymer morphology, cohesive energy density [483] and Young's modulus [479] and of liquid environment [484–486] ("water treeing", see e.g. [484, 485]) have been investigated.

2. Fracto-Emission

Fracto-emission in the narrow sense is understood as the emission of particles (e.g. electrons, ions, ground-state and excited neutrals, and photons) during and following fracture [487–489]. However, since localized fracture events give already rise to an emission at a much earlier stage of polymer deformation, one may as well speak of *mechano-emission* [490–493]. When polymers are deformed in a vacuum electron emission occurs [490]. Such emission takes place during the stretching of carbonchain and heterochain polymers of both oriented and unoriented types. By comparing mechanoemission characteristics with features of the accumulation of molecular products of breakdown one finds a close correlation between mechanoemission and the breakdown of polymers under load [491]. According to Zakrevskii and Pakhotin [490–492], the emitted electrons are formed by autoionization of highly loaded chains possibly involving a tunnel transition of electrons into deep traps which are subsequently destroyed during deformation. The ionized bonds which have a reduced mechanical strength (see Chapter 4 II B 2) should then dissociate rapidly under the applied stress and form a macroion and a macroradical [492].

However, Enikolipian et al. [487] propose as a "more mechanical" mechanism the rupture of the most highly stressed chains which dissipate their stored elastic energy through fragmentation of the newly formed free chain ends and in the form of thermal and electronic excitation of their surrounding. Without any doubt the excellent sensitivity of the methods tracing emission products permits an early detection of localized fracture events. On the other hand, the fact that the events are localized makes it difficult, at this time, to draw valid conclusions with respect to the continuation of the deformation process and the occurence of final fracture.

3. Mechano-Chemistry

In a broader sense mechano-chemistry comprises all aspects of the stress-induced scission of chain molecules. In a narrower sense, however, one speaks of mechanochemical methods if one refers to the *intentional mechanical degradation* of (solid) polymers. The objectives in the latter cases are the comminution or softening of the material or the production of large and highly reactive surfaces as a means to initiate permanent chemical bonding between different polymers. In Table 9.8 an overview on methods and processes is given which may result in a mechanical degradation of chain molecules. The objectives of these processes are indicated with respect to the deformation mechanism. It should be pointed out that in the mechano-chemical methods, in the narrower sense, the degrading solids are subjected to ill defined, complex states of stress leading to a deformation which always includes simultaneously the yielding and flowing of material and the breakage of chains. Table 9.8 lists the deformation mechanisms which are the most important in view of the corresponding objective. Reference is made to those chapters and sections of this book where the particular deformation mechanisms are treated.

In recent years a number of comprehensive review articles on the mechanochemistry of polymers have appeared [221–226, 494, 495]. A competent two-volume handbook by Porter and Casale on this subject [227] is also available. The

Table 9.8. Mechano-chemistry of Solid Polymers: Methods and Objectives

Method, Process	Objective	Deformation Mechanism	Reference to Chapter Section
Grinding, sawing, cutting	comminution	compressive yielding,	8.II. D
Ball or vibromilling Impact loading	comminution shock absorption	elastic compressive and/or tensile deformation	5.II. E/8.II.A.1
Mastication (of elastomers)	softening	stretching and breakage of chain molecules	5, 8.II.E
Extrusion, injection, drawing	forming	shear flow	8.II. D
Mixing, stirring	dispersion of additives	frictional interaction, compressive yielding	
Synthesis through mechanical blending	grafting, copolymerization of incompatible or mechanically different polymers	chain scission and free radical reactions	6.IV. A, 7.III.

reader is referred to these references for any extended information on the effect of experimental variables (type of equipment, temperature, rate of mechanical working, atmosphere) or polymer characteristics (chemical structure, initial molecular weight) on the resulting material properties.

Within this monograph special consideration is given to the loading and breakage of molecular chains. With regard to mechano-chemical processes such as comminution or mechanical synthesis one has to study, therefore,

- the response of the polymer matrix to mechanical working
- the consequences of chain scission: reduction of molecular weight and radical formation
- the results of radical reactions, e.g. oxidative degradation, cross-linking, copolymerization.

The property changes resulting from a mechanical treatment of polymeric materials have been known and utilized for a long time. Thus knowledge of the fact that mechanical working of natural rubber leads to its softening already dates back 120 years [225]. But it was only after the introduction of the concept of macromolecules some fifty years ago that this effect of mechanical treatment was related to the rupture of molecular chains. The fact that the intensive milling of a heterophase mixture (of rubber and maleic acid anhydride) can induce a chemical reaction between the components was first observed in 1941 [224]. Systematic studies of the nature of these chemical reactions and particularly of the role of the formed free radicals began about ten years later [224–225].

The response of polymer matrices and of chain molecules to mechanical working have been intensively discussed in the previous chapters. The ESR investigations of ground polymers (6.IV A and 7.I C 3) have shown that in a comminution process (e.g. milling) chain molecules are broken in large numbers and as a *consequence* of the formation of new surface during the breaking-up of the polymer particles. The

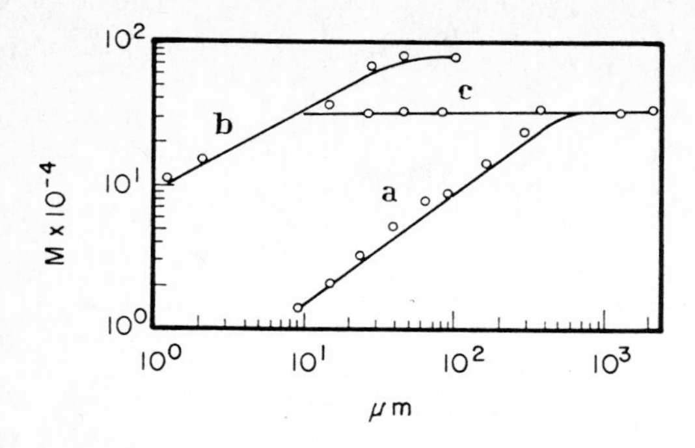

Fig. 9.55. Molecular weight of comminution products as a function of particle size (after [200] from [201]): *a* polystyrene in NO_2, *b* polymethylmethacrylate block copolymer in NO_2, *c* polystyrene in air

chain ruptures have an immediate and notable effect on the sample molecular weight M [221–228]. In Figure 9.55 M is plotted as a function of particle diameter D. The linear sections may be represented by a potential law

$$M(D) = A(D/D_0)^n \tag{9.45}$$

where A and n are material-related constants. Curves *a* and *b* in Figure 9.55 reveal that chain scission is more effective in polystyrene (n = 0.76) than in PMMA (n = 0.50). In both cases comminution was carried out [228] in the presence of radical scavengers (NO_2) to prevent chain recombination or cross-linking. Comminution in air seems to have permitted cross-linking reactions to such an extent that the average molecular weight of a particle size fraction did not change with D (curve *c*). The opposite effect, decrease of M without increase of specific surface was noted by Komissarov et al. [222] in the dispersion of PA and PETP fibers *without cooling*.

It is a general observation that in continuous mechanical working D (and M) do not decrease indefinitely but reach certain limiting values. These limiting average M values are of the order of the entanglement molecular weight M_e. This does not preclude the possibility that in mechanical degradation fractions of low molecular weight and even volatile components can be produced [221–225].

The effect of environmental temperature on the rate of degradation has also been widely studied [221–227]. In view of the composite nature of the degradation process no simple rate equations can be expected. It becomes clear from Eqs. (5.41) and (7.3) that the softening of the matrix (decrease of ζ_0) and the reduced effective bond strength U(T) partly balance each other. According to an extensive review by Casale [226] the temperature effect on matrix rigidity seems to be the dominant one. The slower relaxation times at lower temperatures lead to an increase in mechanical degradation with decreasing temperature (negative temperature coefficient of the overall mechano-chemical reaction).

References for Chapter 9

1. A. A. Griffith: Phil. Trans. Roy. Soc. (London) A 221, 163–198 (1920).
2. Mechanics of Fracture, G. C. Sih, ed., Vol. I-V, Leyden: Noordhoff International Publishing. 1975–1977.
3. R. W. Hertzberg: Deformation and Fracture Mechanics of Engineering Materials, New York: John Wiley 1976.
4. J. G. Williams: Stress analysis of polymers, London: Longman 1973.
5. E. H. Andrews: Fracture in Polymers, Edinburgh: Oliver + Boyd, 1968.
6. F. Kerkhof: Kolloid-Z. u. Z. Polymere *251*, 545 (1973).
7. D. S. J. Dugdale: Mech. Phys. Solids *8*, 100 (1960).

8a. ASTM Book of Standards, Part 31: Physical and Mechanical Testing of Metals (1969).
 b. H. J. G. Blauel, J. F. Kalthoff, E. Sommer: Materialprüfung *12*, 69 (1970).
 c. ASTM Book of Standards, Part 10, E 616 (1981).
 d. Part 35, E 6 (1982).
 e. D. P. Wilhelm: Int. Journ. Fracture *23*, R 77 (1983).
 f. The "Task group on polymers and composites" of the European Group on Fracture prepares a proposal for a nomenclature in fracture mechanics of polymers closely related to [8c–e].
 g. Instead of G_{Ic} the term G_{cc} (corresponding to a K_{cc}) is proposed herewith; it should indicate that energy release rate at which slow crack growth turns into unstable, rapid crack growth (see Fig. 9.5). The generally used term G_{Ic} seems to be less consistant since it has a different meaning in fracture mechanics of metals.

9. J. D. Eshelby: Proc. Royal Soc. A241, 376–396 (1957).
10. G. R. Irwin: Handbuch der Physik 6, Berlin – Göttingen – Heidelberg: Springer 1958, 559.
11. H. H. Kausch: Kunststoffe 66/9, 538–544 (1976).
12. G. P. Marshall, L. H. Coutts, J. G. Williams: Materials Sci. *9*, 1409 (1974).
13. P. W. Beamont, R. J. Young: J. Materials Sci. *10*, 1334–1342 (1975).
14. F. A. Johnson, J. C. Radon: J. Polymer Sci., Polymer Chem. Ed. *11*, 1995–2020 (1973).
15. G. W. Weidmann, W. Döll: Colloid and Polymer Sci. *254*, 205–214 (1976).
16. J. P. Berry, in: Fracture Processes in Polymeric Solids, B. Rosen, ed., New York: Interscience 1964, Chap. II.
17. L. J. Broutman, F. J. McGarry: J. appl. Polymer Sci. *9*, 589 (1965).
18. M. F. Kanninen, A. R. Rosenfield, R. G. Hoagland, in: Deformation and Fracture of High Polymers, Kausch, Hassell, Jaffee Eds., New York: Plenum Press 1974, 471–484.
19. M. Parvin, J. G. Williams: J. Materials Sci. *10*, 1883–1888 (1975).
20. Y. W. Mai: Int. J. Fracture *9*, 349 (1973).
21. R. P. Kambour, S. Miller: General Electric Report No 77CRD009, March 1977.
    R. P. Kambour, A. S. Holik, S. Miller: General Electric Report No 77CRD125, June 1977.
22. S. Arad, J. C. Radon, L. E. Culver: J. appl. Polymer Sci. *17*, 1467–1478 (1973).
23. B. Ellis, R. van Noort: J. Appl. Polym. Sci. *30*, 4517 (1985).

24a. W. Retting: Materialprüfung *5*, 101 (1963).
 b. W. Retting: Kolloid-Z. Z. Polymere *210*, 54 (1966).

25. H. W. Greensmith, L. Mullins, A. G. Thomas: Trans. Soc. Rheol. *4*, 179 (1960).
26. G. J. Lake, P. B. Lindley, A. G. Thomas: Paper 43 (Session IV) Int. Conf. on Fracture 1969, Brighton.
27. A. G. Thomas, in: Deformation and Fracture of High Polymers, New York: Plenum Press 1973, 467–470.
28. A. Ahagon, A. N. Gent: J. Polymer Sci., Polymer Physics Ed. *13*, 1903–1911 (1975).
29. H. K. Mueller; Ph. D. Thesis, California Inst. of Technology, Pasadena, California, June 1968.
30. W. Döll: Energieumsetzung in Wärme beim kritischen Bruchbeginn in PMMA. Report of the Institut für Festkörpermechanik, Freiburg, 1978.
31. D. W. van Krevelen: Properties of Polymers, Amsterdam: Elsevier Publ. Co. 1972, Table 6.6. Polymer Handbook, Brandrup, J., Immergut, E. H., J. Wiley + Sons, New York 1975.

32. J. J. Benbow, F. C. Roesler: Proc. Phys. Soc. B *70*, 201–211 (1957).
33. J. P. Berry: J. Polymer Sci. *50*, 107–115 (1961).
34. J. P. Berry: J. Polymer Sci. *50*, 313–321 (1961).
35. N. L. Svensson: Proc. Phys. Soc. *77*, 876–884 (1961).
36. J. J. Benbow: Proc. Phys. Soc. *78*, 970 (1961).
37. J. P. Berry: J. appl. Physics *34*, 62 (1963).
38. A. van den Boogaert: The Physical Basis of Yield and Fracture, Oxford, 1966, pg. 167.
39. P. I. Vincent, K. V. Gotham: Nature *210*, 1254 (1966).
40. J. P. Berry: in: Fracture VII, H. Liebowitz, Ed., Academic Press, New York 1972.
41. J. O. Outwater, D. J. Gerry: J. Adhesion *1*, 290 (1969).
42. S. Mostovoy, P. B. Crosely, E. J. Ripling: J. Materials, *2*, 661 (1967).
43. G. P. Marshall, L. E. Culver, J. G. Williams: Polymer *37*, 75–81 (1969).
44. R. Griffith, D. G. Holloway: J. Materials Sci. *5*, 302–307 (1970).
45. J. W. Curtis: J. of Physics *3*, 1413–1422 (1970).
46. L. E. Miller, K. E. Puttick, J. G. Rider: J. Polymer Sci. *33*/C, 13–22 (1971).
47. B. Harris, P. W. R. Beaumont, E. Moncunill de Ferran: J. Materials Sci. *6*, 238–251 (1971).
48. A. R. Rosenfield, P. N. Mincer: Polymer Sci. Symposium *32*, New York: John Wiley (1971), 283–296.
49. J. S. Foot, I. M. Ward: J. Materials Sci. *7*, 367–387 (1972).
50. W. Döll: Kolloid-Z. u. Z. Polymere *250*, 1066–1073 (1972).
51. G. P. Marshall, L. E. Culver, J. G. Willliams: Int. J. Fracture *9*, 295 (1973).
52. J. G. Williams, G. P. Marshall, in: Deformation and Fracture of High Polymers, H. H. Kausch, J. A. Hassell, R. I. Jaffee Eds., New York: Plenum Press 1973, 557.
53. H. R. Brown: J. Materials Sci. *8*, 941–948 (1973); G. P. Marshall, J. G. Williams, C. E. Turner, ibid., 949–956 (1973).
54. F. A. Johnson, A. P. Glover, J. C. Radon: Proc. 1973 Symposium on Mechanical Behavior of Materials, 141–148, The Society of Materials Sci., Japan (1974).
55. B. E. Nelson: J. Colloid and Interface Sci. *47*/3, 595–599 (1974).
56. P. W. R. Beaumont: J. Adhesion *6*, 107–137 (1974).
57. G. Goldbach: Kunststoffe *64*/9, 475–481 (1974).
58. R. A. W. Fraser, I. M. Ward: J. Materials Sci. *9*, 1624–1630 (1974).
59. A. D. S. Diggwa: Polymer *15*, 101 (1974).
60. W. Döll: Int. J. Fracture *11*, 184–186 (1975).
61. W. Döll: J. Materials Sci. *10*, 935–942 (1975).
62. R. Ravetti, W. W. Gerberich, T. E. Hutchinson: J. Materials Sci. *10*, 1441–1448 (1975).
63. F. A. Johnson, J. C. Radon: J. Polymer Sci., Polymer Chemistry Ed. *13*, 495–516 (1975).
64. E. Plati, J. G. Williams: Polymer *16*, 915–920 (1975). E. Plati, J. G. Williams: Polymer Engng. and Sci. *15*, 470–477 (1975).
65. R. P. Kusy, M. J. Katz: J. Materials Sci. *11*, 1475–1486 (1976). R. P. Kusy, D. T. Turner: Polymer *17*, 161–166 (1976).
66. W. Döll, G. W. Weidmann: J. Materials Sci., Letters *11*, 2348–2350 (1976).
67. W. Döll: Int. J. Fracture *12*/4, 595–605 (1976).
68. Y. W. Mai, J. G. Williams: J. Materials Sci. *12*, 1376–1382 (1977).
69. J. G. Williams, M. W. Birch: Fracture 1977, Vol. 1, ICF4, Waterloo Canada, June 19–24. 1977, pg. 501–528.
70. W. Döll, E. Gaube: unpublished results.
71. N. J. Mills, N. Walker: Polymer *17*, 335–344 (1976).
72. P. S. Leevers, J. C. Radon, L. E. Culver: Polymer *17*, 627–632 (1976).
73. R. A. Gledhill, A. J. Kinloch: Polymer *17*, 727–731 (1976).
74. G. P. Morgan, I. M. Ward: Polymer *18*, 87–91 (1977).
75. S. Yamini, R. J. Young: Polymer *18*, 1075–1084 (1977).
76. R. P. Kambour: J. Polymer Sci. D (Reviews) *7*, 1–154 (1973).
77. S. Rabinowitz, P. Beardmore: CRC Critical Reviews *1*, 1 (1972).
78. C. C. Hsiao, J. A. Sauer: J. of Appl. Physics *21*/11, 1071–1083 (1950). J. A. Sauer, C. C. Hsiao: Trans. ASME *75*, 895 (1953).
79. J. F. Rudd: J. Polymer Sci. *B1*, 1 (1963).

80. H. A. Stuart, D. Jeschke, G. Markowski: Materialprüfung 6, 77 (1964).
81. A. C. Knight: J. Polymer Sci. A, *3,* 1845 (1965).
82. G. Rehage, G. Goldbach: Die Angewandte Makromolekulare Chemie *1,* 125–149 (1967).
83. R. P. Kambour, R. W. Kopp: General Electric, Report No. 67-C-374 (1967).
84. L. C. Cessna, Jr., S. S. Sternstein: Fundamental Phenomena in the Materials Sci. *4,* (Plenum Press), 45–79 (1967).
85. R. P. Kambour: Applied Polymer Symposia *7,* 215–235 (1968).
86. S. S. Sternstein, L. Ongchin, A. Silverman: Applied Polymer Symposia *7,* 175–199 (1968).
87. R. P. Kambour, A. S. Holik: General Electric, Report No. 69-C-022 (1968).
88. G. M. Bartenev: Vysokomol. soyed. A11/*10,* 2341–2347 (1969). Polymer Sci. USSR 11/*10,* 2663–2671 (1970).
89. R. N. Haward, B. M. Murphy, E. F. T. White: Paper 45 (Session IV), Fracture 1969, Brighton, 45/1–45/13.
90. P. Dunn, G. F. Sansom: J. appl. Polymer Sci. *13,* 1641–1655 (1969).
91. P. Dunn, G. F. Sansom: J. appl. Polymer Sci. *13,* 1657–1672 (1969).
92. P. Dunn, G. F. Sansom: J. appl. Polymer Sci. *13,* 1673–1688 (1969).
93. P. Dunn, G. F. Sansom: J. appl. Polymer Sci. *14,* 1799–1806 (1970).
94. R. S. Moore, C. Gieniewski: J. appl. Polymer Sci. *14,* 2889–2904 (1970).
95. K. C. Rusch, R. H. Beck, Jr.: J. Macromol. Sci.-Phys. B 3(3) 365 (1969), B 4(3), 261 (1970).
96. A. N. Gent: J. Materials Sci. *5,* 925 (1970).
97. L. Morbitzer, R. Holm, U. Brenneisen, K. Heidenreich, H. Röhr: Kunststoffe *60*/11, 861–866 (1970).
98. J. Murray, D. Hull: Polymer Letters *8,* 159–163 (1970), J. Murray, D. Hull: J. Polymer Sci. A-2, *8,* 1521–1543 (1970).
99. Yu. M. Malinskii, V. V. Prokopenko, N. A. Ivanova, V. A. Kargin: Mekh. Polimerov (USSR) *6*/2, 271–275 (1970), Polymer Mech. (USA) *6*/2, 240–244 (1973).
100. Yu. M. Malinskii, V. V. Prokopenko, N. A. Ivanova: Mekh. Polimerov (USSR) *6*/3, 445–448 (1970), Polymer Mech. (USA) *6*/3, 382–384 (1970).
101. F. Fischer: Z. f. Werkstofftechnik 1/2, 74–83 (1970).
102. G. P. Marshall, L. E. Culver, J. G. Williams: Proc. Roy. Soc. Lond. A*319,* 165–187 (1970).
103. G. Menges, H. Schmidt, H. Berg: Kunststoffe *60*/11, 868–872 (1970).
104. G. Menges, H. Schmidt: Plastics & Polymers *38,* 13–19 (1970).
105. T. T. Wang, M. Matsuo, T. K. Kwei: J. appl. Physics *42*/11, 4188–4196 (1971).
106. P. Beahan, M. Bevis, D. Hull: Phil. Magazine *24*/12, 1267–1279 (1971).
107. N. Fujimoto, M. Takayanagi, Y. Yamaguchi: Proceeding of Int. Conf. Mech. Behavior of Materials, Kyoto, 1971, 572.
108. R. N. Haward, B. M. Murphy, E. F. T. White: J. Polymer Sci. A-2/*9,* 801–814 (1971).
109. B. J. Macnulty: J. Materials Sci. *6,* 1070–1075 (1971).
110. J. Pohrt: J. Macromol. Sci.-Phys. B5(2), 299–316 (1971).
111. J. Pohrt: Gummi Asbest Kunststoffe *24,* 594–606, and 700–713 (1971). K. V. Gotham: Plastics and Polymers *40,* 277–282 (1972).
112. Yu. S. Lipatov, F. B. Fabulyak: J. Appl. Polymer Sci. *16,* 2131 (1972).
113. H. Schmidt: Kunststoff-Rundschau *19*/1, 1–7 (1972). H. Schmidt: Kunststoff-Rundschau *19*/2,3, 56–65 (1972).
114. C. B. Bucknall, D. Clayton: J. Materials Sci. *7,* 202–210 (1972).
115. P. Beahan, M. Bevis, D. Hull: J. Materials Sci. *8,* 162–168 (1972).
116. N. Brown, M. F. Parrish: Polymer Letters Ed. *10,* 777–779 (1972).
117. L. Spenadel: J. Appl. Polymer Sci. *16,* 2375–2386 (1972).
118. I. Narisawa, T. Kondo: Int. J. of Fracture Mechanics *8*/4, 435–440 (1972).
119. I. Narisawa: J. Polymer Sci. Part A-2/*10,* 1789–1797 (1972).
120. J. Hoare, D. Hull: Phil. Magazine *26*/2, 443–455 (1972).
121. G. Menges, E. Alf: Kunststoffe *62*/4, 259–267 (1972).
122. H. E. Driesen: gwf-gas/erdgas *113*/6, 265–270 (1972).
123. P. I. Vincent, S. Raha: Polymer *13*/6, 283–287 (1972).
124. E. H. Andrews, L. Bevan: Polymer *13*/7, 337–346 (1972).

125. R. P. Kambour, E. E. Romagosa, C. L. Gruner: Macromolecules *5*/4, 335–340 (1972).
126. E. H. Andrews, G. M. Levy, J. Willis: J. Materials Sci. *8*, 1000–1008 (1973).
127. I. Narisawa, T. Kondo: J. Polymer Sci. – Poly. Phys. Ed. – *11*, 223–232 (1973).
128. R. P. Kambour, C. L. Gruner, E. E. Romagosa: J. Polymer Sci. – Poly. Phys. Ed. – *11*, 1879–1890 (1973).
129. N. Brown: J. Polymer Sci. – Poly. Phys. Ed. – *11*, 2099–2111 (1973).
130. E. Gaube, H. H. Kausch: Kunststoffe *63*/6, 391–397 (1973).
131. J. Lilley, D. G. Holloway: Phil. Mag. (8) *28*/1, 215–220 (1973).
132. K. P. Grosskurth: Kautschuk u. Gummi-Kunststoffe *26*/2, 43–45 (1973).
133. G. P. Marshall, J. G. Williams: J. Appl. Polymer Sci. *17*, 987–1005 (1973).
134. S. Fischer, N. Brown: J. Appl. Phys. *44*/10, 4322–4327 (1973).
135. H. G. Olf, A. Peterlin: Polymer *14*, 78–79 (1973).
136. G. Menges: Kunststoffe *63*, 95–100 a. 173–177 (1973).
137. R. N. Haward, D. R. J. Owen: J. Materials Sci. *8*, 1135–1144 (1973).
138. H. G. Olf, A. Peterlin: Macromolecules *6*, 470–472 (1973). H. G. Olf, A. Peterlin: J. Colloid and Interface Sci. *47*/3, 628–634 (1973).
139. S. S. Sternstein, F. A. Myers: J. Macromol. Sci.-Phys. *8*/3–4, 539–571 (1973).
140. J. R. Kastelic, E. Baer: J. Macromol. Sci.-Phys. *7*/4, 679–703 (1973).
141. L. Camwell, D. Hull: Phil. Magazine *27*/5, 1135–1150 (1973).
142. R. J. Oxborough, P. B. Bowden: Phil. Magazine *28*/3, 547–559 (1973).
143. K. J. Takahashi: Polymer Sci. – Poly. Phys. Ed. – *12*, 1697–1705 (1974).
144. M. Kitagawa, K. Motomura: J. Polymer Sci. – Poly. Phys. Ed. – *12*, 1979–1991 (1974).
145. H. G. Olf, A. Peterlin: J. Polymer Sci. – Poly. Phys. Ed. – *12*, 2209–2251 (1974).
146. J. F. Fellers, B. F. Kee: J. Appl. Polymer Sci. *18*, 2355–2365 (1974).
147. E. H. Andrews, G. M. Levy: Polymer *15*/9, 599–607 (1974).
148. T. L. Peterson, D. G. Ast, E. J. Kramer: J. Appl. Phys. *45*/10, 4220–4228 (1974).
149. K. P. Grosskurth: Kautschuk u. Gummi – Kunststoffe *27*/8, 324–328 (1974).
150. P. Beahan, M. Bevis, D. Hull: Proc. R. Soc. Lond. A *343*, 525–535 (1975).
151. G. W. Weidmann, J. G. Williams: Polymer *16*/12, 921–924 (1975).
152. A. S. Argon: Pure and Appl. Chemistry 43/1–2, 247–272 (1975).
153. D. Hull, L. Hoarse: Plastics and Rubber: Materials and Appl. *5*, 65–73 (1976).
154. P. Filzek, R. Süselbeck, W. Wicke: Kunststoffe *66*/1, 38 (1976).
155. M. J. Doyle: J. Polymer Sci., Polymer Phys. Ed., *13*, 127–135 (1975).
156. N. Verheulpen-Heymans, J. C. Bauwens: J. Materials Sci. *11*, 1–6 (1976). N. Verheulpen-Heymans, J. C. Bauwens: J. Materials Sci. *11*, 7–16 (1976).
157. N. Verheulpen-Heymans: J. Polymer Sci.: Polymer Phys. Ed., *14*, 93–99 (1976).
158. Y. Imai, N. Brown: J. Materials Sci. *11*, 417–424 (1976). Y. Imai, N. Brown: J. Materials Sci. *11*, 425–433 (1976).
159. H. G. Krenz, D. G. Ast, E. J. Kramer: J. Materials Sci. *11*, 2198–2210 (1976).
160. H. G. Krenz, E. J. Kramer, D. G. Ast: J. Materials Sci. *11*, 2211–2221 (1976).
161. D. L. G. Lainchbury, M. Bevis: J. Materials Sci. *11*, 2222–2234 (1976).
162. D. L. G. Lainchbury, M. Bevis: J. Materials Sci. *11*, 2235–2241 (1976).
163. S. Sikka: PhD Dissertation, University of Utah, Salt Lake City, Utah (1976).
164. I. D. Graham, J. G. Williams, E. L. Zichy: Polymer *17*/5, 439–442 (1976).
165. A. S. Argon, M. I. Bessonov: Reports of Research in Mech. Process. of Polymers *12*, MIT (1976).
166. A. S. Argon, J. G. Hannoosh, M. M. Salama: Reports of Research in Mech. Process. of Polymer *17*, MIT (1977); Fracture 1977, Vol. 1, Waterloo, Can., pg. 445. A. S. Argon, J. G. Hannoosh: Reports of Research in Mech. Process. of Polymers *21*, MIT (1977), Phil. Mag. *36*/5, 1195–1216 (1977).
167. A. S. Argon, M. M. Salama: Reports of Research in Mech. Process. of Polymers *22*, MIT (1977), Phil. Mag. *36*/5, 1217–1234 (1977).
168. W. Retting: Die Angew. Makrom. Chemie *58*/*59*, 133–174 (1977).
169. K. Smith, R. N. Haward: Polymer *18*/7, 745–746 (1977).
170. C. J. Singleton, E. Roche, P. H. Geil: J. Appl. Polymer Sci. *21*, 2319–2340 (1977).

171. R. P. Kambour: General Electric, Report No. 77CRD169, August 1977.
172. M. Ishikawa, I. Narisawa, H. Ogawa: J. Polymer Sci. (Polymer Phys. Ed.) *15,* 1791–1804 (1977).
173. J. W. Sauer, E. Foden, D. R. Morrow: Polymer Eng. and Sci. *17*/4, 246–250 (1977).
174. R. Bardenheier: DVM, Vorträge der 9. Sitzung des Arbeitskreises Bruchvorgänge, 29–38, 11. 10. 1977.
175. J. Opfermann: DVM, Vorträge der 9. Sitzung des Arbeitskreises Bruchvorgänge, 39–46, 11. 10. 1977.
176. K. P. Grosskurth: DVM, Vorträge der 9. Sitzung des Arbeitskreises Bruchvorgänge, 47–60, 11. 10. 1977.
177. R. A. W. Fraser, I. M. Ward: Polymer *19*/2, 220 (1978).
178. J. A. Sauer: Polymer *19,* NN (1978). S. Warty, D. R. Morrow, J. A. Sauer: Polymer, to be published.
179. J.-C. Pollet, S. J. Burns: Int. J. of Fracture *13,* 667–679 (1977). J.-C. Pollet, S. J. Burns: Int. J. Fracture *13*/6, 775–786 (1977).
180. A. G. Atkins, C. S. Lee, R. M. Caddell: J. Mat. Sci. *10,* 1381 (1975).
181. K. B. Broberg: Arkiv för Fysik *18,* 159–192 (1960).
182. G. T. Hahn, M. F. Kanninen: Fracture 1977, Vol. 1, ICF4, Waterloo, Canada, June 1977, 193.
183. K. Friedrich: Fracture 1977, Vol. 3, ICF4, Waterloo, Canada, June 1977, pg. 1119. K. Friedrich: Dissertation, Ruhr-Universität Bochum, 1978.
184. K. N. G. Fuller, P. G. Fox, J. E. Field: Proc. R. Soc. A(GB), Vol. 341, No. 1627, 537–557 (1974).
185. R. Weichert, K. Schönert: J. Phys. Mech. Sol. 22, 127–133 (1974).
186. R. Weichert: Dissertation, Universität Karlsruhe, 1976.
187. P. C. Moon, R. E. Barker, Jr.: J. Polymer Sci. (Polymer Phys. Ed.) *11,* 909–917 (1973).
188. E. Roeder, H.-A. Crostack: Kunststoffe *67*/8, 454–456 (1977).
189. K. L. DeVries, D. K. Roylance, M. L. Williams: Report UTEC DO 68-056, Univ. of Utah, Salt Lake City, 1968.
190. L. E. Nielsen, D. J. Dahm, P. A. Berger, V. S. Murty, J. L. Kardos: Polymer Sci. (Polymer Phys. Ed.) *12,* 1239 (1974).
191. M. D. Skibo, R. W. Hertzberg, J. A. Manson: J. Mat. Sci. *11,* 479–490 (1976).
192. J. A. Manson, L. H. Sperling: Polymer Blends and Composites, New York/London 1976: Plenum Press.
193. Mehrphasige Polymersysteme, Angew. Makromol. Chem. 58/59, 60/61 (1977).
194. J. Murray, D. Hull: J, Polymer Sci., Part A-2 *8,* 583–594 (1970).
195. B. E. Nelson, D. T. Turner: Polymer Letters *9,* 677–680 (1970).
196. D. Hull, T. W. Owen: J. Polymer Sci. (Polymer Phys. Ed.) *11,* 2039–2055 (1973).
197. W. Döll: Colloid & Polymer Sci. *252,* 880–885 (1974).
198. J. R. Martin, J. F. Johnson: J. Polymer Sci. (Polymer Phys. Ed.) *12,* 1081–1088 (1974).
199. J. Hoare, D. Hull: J. Materials Sci. *10,* 1861–1870 (1975).
200. R. P. Kusy, D. T. Turner: Polymer *18,* 391–399 (1977).
201. S. Gogolewski: Polymer *18,* 63–68 (1977).
202. S. Gogolewski, A. J. Pennings: Polymer *18,* 647–660 (1977).
203. J. E. Stamhuis, A. J. Pennings: Polymer *18,* 667–674 (1977).
204. S. Bandyopadhyay, H. R. Brown: Polymer *19,* 589–592 (1978).
205. E. S. Clark, C. A. Garber: Int. J. Polymeric Mater. *1,* 31–46 (1971).
206. J. Patel, P. J. Philips: Polymer Letters Ed. *11,* 771–776 (1973).
207. E. Andrews: Pure and Appl. Chem. *39*/1–2, 179–194 (1974).
208. G. Menges, N. Berndtsen: Kunststoffe *66*/11, 735–740 (1976).
209. P. I. Vincent, S. Picknell, G. F. Harding: An Investigation of Fracture Criteria for Anisotropic Polyethylene Terephthalate Film Using Mechanical and Optical Techniques, Div. Pol. Sci. Case Western Reserve Univ., Cleveland, Ohio 44106 (1967).
210. V. E. Korsukov, V. I. Vettegren', I. I. Novak, A. Chmel': Mekh. Polimerov 4, 621–625 (1972). Polymer Mechanics 4, 536–539 (1972).

211. M. Takano, L. E. Nielsen: J. Appl. Polymer Sci. *20*, 2193–2207 (1976).
212. H. Oberst: Kunststoffe *53*, 4 (1963).
213. J. A. Sauer: J. Polymer Sci. C (Polymer Symposia) *32*, 69–122 (1971).
214. P. I. Vincent: Polymer *15*, 111–116 (1974).
215. E. H. Andrews, B. J. Walker: Proc. R. Soc. Lond. A. *325*, 57–79 (1971).
216. J. P. Elinck, J. C. Bauwens, G. Homès: Int. J. Fracture Mech. *7*, 277–287 (1971).
217. W. J. Plumbridge: J. Materials Sci. *7*, 939–962 (1972).
218. J. A. Manson, R. W. Hertzberg: CRC Crit. Reviews in Macromol. Sci. *1* (4), 433–499, August 1973.
R. W. Hertzberg, J. A. Manson: Fatigue of Engineering Plastics, New York: Academic Press 1980.
219. J. C. Radon, L. E. Culver: Polymer *16*/7, 539–544 (1975).
220. M. D. Skibo, R. W. Hertzberg, J. A. Manson: Fracture 1977, Vol. 4, ICF4, Waterloo, Canada, June 1977, 1127–1133.
221. N. K. Baramboim: Mechanochemistry of Polymers (translated from the Russian), W. T. Watson, ed., MacLaren 1964.
222. S. A. Komissarov, N. K. Baramboim: Vysokomol. soyed. A11/5, 1050–1058 (1969). Polymer Sci. USSR 11/*5*, 1189–1198 (1969).
223. N. K. Baramboim, W. G. Protasow: die Technik *30*/2, 73–81 (1975).
224. W. Lauer: Kautschuk + Gummi, Kunststoffe *28*/9, 536–613 (1975).
225. A. Casale, R. S. Porter, J. F. Johnson: Rubber Chem. and Techn. *44*/2, 534–577 (1971).
226. A. Casale: J. of Appl. Polymer Sci. *19*, 1461–1473 (1975).
227. A. Casale, R. S. Porter: Polymer Stress Reactions, New York: Academic Press 1978.
228. T. Pazonyi, F. Tüdös, M. Dimitrov: Plaste und Kautschuk *16*/8, 577–581 (1969).
229. J. G. Williams, Fracture Mechanics of Polymers, Chichester: Ellis Horwood Ltd. 1984.
230. J. R. Rice, in: Fracture, an advanced treatise, Vol. 2, H. Liebowitz, Ed., New York: Academic Press 1968, p. 192.
231. G. R. Irwin, R. de Wit: ASTM-Report, 56 (1983).
232. A. J. Kinloch, R. J. Young: Fracture Behaviour of Polymers, Barking: Applied Science Publishers 1983.
233a. M. K. V. Chan, J. G. Williams: Int. J. Fracture *23*, 145 (1983).
233b. S. Hashemi, J. G. Williams: Polymer Eng. Sci. *26*, 760 (1986) and Polymer *27*, 384 (1986)
234. H. H. Kausch, J. G. Williams, in "Fracture", Encyclopedia of Polymer Science and Engineering, 2nd edition, New York: Wiley-Interscience, to be published in 1986.
235. B. Stalder, Ph. D. Thesis: "Techniques expérimentales d'étude de la rupture fragile: développements et application aux polymères", Ecole Polytechnique Fédérale de Lausanne (1985).
B. Stalder, H. H. Kausch: J. Materials Sci., *20*, 2873 (1985)
236. U. Meier, A. Rösli: Material u. Technik *4*, 148 (1978).
237. J. M. Hodgkinson, A. Savadori, J. G. Williams: J. Mater. Sci. *18*, 2319 (1983).
238. J. F. Mandell, A. Y. Darwish, F. J. McGarry: Polym. Eng. Sci. *22*, 826 (1982).
239. A. C. Moloney, H. H. Kausch, R. Stieger: J. Mater. Sci. *18*, 208 (1983).
240. A. J. Kinloch, S. J. Shaw, D. A. Tod, D. L. Hunston: Polymer *24*, 1341 and 1355 (1983).
241. R. P. Burford: J. Mater. Sci. *18*, 3756 (1983).
242. J. E. Holt, D. Schoorl, ibid., 2017.
243. S. W. J. Boatright, G. G. Garrett, ibid., 2181.
244. E. H. Andrews, N. A. Lockington: idid. 1455.
245. I. Narisawa, T. Murayama, H. Ogawa: Polymer *23*, 291 (1982).
246. B. R. Lawn: J. Amer. Ceram. Soc. *66*, 83 (1983).
247. M. Mullins: Int. J. Fract. *24*, 189 (1984), D. Greenspan: Computers and Structures *22*, 1055 (1986).
248. D. M. Esterling: Int. J. Fract. *14*, 417 (1978).
249. E. J. Kramer: in: Advances in Polymer Sci. 52/53, H. H. Kausch Ed., Berlin-Heidelberg: Springer Verlag 1983, p. 2.
250. W. Döll, L. Könczöl: Kunststoffe *70*, 563 (1980).
251. L. Johansson: Diss., Lund Institute of Technology, Lund, Sweden, 1984

252. J. Clements, I. M. Ward: J. Mater. Sci. *18*, 2484 (1983).
253a. P. S. Leevers: ibid. *17*, 2469 (1982).
b. B. Stalder, H. H. Kausch: ibid. *17*, 2481 (1982).
254. A. Pavan: 6th European Conf. on Fracture, Amsterdam, June 1986 (to be published in the proceedings).
255. J. W. Dally, W. L. Fourney, G. R. Irwin: Int. J. Fract. *27*/3–4, 159 (1985).
256. W. Döll, G. W. Weidmann, M. G. Schinker: in: Non-Crystalline-Solids, G. H. Frischat, ed., Trans. Tech., 612 (1977).
256a. Using a fast one-dimensional IR-microscope Egorov et al. (Acta polymerica *31*, 541 (1980)) have determined local exothermal effects during fracture of PETP leading to a maximum short-time temperature rise of 215 K, which is in excellent agreement with the 230 K measured by Döll [30] for PMMA.
257a. J. G. Williams: Appl. Materials Res., 104 (April 1965).
b. J. G. Williams: Int. J. Fracture Mech. *8*, 393 (1972).
c. E. Q. Clutton, J. G. Williams: J. Mater. Sci. *16*, 2583 (1981).
258a. R. Weichert, K. Schönert: J. Mech. Phys. Solids *22*, 127 (1974).
b. R. Weichert, K. Schönert: 7. Sitzung des Arbeitskreises Bruchvorgänge (DMV), Aachen, 8.–10. 10. 1975.
259. V. G. Ukadgaonker, S. K. Chadda, S. K. Maiti: Int. J. Fracture *24*, R23 (1984).
260. N. Ohtani, A. Kobayashi : J. Appl. Polym. Sci. *289*, 1537 (1984).
261. B. Michel, P. Sommer, H. Gründemann: Experim. Technik Phys. *32*, 71 (1984).
262. International Journal of Fracture *27*/3–4 (1985).
263. K. Ravi-Chandar, W. G. Knauss: Int. J. Fract. *26*, 141 (1984).
264. W. L. Fourney, D. B. Barker, D. C. Holloway: Proceedings of the 3 rd Topical Conference on Shock Waves in Condensed Matter and Exhibits, Amsterdam: North-Holland Phys, Publ. 1984, p. 153.
265. K. Takahashi, K. Matsushige, Y. Sakurada: J. Mater. Sci. *19*, 4026 (1984).
266. T. Murayama, Dynamic Mechanical Analysis of Polymeric Material, Amsterdam – Oxford – New York: Elsevier Scientific Publisher Co. 1978.
267. H. J. Schindler, H. Kolsky, Mechanical properties at high rates of strain, Proceedings, Harding J. (ed.), 3rd International Conference on Mechanical Properties of Materials at High Rates of Strain, Institute of Physics, Oxford (UK), p. 299 (1984).
268a. A. S. Kobayashi, M. Ramulu, J. phys. (Paris), colloque *46*/5, 197 (1985).
b. K. Fujimoto, T. Shioya, ibid., 233 (1985).
c. S. M. Walley, J. E. Field, G. M. Swallowe, S. N. Mentha, ibid., 607 (1985).
d. A. S. Kobayashi, M. Ramulu, M. S. Dadkhah, K.-H. Yang, B. S. J. Kang: Int. J. Fracture *30*, 275 (1986).
269a. P. S. Theocaris, J. Milios, Eng. Fract. Mech. *13*, 599 (1979).
b. idem., Int. J. Fract. *16*, 31 (1980).
c. P. S. Theocaris, H. G. Georgiadis, Int. J. Fract. *29*, 181 (1985).
270. H. H. Kausch (Ed.): Advances in Polymer Science 52/53, Berlin – Heidelberg – New York – Tokyo: Springer 1983.
271. M. Dettenmaier, H. H. Kausch: Polymer *21*, 1232 (1980).
272. M. Dettenmaier, in: Advances in Polymer Science 52/53, H. H. Kausch Ed., Berlin – Heidelberg – New York – Tokyo: Springer 1983, p. 57.
273. W. Döll: ibid, p. 105.
274. K. Friedrich: ibid, p. 225.
275. E. J. Kramer, in: Developments in Polymer Fracture, Vol. 1, p. 55, E. H. Andrews ed., London: Appl. Science Publication 1979
276. B. D. Lauterwasser, E. J. Kramer: Phil. Mag. A*39*/4, 469–495 (1979).
277. H R. Brown: J. Mater, Sci. Letters *14*, 237 (1979).
278. A. M. Donald, T. Chan, E. J. Kramer: J. Mater. Sci. *16*, 669 (1981).
279. T. Chan, A. M. Donald, E. J. Kramer: ibid., 676
280. N. J. Mills: ibid., 1317.
281. W. C. V. Wang, E. J. Kramer: ibid. *17*, 2013.

282. A. M. Donald, E. J. Kramer: Polymer *23,* 461 (1982)
283. C. S. Henkee, E. J. Kramer: J. Polym. Sci., Polym. Phys. Ed. *22*, 721 (1984) and J. Mat. Sci. *21*, 1394 (1986).
284. C. S. Henkee: Ph. D. Thesis, Cornell Materials Science Center, Ithacae, New York, 1984.
285. G. H. Michler: Krist, u. Technik *14,* 1357 (1979).
286. G. H. Michler: Coll. & Polym. Sci. *236,* 462 (1985).
287. G. H. Michler: Halle (to be published).
288. E. Paredes, E. W. Fischer: Makromol. Chem. *180,* 2707 (1979).
289a. H. R. Brown, E. J. Kramer: J. Macromol. Sci. – Phys. B*19*(3), 487 (1981).
b. A. M. Donald, E. J. Kramer: Polymer *23*, 457 (1982).
c. A. M. Donald, E. J. Kramer: ibid. 461.
d. A. M. Donald, E. J. Kramer: J. Appl. Polym. Sci. *27*, 3729 (1982).
290. H. R. Brown, Y. Sindoni, E. J. Kramer, P. J. Mills: Polym. Eng. Sci. *24*, 825 (1984).
291. H. R. Brown, P. J. Mills, E. J. Kramer: Materials Science Center Report No. 5321, Cornell University, Ithaca, New York, 1984.
292. H. R. Brown: Polymer *26,* 483 (1985).
293. A. C. M. Yang, E. J. Kramer: J. Polym. Sci., Polym. Phys. Ed. *23,* 1353 (1985).
294a. N. R. Farrar, E. J. Kramer: Polymer *22*, 691 (1981).
b. D. Previero: Etude du PMMA orienté par la mécanique de la rupture, Travail de diplôme, EPF Lausanne, dec. 1982.
295a. L. Bevan: J. Appl. Polym. Sci. *27*, 4263 (1982).
b. L. Bevan: Polym. Bulletin. *10*, 187 (1983).
296a. R. P. Kambour: Polymer Comm. *24*, 292 (1983).
b. R. P. Kambour, E. A. Farraye: Polymer Comm. *25*, 357 (1984).
297. CRC Handbook of Solubility Parameters and other Cohesion Parameters, A. F. M. Barton, Ed., CRC Press Inc., Boca Raton/USA 1983.
298. E. W. Fischer, M. Dettenmaier: J. Non-Cryst. Solids *31*, 181 (1978)
299. J. T. Seitz: 50th Golden Jubilee Meeting, Soc. of Rheology. Boston, 1979.
300. H. H. Kausch, M. Dettenmaier: Polymer Bulletin *3*, 565 (1980).
301. A. Chudnovsky, I. Palley, E. Baer: J. Mater. Sci., *16*, 35 (1981).
302. J. S. Trent, I. Palley, E. Baer: ibid. 331.
303. S. Masuda, N. Asano: J. Appl. Polym. Sci. *29*, 1309 (1984).
304. I. Narisawa, M. Ishikawa, H. Ogawa: J. Mat. Sci. *15*, 2059 (1980).
305. G. A. Kardomateas, I. V. Yannas: Phil. Mag. *52,* 39 (1985).
306a. A. M. Donald, E J. Kramer: J. Mater Sci. *16,* 2967 (1981).
b. ibid. 2977.
307. C. C. Kuo, S. L. Phoenix, E. J. Kramer: J. Mater Sci. Letters. *4* 459 (1985).
308. A. M. Donald: J. Mater. Sci. *20,* 2630 (1985).
309. H. Breuer, J. Stabenow: Angew. Makromol. Chem. *78* 45 (1979).
310. W. Retting: Berliner Polymertage, Oktober 1982.
311. M. E. J. Dekkers: PhD Dissertation, Technical Highschool Eindhoven, 1985.
312a. L. Bevan: J. Mater. Sci. Letters *13*, 216 (1978).
b. L. Bevan, H. Nugent: Polym. Eng. Sci. *23*, 211 (1983).
c. S. S. Chern, C. C. Hsiao: J. Appl. Polym. Sci. *57*, 1823 (1985).
313a. R. P. Kusy, M. J. Katz: Polymer *19*, 1345 (Nov. 1978).
b. E. J. Kramer: J. of Materials Sci. *14*, 1381 (1979).
c. A. M. Donald, E. J. Kramer: Polymer *24*, 1063 (1983).
d. R. P. Kusy, W. F. Simmons: in: Polymer Alloys III, Daniel Klempner and Kurt C. Frisch Ed., Plenum Publishing Corp. (1983).
e. E. J. Kramer: Polym. Eng. Sci. *24*, 761 (1984).
f. A. C.-M. Yang, E. J. Kramer, Chia C. Kuo, S. Leigh Phoenix: report 5674, Materials Sci. Center Cornell University, Ithaca Macromolecules *19*, 2010 and 2020 (1986).
314. M. Dettenmaier, H. H. Kausch: Polymer. Bulletin *3,* 571 (1980)
315. M. Dettenmaier, H. H. Kausch: Coll. & Polym. Sci. 259, 937 (1981).
316. T. Q. Nguyen, M. Dettenmaier, H. H. Kausch: unpublished resutls.

317a. P. H Müller, A. Entgelter: Kolloid. Z. 160, 156 (1975).
b. G. I. Barenblat: Methods of the Combustion Theory in the Mechanics of Deformation, Flow and Fracture of Polymers, in: Deformation and Fracture of Polymers (H. H. Kausch, J. A. Hassell, R. I. Jaffee, eds.), New York – London: Plenum 1973, p. 91.
c. G. P. Andrianova, B. A. Arutyunov, Yu. V. Popov: J. Polym. Sci., Polym. Phys. Ed. 16, 1139 (1979).
d. T. Pakula, E. W. Fischer: ibid 19, 1705 (1981).
e. A. de Brossin, M. Dettenmaier, H. H. Kausch: Helv. phys. acta 55, 213 (1982).
318a. A. M. Donald, E. J. Kramer, R. A. Bubeck: J. Polymer Sci., Polymer Phys. Ed. *20*, 899 (1982).
b. A. M. Donald, E. J. Kramer: J. of Materials Science *17*, 1871 (1982).
c. J. N. Majerus, A. R. Pitochelli: J. Pol. Sci. A-1, *8*, 1439 (1970).
319a. P. S. King, E. J. Kramer: J. of Materials Science *16*, 1843 (1981).
b. A. M. Donald, E. J. Kramer, R. P. Kambour: J. of Materials Science *17*, 1739 (1982).
c. R. Schirrer, C. Goett: J. Mat. Sci. *16*, 2563 (1981).
d. L. H. Lee, J. F. Mandell, F. J. McGarry: Polymer Eng. Sci. *26*, 626 (1986).
320. Polymer Engineering and Science 24/10 and 11, Cleveland Symposia on Macromolecules, Parts I and III (1984).
321. A. M. Donald, E. J. Kramer: Phil. Mag. A. *43*(4), 857 (1981), W. E. Warren: Polymer *25*, 43 (1984).
322. M. Y. Tang, J. F. Fellers: J. S. Lin, Polym. Preprints *24*/2, 376 (1983).
323. E. Passaglia: Polymer *25*, 1727 (1984).
324a. B. N. Sun, C. C. Hsiao: J. Appl. Phys. *57*, 170 (1985).
b. O. S. Brüller: Polymer *19*, 1195 (1978).
c. J. L. S. Wales: Polymer *21*, 684 (1980).
d. O. Frank, J. Lehmann: J. Colloid & Polymer Sci. *264*, 473 (1986).
325. K. Sehanobish, E. Baer, A. Chudnovsky, A. Moet: J. Mater. Sci. *20*, 1934 (1985).
326a. N. Verheulpen-Heymans: J. Mater. Sci. *11*, 1003 (1976).
b. Polymer *20*, 356 (1979); ibid. *21*, 97 (1980).
327a. J. Opfermann: Ph.D. Thesis D 82, Technische Hochschule Aachen, 1978.
b. M. J. Doyle: J. Mater. Sci. *8*, 1165 (1973).
c. M. J. Doyle, J. G. Wagner: ACS no. 154, Washington, 63 (1976).
d. K. P. Grosskurth, M. Schlagenhauf: Kunstst. *66*, 354 (1976).
328. M. Kitagawa, M. Kawagoe: J. Polym. Sci., Polym. Phys. Ed. *17*, 663 (1979).
329a. W. Döll, G. W. Weidmann: Progr. Colloid & Polymer Sci. *66*, 291 (1979).
b. W. Döll, L. Könczöl, M. G. Schinker: Coll. & Polym. Sci. *259*, 171 (1981).
c. W. Döll, U. Seidelmann, L. Könczöl: J. Mat. Sci. Letters, *15*, 2389 (1980).
330. P. Trassaert, R. Schirrer: J. Mater. Sci. 18, 3004 (1983).
331. E. J. Kramer, E. W. Hart: Polymer *25*, 1667 (1984).
332. N. Verheulpen-Heymans: Bruxelles, personal communication 1984.
333. S. S. Pang. Z. D. Zhang, S. S. Chern, C. C. Hsiao: J. Polym. Sci., Polym. Phys. Ed. *23*, 683 (1985).
334. R. Schirrer, M. G. Schinker, L. Könczöl, W. Döll: Coll. & Polym. Sci. 259, 812 (1981).
335. R. Popli, D. Roylance: Polym. Eng. Sci. 22, 1046 (1982).
336. J. F. Fellers, D. C. Huang: J. Appl. Polym. Sci. *23*, 2315 (1979).
337. E. W. Fischer: Verhandlungen der Deutsch. Phys. Ges. 2/81, Frühjahrstagung Marburg, Physik der Hochpolymeren (1981).
338a. D. E. Morel, D. T. Grubb: Polymer *25*, 417 (1984).
b. R. A. Duckett: J. Mater. Sci. *15*, 2471 (1980).
339a. C. B. Bucknall: Toughened Plastics, London: Appl. Sci. Publ. 1977.
b. Polymer Blends, D. R. Paul, S. Newman (ed.): New York: Academic Press 1978.
340a. A. Noshay, E. McGrath: Block Copolymers, Overview and Critical Survey, Academic Press, NY, S. Francisco, London (1977).
b. L. H. Sperling: Interpenetrating Polymer Networks and Relat. Materials, Plenum Press, New York, London (1981).

341. Polyblends, Montreal, April 1984 (Polym. Eng. Sci. *24*/17, 1984), April 1985 (ibid. *26*/1, 1986), April 1986 (ibid. *28*, 1987).
Polymer Alloys: Structure and Properties, Brugge, June 1984.
Toughening of Plastics II, London, July 1985.
European Symposium on Polymer Blends, Strasbourg, May 1987.
IX. Intern. Macromol. Symposium, Interlaken, Sept. 1987.
342. T. Ricco, A. Pavan, F. Danusso: Polymer Engineering & Sci. *18*, 774 (1978) and Polymer *20*, 367 (1979).
343. G. H. Michler: Plaste & Kautschuk *26*/9, 497; *26*/10, 680 (1979).
344. B. Carlowitz: Kunststoffe *70*/7, 405 (1980).
345. C. B. Bucknall, A. Marchetti: Polym. Engineering & Sci. *24*, 535 (1984).
346. L. V. Newman, J. G. Williams: J. Mater. Sci. *15*, 773 (1980).
347. A. S. Argon, R. E. Cohen, B. Z. Jang, J. B. Van der Sande: J. Polym. Sci., Phys. Ed. *19*, 253 (1981) and in ref. [270], p. 275.
348. A. M. Donald, E. J. Kramer: J. Mat. Sci. *17*, 1765 (1982).
349. A. S. Argon, F. S. Bates, R. E. Cohen, O. Gebizlioglu, B. Z. Jang, C. Schwier: 52nd Annual Meeting Abstracts, Soc. of Rheology, Williamsburg, Virginia (1981).
350. L. Morbitzer: Colloid & Polymer Sci. *259*, 832–851 (1981).
351. H. J. Orthmann: Kunststoffe *73*, 96 (1983).
352. N. Brown, S. Fischer: Journ. of Polymer Sci., Polymer Phys. Ed. *13*, 1315–1331 (1975).
353. W.-C. v. Wang, E. J. Kramer, W. H. Sachse: Report No. 4516, Materials Sci. Center, Cornell University, Ithaca, NY/USA.
354. W. C. V. Wang, E. J. Kramer: Polymer *23*, 1667 (1982).
355. E. Kamei, N. Brown: Journ. of Polymer Sci., Polymer Phys. Ed. *22*, 543–559 (1984).
356. H. R. Brown, E. J. Kramer: Polymer *22*, 687 (1981).
357. M. B. Yaffe, E. J. Kramer: Journ. of Materials Sci. *16*, 2130–2136 (1981).
358. K. Iisaka, K. Shibayama: Journ. of Applied Polymer Sci. *24*, 2113–2120 (1979).
359. D. Pütz, G. Menges: The British Polymer Journ. *10*, 69–73 (1978).
360. I. G. Campbell, D. McCammond, C. A. Ward: Polymer *20*, 122–125 (1979).
361. D. M. Bigg, R. I. Leininger, C. S. Lee: Polymer *22*, 539 (1981).
362. C. C. Chau, J. C. M. Li: Journ. of Mater. Sci. *18*, 3047–3053 (1983).
363. W. Tsai, C. C. Chen, J. A. Sauer: Journ. of Mater. Sci. *19*, 3967–3975 (1984).
364. H. A. El-Hakeem, L. E. Culver: J. Appl. Polym. Sci. *22*, 2691 (1978).
365. A. Lustiger, R. D. Corneliussen, M. R. Kantz: Mater. Sci. & Eng. *33*, 117 (1978).
366. Y. Ohde, H. Okamoto: J. Mater. Sci. *15*, 1539 (1980).
367. M. E. R. Shanahan, J. Schultz: J. Polym. Sci., Polym. Phys. Ed. *16*/5, 803 (1978).
368. M. E. R. Shanahan, J. Schultz: ibid. *18*, 1747 (1980).
369. H. H. Kausch: IUPAC Macromolecules, H. Benoit & P. Rempp, Ed., Pergamon Press, Oxford, New York (1982), p. 211.
370. M. E. R. Shanahan, M. Debski, F. Bomo, J. Schultz: Journ. of Polymer Sci. *21*, 1103–1109 (1983).
371. P. L. Soni, P. H. Geil: J. Appl. Polymer Sci. *23*, 11 (1979) and 1167 (1979).
372. S. Bandyopadhyay, H. R. Brown: Polymer Eng. Sci. *20*, 720 (1980).
373. A. Lustiger, R. L. Markham: Polymer *24*, 1647 (1983).
374. S. Bandyopadhyay, H. R. Brown: ibid. *22*, 245 (1981).
375. H. R. Brown: Polymer *19*, 1186 (1978).
376. R. A. Bubeck: Polymer *22*, 682 (1981).
377. M. Raab, V. Hnát: Int. Journ. of Fracture *24*, R93 (1984).
378. J. Miltz, A. T. Dibenedetto, S. Petrie: Journ. of Materials Sci. *13*, 2037 (1978).
379. C. H. M. Jaques, M. G. Wyzgoski: Journ. of Applied Polymer Sci. *23*, 1153 (1979).
380. A. Priori, L. Nicolais, A. T. Dibenedetto: Journ. of Materials Sci. *18*, 1466 (1983).
381. R. P. Kambour, A. F. Yee: Report No. 80 CRD 195, General Electric, Schenectady/USA (Aug. 1980).
382. N. Brown: in: Failure of Plastics, Hanser Publishers, Munich, 1986, Chapter 15.
383. H. Okamoto, Y. Ohde: ibid., Chapter 17.

384. A. Lustiger: ibid., Chapter 16.
385. R. J. Young: "Developments in reinforced polymers", G. Pritchard, Ed., London: Applied Science Publishers 1980.
386. L. J. Broutman, F. J. McGarry: J. Appl. Polym. Sci. *9*, 609 (1965).
387. S. Yamini, R. J. Young: J. Mater. Sci. *14*, 1609 (1979).
388. D. C. Phillips, J. M. Scott, M. Jones: J. Mater. Sci. *13*, 1609 (1979).
389. W. D. Bascom, R. L. Cottington, R. L. Jones, P. Reyser: J. Appl. Polym. Sci. *19*, 2545 (1975).
390. M. J. Owen, R. G. Rose: J. Mater. Sci. *10*, 174 (1975).
391. G. Pritchard, R. G. Rose, N. Taneja: J. Mater. Sci. *11*, 718 (1976).
392. H. R. Brown, I. M. Ward: J. Mater. Sci. *8*, 1365 (1973).
393. B. E. Nelson, D. T. Turner: J. Polym. Sci. *B9*, 677 (1971).
394. S. Jamini, R. J. Young: Polymer *18*, 1075 (1977).
395. A. J. Kinloch, J. G. Williams: J. Mater. Sci. *15*, 987 (1980).
396. M. Atsuta, D. T. Turner: J. Mater. Sci. Lett. *1*, 167 (1982), Polym. Eng. Sci. *22*, 1199 (1982).
397. C. Y. C. Lee, W. B. Jones: Polym. Eng. Sci. *22*, 1190 (1982).
398. F. W. Preston: J. Amer. Ceram. Soc. *14*, 419 (1931).
399. S. Melin: Int. J. Fract. *23*, 37 (1983).
400. A. C. Moloney: unpublished work.
401. R. E. Robertson, V. E. Mindroiu: Polym. Mater. Sci. Eng. *53*, 301 (1985).
402. G. F. Taylor: Proc. Roy. Soc. *A201*, 192 (1950).
403. P. G. Saffman, G. I. Taylor: Proc. Roy. Soc. *A245*, 312 (1958).
404. A. J. Kinloch, D. Gilbert, S. J. Shaw: Polym. Comm. *26*, 290 (1985).
405. B. W. Cherry, K. W. Thomson: J. Mater. Sci. *16*, 1925 (1981).
406. R. J. Morgan, J. E. O'Neal: J. Mater. Sci. *12*, 1966 (1977).
407. A. C. Moloney, H. H. Kausch, T. Kaiser, H. R. Beer: submitted to J. Mater. Sci.
408. A. F. Yee, R. A. Pearson: NASA Report 3718, "Toughening mechanisms in elastomer modified epoxy resins", Part I, August 1983.
409. W. D. Bolt, D. J. Fuller, D. J. Phillips: J. Mater. Sci. *20*, 3184 (1985).
410. M. F. Mott: Engineering *165*, 16 (1948).
411. A. M. Leksovskii, B. L. Baskin, A. Ya. Gorenberg, G. Kh. Usmanov, V. R. Regel: Sov. Phys. Solid State *25*, 630 (1983).
412. U. W. Gedde, J.-F. Jansson: Polymer *26*, 1469 (1985).
413. W. Rose, Ch. Meurer: Progr. Colloid & Polymer Sci. *67*, 167 (1980).
414. V. A. Marichin, L. P. Mjasnikova, Z. Pelzbauer: J. Macromol. Sci.-Phys. *B22*(1), 111 (1983).
415. V. A. Marikhin: Makromol. Chem., Suppl. *7*, 147 (1984).
416. D. R. Norton, A. Keller: Polymer *26*, 704 (1985).
417. H. U. Ischebeck: Kunststoffe *74*, 153 (1984).
418. F. J. Baltá Calleja: Advances i. Polymer Sci. *66*, 117 (1985).
419. D. Briggs: Polymer *25*, 1379 (1984).
420. J. M. Schultz: Polymer Engineering & Sci. *24*, 770 (1984).
421a. H. Niklas, K. Eifflaender: Kunststoffe *49*, 109 (1959).
b. K. Richard, R. Ewald: Kunststoffe *49*, 8 (1959).
c. G. Diedrich, W. Müller, E. Gaube: Kunststoffe *56*, 228 (1966).
d. E. Gaube, G. Diedrich, W. Müller: Kunststoffe *66*, 2 (1976).
e. R. Martino: Modern Plastics International *8*, 43 (1978).
f. G. Ratschmann: Kunststoffe *75*, 164 (1985).
g. E. Gaube, H. Gebler, W. Müller, C. Gondro: Kunststoffe *75*, 412 (1985).
422 concerning PE:
a. P. Stockmayer, S. Wintergerst: 3R International *20*, 274 (1981).
b. C. S. Lee, M. M. Epstein: Polymer Eng. Sci. *22*(9), 549 (1982).
c. G. H. Michler, E. Brauer: Acta Polymerica *34*(9), 533 (1983).
d. M. B. Barker, J. Bowman, M. Bevis: Journ. of Materials Sci. *18*, 1095 (1983).

e. A. Lustiger, R. L. Markham: Polymer *24*, 1647 (1983).
concerning PP:
f. H. Dragaun, H. Hubeny, H. Muschik, G. Detter: Kunststoffe *65*(6), 311 (1975).
g. A. Sandt: Kunststoffe *72*(12), 791 (1982).
concerning PVC:
h. H. Niklas, H. H. Kausch: Kunststoffe *53*, 839 and 996 (1963).
j. L. Johanson, B. Törnell: Acta Polytechnica Scandinavica, Ch. 142 (1980).
k. L. Johansson, B. Törnell, L. Ågren: Acta Polyt. Scandinavica, Ch. 150 (1982).
l. see ref. [251].
423a. W. Müller, E. Gaube: Kunststoffe *72*, 297 (1982).
b. H. Gebler: Kunststoffe *73*, 73 (1983).
c. G. Menges, D. Kirch, J. Nordmeier, E. Winkel, J. Wortberg: Kunststoffe *73*, 258 (1983).
d. F. Altendorfer, H. Janeschitz-Kriegl: Kunststoffe *74*, 325 (1984).
e. G. Menges, J. Nordmeier, K. Esser, H. Gross, D. Weinand, D. Kirch, A. Mayer: Kunststoffe – Plastics *31*/4, *31*/5, *31*/6 (1984).
424a. J. M. Greig: Plastics & Rubber Processing & Applications *1*(No. 1), 43 (1981).
b. U. Meier, H. Dorn, F. Rupp: Kunststoffe – Plastics *24*/6, 10 (1977).
425a. B. W. Cherry, Teoh Swee Hin: Polymer *24*, 1067 (1983).
b. G. G. Trantina: Polymer Eng. Sci. *26*, 776 (1986).
426concerning several polyolefins:
a. R. D. Goolsby, A. M. Chatterjee: Polymer Eng. & Sci. *23*(3), 117 (1983).
concerning PE:
b. P. Flüeler, D. R. Roberts, J. F. Mandell, F. J. McGarry: Report R79-5, Massachusetts Institute of Technology, Dept. of Materials Sci. and Eng., Cambridge, Mass./USA 1979.
c. C. G. Bragaw: Int. Conf. of Def., Yield and Fracture, Cambridge 1979.
d. E. Gaube, W. F. Müller: Kunststoffe *70*, 72 (1980).
e. A. Gray, J. N. Mallinson, J. B. Price: Plastics & Rubber Processing & Applications *1*, 51 (1981).
f. A. Savadori, M. Bramuzzo, C. Marega: Kunststoffe *73*, 203 (1983).
g. M. K. V. Chan, J. G. Williams: Polymer *24*, 234 (1983).
h. J. F. Mandell, D. R. Roberts, F. J. McGarry: Polymer Engineering & Sci. *23*, 404 (1983).
j. E. Gaube, W. Müller: 3R International *23*, 236 (1984).
concerning PP:
k. A. Savadori: Material & Technik *4*, 212 (1985).
concerning PVC: see ref. [238].
427a. E. Gaube, G. Diedrich: Schweißen + Schneiden *24*, 1 (1972).
b. E. Gaube, W. Müller, F. Falcke: Kunststoffe *64*, 193 (1974).
c. E. Gaube: Kunststoffe *67*, 353 (1977).
d. E. Gaube, G. Diedrich: Chem.-Ing.-Tech. *50*, 155 (1978).
428a. G. Diedrich, E. Gaube: Kunststoffe *60*, 74 (1970).
b. J. Hessel, P. John: Kunststoffe *74*, 385 (1984).
c. M. B. Barker, S. R. Bentley, M. Bevis, J. Bowman: Kunststoffe *74*, 501 (1984).
d. E. Barth, R. Schommer: Kunststoffe *74*, 506 (1984).
429a. K. Richard, E. Gaube, G. Diedrich: Materialprüfung *5*, 213 (1963).
b. E. Kramer, J. Koppelmann: Kunststoffe *73*, 714 (1983).
c. E. Kramer: Österr. Kunststoff-Zeitschrift *15*, 3 (1984).
430a. J. Ehrbar: Kunststoffe *53*, 845 (1963).
b. E. Gaube, W. Müller, G. Diedrich: Kunststoffe *56*, 673 (1966).
c. J. Ehrbar, C.-M. v. Meysenbug: J. of Materials Technology *7*, 429 (1976).
d. G. Diedrich, B. Kempe, K. Graf: Kunststoffe *69*, 470 (1979).
e. B. Kempe: Z. Werkstofftech. *15*, 157 (1984).
f. R. Henkhaus, W. Rau: Werkstoffe und Korrosion *36*, 8 (1985).
431. M. J. Doyle: J. of Materials Sci. *18*, 687 (1983).
432. B. Cotterell: The Intern. Journal of Fracture Mechanics *4*, 209 (1968).
433. H. Hendus, E. Penzel: Chemiefasern/Textilindustrie (1976/6), 527.

434. W. H. Hassell: M. S. Thesis, Dept. of Mech. Engineering, Univers. of Utah, Salt Lake City, 1973.
435. Courtesy of H. Hendus, Ludwigshafen 1976.
436a. J. W. S. Hearle: Proc. Textile Inst./Inst. Text. de France Conf. Paris (1975), 60–75.
b. J. W. S. Hearle: "An Atlas of Fibre Fracture". Text. Mfr. (1972), *99*, Jan., Feb., 14; March, 12; May, 20; Aug., 40; Sept., 16; Oct., 40; Nov., 12; Dec., 36; (1973), *100*, Jan., 24; March, 24; April, 34; May, 54; June, 44.
437. Courtesy to J. W. S. Hearle, Manchester 1977.
438a. L. Konopasek, J. W. S. Hearle: J. of Applied Polymer Sci. *21*, 2791 (1977).
b. A. R. Bunsell, J. W. S. Hearle: J. Materials Sci. *6*, 1303 (1971).
A. R. Bunsell, J. W. S. Hearle: J. Appl. Polymer Sci. *18*, 267 (1974).
c. Ch. Oudet, A. R. Bunsell: J. Applied Polymer Sci. *29*, 4363 (1984).
d. Lu Fu-Min, B. C. Goswami, J. E. Spruiell, K. E. Duckett: J. Applied Polymer Sci. *30*, 1859 (1985).
439. H. H. Kausch: Pure and Appl. Chemistry *55*, 833 (1983).
440. S. Hashemi, J. G. Williams: J. Mat Sci. *20*, 4202 (1985).
441. K. Sehanobish, A. Moet, A. Chudnovsky, P. P. Petro: J. Mat. Sci. Letters *4*, 890 (1985).
442. P. Eriksson, M. Ifwarson: Kunststoffe *76*, 512 (1986).
443. E. Böhme: Plast. Modernes & Elastom. *38*/4, 89 (1986).
444. D. G. Legrand: Polym. Eng. & Sci. *24*, 373 (1984).
445. R. W. Hertzberg: Polym. Communications *26*, 38 (1985).
446. G. Sandilands, P. Kalman, J. Bowman, M. Bevis: Polymer Communications *24*, 273 (1983).
447. H. R. Beer, T. Kaiser, A. C. Moloney, H. H. Kausch: J. Mat. Sci. *21*, 4173 (1986).
448. A. C. Moloney, H. H. Kausch, T. Kaiser, H. R. Beer: ibid. *22* (1987) in press.
449. J. G. Williams: Plastics & Rubber Process. & Applic. *1*, 369 (1981).
450. R. K. Mittal, V. Rashmi: Polymer Engin. & Sci. *26*, 310 (1986).
451. S. Lee, J. de la Vega, D. C. Bogue: J. of Applied Polymer Sci. *31*, 2791 (1986).
452. S. Hashemi, J. G. Williams: J. Mat. Sci. *20*, 922 (1985).
453. J. G. Williams, J. M. Hodgkinson: Proc. R. Soc. Lond. A *375*, 231 (1981).
454. S. A. Umar-Khitab, D. McCammond, R. D. Venter: Polym. Eng. & Sci. *25*, 1035 (1985).
455. T. Vu-Khanh, F. X. de Charentenay: Polym. Eng. & Sci. *25*, 841 (1985).
456. J. G. Williams: J. Mat. Sci. *12*, 2525 (1977).
457. Y. W. Mai, J. G. Williams: J. Mat. Sci. *14*, 1933 (1979).
458. P. E. Bretz, R. W. Hertzberg, J. A. Manson: Polymer *22*, 575 (1981).
459. M. Kitagawa, S. Isobe, H. Asano: Polymer *23*, 1830 (1982).
460. J. Michel, J. A. Manson, R. W. Hertzberg: Polymer *25*, 1657 (1984).
461. M. G. Schinker, L. Könczöl, W. Döll: J. Mat. Sci. Letters *1*, 475 (1982).
462. L. Könczöl, M. G. Schinker, W. Döll: J. Mat. Sci. *19*, 1605 (1984).
463. M. G. Schinker, L. Könczöl, W. Döll: Colloid & Polymer Sci. *262*, 230 (1984).
464. P. J. Mills, H. R. Brown, E. J. Kramer: Materials Sci. Center, Cornell University, Ithaca, N.Y./USA, report No. 5399 (1984).
465. J. W. Teh, J. R. White, E. H. Andrews: Polymer *20*, 755 (1979).
466. J. R. White, J. W. Teh: Polymer *20*, 764 (1979).
467. X. P. Liu, C. C. Hsiao: J. Appl. Phys. *58*, 2837 (1985).
468. J. F. Mandell, J. P. F. Chevaillier: Polym. Eng. & Sci. *25*, 170 (1985).
469a. D. S. Matsumoto, M. T. Takemori: General Electric, Schenectady, N.Y./USA, report No. 82CRD240 (1982).
b. M. T. Takemori: ibid report No. 83CRD294 (1984).
c. T. A. Morelli, M. T. Takemori: J. Mat. Sci. *18*, 1836 (1983).
470. P. E. Bretz, R. W. Hertzberg, J. A. Manson: Polymer *22*, 1272 (1981).
471. H. R. Zeller, W. R. Schneider: J. Appl. Phys. *56*, 455 (1984).
472. J. Schirr, thesis: Beeinflussung der Durchschlagfestigkeit von Epoxidharzformstoff durch das Herstellungsverfahren und durch mechanische Spannungen. Fakultät für Maschinenbau und Elektrotechnik der Technischen Universität Carolo-Wilhelmina zu Braunschweig (1974).

473. Ch. Tobler: travail de diplôme, EPFL, Dépt. Electricité (1985).
474. L. Niemeyer, L. Pietronero, H. J. Wiesmann: Phys. Rev. Letters *52*, 1033 (1984).
475. L. Pietronero, H. J. Wiesmann: J. Stat. Phys. *36* (1984).
476. L. P. Kadanoff: Physics Today *2*, 6 (1986).
477. P. Pfluger, H. R. Zeller: J. Bernasconi *53*, 94 (1984).
478. N. P. Bazhanova, P. N. Dashuk, V. V. Korablev, Y. A. Morosov: International Symposium on Discharges and Electrical Insulation in Vacuum, DDR-Berlin, Sept. 1984.
479. M. Hikita, S. Tajima, I. Kanno, I. Ishino, G. Sawa, M. Ieda: Jap. J. of Appl. Phys. *24*, 988 (1985).
480. A. L. Robezhko, V. F. Vazhov, G. V. Efremova, S. M. Lebedev, V. Ya. Ushakov: Sov. Phys. Solid State *23*, 1950 (1981).
481. V. Ya. Ushakov, A. L. Roberzhko, G. V. Efremova: Sov. Phys. Solid State *26*, 25 (1984).
482. V. Ya. Ushakov, A. L. Robezhko, V. F. Vazhov, V. V. Nikitin, S. A. Lopatkin, V. A. Surnin: Sov. Phys. Solid State *27*, 1416 (1985).
483. J. P. Crine, A. K. Vijh: Appl. Phys. Commun. *5*, 139 (1985).
484. R. Patsch: Colloid & Polymer Sci. *259*, 885 (1981).
485. W. Gölz: Colloid & Polymer Sci. *263*, 286 (1985).
486. S. Masuda, N. Asano: J. Appl. Pol. Sci. *29*, 1309 (1984).
487. N. S. Enikolopian, L. S. Zarkhin, E. V. Prut: J. Appl. Pol. Sci. *30*, 2291 (1985).
488. J. T. Dickinson, L. C. Jensen, A. Jahan-Latibari: J. Vac. Sci. Technol. *A2*, 1112 (1984).
489. J. T. Dickinson, A. Jahan-Latibari, L. C. Jensen: J. Mater. Sci. *20*, 229 (1985).
490. V. A. Zakrevskii, V. A. Pakhotin: Sov. Phys. Solid State *20*, 214 (1978).
491. V. A. Zakrevskii, V. A. Pakhotin: Pol. Sci. USSR *25*, 3047 (1983).
492. V. A. Zakrevskii, V. A. Pakhotin: Pol. Sci. USSR *23*, 741 (1981).
493. J. Fuhrmann: Spring Meeting of the German Physical Society, Lausanne, 18–20 March 1985.
494. P. Y. Butyagin: Uspekhi Khimii *53*, 1769 (1984), translated in: Russian Chemical Reviews *53*, 1025 (1984).
495. R. S. Porter, A. Casale: Pol. Eng. & Sci. *25*, 129 (1985).

Chapter 10

Fracture Mechanics Studies of Crack Healing

I. Introduction

The formation of cracks, i.e. fracture, and the healing of cracks in polymer solids are two complementary processes. Some of the molecular mechanisms described in the previous chapters on fracture also apply – by simply changing direction – to crack healing, such as relaxation of chains towards an equilibrium conformation, establishment of network isotropy, closure of voids and formation of entanglements. The role of these mechanisms in *fracture* has become clearer through the study of *crack healing*. It is for this reason that a chapter on this subject is included in a book on fracture.

The term crack healing as it is used today designates the build-up of mechanical strength between two (crack-) surfaces of a (fracture mechanics) specimen which are brought into contact (above T_g). Through determination (at room temperature) of the stress intensity factor K_{Ii} at which the crack begins to reform and propagate through the healed contact zone, a quantitative measure of the healing effect is obtained. This method was principally developed in Lausanne [1, 2, 4] and Urbana [3, 5]; it has recently been discussed in some detail by Kausch et al. [6].

The physical mechanisms involved in crack healing are the following (Fig. 10.1):
- Establishment of contact and adhesion between the surfaces
- Viscoelastic deformation of surface irregularities
- Interdiffusion of molecular segments across the interface
- Formation of new entanglements in the interfacial region.

The concentration and stability of these new entanglements determine the material resistance during reopening of the crack. The principal mechanisms had already been noted in 1949 [7] when it had been observed that the coalescence of thermoplastic particles only occurred well above the (glass-) transition temperature and by "interpenetration of the chains in the two surfaces".

Adhesion, viscoelastic deformation, diffusion, and entanglement formation are obviously of quite general importance and have been studied extensively (see for instance [6–58]). They are also the basis of technical welding [8] which, however, will not be covered in this monograph.

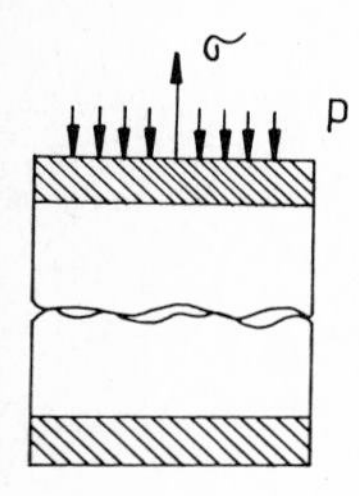

Fig. 10.1. Stages of crack healing: 1, Pointwise physical contact; 2, Viscoelastic deformation and wetting; 3, Chain interdiffusion; 4, Establishment of equilibrium. – A slight normal pressure ($p \sim 1$ bar) is needed to obtain physical and thermal contact during the healing procedure. The tensile stress σ necessary to separate the surfaces after the healing procedure can be used as a qualitative measure of the healing effect (from [4])

II. Models of Adhesive and Cohesive Joint-Strength

A. Adhesion and Tack

A measure of the reversible, purely physical attraction of two solids across an interface by Van der Waals forces [9] is the work of adhesion W_{adh}. This quantity is for instance given by the relationship of Dupré [10]:

$$W_{adh} = \gamma_1 + \gamma_2 - \gamma_{12} \sim 2[(\gamma_1^d \gamma_2^d)^{1/2} + (\gamma_1^h \gamma_2^h)^{1/2}] \tag{10.1}$$

where the superscripts d and h refer to dispersion forces and hydrogen bonds respectively. Adhesion and autohesion (the adhesion between alike bodies) have been treated comprehensively by Voyutskii [11]. In this respect it is useful to distinguish between *adhesion* as defined above and *adherence* which designates the energy measured experimentally when separating two surfaces in contact. The latter term includes the energy of possible viscoelastic deformations occurring in the surface regions (or in even larger parts of the contacting bodies).

The adherence between identical elastomers, known as *tack*, has been studied for a long time (see e.g. [12–16]). Bister, Borchard, Rehage et al. [13, 14] have identified four stages in the build-up of strength between surfaces in contact (Fig. 10.2). The influence of thermodynamical and viscoelastic surface properties on the adhesive joint strength had been further analyzed by Koszterszitz [15], Wool et al. [3] and Zosel [16]. In most of these studies fracture of the adhesive bonds occurred in the plane of contact.

A different and important concept, the *fracture mechanics analysis of adhesive failure* had been introduced by Williams [17] and was further developed and applied subsequently [18–20]. As becomes clear from the previously cited references the *viscoelastic behavior* of the contacting surfaces and of the molecules forming them is of prime importance for the establishment of interfacial contact and for the build-up of strength through the interdiffusion of chains and the formation of entanglements. From the very large body of literature on viscoelasticity only a few papers may be cited [21, 22].

Generally the strength of a joint A/B is *cohesive* in nature whenever the failure does not occur exactly at the interface *between* A and B but at some (small) distance from it and *within* A or B. In the case of joints between identical or compatible polymers it seems to be adequate to speak of cohesive failure whenever the interaction between entangled A and B chains is so strong that their separation gives rise to large scale local deformation. In the studies of Bister and Borchard (Fig. 10.2) on elastomers cases I and II must be classified, therefore, as adhesive failure, case IV as cohesive failure. Case III fulfills both criteria: separation occurs exactly at the interface between A and B but the interaction between the just entangled molecules is sufficiently strong so as to provoke notable deformation of the interface.

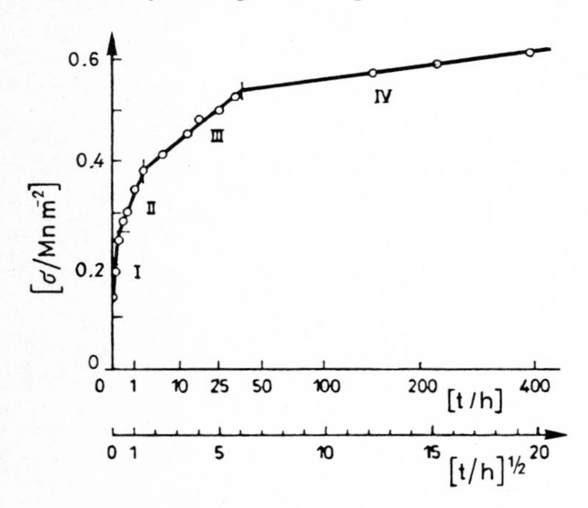

Fig. 10.2. Strength of adherence of polyisobutylene (Oppanol B 200) as a function of contact time; in region I physical contact is gradually established, in region II chain ends penetrate the interface, in III they form entanglements (which are unstable in subsequent fracture); in IV the separation of the adhering surfaces causes large scale deformation of the whole sample, fracture occurs outside of the former interface (after Bister, Borchard, and Rehage [13])

B. Joint Strength of Thermoplastic Polymers

The joint strength due to purely physical attraction is low and is of the order of the "true surface energy", γ. To achieve practical strength, there must be an interchange of material across the interface by some sort of chain diffusion process. As later discussed in detail these diffusion processes can only occur at temperatures equal to or above T_g. This condition was easily fulfilled for the studies at room temperature of elastomer adherence [11–16]. In the case of thermoplastic materials, however, strong attention must be given to the fact that glass transition temperature T_g, temperature T_h of bond formation, and temperature of bond strength measurement (normally ambient temperature) will be quite different. If $T_h \gg T_g$ one has to do with polymer melts and with the formation of *weld or knit lines.* This important processing effect has more recently been studied for injection molded thermoplastics such as PS and PC [23a–g] and blends of EPDM and PP [23h]. Depending on the time of chain interpenetration and the molecular mobility a distinct effect on the strain at break of weld-line containing samples or on their strength at break had been observed (see Fig. 10.3). Qualitative evidence of the effect of chain diffusion was also obtained from studies of the *coalescence and fusion behavior* of particles in contact [24].

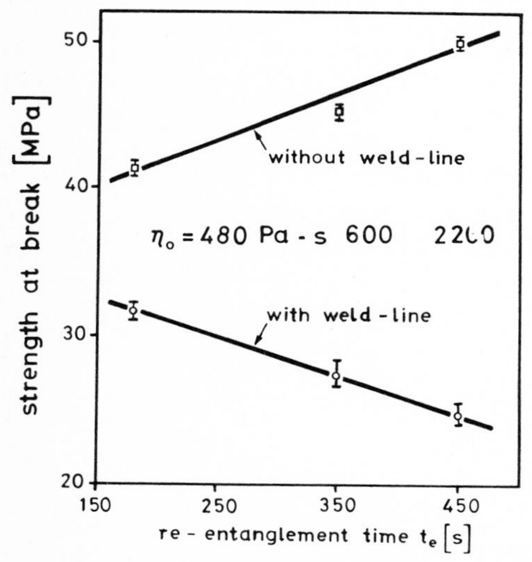

Fig. 10.3. Effect of weld-lines on the breaking strength of PS bars (the re-entanglement time t_e is determined in a double step shear test at 180 °C as the time necessary to restore the entanglement density in the melt after a first shear deformation (after Baird and Pisipati [23e])

Before analyzing the relation between joint strength and chain interdiffusion at solid surfaces any further it is necessary to consider the kinetics of *contact establishment.* It is reasonable to assume that the strength of two surfaces brought into contact will firstly be proportional to the contact surface area ratio A/A_0, where A is the contact area and A_0 the total cross-sectional area of the joint, and secondly to the proportion N/N_0 of the links formed across that area, where N is the number of links actually formed and N_0 the concentration of links at "full strength", G_{co}. Healing would then be described by the following equation:

$$\frac{G_{c(t)}}{G_{co}} = \frac{A(t)}{A_0}\,\frac{N(t)}{N_0}. \qquad (10.2)$$

The achievement of complete contact (i.e. when $A/A_0 = 1$) for a surface with some initial roughness requires the distortion of that roughness under the action of interfacial pressure, both external and from surface work. In polymers, the time-dependence of this process will arise from viscoelasticity and the duration will be governed by the current material stiffness. Clearly, for high temperatures ($T > T_g$), the modulus is low and, even under small external pressures (pressure $p \simeq 0.8$ bar), complete contact will be achieved rapidly.

Some estimate of the time-dependence of contact area changes can be made by assuming that the modulus of the material has the form

$$E = E_0(t/t_0)^{-m}, \qquad (10.3)$$

where E is the Young's modulus at time t, E_0 is a reference modulus at time t_0, and m is a constant. Then, Hertzian contact theory [25] gives

$$A \propto \left(\frac{\sigma}{E}\right)^{2/3}, \qquad (10.4)$$

where σ is the applied compressive stress. Surface energy may contribute to the effective σ. By now assuming an extreme condition, namely that the development of strength is exclusively due to the increase in contact area, i.e. that $N(t) = N_0$, Eq. 10.2 gives $G_c \propto A$. Thus, it would be expected that the time-dependence of the corresponding stress intensity factor after healing, K_{Ii}, would be of the form

$$K_{Ii} \propto \left(\frac{\sigma}{E_0}\right)^{1/3} t^{m/3}. \qquad (10.5)$$

For low temperatures, i.e. for elastic behavior, m goes to zero and E_0 is much larger than σ, thus the degree of contact is small and remains almost constant. For $T > T_g$, σ/E_0 can be significant even for modest contact pressures and m approaches 1 for viscous type behavior, leading to a rapid establishment of complete contact. The constant of proportionality will, of course, be governed by the roughness of the two surfaces. If the extreme condition were valid one would expect, therefore, a dependency of K_{Ii} on $t^{1/3}$.

The links at the interface are formed by the crossing of chain ends over the interface and the interpenetration of chains. This occurs at $T > T_g$. With increasing healing time all traces of the original interface will gradually dissappear. If the healing is interrupted at a very early stage and the crack is reopened, then apparently adhesive failure will occur at a plane corresponding to the former interface and at unmeasurably small values of K_{Ii} (for PMMA this value will be below 0.2 MPa $m^{1/2}$). In the

K_{Ii} region, however, where the progressive healing of cracks can be followed quantitatively ($0.3 < K_{Ii} \leq 1$ MPa $m^{1/2}$ for PMMA) one clearly observes extensive fibrillation of the partially healed material (see Fig. 10.4); the ensuing fracture of the sample must, therefore, be classified as cohesive even if it is still the region of the former interface which remains the weakest part within the structure.

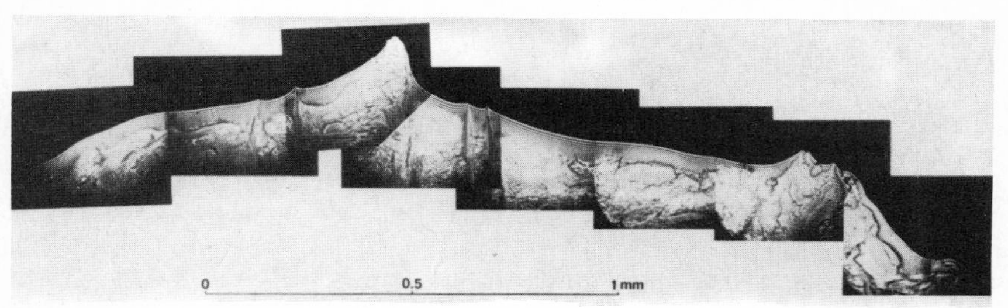

Fig. 10.4. Crack front in a partially healed, dried PMMA 7H (t_h = 10 min, K_1 = 0.82 MN $m^{-3/2}$, from Ref. [59]). The interference micrograph shows well that the crack front (irregular, cloudy fringe pattern) is preceded on its whole width by at least a few narrow fringes which are characteristic for a craze zone

This shows that the condition $N(t) = N_0$ has been too extreme and that there will be a separate time-dependence of K_{Ii} associated with the self-diffusion of the polymer. The time-dependence observed in G_c will depend on which of the two mechanisms is the slower, and thus the controlling process. It is, of course, possible to visualize a rather complex interactive process when the two time scales are similar with each section of the area having a different diffusion time depending on its time in contact. Such a mechanism could be modelled but as a first approach the two time scales will be considered to be distinctly different. For smooth surfaces of "normal" molecular weight samples at temperatures $T > T_g$, it is likely that intimate contacts are achieved very quickly and that diffusion is the governing process. Some consequences of slow wetting of very high molecular weight material as observed by Wool et al. [3] will be discussed later. In any event the actual diffusion rates may be dependent on the state of the surface.

In the following section a model description of crack healing is given which evidently must try to relate the interdiffusion of chains achieved during the healing process with the material resistance measured in a subsequent experiment.

C. Model Description

The model used to interprete crack healing studies is shown in Figure 10.5. An ideally plane crack in a (compact tension) specimen is visualized. The "strength" of the virgin material is characterized by its stress intensity factor K_{I0}. After crack closure and healing the crack is reopened (at room temperature); the thus determined stress intensity factor K_{Ii} is a measure of the healing effect. The healing is brought about by the interpenetration of molecular coils from either side of the interface. It is absolutely admitted that the real interface is not an ideal plane. This will not change the evaluation of the data very much since in any event it must be assumed that the

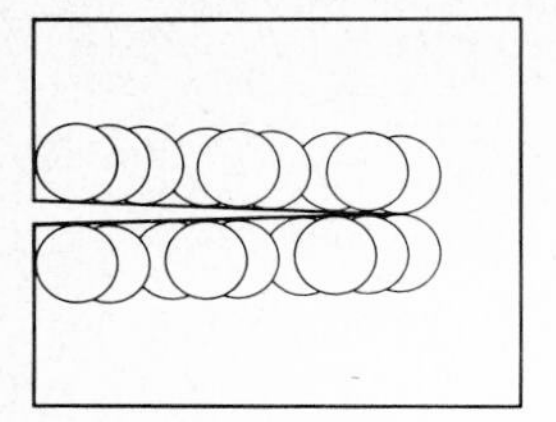

Fig. 10.5. Compact tension or double cantilever beam specimens are generally used for crack healing experiments; basic molecular analysis of these experiments mostly refers to molecular coils situated at an ideally plane interface before interdiffusion

molecular displacements in a certain surface region contribute to the healing process. Important, however, are the composition, the density and the state of orientation of the surface layers and the influence of these parameters on chain diffusion.

The study of *diffusion* and especially of the *self-diffusion* of entangled systems has received growing interest in the last decade, particularly due to the progress in theoretical concepts [26–28] and experiment methods such as radioactive tracer technique [29–31], nuclear magnetic resonance [32–40], infrared microdensitometry [41–45], scanning electron microscopy combined with energy-dispersive X-ray analysis [46], frustrated multiple internal reflection spectroscopy [47], forced Rayleigh scattering [48, 49], neutron scattering [50, 51], fluorescence redistribution after pattern photobleaching [52, 53], forward recoil spectrometry [54] and Rutherford backscattering spectrometry [55–57]. An excellent review on polymer selfdiffusion in entangled systems has just been given by Tirrell [58].

The different modes of chain diffusion (center of mass diffusion, reptation) and their dependence on scaling laws and on the thermodynamic interaction between the diffusing species are extensively discussed in the cited literature [24–58]. With respect to the healing of interfaces between identical or compatible polymers as discussed in the following, it can be said that reptation is the predominant mode of chain diffusion. This random molecular motion leads to the crossing of chain ends over the interface, to a gradual interpenetration of chains and to the formation of new entanglements. Due to the statistical nature of this process there will be a large local variation of the number of diffusing chain ends and of their curvilinear depth $\Delta\ell$ of interpenetration. This means that even at small values of $\langle\Delta\ell\rangle^{1/2}$ there are some chains which have penetrated much further than the minimum distance $\Delta\ell_{min}$ necessary to form an entanglement. At the end of the crack healing process $n(t_h)$ new entanglements per unit surface area will have formed in the interfacial region near the former crack surfaces. Unless the healing has been absolutely complete, $n(t_h)$ will be smaller than n_e, the corresponding concentration of entanglements in the virgin material.

In the cited healing studies [1–5] cracks are reopened subsequent to healing in order to measure quantitatively the healing effect. An analysis of the microscopic fracture process of glassy polymer such as PMMA or SAN showed [59, 60] that fracture at room temperature proceeds through the formation and rupture of craze fibrils (see Fig. 10.4).

In the interfacial regions of incompletely healed specimens there are fewer entanglements than elsewhere and possibly they are found close tot the chain ends. It must then be assumed that in the process of fibrillation especially the newly entangled chains are highly stressed and that they are gradually loosing those entanglements which are close to the chain ends. Evedently the complete dissolution of the new entanglements in a particular cross-section of a fibril will lead to the rupture of that fibril (see Fig. 10.6). The above assumption would conveniently explain the observed correlation between the extent of fibril deformation and rehealing [1–5]. In order to establish a quantitative relation between healing times and fracture energy G_{Ii} it is necessary to know how $n(t_p)$ is built up through interdiffusion during crack healing and how this same concentration of entanglements is reduced "to zero" (or to some critical value) in the subsequent fibrillation process.

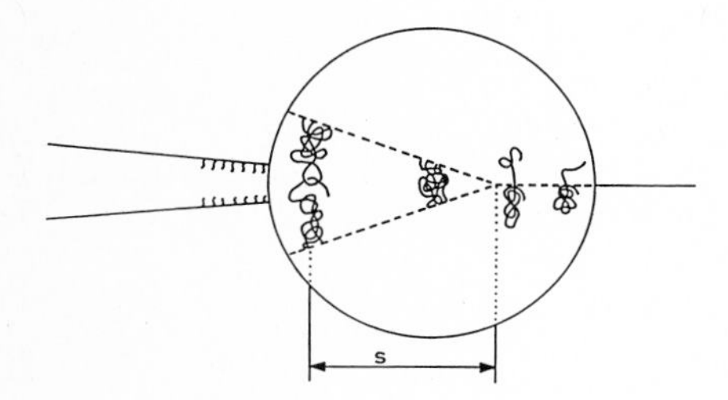

Fig. 10.6. Schematic representation of the possible relation between depth of chain interpenetration ($\Delta \ell$), entanglement formation, craze length (s) and stress intensity factor K_I:

$$s = \frac{\pi}{8} \frac{K_I^2}{\sigma_F^2} \sim \Delta \ell$$

The diffusion of a particular chain molecule in the melt is evidently subjected to the constraints imposed on it by its neighbors. As proposed by de Gennes [26a] and Doi and Edwards [27] these constraints can be considered to form a *tube* within which the molecular segments can perform wriggling and translational motions. The latter are described by the Einstein relation:

$$\langle \Delta \ell^2 \rangle = 2 D_t t_h \tag{10.6}$$

where the "tube diffusion coefficient D_t" is related to the monomeric friction coefficient ζ_0:

$$D_t = kTM_0/(\zeta_0 M). \tag{10.7}$$

A chain of length L will have completely *escaped* from its original tube at time t_R:

$$t_R = L^2/2D_t = (\zeta_0 L_0^2/2kTM_0)M^3. \tag{10.8}$$

At this time the center of mass of the chain will have moved by a distance $\langle \Delta y^2 \rangle^{1/2}$ approximately equal to the radius of gyration, R_g. The self-diffusion coefficient, D_s, then is derived as

$$D_s = \frac{\langle \Delta y^2 \rangle}{2t} = \frac{Rg^2}{2t_R} \sim M^{-2} \tag{10.9}$$

The molecular weight dependence predicted for the self-diffusion coefficient D_s of (long) chains by reptational motion is stronger, therefore, than that of chains in (dilute) solutions for which the Rouse theory gives

$$D \sim M^{-1}. \tag{10.10}$$

Most of the cited studies of self-diffusion in entangled systems [33–58] have confirmed the M^{-2} dependency of D_s (see [58] for an extended discussion). In the following it is assumed that the relations 10.6 to 10.9 also describe the interdiffusion of chains in the crack healing experiments.

Kausch, Jud et al. [1, 2, 4, 6] used an *entanglement model* for the analysis of their crack healing data. They assumed that the number, n, of entanglements formed per unit surface area is proportional to the "distance $\Delta \ell$" travelled by the molecular chains

$$\frac{n(t_h)}{n(\tau_0)} = \left(\frac{2D_t}{2D_t} \frac{t_h}{\tau_0} \right)^{1/2}, \tag{10.11}$$

where τ_0 is the time necessary for "complete healing". It follows from the above criterion that D cancels out, i.e. that the *time dependency* of n(t) is the same no matter which diffusion mechanism applies. These authors then have further assumed that the fracture energy $G_c(t)$ of the healed interface is proportional to the number of entanglements formed [1, 2, 4, 6, 64–66]. In terms of the stress intensity factor, K_{Ii}, this gives

$$\frac{K_I}{K_{I_0}} = \left(\frac{G_c}{G_{c_0}} \right)^{1/2} = \frac{n(t_h)}{n(\tau_0)} = \left(\frac{t_h}{\tau_0} \right)^{1/4}. \tag{10.12}$$

It is the merit of the Kausch-Jud model of crack healing to have explained for the first time and in a straightforward manner the $t^{1/4}$ dependence of the increase of the stress-intensity factor. Nevertheless, a more detailed discussion of the two basic assumptions (Eqs. 10.11 and 10.12) is necessary in relation with the experimental data.

The crack healing models of Wool et al. [3, 5, 67–69] and of Prager and Tirell et al. [70–72] are based on very similar considerations though their final fracture criteria are slightly different. Wool et al. [67–69] look at the average interpenetration distance of the χ segments belonging to those parts of the chains which have completely disengaged themselves from their original tubes (they call those parts "minor chains"; cf. Fig. 10.7). They then assume that the fracture stress σ is proportional to χ which gives [5, 67]:

$$\sigma \sim (t_h/M)^{1/4}. \quad (10.13)$$

Wool et al. [3] point out, however, that the time dependency of σ will be different in the case of notable contributions of the wetting stress to σ and in the case of slow wetting. This could lead to a stress intensity factor after healing of:

$$K_{Ii}(t) \sim at^{1/4} + b$$

and to a corresponding energy release rate

$$G_c(t) \sim ct^{1/2} + dt^{1/4} + e.$$

In their analysis they consider the frictional energy losses during pull-out of minor chains to be the principal contribution to the fracture energy. In the case of long chains chain scission would become increasingly important.

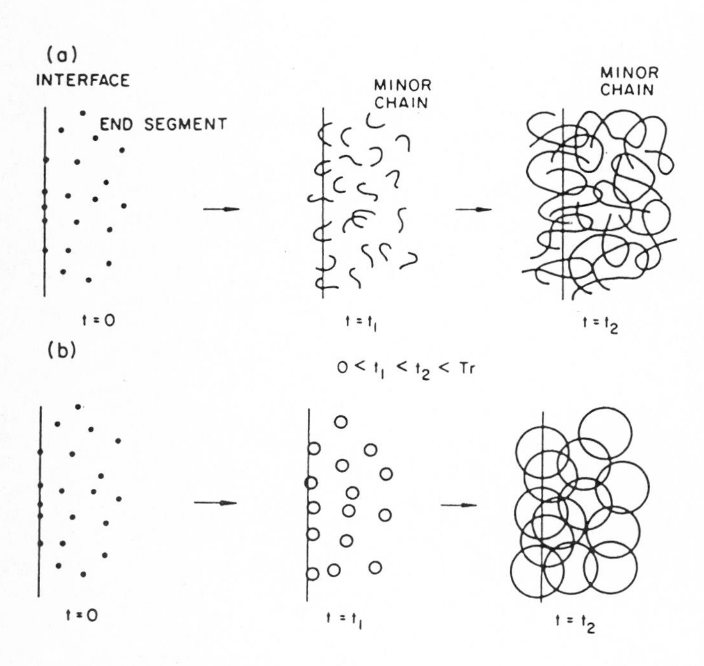

Fig. 10.7. **a** Model of the growth of minor chains that have emerged from one side of an interface, **b** Growth of the spherical envelopes of those minor chains (from Whool and O'Connor [3])

Prager and Tirrell [70–72] have carried out computer simulation studies of chain interdiffusion (see Fig. 10.8). They find that in a system describing freshly broken samples which contain many chain ends at the contacting surfaces *the number of molecular bridges* spanning the original interface increases with $t_h^{1/4}\ M^{-1/4}$; in equilibrated samples they except that this number increases with $t^{1/2}\ M^{-3/2}$.

Whereas all the above studies [1–6, 26, 47, 64–72] lead to the same conclusion, namely that chain reptation is the essential healing mechanism, they give rise to different interpretations as to how the "minor chains" influence the fracture energy. This point will, therefore, be taken up again in the discussion of the experimental results.

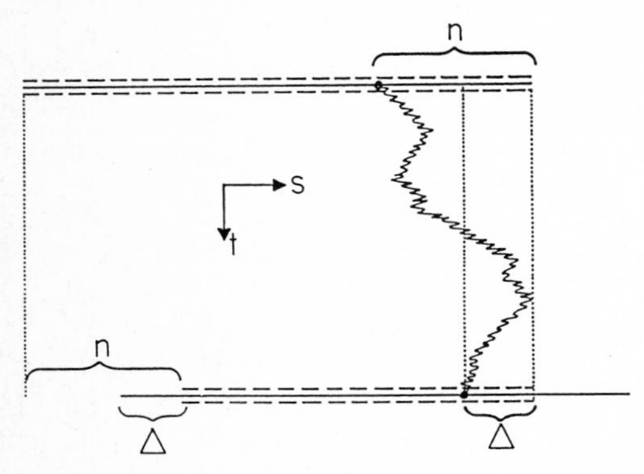

Fig. 10.8. One-dimensional computer simulation of the displacement by reptation of a given segment (•) of a chain (——) in its tube (= = =); the number n of segments which the original tube has lost in time t is replaced by Δ new ones (from Prager and Tirrell [70])

III. Experimental Studies

A. Uncrosslinked Thermoplastics

1. Experimental Procedure

Representative results have been obtained [1, 2, 47] with poly(methyl methacrlate), PMMA 7H (molecular weight $M_w = 1.20 \times 10^5$; molecular weight to molecular number ratio $M_w/M_n = 2$) from Röhm GmbH., and styrene acrylonitrile (SAN), Luran 368R co-polymer (25 mol % AN, of $M_w = 1.20 \times 10^5$) from BASF. Both have the same glass transition temperature, $T_g = 375$–377 K, and are completely compatible [73]. Compact tension (CT) specimens were compression moulded from granules. The specimens were of length 26 mm, width 26 mm and depth 3 mm; a notch was introduced into the sample by sawing, it was subsequently sharpened by a razor blade. The specimens were fractured in tension and their original stress intensity factor, K_{I_0} was determined (cf. Sect. 9 I A). For one series of specimens, the smooth fracture surfaces were brought in contact again and very lightly pressed together.

This sandwich-like preparation was then placed in a pre-heated hot-press, and a slight pressure of about 1 bar was applied normal to the major specimen surface. The specimen temperature, T_h, was set to between 1 and 15 K above the glass transition temperature, T_g. The pressure was needed to establish full contact between the fracture surfaces but it was so small that the specimen geometry was not affected. The samples for this experiment, which truly can be called "crack healing", have always been held at ambient atmosphere. They had a humidity of about 1%.

As indicated by the authors [2], vacuum-dried samples were taken and treated as above in a second series of experiments. For a third series, flat surfaces were prepared by polishing the fracture surfaces of dried specimens. Arbitrarily-selected surfaces were then welded together as described above. In each case the re-healed or welded specimens were completely uniform and transparent and did not show any trace of the former surfaces.

2. Results and Observations

The fracture toughness at crack initiation, K_{Ii}, of rehealed and welded CT samples clearly is a function of healing time, t_h, and healing temperature, T_h. The fairly large scatter involved in these measurements requires a large number of experiments. In Figure 10.9 the results of series 1 (curves 1 to 4) and series 3 (curve 5) are presented. In a K_{Ii} against $t_h^{1/4}$-plot, all the lines are straight and go through the origin. The results from series 2 (vacuum dried samples) are also represented by straight lines having a smaller slope, however.

The experimental data are in full agreement with the model representation; they confirm that crack healing undoubtedly progresses with chain interpenetration and that entanglements must have formed in a completely healed specimen. The data comply with the assumption that the concentration $n(t_h, T_h)$ of entanglements per unit surface area coupling chains from opposite surface sides increases *in proportion to the average length* $(\langle \ell^2 \rangle)^{1/2}$ of chain interpenetration (Eq. 10.11). It follows that

$$\frac{n(t_h)}{n_0} = \frac{(2D_t t_h)^{1/2}}{\langle \ell_0{}^2 \rangle^{1/2}} \tag{10.11a}$$

where $(\langle \ell_0{}^2 \rangle)^{1/2}$ is the average length of interpenetration necessary to obtain full healing. Equation (10.11a) does not permit the determination of D_t and $\langle \ell_0{}^2 \rangle$ independently, but it explains very well the experimental observation that K_{Ii} depends linearly on $t_h^{1/4}$.

From Figure 10.9 one can determine the temperature dependence of D_t. If it is assumed that the critical number of links, n_0, and their volume density do not depend on temperature, then it follows from Equation 10.11 that

$$\frac{D(T)}{D_0} = \frac{\tau_0(T_0)}{\tau_0(T)}. \tag{10.14}$$

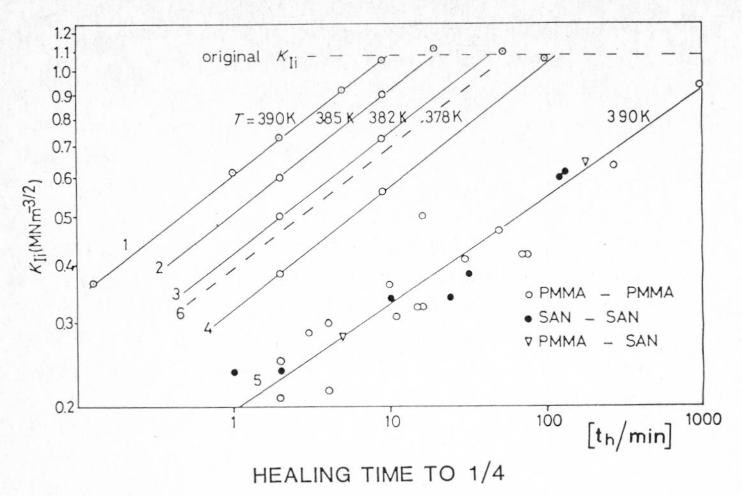

Fig. 10.9a. Double logarithmic plot of fracture toughness K_{Ii} against penetration time. Curves 1–4: healing of broken PMMA specimens immediately after fracture (the points are average values taken from 20 to 30 measurements). Curve 5: surfaces welded after vacuum drying and polishing (all data points are indicated individually). Curve 6: healing at 390 K immediately after fracture of dried PMMA samples (from [2])

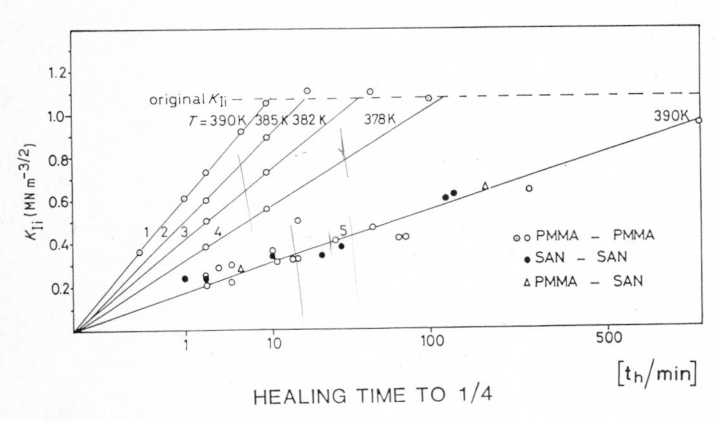

Fig 10.9b. Plot of the same data against $t_h^{1/4}$

The experimental values obtained from Figure 10.9 are well represented by an Arrhenius law of D

$$D(T) = D_0 \exp\left(-\frac{E_a}{RT}\right), \qquad (10.15)$$

with an activation energy, E_a, equal to 274 kJ mol^{-1}. This value of E_a is of a magnitude similar to that given for PMMA in [74] for the α-transition ($\simeq$ 400 kJ mol^{-1}) which describes main-chain motion. It is much larger than the activation energies thus far reported for the self-diffusion coefficients which have been referrenced by Jud [2, 47] and Tirrell [58] for experiments carried out at temperatures T distinctly higher than T_g.

3. Estimation of Diffusion Coefficients

The diffusion model predicts very well the time-dependence of the rehealing experiments but it leaves a number of questions open concerning the physical interpretation of the observed diffusion mechanism, such as the absolute values of D_t, and D_s, the role of chain-ends and of molecular weight, the influence of relaxing fibrils, formed in the first fracture event, and the nature of the established physical links. Absolute values of a diffusion coefficient, D_s, can be estimated, however, using a theory of Graessley [21b], based on the work of Doi and Edwards [27]. D_s can be related to measurable viscoelastic and structural parameters such that

$$D_s = \frac{G_0}{135}\left(\frac{\rho RT}{G_0}\right)^2 \left(\frac{R_g^2}{M}\right) \frac{M_c}{M^2 \eta_0(M_c)}, \tag{10.16}$$

the definitions and values of the quantities used in this equation are given in Table 10.1. The calculated values of D also permit the determination of the depth of interpenetration, $\langle \Delta y^2 \rangle^{1/2}$; for fully rehealed specimens Jud et al. [2] found $\langle \Delta y^2 \rangle^{1/2}$ to have a value of 2.5 nm. This seems to be a reasonable depth, since the radius of gyration for PMMA of the critical molecular weight is about 5 nm.

Table 10.1. Significance and Values of Quantities used for the Calculation of the Diffusion Constant of PMMA 7H by Jud et al. [2]

$G_0(T)$	Plateau modulus	$6.36 \cdot 10^4$ Nm^{-2} $(0.52\,\rho RT/M_c)$
$\rho(T_0)$	Density	$1.14 \cdot 10^3$ kg m^{-3}
R, T	Universal gas constant, absolute temperature	
M	Molecular weight	120000
R_e^2/M	Mean square end to end distance/M	$0.456 \cdot 10^{-20}$ m^2 mol g^{-1}
M_c	Critical M for entanglement	30000 g mol^{-1}
$\eta_0(T, M_c)$	Zero shear viscosity for M_c	$3.78 \cdot 10^8$ Ns m^{-2} at T = 387 K $2.14 \cdot 10^7$ Ns m^{-2} at T = 390 K

For a PMMA 7H, polished and dried, one obtains a diffusion coefficient D(390 K) = $8.33 \cdot 10^{-22}$ m^{2}s^{-1} and a t_h = 60000 s. Thus, $\sqrt{4Dt_h}$ amounts to 14 nm. Under these conditions only 4% of all the chains in a 20 nm layer on either side of the interface have interpenetrated into the opposite matrix by at least R_g [4]. If one assumes that they are *fully entangled* they then contribute 4% of the equilibrium entanglement density of $20 \cdot 10^{24}$ m^{-3}. The contribution of those chains having only partially intersected may be roughly estimated to correspond to 5%. In all one has, therefore, at τ_0 a cconcentration of newly formed entanglements which amounts to 9% of the equilibrium concentration. It is thus quite understandable that the $K_{Ii}(t_h^{1/4})$ curves are linear all the way up to the K_{I_0}-level.

The above estimate should be taken with all due precaution in view of the simplifications introduced into this model, particularly concerning the exact diffusion mechanism in the early stage of crack healing and the relation between depth of interpenetration and number of entanglements formed per chain. It may serve, however, as a first quantitative basis for ensuing discussions.

4. Reopening of Healed Cracks

As discussed above, the mechanical measurement of K_{Ii} at the moment of the reopening of a healed crack forms the basis of the healing studies. In order to obtain further information on the healing mechanisms, some interference microscopical studies were undertaken by the Lausanne group jointly with the Fraunhofer-Institut für Werkstoffmechanik in Freiburg [59]. From these investigations it was learned that the crack front in healed specimens is as irregular as the length, s, of the craze zone at the crack tip (Fig. 10.4). Both irregularities must be ascribed to the fluctuation during rehealing. In those areas where the chains have penetrated less than the everage distance, $(2D_t t_h)^{1/2}$, only short craze zones will be formed. Thus, according to the Dugdale model the local K value will be smaller:

$$K(s) = (8s\sigma_F^2/\pi)^{1/2} \tag{10.17}$$

where σ_F is the yield stress.

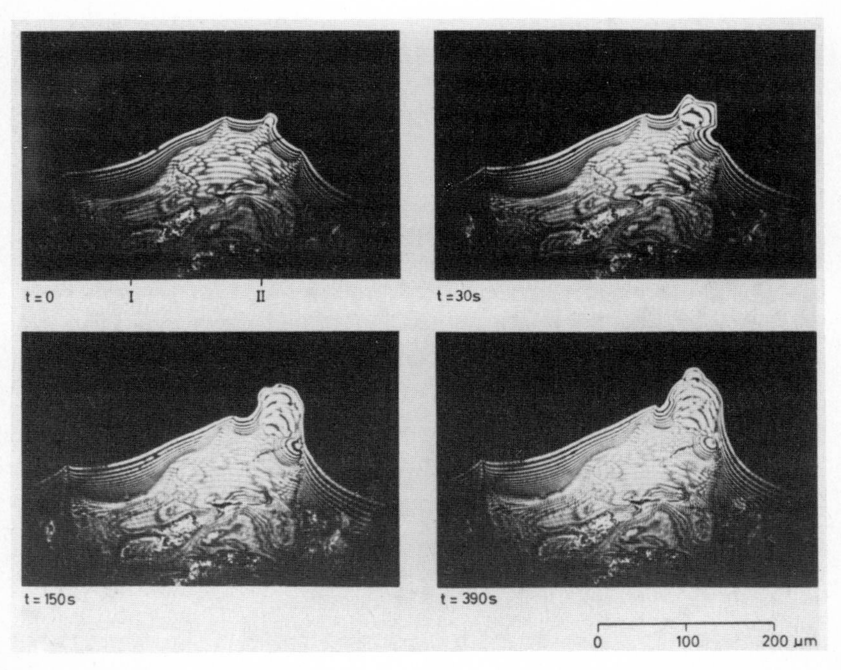

Fig. 10.10. Different stages of slow crack growth in a partially healed PMMA sample, K_I = 0.82 MN $m^{-3/2}$ (from [59])

By determining K(s) along the crack front and subsequently averaging Könczöl et al. [59] obtained a value of K(s) = 0.81 MN $m^{-3/2}$. This value was in excellent agreement with the applied K value during slow crack propagation of 0.82 MN $m^{-3/2}$.

The slow crack growth in rehealed PMMA samples is exemplified in Figure 10.10. The crack grows preferentially through a widening of the weakest regions. Obviously the crack front does no longer present the slightly bent profile observed in virgin samples but a real "alpine landscape". This fact may influence the evaluation of any crack rate effects but does not preclude the use of K_{Ii} as a measure of the healing process.

B. Crosslinked Thermoplastics

1. Experimental Procedure

In the studies of Nguyen et al. [64] compact tension (CT) specimens were compression moulded after careful drying in vacuo to remove any adsorbed water from the same SAN material described in Sect. III A1. To cross-link the samples, unbroken CT specimens, 26 x 26 x 3 mm^3 in size, were irradiated with γ-rays at room temperature in vacuum-sealed Pyrex tube to a predetermined dose up to 800 Mrad (Sulzer source of 60 Co of 50 kCi). SAN belongs to the group of polymers which mainly undergo cross-linking during irradiation [75]. It is well known that reactive species, ions and free radicals, are still present in appreciable amounts after irradiation in polymeric materials. In order to eliminate any post-irradiation effect, all the samples were annealed under vacuum before measurements for several days at a temperature T_{irr} some 5 K above the glass transition temperature, T_g. The irradiated and annealed CT specimens were subsequently broken at room temperature and submitted to rehealing treatment as described before.

In order to give a quantitative interpretation to the rehealing experiments, the polymer network must be well characterized particularly concerning the average distance between two points of cross-linking (see [64]).

2. Results of Mechanical Tests and Observations

In the mechanical tests the fracture toughness, K_{Ii}, at crack initiation was thus measured on the rehealed CT samples as a function of contact time, t_h. The samples had been exposed to various degrees of cross-linking before the fracture and rehealing treatment. As shown in Figure 10.11, an initial linear dependence between the square of K_{Ii} and $t_h^{1/2}$ is observed both with cross-linked samples and non-irradiated samples. The linear increase of K_{Ii}^2 is followed by a plateau region (K_{Ip}^2). However, whereas K_{Ii}^2 of an uncrosslinked sample reaches a plateau value only when the original fracture toughness K_{Ip}^2 is fully restored, the irradiated samples show a plateau at a value K_{Ip}^2 smaller than K_{I0}^2 (Fig. 10.11). The plateau value decreases with the dose received and there is a good correlation between K_{Ip}^2/K_{I0}^2 and the square root of $\overline{M_x}$ (Table 10.2).

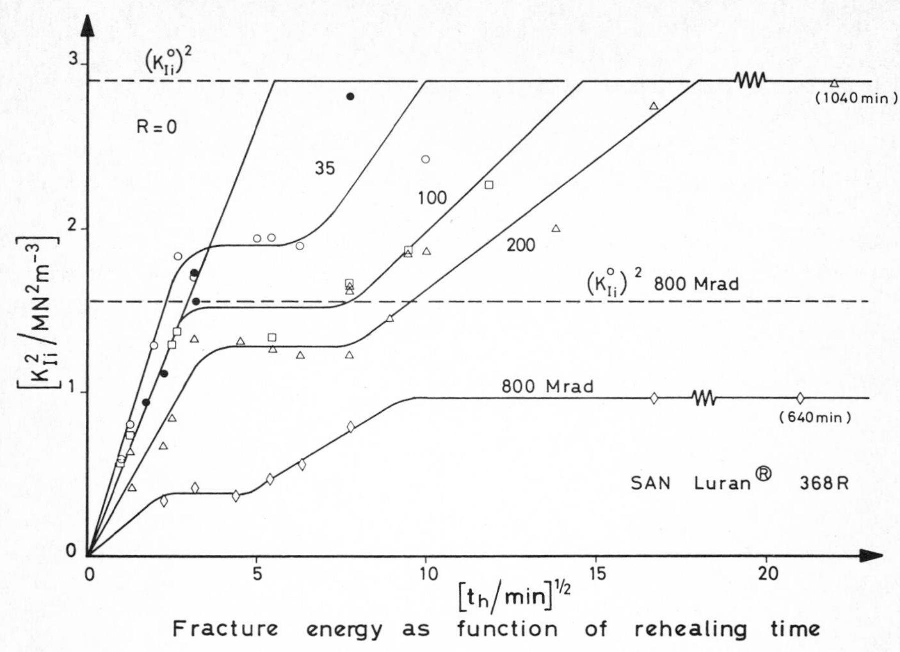

Fig. 10.11. Fracture energy of crosslinked SAN as a function of rehealing time t_h (from [64]); parameter : irradiation dose in Mrad

Another striking fact in irradiated samples is the occurrence of a second recovery in fracture toughness which begins some minutes to one hour after the first plateau was reached. With the exception of the highest dose of 800 Mrad, the original fracture toughness of the virgin material is eventually fully restored. The second recovery gives a straight line on a K^2_{Ii} vs. $t_h^{1/2}$ plot, suggesting that a diffusion process is also responsible for this stage. The modest variation of the slopes for the initial recovery of fracture energy for all samples, regardless of the irradiation dose, indicates that cross-linkage before fracture influences little the kinetics of the first stage of rehealing.

3. Discussion

Specimens which had received a dose of 200 Mrad *after* fracture did not show any measurable rehealing at 390 K, even after several days of penetration. The creation of a fracture surface seems to be necessary, therefore, before healing can occur in cross-linked samples. It is known that the superficial layer next to the fracture surface is highly affected by the fracture process resulting in disentanglements of the molecular coils and chain scissions. In highly cross-linked polymers, chain scission will be prevalent, a fact which has been well documented in the past using ESR spectroscopy. The net effect in the used samples may well be the formation of two layers: a first layer (region A in Fig. 10.12) which consists essentially of free chain ends liberated by slippage from the matrix or created by breakage of segments between cross-linking points. The width of this layer can be estimated to be given by ($\langle R_e^2 \rangle^{1/2}$) the average distance between a chain end and the nearest cross-link along the chain. Adjacent to this region A of highly mobile chain ends one finds a region B (cf. Fig. 10.12) plastically deformed by the fracture event. In order to explain the second recovery it is required that this layer B also contains mobile chains but that their degree of mobility is much smaller than that of the A-chains. The lower mobility could be ascribed, for instance, to the greater length of the diffusing units and to the presence of cross-links in the matrix hindering the diffusion even of the uncross-linked chains.

The two-layer model presented in Figure 10.12 certainly is a simple one and its limits will become apparent in the later discussions.

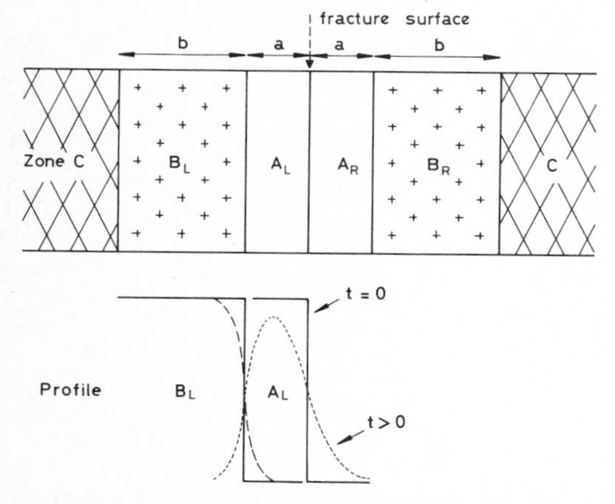

Fig. 10.12. Two-layer model of interdiffusion (from [64])

Magnitude of the Plateau Fracture Toughness K_{Ip}^2

It is evident that only chains with their ends within a certain distance of the interface can diffuse easily across this interface and give effective contributions to the transmission of mechanical stress [1–6]. In the present case a chain segment of molecular weight M_1 may be considered which is attached to a cross-link point at a distance d_1 from the interface. If d_1 is no larger than the radius of gyration, R_{g1}, then the end of this particular segment has a good chance to cross the interface. The width *a* of region A in Figure 10.12 should be taken, therefore, as the radius of gyration of the free chain segments found in region A. No matter whether those chain segments are liberated by slippage or created by breakage, they will have a number average molecular weight of $\overline{M_x}/2$. If it can be assumed that the free chain segments adopt a random coil configuration then the width *a* of the highly mobile layer will be given by:

$$a = \sqrt{\frac{\overline{M}_x}{2}} \cdot \left\langle \frac{R_g^2}{M} \right\rangle_0^{1/2} . \qquad (10.17)$$

For the molecular-weight independent term $\left\langle \frac{R_g}{M} \right\rangle_0^{1/2}$ one may take the value of $7 \cdot 10^{-4}$ [$nm^2 mol/g$] determined by Kirste et al. [76]. During the healing treatment a fracture energy $G_{Ii} \sim K_{Ii}^2$ will be built up first by interdiffusion of the mobile segments quite in analogy to the observations with uncross-linked polymers [1, 3, 4]; K_{Ii}^2 is, therefore, proportional to the distance of interpenetration. Once this distance has reached the value of a, the chain ends are completely interdiffused. At this moment one has:

$$K_{Ip}^2/K_{I_0}^2 = a/a_0 \tag{10.18}$$

where a_0 is the interpenetration distance necessary to establish the full strength of the virgin material. If a_0 is taken to be the radius of gyration of an average original molecule ($\overline{M_{n_0}} = 100000$), one should have:

$$G_{Ip}/G_{I_0} = K_{Ip}^2/K_{I_0}^2 = \sqrt{\overline{M}_x/200000}. \tag{10.19}$$

As is revealed by Table 10.2, there is indeed a good linear correspondence between K_{Ip}^2 and $\sqrt{\overline{M}_x}$. The fact that at high degrees of cross-linking $K_{Ii}^2/K_{I_0}^2$ is found slightly larger than $\sqrt{\overline{M}_x/200000}$ might be due to a non-random cutting of chain segments in the first fracture process and to an underestimation of $\overline{M}_x$. In view of the very wide molecular weight distribution of the original sample this aspect will not be pursued any further.

Table 10.2. Relationship between the relative value G_{Ip}/G_{I_0} of the plateau and $\overline{M}_x$

Dose (Mrad)	$\overline{M}_x$	G_{Ip}/G_{I0}	$\sqrt{\overline{M}_x/200000}$
0	–	1	–
35	88650	0.65	0.64
100	57700	0.52	0.56
200	41000	0.45	0.40
800	13500	0.25	0.20

The Second Recovery in Fracture Toughness and the Double Layer Model

If chain ends were the only diffusing species, no increase in G_I should be observed after the plateau energy G_{Ip} was reached. The occurrence of the second recovery in fracture toughness on a long time scale indicates that less mobile species, able to give rise to additional entanglements, are also present near the fracture surfaces. These may be formed from the remains of the broken network disentangled during the fracture. The presence of cross-link points lowers the diffusion coefficient so that linear or very lightly branched portions of these molecules start to intermingle only after complete interdiffusion of the dangling chain ends.

Nguyen et al. [64] have presented a mathematical model with two diffusing layers to produce the two-step diffusion process mentioned above. The model is described by two diffusion coefficients: the self-diffusion coefficient D_A of A_L into A_R, and the mutual diffusion coefficient D_B of A_L into B_L (or A_R into B_R). The number of entanglements formed and thus G_c is assumed to be proportional to the number of species A and B which have diffused across the boundary at x = 0:

$$G_c(t) \sim \int_0^{a+b} \{C_{A_L}(x, t) + C_{B_L}(x, t)\} dx. \tag{10.20}$$

The movements of A and B will be correlated if the two species are bound together and belong to the same molecule. However, to make the mathematics more tractable, Nguyen et al. assumed that they are independent and that the diffusion of each species follows Fick's laws. This need not be true for the actual system under investigation, however, the simplified model allowed to study some basic characteristics of a two-step diffusion process. The best fit of this model to the experimental data for specimens irradiated to 200 Mrad is obtained with $D_A/D_B = 12$. The authors note, however, that the model used so far does not lead to a well pronounced plateau. It will be necessary, therefore, to consider two additional parameters: the oriented state of the fibrillar matter in the surface regions and the attachment of the dangling chain sections to the crosslinked network.

C. Polymer Mixtures

1. Experimental Results

In analogy to the above observations it was expected that healing of compatible blends AB would occur through interdiffusion of the constituent molecules A and B across the interface. Depending on the differences in mobility of A and B it would have to be expected that the more mobiles species would first contribute to the healing process.

In the experiments of Petrovska, Kausch et al. [6, 65, 66] blends of polystyrene (PS) as dominant matrix with either a stiff component (poly-2,6-dimentyl-1,4-phenylene oxyde – PPO) or a mobile component (polyvinylmethyl ether – PVME) were used. As to be expected the glass transition temperature depended on the composition of the blend. In each case the healing experiments were carried out at a healing temperature $T_h = T_g + 10$ K.

The increase in stress intensity factor with interdiffusion time t_h ("healing curves") are shown in Figures 10.13 and 10.14. It is readily seen that

- the presence of 5% PPO retards the healing process in its later stages
- retardation is the more effective the higher the molecular weight of the PPO
- in the 5% PPO blend there is an indication of a plateau region althought is not very pronounced
- healing of the PVME/PS blends is also slower (at 69 °C) than healing of pure PS (at 117 °C)

– PPO seems to slow down the healing more effectively than PVME (adding of 5% PPO of M_w = 16000 retards the healing about as much as adding 20% PVME of M_w = 90000, see Figures 10.13 and 10.14).

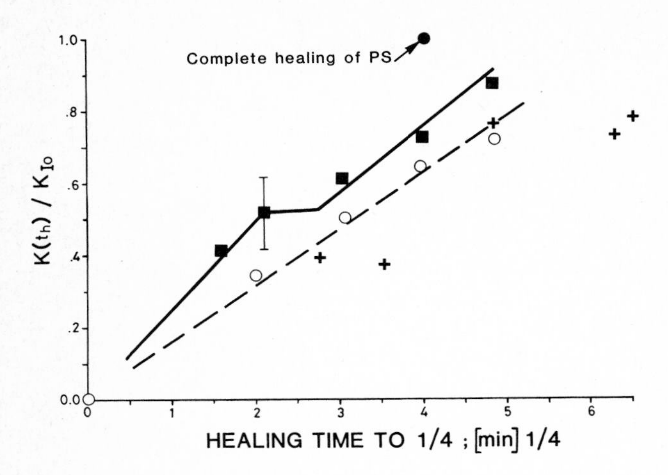

Fig. 10.13. Healing of 5% PPO/95% PS blends with different molecular weight of PPO (molecular weight of PS 240000):

(■) M_w(PPO) = 16000, T_h = 119 °C
(○) M_w(PPO) = 50000
(+) M_w(PPO) = 103000

(The experimental error is indicated only for one point, the others being approximately of the same order of magnitude). From [6]

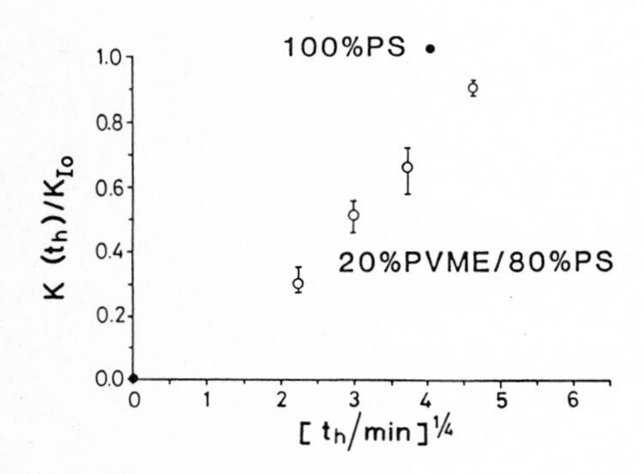

Fig. 10.14. Healing of PVME/PS blend (the time of complete healing of pure PS is also indicated). From [6]

2. Conclusions

Analysis of the thermal, orientational, and viscoelastic behavior of all these blends shows that the thermally determined glass transition temperatures do not describe mechanically equivalent states. This is demonstrated by the two DSC curves obtained for pure PS and for a PS/PVME blend respectively (Fig. 10.15). In the case of pure PS a transition of notable magnitude is achieved within a temperature interval of 22 K. In the blend the magnitude is comparable but the transition is spread out over an interval of 49 K. Whatever the way of determining a T_g from DSC measurements, it indicates a reference temperature at which additional modes of vibration are liberated. These modes will aid the reptational diffusion of the chain molecules in the healing experiments as well as the stress- and entropy-relaxation processes studied by Lefebvre et al. [77–79] who investigated how the degree of segmental orientation obtained in hot-drawn PS blends decreased with increasing temperature and decreasing rate of deformation.

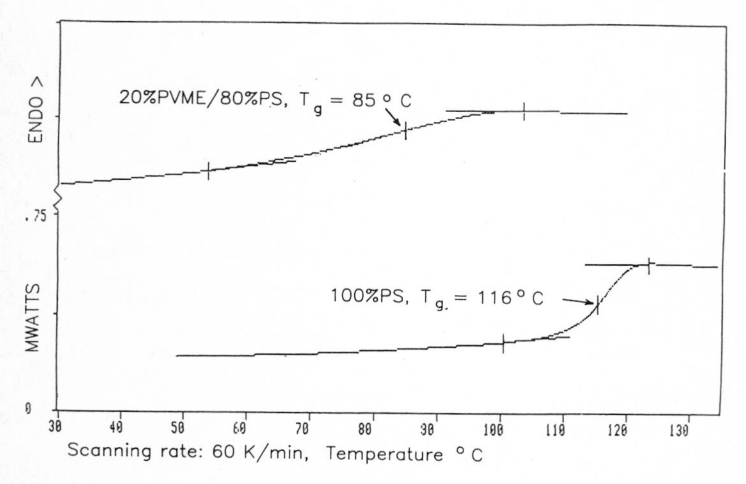

Fig. 10.15. DSC curves used for the determination (by extrapolation to "zero" scanning rate) of the glass transition temperatures of PS and PS/PVME blend. From [6]

Moreover, the rate of loss of orientation of PS after a fixed strain appears to be retarded by the presence of PPO [77], and the stress relaxation response at T_g +40 °C is shifted to longer times by the presence of PVME [78]. Thus, the PS orientation was enhanced by both the presence of PPO [77] and PVME [78]. The authors explained this fact by the energetic interaction between the different components in the blend which is equivalent to increasing the intersegmental friction. The PVME chain segments did not show any orientation after hot drawing at T_g + 11.5 K since their mobility was obviously sufficiently large so as to relax any preferred orientation within the drawing period.

The high mobility of the PVME chains did not manifest itself in the crack healing experiments. Neither a two-stage healing process nor accelerated healing has been observed. The apparently contradictory results obtained by the different methods (DSC, degree of orientation of uniaxially stretched samples, viscoelastic behavior, mechanical effects of interdiffusion) is due to the fact that segmental motion and chain reptation obviously depend to a different degree on the form of the relaxation time spectrum and on which portion of the spectrum is being sampled by the experiments. The difference between segmental and whole chain motion has long been ascribed to the retarding influence of entanglements. This concept was perhaps first given a molecular interpretaiton by the FLW modelling of pure polymer response by assuming that a Rouse treatment would apply [80]. Here one takes the monomeric friction factor to be the quantity which characterizes the mobility of short segments; it is this friction factor which locates the glass-to-rubber transition of a given polymer on the time scale, at a given temperature above T_g [81, 82]. This factor would also characterize the steady flow viscosity of a polymer of $M = 2 M_e$, measured at the same effective $T - T_g$ temperature.

Above the entanglement point, the terminal relaxation zone was taken to be described by the same Rouse formulation, but with a viscosity characterized by the actual steady flow value for the polymer investigated. This corresponds to an increase in the friction factor by the ratio $(M/M_e)^{2.4}$. The manner in which the chains went from segmental response to total chain response was admittedly not clear, and so response in the plateau region was not well described [6].

Both points are now covered in the de Gennes, Doi-Edwards and Graessley treatment of undiluted polymers [26–28]. The long time response of the "entangled" chain is controlled by the reptation along the tube (though with an M^2 dependence leading to an M^3 dependence of viscosity), but motion *is* allowed within the tube and is specifically modeled as "Rouse-like motion" at short time. These two parts of the modern theory therefore correspond formally to the early FLW assumption of the applicability of the Rouse treatment.

For present purposes then, segmental motion is governed by the monomeric friction factor. However, in a widely spread transition (PS/PVME) the long range modes of PVME chain motion, essential for reptation, are still blocked at $T_g + 10$ K. Consequently one may find notable segmental relaxation well before a displacement of whole chains.

It has to be taken into account that the addition of small amounts of a "narrow-transition material" (PS) to a "wide-transition material" (PVME) may widen the transition even further. An excellent demonstration of this (unexpected) fact has been given by Brekner, Cantow and Schneider [84] whose results are reproduced in Figure 10.16. The addition of small amounts of (any molecular weight) PS to PVME widens the relaxation time spectrum.

It is also of interest to consider the location of the healing time on the time scale provided by the stress relaxation measurements. For this purpose one notes the healing time of pure PS at $T = T_g + 11.5$ °C to be $4^4 = 256$ min. At $T_g + 40$ °C, this corresponds to $10^{0.7}$ sec. as calculated from the WLF equation using $C_1 = 13.7$ deg. and $C_2 = 50$ deg [83]. This is just into the terminal zone as seen in the stress relaxation modulus data of Faivre, Jasse and Monnerie (Fig. 7 of ref. 78b). Hence, the healing

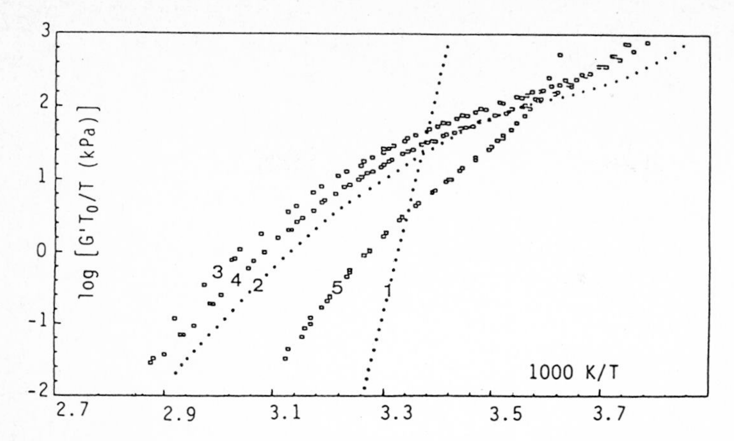

Fig. 10.16. Isochrone master curves of G′ of the mixture of PVME (M_w= 73000) with oligomer PS_{800} (M_w = 800), 1. PS_{800}, 2. PVME, 3. mixt. w. 8.6%, 4. mixt. w. 24%, 5. mixt. w. 75.4% PS_{800} (taken from ref. [84])

time is evidently relative to the longest relaxation time, or tube renewal time. The shortes measurable healing time is about five minutes because of the time required for sample warm-up and cool-down. This corresponds to about 10^{-1} sec. at Tg +40 °C, which is midway along the plateau. Thus, even at the shortest healing time available, all Rouse submolecules shorter than M_e will have relaxed out.

D. General Conclusions Derived from Crack Healing Studies

So far, the most interesting results have been obtained from glassy amorphous polymers such as PMMA, SAN, or PS, used in the form of homopolymers or blends (PMMA/SAN; PS/PPO; PS/PVME). In these cases "interfacial strength" (as measured through K_{Ii}) was gradually restored with healing time t_h. The most remarkable fact certainly is that K_{Ii} increases with $(t_h)^{1/4}$; this seems to be brought about by the strong dependence of the fracture energy G_c of glassy polymers on fibril stability which, in turn, depends on the degree of reentanglement around the former interfacial zone. The assumption that reentanglement increases with $(t_h)^{1/2}$ would then tie in a straightforward manner chain diffusion to healing.

The available information on the molecular weight dependence of healing [47, 68, 70] suggests that chain diffusion occurs through reptation. However, more extensive studies of this aspect would be welcome in order to better separate the molecular weight dependence of the calculated diffusion coefficient from other effects caused for instance by changes of T_g and K_{I0} with M_w or by a non-even distribution of chain ends in the interfacial zone.

A final remark concerns semicrystalline polymers. Studies undertaken in Lausanne [85] with HDPE, PA6 and PVC or PVC/PMMA blends have consistently failed to achieve gradual healing, even when using rather elevated temperatures. In these cases, the existence of a network with crystalline junction points [86] seems to effectively prevent sufficient interdiffusion of chain molecules and thus application of the previously described "static" crack healing method based on fracture mechanics specimens.

References for Chapter 10

1. K. Jud, H. H. Kausch: Polymer Bulletin *1*, 697 (1979).
2. K. Jud, H. H. Kausch, J. G. Williams: J. Mater. Sci. *16*, 204 (1981).
3. R. P. Wool, K. M. O'Connor: J. Appl. Phys. *52*/10, 5953 (1981).
4. H. H. Kausch, K. Jud: Plastics & Rubber Proc. & Appl. *2*, 265 (1982).
5. R. P. Wool, K. M. O'Connor: J. Polymer Sci., Polym. Lett. Ed. *20*, 7 (1982).
6. H. H. Kausch, D. Petrovska, R. F. Landel, L. Monnerie: to be published Polym. Eng. + Sci. (1987).
7. H. P. Meissner, E. W. Merrill: Modern Plastics, April, 104 (1949).

8a. G. Diedrich, E. Gaube: Kunststoffe *60*, 74 (1970).
 b. T. Hadick: Schweissen von Kunststoffen. Krefeld: Deutscher Verlag für Schweißtechnik 1972.
 c. Federal Republic of Germany: DVS 2207, 2209, 2211, Switzerland: 0 0 0

9. D. Langbein: J. Adhesion *1*, 237 (1969)
10. F. M. Fowkes, in: Contact Angle, Wettability and Adhesion. Adv. Chem. Ser. *43*, Am. Chem. Soc. 1964, 108.
11. S. S. Voyutskii: Autohesion and Adhesion of High Polymers, New York: Wiley-Interscience 1963.

12a. W. G. Forbes, Mcleod: Trans IRI *30*, 154 (1958).
 b. J. D. Skewis: Rubber Chem. Tech. *39*, 217 (1966).
 c. M. L. Studebaker, J. R. Beatty, in: Science and Technology of Rubber. F. R. Eirich, Ed., Academic Press 1978.

13a. E. Bister: Dissertation, T. U. Clausthal (1977).
 b. E. Bister, W. Borchard, G. Rehage: Kautschuk + Gummi *29*/9 527 (1976).

14. L. Bothe, G. Rehage: Angew. Makromol. Chem. *100*, 39 (1981).

15a. G. Koszterszitz: Coll. & Polym. Sci. *258*, 685 (1980).
 b. G. Koszterszitz: Kautsch. + Gummi Kunstst. *35*, 11 (1982).

16. A. Zosel: Coll. & Polym. Sci. *263*, 541 (1985).
17. M. L. Williams: J. Adhesion *4*, 307 (1972).

18a. M. L. Anderson, K. L. DeVries, M. L. Williams: Int. J. Fract. *9*, 421 (1973).
 b. D. H. Kaelble: J. Appl. Polym. Sci. *18*, 1869 (1974).

19a. A. N. Gent, R. P. Petrich: Proc. Roy. Soc. A*310*, 433 (1969).
 b. K. Kendall: J. Phys. D. Appl. Phys. *5*, 1782 (1973).
 c. J. Cognard: J. Adhesion *20*, 1 (1986).

20a. D. Maugis: Le Vide *186*, 1 (1977).
 b. D. Maugis, M. Barquins: J. Phys. D: Appl. Phys. *11*, 1989 (1978).
 c. D. Maugis: J. Mater. Sci. *20*, 3041 (1985).

21a. W. W. Graessley: J. Chem. Phys. *43*, 1696 (1965).
 b. W. W. Graessley: J. Polym. Sci., Polym., Polym. Phys. *18*, 27 (1980).

22a. H. Watanabe, T. Kotaka: Macromol. *16*, 769 (1983).
 b. H. Watanabe, T. Kotaka: ibid. *17*, 342 (1984).

23a. S. C. Malguarnera, D. C. Riggs: Polym.-Plast. Technol. Eng. 17, 193 (1981).
 b. S. C. Malguarnera: Polym.-Plast. Technol. Eng. *18*, 1 (1982).
 c. H G. Moslé, R. M. Criens: Kunststoffe *72*, 222 (1982).
 d. R M. Criens, H. G. Moslé: Polym. Eng. Sci. *23*, 591 (1983).
 e. R. Pisipati, D. G. Baird: European Meeting on "Polymer Processing and Properties", Capri, Italy, 12–16. 6. 1983.
 f. D. G. Baird, R. Pisipati: Polym. News *8*, 301 (1983).
 g. H. H. Winter, K. H. Wei: Interrelations Between Structure and Properties of Polymeric Materials, J. C. Seferis ed., Amsterdam: Elsevier 1983.
 K. H. Wei, M. F. Malone, H. H. Winter: Polym. Eng. + Sci. *26*, 1012 (1986).
 h. R. C. Thamm: Rubber Chem. Technol. *50*, 24 (1979).

24a. T. Hattori, K. Tanaka, M. Matsuo: Polym. Eng. Sci. 12, 199 (1972).
 b. E. M. Katchy: Kunststoffe *74*, 755 (1984).
 c. N. Rosenzweig, M. Narkis: Polymer, Polym. Comm. *21*, 988 (1980).

25. S. P. Timoshenko, J. N. Goodier: Theory of Elasticity, 3rd Edn., Tokyo, McGraw Hill Kogakusha Ltd. 1970, p. 412.
26a. P. G. de Gennes: J. Chem. Phys. *55*, 572 (1971).
b. P. G. de Gennes: C. R. Acad. Sc. Paris, *291*, B 219 (1980).
c. P. G. de Gennes: C. R. Acad Sc. Paris, *292*, II 1505 (1981).
d. G. B. McKenna: Polymer Comm. *26*, 324 (1985).
27. M. Doi and S. F. Edwards: J. Chem. Soc. Faraday Trans. II, *74*, 1789 (1978).
28. W. W. Graessley: Advances in Polymer Science *47* (1982).
29a. F. Bueche, W. M. Cashin, P. Debye: J. Chem. Phys. *20*, 1956 (1952).
b. J. D. Ferry: J. Phys. Chem. *66*, 2699 (1962).
c. G. P. Chen, J. D. Ferry: Macromol. *1*, 270 (1968).
30. F. Bueche: J. Chem. Phys. Lett. *48*, 1410 (1968).
31. Y. Kumagai, H. Watanabe, K. Miyasaka, T. Hata: J. Chem. Eng. of Japan *12*, 1 (1979).
32. D. W. McCall, C. M. Huggins: Appl. Phys. Lett. *7*, 153 (1965).
33. J. E. Tanner: J. Chem. Phys. *52*, 2523 (1970).
34. J. E. Tanner, K. J. Liu, J. E. Anderson: Macromol. *4*, 586 (1971)
35. J. E. Tanner: Macromol. *4*, 748 (1971).
36. R. Kimmich, R. Bachus: Coll. Polymer Sci. *260*, 911 (1982).
37. R. Bachus, R. Kimmich: Polymer *24*, 964 (1983).
38a. G Fleischer: Polymer Bull. *9*, 152 (1983).
b. G. Fleischer: Polymer Bull. *11*, 75 (1984).
39. G. Fleishcer: Polymer *26*, 1677 (1985).
40. T. Cosgrove, R. F. Warren: Polymer *18*, 255 (1977).
41. J. Klein, B. J. Briscoe: Proc. Roy. Soc. London, Ser. A *365*, 53 (1979).
42. J. Klein: Nature *271*, 143 (1978).
43. J. Klein: Macromol. *11*, 852 (1978).
44. J. Klein: Phil. Mag. A *43*, 771 (1981).
45. J Klein, D. F. Fletcher, L. J. Fetters: Nature *304*, 526 (1983).
46. P. T. Gilmore, R. Falabella, R. L. Laurence: Macromol. *13*, 880 (1980).
47. K. Jud, PhD Thesis 413, Départment des Matériaux, Ecole Polytechnique Fédérale de Lausanne, 1981.
48. J. Couttandin, H. Sillescu, R. Volkel: Makromol. Chem., Rapid Commun. *3*, 649 (1982).
49a. M. Antonietti, J. Coutandin, R. Grüter, H. Sillescu: Macromol. *17*, 798 (1984) and ibid. *19*, 793 and 798 (1986).
b. M. Antonietti: Dissertation, Inst. f. Physikalische Chemie, Universität Mainz 1985.
50. C. P. Bartels, W. W. Graessley, B. Crist: J. Polym. Sci.: Polym. Lett. *21*, 495 (1983) and Macromolecules *19*, 785 (1986).
51. M. Stamm: ACS Polym. Div. Preprints *24*, 380 (1983).
52. B. A. Smith: Macromol. *15*, 469 (1982).
53. B. A. Smith, E. T. Samulski, L. P. Yu, M. A. Winnik: Phys. Rev. Lett. *52*, 45 (1984).
54a. P. J. Mills, P. Green, C. Palmstrom, J. W. Mayer, E. J. Kramer: Appl. Phys. Lett., *45* (9), 957 (1984).
b. R. J. Composto, J. W. Mayer, E. J. Kramer: Fast Mutual Diffusion in Polymer Blends (to be published)
55. J. F. Romanelli, J. W. Mayer, E. J. Kramer, P. Russell: MSC Report 5289, Dept. Mater, Sci. and Eng., Cornell University, Ithaca, 1984.
56. E. J. Kramer, P. Green, C. Palmstrom: Polymer *25*, 473 (1984).
57. P. Green, C. Palmstrom, J. W. Mayer, E. J. Kramer: Macromol. *18*, 501 (1985).
58. M. Tirrell: Rubb. Chem. Techn. *57*, 523 (1984).
59. L. Könczöl, W. Döll, H. H. Kausch: K. Jud, Kunststoffe *72*, 46 (1982).
60. W. Döll, Crazing in Polymers, H. H. Kausch (Ed.): Advances in Polymer Science Vol. 52/53, Heidelberg – New York: Springer-Verlag 1983.
61. S. Fakirov: J. Polym. Sci., Polym. Phys. *22*, 2095 (1984).
62. S. Fakirov: Makromol. Chem. *185*, 1607 (1984).
63. S. Fakirov: Polymer Comm. *26*, 137 (1985).

64. T. Q. Nguyen, H. H. Kausch, K. Jud, M. Dettenmaier: Polymer *23,* 1305 (1982).
65. D. Petrovska, M. Dettenmaier, H. H. Kausch: Helv. Phys. Acta *57,* 244 (1984).
66. D. Petrovska, H. H. Kausch, J. P. Faivre, B. Jasse, L. Monnerie: ibid 754.
67. Y. H. Kim, R. P. Wool: Macromolecules *16,* 1115 (1983).
68a. R. P. Wool: Proc. IV. Int. Congress on Rheology, Mexico, 1984, p. 573.
69a. R. P. Wool: J. Elastomers & Plast. *17,* 106 (1985).
b. R. P. Wool: Rubb. Chem. & Techn. *57,* 307 (1984).
70. S. Prager, M. Tirrell: J. Chem. Phys. *75,* 5194 (1981).
71a. D. Adolf, S. Prager, M. Tirrell: J. Chem. Phys. *79,* 7015 (1983).
b. D. Adolf, M. Tirrell: J. Polym. Sci., Polym. Phys. Ed. *23,* 413 (1985).
72. S. Prager, D. Adolf, M. Tirrell: J. Chem. Phys. *84,* 5152 (1986).
73. S. Krauss in: Polymer Blends Vol. 1, New York – London: Academic Press 1978, Ch. 2.
74. N. G. McCrum, B. E. Read, J. G. Williams: Anelastic and Dielectric Effects in Polymeric Solids, New York: Wiley 1976, p. 258.
75. M. Dole: The Radiation Chemistry of Macromolecules, New York – London: Academic Press 1973, p. 91.
76. J. Jelenic, R. G. Kirste, B. J. Schmitt, S. Schmitt-Strecker: Makromol. Chem. *180,* 2057 (1979).
77a. D. Lefebvre, B. Jasse, L. Monnerie: Polymer *22,* 1616 (1981).
b. D. Lefebvre, B. Jasse, L. Monnerie: Polymer *25,* 318 (1984).
78a. J P. Faivre: Thèse de docteur-ingénieur, Université Pierre et Marie Curie, Paris, 1985.
b. J P. Faivre, B. Jasse, L. Monnerie: Polymer *26,* 879 (1985).
79. Y. Zhao, B. Jasse, L. Monnerie: Polymer Bulletin *13,* 259 (1985).
80. J. D. Ferry, R. F. Landel, M. L. Williams: J. Appl. Phys. *26,* 359 (1955).
81. R. F. Landel, R. F. Fedoux: Proc. Int. Conf. on Mech. Behav. Matls., 1971, Soc. Matls. Sci., Tokyo, Japan, 1972, Vol. III., p. 496.
82. J.-D. Ferry: Viscoelastic Properties of Polymers. New York: Wiley, chapt. 10, 3rd ed. 1980.
83. ibid. chapt. 11.
84. M.-J. Brekner, H.-J. Cantow, H. A. Schneider: Polymer Bulletin *14,* 17 (1985).
85. D. Petrovska, H. H. Kausch: unpublished results, Laboratoire de polymères, EPFL, Lausanne 1986.
86. M. Theodoru, B. Jasse: J. Polymer Sci., Polymer Phys. Ed., to be published 1987.

Appendix

Table A.1. List of Abbreviations of the most Important Polymers

(According to DIN 7723, DIN 7728, ASTM D 1600-64 T, ASTM 1418-67, ISO/R 1043-1969)

ABS	Acrylonitrile-butadiene-styrene copolymer	NBR	Acrylonitrile-butadiene rubber
ABR	Acrylonitrile-butadiene rubber	NCR	Acrylonitrile-chloroprene rubber
AMMA	Acrylonitrile-methyl methacrylate copolymer	NR	Natural rubber
ASA	Acrylester-styrene-acrylonitrile copolymer	PA	Polyamide
BR	Polybutadiene rubber	PA 6	Polycaprolactam
CA	Cellulose acetate	PA 66	Poly(hexamethylene adipamide)
CAB	Cellulose acetobutyrate	PA 6.10	Poly(hexamethylene sebacamide)
CAP	Cellulose acetopropionate	PA 11	Polyamide of 11-amino-undecanoic acid
CN	Cellulose nitrate	PA 12	Polylaurolactam
CP	Cellulose propionate	PAA	Poly (acrylic acid)
CPE	Chlorinated polyethylene	PAN	Polyacrylonitrile
CPVC	Chlorinated poly(vinyl chloride)	PB	Polybutylene
EC	Ethyl cellulose	(PBA)	Polybutyl acrylate
EP	Epoxide resin	PBTP	Poly(butylene glycol terephthalate)
EPDM	Ethylene propylene terpolymer rubber	PC	Polycarbonate
EPR	Ethylene-propylene rubber	PCTFE	Poly(chlorotrifluoroethylene)
EPTR	Ethylene-propylene terpolymer rubber	PDAP	Poly(diethyl phthalate)
ETFE	Ethylene-tetrafluorethylene copolymer	PE	Polyethylene
EU	Polyether-urethane	PEC	Chlorinated polyethylene
EVA	Ethylene-vinyl-accetate copolymer	PED	Deuterated polyethylene
HDPE	High-density polyethylene	PEH	PE (as opposed to PED)
HiPS	High impact polystyrene	PEO	Polyethylene oxide
IR	Isoprene rubber	PEOB	(Poly[p-(2-hydroxyethoxy) benzoic acid])
IIR	Butyl rubber, isoprene-isobutylene copolymer	(PES)	Polyethersulfone
MF	Melamin-formaldehyde resin	PETP	Poly(ethylene terephthalate)
		PF	Phenol-formaldehyde resin
		PI	Poly-trans-isoprene
		PIB	Polyisobutylene
		PMMA	Poly(methyl methacrylate)

Table A. 1. (continued)

PMP	Poly(4-methylpentene-1)	PVDF	Poly(vinylidene fluoride)
PαMS	Poly-α-methyl styrene	PVF	Poly(vinyl fluoride)
POM	Polyoxymethylene = poly-acetal = polyformaldehyde	PVFM	Poly(vinyl formal)
		PY	Unsaturated polyester resin
PP	Polypropylene	RF	Resorcinol-formaldehyde resin
PPO	Poly(2,6-dimethyl-1,4-phenylene oxide)	RP	Reinforced plastic
		SAN	Styrene-acrylonitrile copolymer
PS	Polystyrene		
PSU	Polysulfone	SB	Styrene-butadiene copolymer
PTFE	Poly(tetrafluoroethylene)	SBR	Styrene-butadiene rubber
PTMT	Poly(tetramethylene terephthalate)	SBS	Styrene-butadiene block copolymer
PUR	Polyurethane	SI	Silicone rubber
PVA(L)	Poly(vinyl alcohol)	SIR	Styrene-isoprene rubber
PVAC	Poly(vinyl acetate)	SMS	Styrene-α-methylstyrene copolymer
PVB	Poly(vinyl butyral)		
PVC	Poly(vinyl chloride)	UE	Polyurethane rubber
PVCA	Vinyl chloride-vinyl acetate copolymer	UF	Urea-formaldehyde resin
		UP	Unsaturated polyester resin
PVDC	Poly(vinylidene chloride)	UR	Polyurethane rubber

Table A. 2. List of Abbreviations not Referring to Polymer Names

CED	Cohesive energy density
CN	Center notched specimen (fracture mechanics)
COD	Crack opening displacement (fracture mechanics)
CT	Compact tensile specimen (fracture mechanics)
DCB	Double cantilever beam (fracture mechanics)
DT	Double torsion (fracture mechanics)
DZ	Deformation zone
ESC	Environment stress cracking
ESR	Electron spin resonance
GPC	Gel permeation chromatography
IR	Infra red
NMR	Nuclear magnetic resonance
SANS	Small angle neutron scattering
SAXS	Small angle X-ray scattering
SEM	Scanning electron microscopy
SEN	Single edge notch specimen (fracture mechanics)
TPB	Three point bending (fracture mechanics)
UV	Ultra violet
WAXS	Wide angle X-ray scattering
WOL	Wedge open loaded specimen (fracture mechanics)

Table A. 3. List of Symbols

Symbol	*General Meaning* (if not indicated otherwise in the section where the symbol appears)	*Unit Used*
C	Concentration of end groups	mol m^{-3}
C	Bending compliance	m N^{-1}
C_p	Specific heat	J kg^{-1} K^{-1}
D	Bond dissociation energy	kJ mol^{-1}
D	Average spherulite diameter	μm
E	Young's modulus	GN m^{-2}
E'	Storage modulus	GN m^{-2}
E"	Loss modulus	GN m^{-2}
E*	Complex modulus	GN m^{-2}
E_A	Activation energy	kJ mol^{-1} or kcal mol^{-1}
E_c	Elastic modulus of crystalline regions	GN m^{-2}
E_k	Young's modulus of chains	GN m^{-2}
E_t	Creep modulus	GN m^{-2}
ΔF	Activation energy	kJ mol^{-1}
G	Shear modulus	GN m^{-2}
G'	Storage modulus	N m^{-2}
G"	Loss modulus	N m^{-2}
G*	Free enthalpy of activation	kcal mol^{-1}
G_A	Energy release rate at crack arrest	kJ m^{-2}
G_c	Energy release rate corresponding to K_c	kJ m^{-2}
G_{cc}	Energy release rate corresponding to K_{cc}	kJ m^{-2}
G_{c1}	Plain-strain energy release rate	kJ m^{-2}
G_{c2}	Plain-stress energy release rate (plastic component of G_c)	kJ m^{-2}
G_d	Dynamic energy release rate	kJ m^{-2}
G_{Ic}	Critical energy release rate	kJ m^{-2}
J	J-contour integral	kJ m^{-2}
K	Compressive modulus	GN m^{-2}
K_o	Initial stress intensity factor	MPa $m^{1/2}$
K_I	Stress-intensity factor in crack opening mode I (as a general rule, it is assumed that K-values refer to mode I unless modes II or III are specifically indicated)	MPa $m^{1/2}$
K_{Ic}	Critical stress intensity factor	
K_c	The value of stress-intensity factor that exists in a given situation, which leads to non-critical crack extension	MPa $m^{1/2}$
K_{cc}	Critical stress intensity factor (stress intensity factor leading to unstable crack propagation)	MPa $m^{1/2}$
K_i	Stress-intensity factor at initiation of crack growth (-threshold)	MPa $m^{1/2}$
K_m	Stress intensity factor below which no crazing occurs	MPa $m^{1/2}$
L	Contour length of a segment	nm or Å
L_o	End-to-end distance of a segment	nm or Å
M	Molecular weight	(g mol^{-1})
M_e	Chain molecular weight between crosslinks or entanglement points	(g mol^{-1})

Table A. 3. (continued)

Symbol	*General Meaning* (if not indicated otherwise in the section where the symbol appears)	*Unit Used*
M_n	Number average molecular weight	$(g\ mol^{-1})$
M_w	Weight average molecular weight	$(g\ mol^{-1})$
N	Number of chains between cross links	
N_F	Number of cycles to failure	
N_L	Avogadro number	$6.022 \cdot 10^{23}\ mol^{-1}$
Q_b	Amount of heat dissipated	kJ
R	Molar gas constant	$8.314\ J\ mol^{-1}\ K^{-1}$
R	Material resistance	
S	Internal entropy	$kJ\ K^{-1}$
S_v	Volume of liquid absorbed/unit vol. of polymer	–
T	Absolute temperature	K
T_g	Glass transition temperature	°C
U	Internal energy	$kJ\ mol^{-1}$
U_o	Activation energy for bond scission	$kJ\ mol^{-1}$ or $kcal\ mol^{-1}$
y	Geometric correction factor	
a	Crack length	m
a_n	Impact strength	$kJ\ m^{-2}$
a_T	Shift factors of relaxation times	
$\bar{f}$	Fractional unoccupied volume	cm^3
g	Spectroscopic splitting factor	
h	Planck's constant	$6.626 \cdot 10^{-34}\ Js$
k	Boltzmann's constant	$1.380 \cdot 10^{-23}\ JK^{-1}$
k_c	Rate of thermomechanical chain scission	s^{-1}
k_S	Notch sensitivity factor of strength	
k_T	Notch sensitivity factor of energy to break	
m_I	Spin quantum number	
n	Number of chain atoms	
Δn	Birefringence	
p_c	Craze yield stress	$MN\ m^{-2}$
q	Molecular cross-section	m^2
r	Chain end-to-end distance	nm or Å
r_p	Plastic zone size	μm
t_b	Time to break	s
α	Activation volume	$m^3\ mol^{-1}$ or nm^3
β	Activation volume for chain scission	$m^3\ mol^{-1}$ or nm^3
γ	Activation volume of fracture under tensile stress	$m^3\ mol^{-1}$ or nm^3
δ	Solubility parameter	$cal^{1/2}\ cm^{-3/2}$
$\tan \delta$	Loss tangent	–
ϵ	Strain	–
ϵ_c	Critical strain	–
ϵ_y	Yield strain	–
η	Viscosity (dynamic)	$Ns\ m^{-2}$
λ	Uniaxial extension ratio	–
λ	Heat conductivity	$W\ K^{-1}\ m^{-1}$
λ_s	Linear swelling ratio	–

Table A. 3. (continued)

Symbol	*General Meaning* (if not indicated otherwise in the section where the symbol appears)	*Unit Used*
μ	Coefficient of friction	–
ν	Poisson's ratio	–
ρ	Polymer density	$g \cdot cm^{-3}$
ρ_a	Radius of crack tip	μm
σ	Macroscopic stress	$GN\ m^{-2}$
σ_b	Breaking strength	$GN\ m^{-2}$
σ_c	Critical stress	$GN\ m^{-2}$
τ	Relaxation time	s
τ^*	Critical shear stress	$GN\ m^{-2}$
ψ	Axial chain stress	$GN\ m^{-2}$
ψ_c	Chain strength	$GN\ m^{-2}$
ω	Loading frequency	Hz
ω	Angular frequency	Hz

Table A. 4. Conversion Factors

Frequently found non-SI Units	SI-Units	Conversion
atmosphere, at	Nm^{-2} or Pascal, Pa	$1\ at = 98.07 \cdot 10^3\ N\ m^{-2}$
	bar	$1\ at = 0.9807\ bar$
calorie, cal	Joule, J	$1\ cal = 4.187\ J$
centipoise, cP	Pascal second, Pa · s	$1\ cP = 1.000 \cdot 10^{-3}\ Pa \cdot s$
dyn	Newton, N	$1\ dyn = 1 \cdot 10^{-5}\ N$
erg	Joule, J	$1\ erg = 10^{-7}\ J$
foot, ft	meter, m	$1\ ft = 3.048 \cdot 10^{-1}\ m$
foot pound-force, lbft	Joule, J	$1\ lb.ft = 1.356\ J$
inch, in	meter, m	$1\ in = 2.540 \cdot 10^{-2}\ m$
kilopond, kp	Newton, N	$1\ kp = 9.807\ N$
kilogram-force per square centimeter	Nm^{-2} or Pascal, Pa	$1\ kgf\ cm^{-2} = 9.807 \cdot 10^4\ N\ m^{-2}$
	–	
mil (= 0.001 in)	micrometer, μm	$1\ mil = 25.4\ \mu m$
pound per square inch, PSI	Nm^{-2} or Pascal, Pa	$1\ PSI = 6.897 \cdot 10^3\ N\ m^{-2}$

Subject Index

Page numbers set in *italics* refer to the headings.